PLANT PROPAGATION
Concepts and
Laboratory Exercises

PLANT PROPAGATION
Concepts and Laboratory Exercises

Edited by

Caula A. Beyl
Robert N. Trigiano

CRC Press is an imprint of the
Taylor & Francis Group, an **informa** business

CRC Press
Taylor & Francis Group
6000 Broken Sound Parkway NW, Suite 300
Boca Raton, FL 33487-2742

© 2008 by Taylor & Francis Group, LLC
CRC Press is an imprint of Taylor & Francis Group, an Informa business

No claim to original U.S. Government works
Printed in the United States of America on acid-free paper
10 9 8 7 6 5 4 3 2 1

International Standard Book Number-13: 978-1-4200-6508-4 (Hardcover)

This book contains information obtained from authentic and highly regarded sources Reasonable efforts have been made to publish reliable data and information, but the author and publisher cannot assume responsibility for the validity of all materials or the consequences of their use. The Authors and Publishers have attempted to trace the copyright holders of all material reproduced in this publication and apologize to copyright holders if permission to publish in this form has not been obtained. If any copyright material has not been acknowledged please write and let us know so we may rectify in any future reprint

Except as permitted under U.S. Copyright Law, no part of this book may be reprinted, reproduced, transmitted, or utilized in any form by any electronic, mechanical, or other means, now known or hereafter invented, including photocopying, microfilming, and recording, or in any information storage or retrieval system, without written permission from the publishers.

For permission to photocopy or use material electronically from this work, please access www.copyright.com (http://www.copyright.com/) or contact the Copyright Clearance Center, Inc. (CCC) 222 Rosewood Drive, Danvers, MA 01923, 978-750-8400. CCC is a not-for-profit organization that provides licenses and registration for a variety of users. For organizations that have been granted a photocopy license by the CCC, a separate system of payment has been arranged.

Trademark Notice: Product or corporate names may be trademarks or registered trademarks, and are used only for identification and explanation without intent to infringe.

Library of Congress Cataloging-in-Publication Data

Plant propagation concepts and laboratory exercises / editors, Caula A. Beyl and Robert N. Trigiano.
    p. cm.
  Includes bibliographical references and index.
  ISBN 1-4200-6508-4 (alk. paper)
  1. Plant propagation. 2. Plant propagation--Laboratory manuals. I. Beyl, Caula A. II. Trigiano, R. N. (Robert Nicholas), 1953-

SB119.P56 2008
631.5'3078--dc22
                                                                2007048934

Visit the Taylor & Francis Web site at
http://www.taylorandfrancis.com

and the CRC Press Web site at
http://www.crcpress.com

Contents

Acknowledgments xi
The Editors xiii
Contributors xv

Part I
Introduction

Chapter 1   Introduction to Plant Propagation 3

Caula A. Beyl and Robert N. Trigiano

Part II
Structures for Plant Propagation

Chapter 2   Propagation Structures: Types and Management 15

Gerald L. Klingaman

Chapter 3   Holistic Thought Process for the Design of Propagation Facilities 29

Milton E. Tignor

Chapter 4   Intermittent Mist Control for Plant Propagation 37

David W. Burger

Part III
Plant Propagation Media and Containers

Chapter 5   Media and Containers for Seed and Cutting Propagation and Transplanting 43

Calvin Chong

Chapter 6   Physical Properties and Other Factors to Consider when Selecting Propagation Media 57

Calvin Chong, John E. Preece, Caula A. Beyl, and Robert N. Trigiano

Chapter 7   Media for Cutting Propagation 63

Patricia S. Holloway

Part IV
Plant Propagation Diseases and the Importance of Sanitation

Chapter 8   Disease Management 75

Alan S. Windham

Chapter 9 Disinfestation of Soil and Planting Media 87
James J. Stapleton

Chapter 10 *Botrytis* and Other Propagation Pathogens 91
Mark P. McQuilken

Chapter 11 Crop Certification Programs and the Production of Specific Pathogen-Free Plants 99
Danielle J. Donnelly and Adam Dale

Chapter 12 Integrated Pest Management for Plant Propagation Systems 113
William E. Klingeman

Part V
Evaluation of Propagated Plants

Chapter 13 Evaluation of Data from Propagation Experiments 127
Michael E. Compton

Part VI
General Concepts for Successful Vegetative Propagation

Chapter 14 Plant Growth Regulators Used in Propagation 143
Zong-Ming Cheng, Yi Li, and Zhen Zhang

Chapter 15 Juvenility and Its Effect on Macro- and Micropropagation 151
Caula A. Beyl

Chapter 16 Chimeras 163
Robert M. Skirvin and Margaret A. Norton

Part VII
Propagation by Stem Cuttings

Chapter 17 Cloning Plants by Rooting Stem Cuttings 177
John M. Ruter

Chapter 18 Anatomical and Physiological Changes That Occur during Rooting of Cuttings 189
Ellen T. Paparozzi

Chapter 19 Use of Auxins for Rooting Cuttings 195
Ellen G. Sutter and David W. Burger

Contents

Chapter 20 Adventitious Rooting of Woody and Herbaceous Plants 201

Lori A. Osburn, Zong-Ming Cheng, and Robert N. Trigiano

Chapter 21 Rooting Cuttings of Tropical Plants 213

Richard A. Criley

Chapter 22 Care and Management of Stock Plants 225

Gary R. Bachman

Part VIII
Propagation by Leaf and Root Cuttings

Chapter 23 Adventitious Shoot and Root Formation on Leaf and Root Cuttings 233

Caula A. Beyl

Chapter 24 Propagation by Leaf Cuttings 245

John L Griffis, Jr., and Malcolm M. Manners

Chapter 25 Propagation of Sumac by Root Cuttings 253

Paul E. Read

Part IX
Layering

Chapter 26 Layering 261

Brian Maynard

Part X
Grafting and Budding

Chapter 27 Grafting: Theory and Practice 273

Kenneth W. Mudge

Chapter 28 Grafting and Budding Exercises with Woody and Herbaceous Species 293

Garry V. McDonald

Part XI
Bulbs and Plants with Special Structures

Chapter 29 Storage Organs 303

Jeffrey A. Adkins and William B. Miller

Chapter 30 Propagating Selected Flower Bulb Species 311

William B. Miller, Jeffrey A. Adkins, and John E. Preece

Part XII
Micropropagation

Chapter 31 Micropropagation 319

Michael E. Kane, Philip Kauth, and Scott Stewart

Chapter 32 Getting Started with Tissue Culture: Media Preparation, Sterile Technique, and Laboratory Equipment 333

Caula A. Beyl

Chapter 33 Micropropagation of Mint (*Mentha* spp.) 347

Sherry L. Kitto

Chapter 34 Micropropagation of Tropical Root and Tuber Crops 355

Leopold M. Nyochembeng

Chapter 35 Micropropagation of Woody Plants 365

Robert R. Tripepi

Part XIII
Seed Production and Propagation

Chapter 36 Sexual Reproduction in Angiosperms 379

Robert N. Trigiano, Renee A. Follum, and Caula A. Beyl

Chapter 37 Breeding Horticultural Plants 391

Timothy A. Rinehart and Sandra M. Reed

Chapter 38 Seed Production, Processing, and Analysis 401

J. Kim Pittcock

Chapter 39 Environmental Factors Affecting Seed Germination 407

Emily E. Hoover

Chapter 40 Producing Seedlings and Bedding Plants 411

Holly L. Scoggins

Chapter 41 Practices to Promote Seed Germination: Scarification, Stratification, and Priming 421

Caula A. Beyl

Part XIV
In Conclusion: Special Topics

Chapter 42 Myths of Plant Propagation 435

Jeffrey H. Gillman

Chapter 43 Intellectual Property Protection for Plants 441

Christopher Eisenschenk

Index 449

Acknowledgments

Foremost, we wish to recognize the extraordinary efforts and talents of all the contributing authors—their creativity, support, advice, understanding, and patience throughout the conception and long-delayed development of this book was nothing less than phenomenal; the Institute of Agriculture at the University of Tennessee and especially Dr. Carl Jones, head of Entomology and Plant Pathology for providing the time and financial support necessary for RNT to complete the book. We wish to gratefully acknowledge the contributions of Dr. John E. Preece; we value him as a colleague and a friend. We value the extra contributions of our colleagues, Margery L. Daughtrey, Alan S. Windham, and Frank A. Hale, who provided many fine photographs included in the text and CD ROM. We also extend very special thanks to John Sulzycki, senior editor at CRC Press, who worked tirelessly to see this project through to completion and never lost confidence in its editors and contributors to finish the task. We thank Pat Roberson, Gail Renard, and Randy Brehm at CRC Press, whose constant encouragement and work were essential for the completion of this textbook. Lastly, RNT sincerely thanks CAB, RLB, SAB, DJG, BHO, JS, ANT, and CGT for their advice, insight, friendship, and constant support—you've all made my life much richer.

The Editors

Caula A. Beyl, dean of the College of Agricultural Sciences and Natural Resources, received a B.S. degree in biology with majors in botany and zoology in 1974 from Florida Atlantic University, Boca Raton, and an M.S. degree in horticulture from Purdue University in West Lafayette, Indiana, in 1977. In 1979, she obtained a Ph.D. from Purdue University in the area of stress physiology. She began her academic career as a postdoctoral researcher at Alabama A&M University, Huntsville, and after one year, she joined the faculty in horticulture in the Department of Natural Resources and Environmental Studies. She was promoted to associate professor in 1987 and professor in 1992. In more than 27 years as a researcher, teacher, and administrator, she has served as principal investigator or coinvestigator on 41 funded research projects in various areas of horticulture, stress physiology, and space biology. She has been major advisor to 8 doctoral candidates and 28 master's students, half of whom were minority students. Her research has resulted in 37 refereed research publications, 7 book chapters, and 115 abstracts, presentations, or proceedings, 17 of which were on institutional research topics. Dr. Beyl is a member of the American Society for Horticultural Science and the honor society of Gamma Sigma Delta. She served as the editor for the *Plant Growth Regulation Quarterly* from 2001 to 2005 and associate editor in the area of environmental stress physiology for the *Journal of the American Society for Horticultural Science* from 2000 to 2002. In 1995, she received the School of Agriculture Outstanding Researcher Award and, in 1998, the Abbott Award for outstanding research paper from the Plant Growth Regulation Society. In 2005, she won the AAMU Researcher of the Month Award. As an undergraduate and graduate educator, Dr. Beyl has taught 14 different courses including agricultural leadership and a graduate-level scientific writing course and was honored for outstanding teaching with the Alabama A&M University Outstanding Teacher Award in 1998 and School of Agriculture Outstanding Teacher Awards in 1998 and 1991. In 2003, she was the recipient of the Distinguished Alumna Award from the horticulture program at Purdue University. From 1998 to 2002, she served as the director of the Plant Science Center and guided curriculum revision, scholarship funding, and recruitment efforts for the center.

In 2002, she assumed the role of director of the Office of Institutional Planning, Research, and Evaluation. As such, she revitalized the office, helped to revise and write new strategic and effectiveness documents for the university, and was essential in guiding Alabama A&M University through its reaffirmation of accreditation process including the development of a Quality Enhancement Plan. The planning documents that she helped to develop were honored by the Southern Association of Institutional Researchers with the Outstanding Planning Document Award for 2004. She has served as a consultant to Bethune Cookman College, Bennett College for Women, and Southern University at Baton Rouge for SACS, QEP, and learning outcome development. She also served as the Quality Enhancement Plan evaluator for SACS during its onsite visit to University of Central Florida. She has presented on a variety of retention, institutional planning, learning outcomes, and QEP development topics to NACDRAO, SAIR, SEF, and SACS, among others. She served as an Alabama Quality Award Examiner in 2006. In 2007, she was appointed interim dean of graduate studies at Alabama A&M University. Since joining the University of Tennessee in June of 2007 as the dean of the College of Agricultural Sciences and Natural Resources, she has been active implementing strategies to focus on retention, recruitment, and diversity issues as well as integrating technological paradigms into the classroom to enhance learning of the millennial generation.

Robert N. Trigiano received his B.S. degree with an emphasis in biology and chemistry from Juniata College, Huntingdon, Pennsylvania, in 1975 and an M.S. in biology (mycology) from the Pennsylvania State University in 1977. He was an associate research agronomist working with mushroom culture and plant pathology for Green Giant Co., Le Sueur, Minnesota, until 1979 and then a Mushroom Grower for Rol-Land Farms, Ltd., Blenheim, Ontario, Canada, during 1979 and 1980. He completed a Ph.D. in botany and plant pathology (co-majors) at North Carolina State University at Raleigh in 1983. After concluding postdoctoral work in the Plant and Soil Science Department at the University of Tennessee, he was appointed an assistant professor in the Department of Ornamental Horticulture and Landscape Design at the same university in 1987, promoted to associate professor in 1991, and to professor in 1997. He served as interim head of the department from 1999 to 2001. He joined the Department of Entomology and Plant Pathology at the University of Tennessee in 2002.

Dr. Trigiano is a member of the American Phytopathological Society (APS), the American Society for

Horticultural Science (ASHS), and the honorary societies of Gamma Sigma Delta, Sigma Xi, and Phi Kappa Phi. He received the T. J. Whatley Distinguished Young Scientist Award (The University of Tennessee, Institute of Agriculture) and the Gamma Sigma Delta Research individual and team Award of Merit (The University of Tennessee). He has received ASHS publication award for the most outstanding educational paper and the Southern region ASHS L. M. Ware distinguished research award. In 2006, he was elected Fellow of the American Society for Horticultural Science. He was awarded the B. Otto and Kathleen Wheeley Award of Excellence in Technology Transfer at the University of Tennessee in 2007. He has been an editor for the ASHS journals, *Plant Cell, Tissue and Organ Culture* and *Plant Cell Reports* and is currently the co-editor of *Critical Reviews in Plant Sciences* and senior editor of *Plant Disease*. Additionally, he has co-edited six books, including *Plant Tissue Culture Concepts and Laboratory Exercises, Plant Pathology Concepts and Laboratory Exercises,* and *Plant Development and Biotechnology.*

Dr. Trigiano has been the recipient of several research grants from the U.S. Department of Agriculture (USDA), Horticultural Research Institute, and from private industries and foundations. He has published more than 200 research papers, book chapters, and popular press articles. He teaches undergraduate/graduate courses in plant tissue culture, mycology, DNA analysis, protein gel electrophoresis, and plant microtechnique. His current research interests include diseases of ornamental plants, somatic embryogenesis and micropropagation of ornamental species, fungal physiology, population analysis, DNA profiling of fungi and plants, and gene discovery.

Contributors

Jeffrey A. Adkins
Department of Plant Sciences and Entomology
University of Rhode Island
Kingston, Rhode Island

Gary R. Bachman
Coastal Research and Extension Center
Mississippi State University
Biloxi, Mississippi

Caula A. Beyl
Department of Plant Sciences
University of Tennessee
Knoxville, Tennessee

David W. Burger
Department of Plant Sciences
University of California
Davis, California

Zong-Ming Cheng
Department of Plant Sciences
University of Tennessee
Knoxville, Tennessee

Calvin Chong
Department of Plant Agriculture
University of Guelph
Guelph, Ontario, Canada

Michael E. Compton
School of Agriculture
University of Wisconsin-Platteville
Platteville, Wisconsin

Richard A. Criley
Department of Tropical Plant and Soil Sciences
University of Hawaii
Honolulu, Hawaii

Adam Dale
Department of Plant Agriculture
University of Guelph
Simcoe, Ontario, Canada

Danielle J. Donnelly
Department of Plant Science
McGill University
Ste-Anne-de-Bellevue, Quebec, Canada

Christopher Eisenschenk
Saliwanchik, Lloyd & Saliwanchik
Gainesville, Florida

Renee A. Follum
Entomology and Plant Pathology
University of Tennessee
Knoxville, Tennessee

Jeffrey H. Gilman
Department of Horticultural Science
University of Minnesota
St. Paul, Minnesota

John L Griffis, Jr.
Department of Tropical Plant and Soil Sciences
University of Hawaii at Manoa
Honolulu, Hawaii

Patricia S. Holloway
Department of Plant, Animal and Soil Sciences
University of Alaska–Fairbanks
Fairbanks, Alaska

Emily E. Hoover
Department of Horticultural Science
University of Minnesota
St. Paul, Minnesota

Michael E. Kane
Environmental Horticulture Department
University of Florida
Gainesville, Florida

Philip Kauth
Environmental Horticulture Department
University of Florida
Gainesville, Florida

Sherry L. Kitto
Department of Plant and Soil Sciences
University of Delaware
Newark, Delaware

Gerald L. Klingaman
Department of Horticulture
University of Arkansas
Fayetteville, Arkansas

William E. Klingeman
Department of Plant Sciences
University of Tennessee
Knoxville, Tennessee

Yi Li
Department of Plant Science
University of Connecticut
Storrs, Connecticut

Malcolm M. Manners
Horticulture Department
Florida Southern College
Lakeland, Florida

Brian Maynard
Department of Plant Sciences and Entomology
University of Rhode Island
Kingston, Rhode Island

Garry V. McDonald
Department of Horticulture
University of Arkansas
Fayetteville, Arkansas

Mark P. McQuilken
Life Sciences Teaching Group
The Scottish Agricultural College
Ayr Campus, Auchincruive Estate, Ayr
Scotland, United Kingdom

William B. Miller
Department of Horticulture
Cornell University
Ithaca, New York

Kenneth W. Mudge
Department of Horticulture
Cornell University
Ithaca, New York

Margaret A. Norton
Department of Natural Resources and Environmental Sciences
University of Illinois at Urbana-Champaign
Urbana, Illinois

Leopold M. Nyochembeng
Department of Natural Resources
Alabama A&M University
Normal, Alabama

Lori A. Osburn
Department of Plant Sciences
University of Tennessee
Knoxville, Tennessee

Ellen T. Paparozzi
Department of Agronomy and Horticulture
University of Nebraska–Lincoln
Lincoln, Nebraska

J. Kim Pittcock
Dean of Agriculture
Arkansas State University
Jonesboro, Arkansas

John E. Preece
Plant, Soil, and Agricultural Systems
Southern Illinois University
Carbondale, Illinois

Paul E. Read
Department of Agronomy and Horticulture
University of Nebraska–Lincoln
Lincoln, Nebraska

Sandra M. Reed
USDA, Agricultural Research Service
Floral and Nursery Plants Research Unit, U.S. National Arboretum
McMinnville, Tennessee

Timothy A. Rinehart
USDA-ARS MSA,
Southern Horticultural Laboratory
Poplarville, Mississippi

John M. Ruter
Department of Horticulture
University of Georgia
Tifton, Georgia

Contributors

Holly L. Scoggins
Department of Horticulture
Virginia Tech
Blacksburg, Virginia

Robert M. Skirvin
Department of Natural Resources and Environmental Sciences
University of Illinois at Urbana-Champaign
Urbana, Illinois

James J. Stapleton
Kearney Agricultural Center
University of California
Parlier, California

Scott Stewart
Environmental Horticulture Department
University of Florida
Gainesville, Florida

Ellen G. Sutter (deceased)
Department of Plant Sciences
University of California
Davis, California

Milton E. Tignor
Haywood Community College
Clyde, North Carolina

Robert N. Trigiano
Department of Entomology and Plant Pathology
University of Tennessee
Knoxville, Tennessee

Robert R. Tripepi
Plant, Soil and Entomological Sciences
University of Idaho
Moscow, Idaho

Alan S. Windham
Department of Entomology and Plant Pathology
University of Tennessee
Knoxville, Tennessee

Zhen Zhang
College of Horticulture
Nanjing Agricultural University
The People's Republic of China

Part I

Introduction

1 Introduction to Plant Propagation

Caula A. Beyl and Robert N. Trigiano

CHAPTER 1 CONCEPTS

- Early Greeks, Romans, and Chinese knew and used various techniques of plant propagation including rooting cuttings, air layering, and graftage.

- Knowledge of plant growth hormones and the role of the endogenous growth regulator, auxin, in promoting the initiation of roots was a major milestone in plant propagation development.

- Various methods are used to propagate plants, but ultimately the methodology used depends upon whether a plant is desired that is genetically identical to the original.

- Sexual processes result in propagation by seed, particularly important to the vegetable, bedding plant, and nursery industries.

- Asexual methods include division, cuttage using various parts of the plant, such as stems, leaves, and roots, and budding and grafting.

- There are many challenges to successful plant propagation including lack of knowledge, recalcitrance associated with the phase change from juvenility to maturity, and various pathogens that thrive in the propagation environment.

- Future approaches to plant propagation may include genetic and physiological studies of rooting, manipulation through tissue culture to stabilize the juvenile phenotype, adjusting conditions in which stock plants are held to optimize rooting of propagules taken from them, and use of special properties of certain bacteria, such as *Agrobacterium rhizogenes*, to induce roots, in this case, hairy roots.

INTRODUCTION

Man's dependence on plants as the most significant source of food depends predominately on the ability not only to cultivate plants and utilize them, but also to propagate them. This knowledge of plant propagation, the multiplication or making of more plants, is both a fascinating art and an exciting science. Merely walking through the produce department of a grocery store and examining the array of fruits, vegetables, and flowers available should engender an appreciation for all the various techniques of plant propagation used. If the plants themselves could reveal how they were propagated, an impressive array of techniques would be described, from the relatively simple methods of seed germination and grafting to the more cutting-edge techniques of tissue culture, which often include molecular genetics.

HISTORY OF PLANT PROPAGATION

The origins of plant propagation are hard to document, but it is reasonable to believe that plant propagation co-developed with agriculture approximately 10,000 years ago. The earliest propagation may have been an inadvertent sowing of seeds gathered during collection and harvesting activities, which evolved into deliberate agriculture. Descriptions of early horticulture in Egypt, Babylon, China, and other countries suggest that the culture of ornamental and food crops was fairly well understood and that they could be propagated easily. The Old Testament contains the following passage from Ezekial 17:22: "... I myself will take a shoot from the very top of a cedar and plant it. I will break off a tender sprig from its topmost shoots and plant it on a high and lofty mountain ... it will produce branches and bear fruit and become a splendid cedar...." This indicates that the concept of taking cuttings was well known at that time. Babylon and Assyria were known for terraced gardens and parks. Such deliberate and extensive cultivation of ornamentals required knowledge of how they could be propagated. The Greek philosopher Theophrastus (371 to 287 BC), a student of Aristotle, made observations on suckering of olive, pear, and pomegranate. With respect to grafting, he even wrote

FIGURE 1.1 Mosaic from the south of France at St. Romain en Gal near Montpellier, dating from the first half of the first century of the present era and depicting topworking of trees.

in *De Causis Plantarum*, "It is also reasonable that grafts should best take hold when scion and stock have the same bark, for the change is smallest between trees of the same kind...." Roman writings contain references to budding and grafting, and Roman mosaics depict grafting (Figure 1.1). Cato, a Roman statesman in the second century, described cuttings or scions of apple grafted onto sturdy rootstocks. Graftage and topworking were illustrated in medieval treatises (Figure 1.2).

The Wardian case, an interesting invention by Dr. Nathanial Ward in the early 1800s, enabled not only germination of fern spores and orchids, but also transportation of newly germinated and delicate plants across long distances. This enabled the plants to survive and arrive at their destinations in good condition. Wardian cases became quite popular for growing orchids, but were also instrumental in helping establish the tea plantations of Assam and rubber trees in Ceylon. Today, replicas of Wardian cases are still used by garden enthusiasts for propagating and protecting sensitive plants (Figure 1.3).

Much of modern plant propagation is unchanged from practices in use before the 1900s. The types of cuttings that we use today are the following: softwood, semihardwood, hardwood, herbaceous, leaf, and root. All were used at that time. Propagators were aware that many cuttings root at nodes and were maximizing cutting production by using single-node (leafbud) cuttings. Growers were propagating plants using bulbs, rhizomes, stolons, with techniques such as separation and division. Both layering and

FIGURE 1.2 A sixteenth-century plate depicting steps to be used for successfully topworking trees. (From Huxley, Anthony, *An Illustrated History of Gardening*, Lyons Press, New York. With permission.)

Introduction to Plant Propagation

FIGURE 1.3 Modern replica of the traditional Wardian case commonly used for propagation of ferns and orchids and culture of plants needing protection.

grafting were used commercially before the 1900s, and methods used then are largely unchanged today.

Transpiration of cuttings was controlled using high-humidity environments, such as glass-covered rooting chambers, some of which had supplemental bottom heat provided by lamps or recirculating water systems. Diseases (likely caused by *Botrytis* and other pathogens) were a problem in these rooting chambers. Mist and fog for propagation were not developed until the twentieth century (Figure 1.4).

The major discoveries and advancements in plant propagation in the twentieth century are the developments of intermittent mist systems and fog systems, discovery of plant growth substances, sanitation and disease control, knowledge of juvenility and chimeras, micropropagation, and the use of micropropagation techniques to escape pathogens, such as viruses, mycoplasms, and bacteria.

HEAT (STEAM) DISINFESTATION OF MEDIA

Many losses were incurred because of the presence of weeds and pathogens in the soil before the twentieth century. In the early 1900s, growers were beginning to use steam to disinfest soil prior to planting. By the late 1930s to early 1940s, people began partially disinfesting (pasteurizing) greenhouse and propagation media because the plants grew better than when the soils were sterilized by overheating. By the 1950s, people were aware that nitrifying and other beneficial microbes were killed by steam sterilization, and that has led to modern steaming methods aimed at killing insects, pathogens, and weed seeds

FIGURE 1.4 Mist being used over cuttings in a greenhouse. (Photo coutesy of John Ruter, University of Georgia.)

DEVELOPMENT OF MIST AND FOG PROPAGATION SYSTEMS

Prior to the twentieth century, transpiration from cuttings was reduced by use of high-humidity chambers, shading, cutting leaves in half and otherwise reducing leaf area, keeping the soil moist, and frequently sprinkling cuttings with water. By the 1940s and 1950s, intermittent mist and fog were being used for plant propagation. Commercial adoption of fog lagged behind mist because of disease problems. By the 1970s, it was well known that ventilation combined with fog reduced incidence of disease, and therefore fog propagation has become more commonplace at commercial nurseries. Fog and intermittent mist are now indispensable in cutting propagation of plants.

AUXINS

One of the most important discoveries in the history of plant propagation was the auxins and their role in root initiation. It was known for some time prior to the 1930s that the presence of leaves on cuttings was critical for adventitious rooting. In fact, it was known that concentrated juice from leaves could stimulate rooting of cuttings. In the early 1930s, the auxin indole-3-acetic acid (IAA) was isolated and shown to stimulate root formation. This hormone is produced in the growing points of shoots, especially in the young, expanding leaves. By the late 1930s, indole-3-butyric acid (IBA), naphthaleneacetic acid (NAA), and their commercial formulations Auxan® and Rootone® were available and were being tested for rooting cuttings of many different species of plants. The use of auxin has increased the percentage of cuttings that root and the number and distribution of roots on cuttings. The quick commercial adoption of auxins for rooting speaks to the great need for root-inducing substances.

MICROPROPAGATION

Early in the twentieth century, pioneers in plant tissue culture first developed *in vitro* techniques for germination of orchid seeds. Later, a medium was developed that supported the growth of roots floating on its surface. However, *in vitro* culture was far from being commercially viable because of the lack of a plant growth regulator that would stimulate cell division.

By the early 1950s, researchers had shown that water extracts from vascular tissue of tobacco, malt extract, liquid *Cocos nucifera* L. (coconut) endosperm, and an extract from solid coconut endosperm all induced cell division. Additionally, autoclaved DNA from herring sperm and from calf thymus stimulated cell division in tobacco wound callus tissue cultured *in vitro*. This activity was not present if the DNA was not autoclaved.

Cytokinins are a class of growth regulators that induce cell (*cyto*) division (*kinin*). The first cytokinin, kinetin, was isolated, purified, and crystallized from autoclaved animal DNA and was shown to stimulate cell division and shoot multiplication. Since then, benzyladenine, thidiazuron, and others have been developed. These compounds are much more potent (effects expressed at much lower concentrations) cytokinins than kinetin and are in wide use in the commercial micropropagation industry today.

By manipulating the plant growth regulators, medium formulation, and plant materials, it is now possible to produce somatic (vegetative) embryos, adventitious shoots, axillary shoots, and adventitious roots using *in vitro* techniques. As a result, commercial micropropagation is an important part of the plant propagation industry.

ESCAPING PATHOGENS

Plant diseases, especially those caused by viruses, are especially problematic because plants with viruses cannot be "cured." During the late 1940s, there was an outbreak of spotted wilt disease of *Dahlia pinnata* (Dahlia), which is caused by a virus. Tip cuttings rooted from infected plants, however, showed no spotted wilt symptoms. By the early 1950s, scientists were excising very small 250 µm apical meristematic domes from dahlia plants with symptoms of dahlia mosaic and placing them *in vitro* where they elongated. When the shoots were 1 to 2 cm long, they were grafted onto young virus-free plants, where they grew normally with no dahlia mosaic symptoms.

By the mid- to late-1950s, scientists were combining thermotherapy with tissue culture. Virus-infected plants were grown at 25°C to 40°C prior to excising meristematic domes and placing them *in vitro*. A percentage of the resulting plants were shown to be free from certain viruses by grafting onto indicator plants or using modern immunoassay techniques, such as enzyme-linked immunosorbent assay (ELISA), which was developed in the 1970s for the detection of viruses. Therefore, plant pathologists have helped plant propagators in the twentieth century by developing methods for steaming and disinfestation of media, benches, and planting containers. In turn, plant propagators have helped plant pathologists by developing techniques including rooting of cuttings, grafting, and tissue culture that can be used to escape viruses. Escaping viruses is an important part of the fruit, nut, and vegetable industries of the twenty-first century.

ROLE AND IMPORTANCE OF PLANT PROPAGATION TODAY

Everything that is dependent on plants in any way is also dependent on the ability of plants to be propagated.

Large-scale agriculture of crops, such as for grains and certain vegetables, depends on successful seed germination. Looking at it another way, you could say that seeds are the first necessary step in the world's food chain, and this dependency has resulted in a huge industry. The seed industry represents a total worldwide market of $21 billion per year (ETC Group, 2005). The requirements to germinate can be as simple as appropriate moisture and temperature conditions, or they can be complex, requiring several months of moist, cold chilling treatment (stratification) or other specialized environmental conditions. Sometimes, the seed coat with its barrier to water penetration must be broken or weakened (scarification). Large numbers of seed can be planted mechanically, but in some cases, they must be planted by hand to avoid damage and obtain a better stand, such as for buckeye (*Aesculus parviflora* Walt.) (Figure 1.5). Nursery industries all over the world use a variety of plant propagation techniques from T-budding dogwoods in Tennessee (Figure 1.6) and June-budding peaches in Georgia to stool-bedding apple rootstocks in England. Nurserymen propagating ornamental trees, shrubs, and grasses or specializing in fruit trees must know the best way to propagate each species and have a thorough knowledge of the best "window of opportunity" that works for each species.

Floral and nursery business are two segments of horticulture that have been recognized as the fastest growing in recent years. The growth and demand for bedding plants has probably been the most significant factor in this growth over the past twenty years. Establishment of bedding plants is dependent upon successful and uniform seed germination. Another area of horticulture dependent

FIGURE 1.5 Seeds of Ohio Buckeye (*Aesculus parvifolia*) being planted by hand at Shadow Nursery, Winchester, TN, to prevent damage and increase the number that will germinate successfully. (Photo courtesy of Don Shadow, Shadow Nursery, Winchester, TN.)

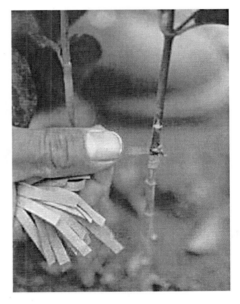

FIGURE 1.6 Fall budding dogwoods in the field at Shadow Nursery in Winchester, TN. The bud stick from which the bud of the desired cultivar was taken can be seen in the propagator's right hand. Notice that the budding rubber used to wrap the T-bud is wrapped from below the insertion point upward. (Photo courtesy of Don Shadow, Shadow Nursery, Winchester, TN.)

upon successful seed germination and production of transplants is the vegetable production industry. The Chinese described grafting of vegetable crops as early as the fifth century, and today both Japan and Korea have developed robots and grafting machines to produce large numbers of grafted tomato, eggplant, and peppers.

Even the forest industry is very interested in plant propagation techniques being used to improve performance of the trees, from establishment of seed orchards containing superior tree selections to use of cuttings of superior clones. Some approaches have focused on somatic embryogenesis or creating an embryo identical to the original from the somatic portion of the plant. Clonal approaches are being studied all over the world for species such as eucalyptus and pine, and molecular genetics techniques are being explored to boost the genes for auxin production in aspen to increase how rapidly the transgenic tree will grow.

The craft and science of plant propagation is particularly important for the preservation of rare and endangered plants. Both traditional and specialized propagation techniques are used to propagate species recognized as either rare or endangered so that they may be preserved and, in some cases, reintroduced into new and old habitats. Sometimes, these techniques involve *in vitro* seed germination, and in other cases, various tissues may be used in tissue culture to create a large number of plants very rapidly for species that respond well to the *in vitro* environment.

Plant propagators working for arboreta and botanical gardens must have a thorough knowledge of plant propagation to be able to germinate seeds sent to them from all parts of the world or to increase numbers of rare or highly valued specimen plants to share with other arboreta and botanical gardens.

HOW THIS BOOK IS ORGANIZED

This book is based on a successful model of organization used in *Plant Tissue Culture Concepts and Laboratory Exercises,* Second Edition, and *Plant Pathology Concepts and Laboratory Exercises,* Second Edition, which combines concept chapters with accompanying laboratory exercises to provide additional in-depth information and hands-on tasks that illustrate the principles found in the concept chapters. This not only aids those using the book for teaching, but also provides examples of techniques that can be used as models for other species by those using the book as a reference. The various chapters are contributed by horticultural scientists with expertise in the disciplines of plant propagation, breeding, pathology, tissue culture, and seed technology, among others, who have had extensive experience teaching plant propagation.

The concept chapters always begin with a list of some of the more important ideas, the concept box, contained in the section. These bullet lists are intended to be a type of "executive summary" for the chapter as well as to alert students to major points of the topics. The laboratory exercises are organized in a standard format throughout the book. Teachers and students are provided a list of materials necessary to complete the experiments, followed by step-by-step procedures that detail exact methodology for each exercise. This book is unique in propagation teaching aids, as it provides, in general terms and descriptions, what the anticipated results of the laboratory exercises should be. The last section of the laboratory lists a set of questions that are intended to stimulate discussion and thought about the experiments. The laboratory exercises included in the book have been used repeatedly and proven to be reliable for classroom usage. They provide a broad exposure to students of hands-on application of the concepts presented.

After the Introduction in chapter 1, Part II of the book leads readers through a progression beginning with propagation structures of different types and how these can be managed for more effective propagation in chapter 2. The next two chapters are laboratory exercises that delve into the design and use of various propagation systems. The rationale for the design of mist or fog propagation systems for the production of cuttings or for germinating seed is explored in chapter 3. Chapter 4 compares the rooting performance of cuttings in a propagation tent versus those rooted under mist. Intermittent mist/fog systems are one of the most significant innovations in asexual propagation, enabling successful propagation of many species by providing a high-humidity environment conducive to rooting. Chapter 4 allows study of the various control systems that determine misting frequency and how to determine whether the mist is being distributed uniformly across the bench.

A complete book on plant propagation cannot neglect the containers and media used for seed germination, cutting propagation, and acclimatization of cuttings. Part III, chapter 5 details the types of media and their characteristics that take the place of soil in today's propagation media, since they provide many advantages over soil including lightness and ease of handling as well as freedom from pathogens, pests, weed seed, or other contaminants. Chapter 6 deals with the factors that must be considered before an appropriate medium can be chosen and gives procedures for determining bulk density and pore space, important characteristics of media. This chapter is followed by laboratory exercises used for seed propagation (chapter 7).

Sanitation can determine the success or failure of seed or cutting propagation and can make the difference whether a commercial propagation establishment is economically viable. Part IV deals with the major threats to successful plant propagation offered by plant diseases and the importance of sanitation. In chapter 8, the characteristics of various common pathogens are presented, along

Introduction to Plant Propagation

with information on how they are typically spread. These concepts are reinforced by laboratory exercises on how to disinfest soil and planting media in chapter 9 and an exercise that illustrates the impact of *Botrytis* and other propagation pathogens in chapter 10. It also contains laboratory exercises on chemical control of gray mold, effect of ventilation on its development, and biological control of *Pythium*. Chapter 11 provides important information on the concept of producing specific pathogen-free plant and crop certification programs in Canada. Integrated pest management, which has almost become ubiquitous in literature on greenhouse and nursery management, is detailed in chapter 12 with a focus on its application to propagation systems.

Part V contains only one chapter, but it is a very important chapter for anyone who conducts plant propagation research, whether an undergraduate or graduate student, or a professional in the discipline. Chapter 13 concerns how to evaluate propagation experiments, collect, handle, and analyze the data. In most books on plant propagation, this issue is ignored completely or mentioned only briefly.

In Part VI, there are three chapters, each a concept with its associated laboratory exercises, that explore general concepts for successful vegetative propagation including plant growth substances used in propagation (chapter 14), the role of juvenility and phase change in propagation success (chapter 15), and how chimeras affect the outcome of vegetative propagation (chapter 16). The use of plant growth substances, such as indolebutyric acid and naphthaleneacetic acid, has been one of the most important discoveries for plant propagators, having a major impact on the number of species that can now be clonally propagated via stem cuttings. These auxin-type growth regulators stimulate the induction and formation of roots on cuttings. The change of phase of a plant from juvenile to physiologically mature also greatly impacts the likelihood of propagation success. Cuttings of some species are very difficult to impossible to root once they have become mature, and propagators have used a number of different techniques to "restore" juvenility, such as serial propagation, severe heading back, and *in vitro* culture. Chimeras are plants composed of layers or sectors that have different genotypes, so depending on where new shoots arise to become new plants, they may have a different phenotype than the parent plant. The genotype is the heritable genetic blueprint for the plant, and the phenotype is the manifestation of the plant that results from putting that genetic blueprint to work—its structure, function, and behavior.

Parts VII through XIII explore the various types of propagation, starting with the use of stem cuttings in Part VII. For many, the idea that you can make a new plant genetically identical to the first by taking a cutting, inserting it into a rooting medium, keeping it moist and humid, and then watching new roots form is tremendously exciting. Home gardeners share cuttings this way quite frequently. The types of cuttings and the kinds of conditions that help them to root successfully are explored in chapter 17. From a scientific standpoint, it is also exciting to follow the anatomical and physiological changes that occur in rooting that are detailed in chapter 18. The laboratory exercises described in chapter 19 and chapter 20 demonstrate the effects of different concentrations and formulations of auxins on rooting of cuttings. Chapter 21 goes a bit farther with variations on how the cuttings are made, using tropical foliage plants as the experimental subject. The previous chapters in this section have focused on the propagule used, in this case stem cuttings, but chapter 22 focuses on how the care and management of the stock plant from which the cuttings are taken can influence how successfully the cuttings will root.

Many people are familiar with the use of stem cuttings, but leaves and roots can also serve as very successful propagules (Part VIII). Many plants can be successfully rooted from just the leaf blade and others from the leaf and petiole (chapter 23). Many home horticulturists do this with leaf and petiole cuttings of African violets rooted in water. Many other techniques with leaf cuttings are described for students to try in the laboratory exercises in chapter 24. For other plants, particularly those that sucker readily, roots can serve as a source of cuttings or can be forced to produce shoots that, in turn, can be used as cuttings (chapter 25). There are many examples among woody trees of new trees developing from the roots of older ones until dense stands are formed (aspen, sumac). Chapter 25 uses one of these, sumac, to demonstrate propagation by root cuttings.

Part IX contains chapter 26, which describes the time-honored technique of layering to produce new clonal plants on their own roots. Layering involves techniques that encourage roots to form before the cutting is severed from the parent plant. Layering techniques probably were adopted through close observation of what occurs in nature with some shrubs or trees. Low-lying branches that come into prolonged contact with the ground may form roots at that point. The technique, although slow, is still valuable with some species that are extremely difficult to root with cutting propagation.

Techniques of grafting and budding in Part X do require skill as well as science. Depictions of early knowledge and application of grafting has occurred in historical depictions from the sixteenth century (Figure 1.2). Grafting and budding both involve combining two plants, one serving as the rootstock or bottom portion of the plant, and the other serving as the top portion of the plant (chapter 27). With grafting, the scion or a piece of stem containing several buds is cut so that it can be joined with the rootstock. With budding, the scion piece consists of only one bud. These techniques are extremely important in the propagation of many fruit tree species including apple and peach. Chapter 28 offers a variety of very easy-to-complete laboratory exercises on grafting techniques using rose, coleus, and boxwood.

Part XI deals with bulbs and plants with special structures. Chapter 29 describes the use of underground storage structures to propagate plants asexually. We take advantage of the underground storage organs, such as bulbs and corms, and use various techniques to encourage the bulbs to form more propagules. The laboratory exercise in chapter 30 includes experiments with the propagation of lily, a nontunicate bulb (unlike onion, for example, which is protected by a dry papery sheath—a modified leaf, around continuous concentric layers). It also includes an experiment with hyacinth, a nontunicate bulb. Propagation using tubers (Irish potato and dahlia) is covered in an experiment, and the chapter ends with a general observational activity of many bulbous and tuber crops in the field.

In the next section (Part XII), the chapters describe the exciting world of micropropagation, where small pieces of tissue or even single cells can be grown in a sterile environment on medium containing all the nutrients, minerals, vitamins, growth regulators, etc. that the tissue (or explant) needs to proliferate more cells (chapter 31). The cells that grow from that original explant are called callus. Depending upon what growth regulators are used, explants can be tweaked to form shoots and roots or, via another route, embryo-like structures. Because a sterile (axenic) environment is needed and explants have such precise requirements for what is contained in the tissue culture medium, micropropagation laboratories have certain requirements in common (chapter 32). Many of the operations manipulating the explants and placing them on sterile autoclaved medium are conducted in sterile transfer hoods. In chapter 33, the procedures for micropropagation of mint are described, from disinfesting the shoots for use as explants and preparing the medium to placing them in the culture vessels. Chapter 34 describes micropropagation of tropical plants, and chapter 35 describes special procedures used with woody plants.

The most fundamental classification of plant propagation systems is not by the technique used, but by whether they are the result of sexual or asexual processes. The previous parts of the book depicting types of plant propagation have all dealt with asexual or clonal propagation, in which the propagule is an identical genetic copy of the original plant. Part XIII of the book deals with seed propagation. Seeds, unless they are apomictic, are the result of a sexual reproduction of plants. Apomixis is a special case of seed propagation that is asexual. This occurs with some species of citrus and Kentucky bluegrass. The embryo results either from cells of the ovule or from an egg with 2n chromosomes that can develop without fertilization. The features of sexual reproduction in plants are featured in chapter 36, which includes examples of higher plants, ferns, and mosses. Not only understanding the processes involved in developing a seed, but also knowing how to produce and germinate seed are important. Chapter 37 explores plant breeding, a process that is essential to transferring traits between cultivars and development of new cultivars with enhanced traits. Chapter 38 deals with seed production, analysis, and processing and provides a laboratory exercise to emphasize the topic. In chapter 39, the factors that affect seed germination are investigated in experiments that demonstrate how important each of those factors is. Chapter 40 contains information on producing seedlings and bedding plants, without which, garden centers in spring would be much less colorful, and gardens would not get a head start. Seeds have the potential to germinate into seedlings, which then develop into plants. Some seeds only require an appropriate temperature and moist conditions to germinate readily. Others require much more complex conditions before they will germinate. Scarification treatments are used to weaken or remove hard seed coats, which are impediments to germination. Some seeds also require a period of cold moist stratification. In chapter 41, the laboratory exercises acquaint students with various techniques used for scarification and the effectiveness of scarification coupled with stratification. Finally, some seeds are allowed to imbibe or take up water, which allows the initial events in germination to occur, but they are prevented from germinating fully by imbibing them in a germination solution containing solutes or an osmoticum, which keeps germination from proceeding. When the seeds are dried and then imbibed again, they germinate much faster for having had the head start.

The last section (Part XIV) contains chapters that are best described as "in conclusion: special topics." Chapter 42 takes a light-hearted, often humorous look at some propagation myths—those "truths" handed down to you from your grandmother. This chapter may also be described as "Myth Busters." Plant protection (chapter 43) has become increasing important function for the horticultural industry in the past several decades. Horticulture is big business, and tremendous resources are invested in developing new plants. These investments are now typically protected by legal processes, which may include national and international jurisdictions. This chapter provides an overview of plant protection instruments, which may take the form of trademarks, copyrights, or patents, describes plant breeder's rights, outlines the process for awarding of exclusive rights to plants, and briefly addresses remedies or enforcement of these rights.

Lastly, we've included a CD with the book that contains most of the figures appearing in the book. The figures are presented in color with captions and/or notes in PowerPoint formats, which are amenable to classroom or individual study. We have intentionally "mounted" the figures on a blank background so that they may be easily incorporated into your own presentations. Supplemental photographs have been added to some of the chapters to further illustrate important concepts.

FUTURE OF PLANT PROPAGATION

Many different approaches are being pursued to develop new methods of plant propagation, and these may include a better understanding of the genetic and physiological bases of rooting, manipulation through tissue culture to stabilize the juvenile phenotype, adjusting conditions that stock plants are held in to optimize rooting of propagules taken from them, and use of special properties of certain bacteria, such as *Agrobacterium rhizogenes*, to induce roots, in this case, hairy roots. Another approach may employ genetic engineering techniques to identify genes that contribute to ease of rooting and insert those genes into more recalcitrant species.

Just examining the scientific literature devoted to plant propagation and its various techniques, such as tissue culture, can be overwhelming. For someone wanting the latest information about plant propagation and the physiology associated with induction of rooting, a membership in the American Society for Horticultural Science (ASHS) would be an excellent approach. ASHS provides outlets for peer-reviewed research in its three journals, which range from predominately applied research to somewhat basic. Another excellent source of current information about the latest techniques and how they have been applied to various species that are difficult to propagate is the International Plant Propagator's Society. Since 1951, this society has been sharing information about plant propagation. It has eight regions, including three in the United States—Eastern Region, Southern Region, and Western Region. Attending its meetings is a delight for anyone who wants to focus specifically on propagating plants.

As you learn more about plant propagation in the following chapters, you will appreciate the variety of techniques used to propagate plants and the diversity of the kinds of propagules used. We hope that this knowledge will prepare you to be successful in your future plant propagation endeavors. We look forward to receiving any comments or suggestions that you may have that would improve either the content or presentations in the book and CD.

SUGGESTED LITERATURE

ETC Group. 2005. *Global Seed Industry Concentration—2005*. ETC Group Communiqué, September/October, 2005. Issue 90.

Janick, J. 2002. Ancient Egyptian agriculture and the origins of horticulture. *Acta Hort*. 582: 23–39.

Janick, J. History of Horticulture lecture series. http://newcrop.hort.purdue.edu/newcrop/history/default.html. Accessed August 27, 2006.

Theophrastus (ca. 300 BC). The modes of propagation in woody and herbaceous plants. Propagation in another tree: grafting, in *Classic Papers in Horticultural Science*. Janick, J. (Ed.), American Society for Horticultural Science, Prentice Hall, Englewood Cliffs, NJ, 1989.

Part II

Structures for Plant Propagation

2 Propagation Structures: Types and Management

Gerald L. Klingaman

CHAPTER 2 CONCEPTS

- A variety of structures, such as greenhouses, cold frames, poly tunnels, and hot beds are used to control the propagation environment.

- Facilities should be designed to allow for maximum flexibility in terms of usage and potential expansion.

- Propagation sites should have access to large volumes of high-quality water, which should be tested for salinity, alkalinity, pH, hardness, electrical conductivity, and mineral content.

- Greenhouse coverings (roofs) are made from polyethylene or other films of various compositions, fiberglass-reinforced plastic, rigid acrylic or polycarbonate resins, and glass. Each of these types of coverings offers advantages and disadvantages.

- Bench systems can be arranged to maximize space utilization—rolling benches use as much as 90% of the usable space, whereas peninsular arrangements are about 75% efficient. Many growers also use floor space for growing plants.

- Heating of greenhouse structures can be accomplished using gas and oil as fuels and passive solar irradiation; electrical heating systems are typically only used to power heat mats and/or heating cables.

- Greenhouses are cooled by ventilation (open roofs or side vents), wet pads and cooling fans (evaporative cooling), fog systems, and most commonly by reducing solar heating with shade cloth.

An environment suitable for plant growth must be provided during propagation, taking into account the reduced capacity of the plant to take up water without a root system. Of primary concern is that desiccation must be prevented until the plant becomes sufficiently established to endure the rigors of the real-world environment. By thoroughly understanding seasonal growth patterns and weather cycles, and with a good measure of luck, people have propagated plants for centuries with little or no special equipment. In today's economy, a variety of structures, such as greenhouses, cold frames, and hot beds, are employed to increase control over the environment to maximize the likelihood of success.

SITE SELECTION

LAYOUT

The type of propagation dictates the facilities that are necessary. Facilities should be designed to allow for maximum flexibility in terms of usage and potential for expansion because businesses may grow, and they may often have to respond to changes in market demand. Because propagation is the most basic part of a greenhouse or nursery operation, these areas should be easily accessible to the main growing facilities. If additional space is needed for propagation, it can be taken from the growing facilities. For example, a container nursery operation could give up some outdoor growing area for additional propagation facilities, whereas a greenhouse operation could turn over some of its production ranges for propagation. For efficient management, propagation areas should be adjacent to each other and not scattered throughout a facility.

Many nursery and greenhouse businesses now have regional production facilities situated in several locations, enabling the operation to take advantage of unique climatic conditions or improved access to major or new emerging markets. In most cases, it is advantageous for each facility to have its own propagation area. Because climates and local weather conditions vary, even within a given hardiness zone, it is easier to manage propagation schedules and have liners or transplants ready to move into the production mode if they are grown locally.

Topography

The physical topography of the site must be considered for both propagation and production purposes. Ideally, gently sloping but essentially level land is preferred for nursery and greenhouse purposes, however, there are many examples of producers that have made do with less than ideal sites, often using extremely sloping and rolling land.

Propagation facilities use a lot of water, and some provision must be made to assure adequate water drainage away from the site. Drainage issues are easy to address before structures are built, but difficult and expensive to deal with following construction. Excess surface water not only creates an unpleasant work environment, it serves as a breeding ground for insects, such as shore flies and fungus gnats, pests such as slugs and snails, and root rot pathogens, such as *Phytophthora, Rhizoctonia,* and *Pythium* species (chapter 8). Ideally, greenhouses should have approximately 1% slope along the length of the span, with drainage swales outside to move rainwater away from the site. The floor of single greenhouses should be crowned (higher in the center) to move surface water to the outer edge of the house. Areas with less than 1% slope should have in-ground tile installed to assure adequate water removal. For gutter-connected structures with concrete floors, the floor should be sloped into drainage catch basins to move excess water. Some governments now regulate water discharge from nursery and greenhouse ranges. Water catchments are designed to control storm water runoff and reduce the amount of nonpoint source pollution leaving the site. Most nurseries and greenhouses use the water in these catchment ponds to irrigate their growing crops, but not for propagation.

WATER

Perhaps the most important consideration before locating any horticultural operation is to assure that the site has access to large volumes of high-quality water. Propagation is no exception. In fact, water used for propagation must be of higher quality than that used to grow the finished crop because young tender plants are more susceptible to stresses, such as high salts and attack by pathogens. Before a site is selected for a propagation facility, the quality of water should be tested. The most important water quality characteristics are salinity, alkalinity, pH, hardness, electrical conductivity, and mineral content (Table 2.1).

Well water is preferable to surface water, provided the quality parameters are the same. Surface water supplies are more variable than well water during the course of a growing season because they are affected by heavy rains, droughts, point source and nonpoint source pollutions, and complications caused by algae and disease organisms. A multistage treatment facility may be needed to ensure high-quality water for propagation if surface water is the only available source. This is accomplished by passing the water through a sand filter to remove as many of the particulates as possible. In-line canister filters in the propagation house may also be required to prevent plugging of mist and fog nozzles. Surface water destined for use in propagation can be disinfested using either a chlorination system or UV lights.

Of these, chlorination is currently most common and effective. Many municipalities use chlorine gas as a disinfestant, but this corrosive, deadly gas is difficult to han-

TABLE 2.1
Water Quality Guidelines for Use in Propagation

Water Characteristics	Optimum	Maximum Usable
Electrical Conductivity (EC)	0–0.3 dS/m	1.5 dS/m
Alkalinity	0–60 ppm as $CaCO_3$	150 ppm as $CaCO_3$
Hardness	20–60 ppm	200 ppm
pH	5–7	8.5 upper limit; 3.5 lower
Calcium (Ca)	20–40 ppm[a]	120 ppm[a]
Magnesium (Mg)	5–10 ppm[a]	25 ppm[a]
Sodium (Na)	0–10 ppm	50 ppm
Chloride (Cl)	0–40 ppm	140 ppm
Iron (Fe)	1–3 ppm	5 ppm
Manganese (Mn)	0.2–0.5 ppm	1 ppm
Zinc (Zn)	0–0.1 ppm	0.2 ppm
Copper (Cu)	0–0.05	0.2 ppm

[a] The ratio of calcium to magnesium should remain between 4:1 and 6:1.

dle, so most propagators use either calcium hypochlorite [$Ca(OCl)_2$]—the material used to treat swimming pools—or sodium hypochlorite ($NaOCl$)—the component found in household bleach products. The amount of chlorine required to treat the water depends on the amount of organic residue, the temperature, pH, and the amount of time allowed for the treatment. Most water supplies have a free residual level of about 1 ppm chlorine at the faucet, which is achieved by adding about 10 ppm at the treatment plant.

Well water that does not meet all of the quality parameters can still be used, provided a means of correcting the specific problem can be found. Some problems, such as a high level of alkalinity, can easily be corrected by injecting acid into the water line. High electrical conductivity and high total dissolved solids, common problems in coastal areas or arid regions, can be alleviated by processing through reverse osmosis treatment plants. Because of the importance of a sustainable, high-quality supply of water, it is critical to explore the economics of water treatment before selecting a site for propagation.

Municipal water supplies can be used for propagation and may be cost competitive if extensive water treatment is necessary. Chlorine is added by most municipalities to disinfest the water supply and is perfectly safe for use in propagation. Most communities now fluoridate their water supply to help reduce tooth decay. The fluoride added does no damage to most plants, but a few species in the Agavaceae (*Dracaena* and *Cordyline*) and Liliaceae (*Chlorophytum* and *Lilium*) accumulate fluorides in their leaf tissues, and the leaf tips may become necrotic. This problem usually only occurs when the fluoride concentration in the water is above 1 ppm.

As populations have grown, water demand has increased, resulting in restrictive ordinances during periods of drought. Some communities treat agricultural usage favorably, whereas others equate it with lawn watering, which frequently goes on even–odd watering cycles at the first hint of drought. Entire days without water could result in stress or death of young propagules.

FACILITY NEEDS

The size and nature of the propagation facility depends on the kind of propagation to be conducted. In almost all instances, greenhouses will satisfy most of the needs of the propagator, but support facilities are also needed. Headhouse facilities for propagation ranges usually amount to about 5 to 10% of the allocated greenhouse space. The headhouse facilities might be used for offices, cutting preparation, potting, cold storage, general storage, and restrooms. Some of the functions of the headhouse can be done in the field, such as trimming the cuttings as they are removed from the stock plants. Transplanting can sometimes be accommodated by using potting wagons, thus facilitating the flow of liners from propagation to the next growing stage. As the size of the propagation area increases, the percentage of area devoted to headhouse facilities will decrease. If tissue culture propagation is anticipated, an isolated area that can be kept clean should be used.

The amount of space required for propagation depends on the number of plants to be produced, the number of cuttings stuck (planted) per given area (plant density), and the length of time that the cuttings will occupy the propagation space. Additional propagules may be planted to compensate for losses caused by poor rooting or low germination percentages and myriad other biotic and environmental factors. Planning for the required amount of space requires a thorough understanding of plant propagation characteristics and careful attention to the details that can make or break a crop. Good propagators are detail-oriented, observant, and consistently looking for ways to minimize loss during propagation.

In a perfect world, one would stick the number of cuttings desired for a particular crop and it would be easy to estimate space requirements. For example, only 50% of the cuttings of *Juniperus virginiana* L. 'Skyrocket' (Skyrocket Juniper) usually root, and of those, only half of the cuttings that root will be vigorous enough to make a good liner. So, if you wanted 1000 usable liners, you would have to stick 4000 cuttings [number of cuttings to stick = number of liners desired / (percent rooting × percent vigorous liners) or 4000 = 1000 / (0.5 × 0.5)]. Not only does the propagator have to acquire and stick the extra cuttings, room must also be provided for them in the propagation area. Obviously, anything that can be done to improve the rooting and vigor of cuttings will increase the efficiency of the operation.

To determine the amount of space required for a given crop, you must decide on density of the propagules in the propagation area and whether they will be in open beds or in flats. Open bed systems are often used for the most difficult species and cultivars because they provide optimum drainage and aeration for the propagules.

Higher plant densities are desirable; however, if planting densities are too high, lower quality plants could result. The two most common problems with high plant densities are legginess (stretching) and *Botrytis* infection. Density is also determined by the length of time the cuttings will be in the propagation area, the leaf size and shape of the species, and the inherent characteristics of the plant that allow it to grow close together in a mist bed and still produce a high-quality liner. Plants such as some cultivars of chrysanthemum root quickly and can be placed at high densities with as many as 800 cuttings per square meter (72 per sq. ft.), the same density that can be used for slow-to-root *Juniperus horizontalis* Moench 'Wiltonii' (Blue Rug Juniper). The cutting spacing required to achieve these densities is 2.5 by 5 cm (1 × 2 in.). Chrysanthemum cuttings will remain in the propa-

FIGURE 2.1 A poly tunnel propagation house filled with 25,000 rooted barberry cuttings in rooting trays.

gation area about two weeks, whereas the juniper cuttings may occupy their space in the propagation bench for 8 to 12 months, depending on the production schedule. Plants with larger leaves, such as *Euphorbia pulcherrima* Willd. ex Klotzsch (Poinsettia) and *Ilex cornuta* Lindl. & Paxt. 'Burfordii' (Burford Holly), are usually spaced 5 by 5 cm (2 × 2 in.) and have a density of 400 cuttings per square meter (36 per sq. ft.). Poinsettias will be under mist for about three weeks, while the holly may occupy the propagation area for three or more months.

For easy-to-root species and for bedding plants, various containers have become popular. Plug production of bedding plants permits growers to produce plants at very high densities during the initial stages of seedling growth when the rate of growth is slow, thus increasing the efficient use of the greenhouse space. Propagation containers vary dramatically from crop to crop, but they all share one feature, namely, a specific spacing arrangement per given area. In the open bed, the propagator can determine the spacing to be used, but when rooting or plug trays are used, the choice of container establishes the density of propagules per unit area.

Rooting trays are plastic inserts designed to fit into a nursery carrying tray (Figure 2.1). These have a given number of planting cells per sheet—often 18, 36, or 72—giving 102, 205, or 411 planting cells per square meter (13, 26, or 52 per sq. ft.). Plug trays are even smaller, with individual trays having from 188 to 480 cells per tray and plant densities of 800 to 4000 seedlings per square meter (100 to 500 per sq. ft.). Individual plastic pots may also be used for propagation with densities per square meter similar to those given for the rooting trays. For easy-to-root species, such as poinsettia, growers often use a technique called direct sticking, where they root directly in the pot in which the plant will be sold.

As you note from the above discussion, propagation facilities may be in use for a long time or for only a matter of weeks. For crops, such as chrysanthemum, that have a year-round demand, propagators use the same space over and over during the course of a year. A chrysanthemum propagator may have as many as 25 "turns" or cycles of the bench space per year. On the other hand, woody plants may have only one turn per year because of the longer times required for rooting and subsequent growth. Easy-to-root woody nursery crops, such as *Forsythia x intermedia* Zab. or *Weigela florida* (Bunge) A. DC., often occupy propagation space for a few months, but seasonally related planting cycles may dictate that the plants be held for several more months before they are planted. One of the principal roles of the propagator is to take the mosaic of several hundred to a thousand or more species or cultivars that are grown by the nursery and schedule the time and space allotment for the crop in propagation. A spreadsheet can be used to calculate the space requirement for each crop. Data in the spreadsheet can be based on the propagation requirements for the species as well as information from past years' performances, such as plant spacing, rooting percentages, and other modifying factors. Successful propagators maintain accurate records of propagation performances from past years, and they use the experience gained to increase their efficiency in future years.

ENVIRONMENTAL REQUIREMENTS FOR PROPAGATION

In this book, a number of different propagation techniques are discussed, but central to all of them is the basic requirement that the plant tissue be maintained in a turgid condition until the new propagule is able to support itself in an open environment. Preventing desiccation is the most crucial factor involved in propagation, and a number of techniques, such as intermittent mist and high-humidity chambers, have been developed to forestall this possibility. Water loss from tissue increases as sunlight, air movement, and temperature increase. At 100% humidity, water loss through transpiration stops.

Growth and rooting, though, are temperature-related phenomena occurring between approximately 10°C and 32°C (50°F to 90°F), depending on the sensitivity of the crop. Most greenhouse and nursery plants root best when the rooting medium is between 21° and 27°C (70° to 81°F). These kinds of environmental conditions can be satisfied without much trouble in a steamy tropical jungle, but in temperate climates, we need help—a greenhouse.

GREENHOUSE STRUCTURES

Greenhouses are a mainstay in plant propagation, permitting easy manipulation of all of the environmental parameters that affect tissue desiccation. Small growers may use just a portion of a greenhouse for propagation, with the rest of the house devoted to production. Large growers have ranges of greenhouses devoted exclusively to propagation. In the United States, approximately 63% of all greenhouse space is covered with polyethylene sheeting, glass is approximately 14%, and the remainder is about equally divided between fiberglass-reinforced plastic and rigid plastic panels. Polyethylene-covered greenhouses are the most common propagation structures because they are inexpensive, double polyethylene offers higher insulation values, and although light levels are lower than under glass, they are more than adequate for propagation.

Greenhouse frames were initially built from decay-resistant woods, such as redwood or cypress. These eventually gave way to steel pipe frames supporting wooden sashes to hold the glass. This technology lasted through World War II when new construction techniques and materials began to enter the market. Extruded aluminum structural components began to replace wooden frames, providing a low-maintenance alternative to the laborious task of stripping wooden sills, repainting, and reglazing every few years. When polyethylene sheeting first became available in the mid-1950s, wooden A-frame houses became popular. These all-wood houses had a short useful life and were quickly phased out as the growers' financial status improved. The wood houses were replaced primarily with tubular galvanized metal Quonset houses. Today, greenhouse frames are most commonly made from galvanized steel that is fabricated in a variety of shapes and sizes. Extruded aluminum is still available and commonly used for sash bars to support glass or rigid plastic panels.

COLD FRAMES AND HOT BEDS

At the turn of the twentieth century, cold frames and hot beds were the primary structures used to propagate plants. These frames were built low to the ground, with the north side slightly higher so that the glass sash covering the structure had about a 15° slope to the south to intercept more sunlight in the winter months. While they served their purpose, they were difficult to automate. Because the air space inside the structure was small, cold frames and hot beds heated or cooled quickly as weather conditions changed, necessitating constant attention. Because of these limitations and the amount of hand labor required to check constantly and maintain the frames, they are used infrequently today for commercial purposes. Cold frames and hot beds are easy to build, inexpensive, and unobtrusive in the landscape, so hobby gardeners still find uses for them.

POLY TUNNELS

Today, the term cold frame is being used more often to describe an unheated poly tunnel or Quonset greenhouse (Figure 2.2). These structures are usually about 2.5 m (7 ft.) tall, 4 m (13 ft.) wide, and 30 m (97 ft.) in length, but

FIGURE 2.2 Poly tunnel greenhouses that will be used for in-ground propagation of woody nursery stock. Each greenhouse can hold in excess of 50,000 cuttings.

FIGURE 2.3 The basic Quonset design is versatile and can be adapted to a wide number of uses. In this small propagation frame PVC tubing is used to support the plastic that will be added as winter approaches.

they can be much longer if need dictates. The structures are made from pieces of pipe tubing bent in a bow placed 1.3 m (4 ft.) to 2 m (6 ½ ft.) apart, depending on the anticipated snow load for the area. There is usually little or no internal bracing, except for a purlin that runs the length of the house at the ridge line. They are anchored by inserting the bows into a larger piece of pipe that is driven into the ground. Most growers use only a single layer of polyethylene when covering unheated tunnels.

The preferred orientation of unheated poly tunnels is north and south. With a north–south orientation, the sun passes over the house and uniformly heats the structure as it passes in its arc through the sky. If the house is oriented east and west, especially during the late winter months, the south side of the house will be much warmer than the north side. This can cause plants to come out of dormancy at different times. Environmental control in poly tunnel structures is minimal, usually limited to use of shade cloth or ventilation by opening the doors and/or rolling up the sides. Many growers in warm-temperate climates use the protection afforded by the poly during the cold winter months, but then cut progressively larger vent holes in the plastic as the temperature increases during the summer. Propagators sometimes install heaters in these houses, but keep them set just above freezing to prevent the plumbing from freezing and reduce the possibility of winter burn on broadleaf plants, but the houses are not warm enough to encourage midwinter growth.

A scaled-down version of these hoop houses, only wide enough to cover an outdoor bed, can be constructed for small producers or hobbyists (Figure 2.3). In Europe, the term cloche, originally used to describe a glass bell jar to cover a single plant, is used to describe these small poly structures, especially when they are used in the vegetable garden. Unheated hoop houses (high tunnels) are also used for the production of vegetables, flowers, and fruits directly in the ground during cooler seasons.

STAND-ALONE GREENHOUSES

A number of different greenhouse designs have been used over the years. The first to be employed on a large scale by commercial growers were glass houses using typical gable construction, and looking much like a contemporary ranch style house, but built of glass. These first houses were built with either cypress or redwood rafters supported by a steel pipe frame. This system remained largely unchanged in design from the end of the Victorian era until the 1950s, when new materials, especially plastic coverings, extruded aluminum sashes, and galvanized tubing, became more readily available. Few of these stand-alone glass houses are used today, except for hobby greenhouses and some institutional construction. Curved surfaces, including the gothic-arch style, are popular for these high-end greenhouses.

The workhorse of the stand-alone greenhouse is a larger version of the poly tunnel, the Quonset greenhouse, which became popular after 1970 because of its low cost and relatively simple construction (Figure 2.4). The Quonset design was developed by American forces during World War II as a quick way of building portable buildings and hangers. As greenhouses, these free-standing structures are usually about 10 m (32 ft.) wide and 4 m (10 ft.) tall with the length usually 30 to 120 m. They are more substantially built than the poly tunnels, with raised sidewalls, internal bracing, heating and cooling systems, and various environmental control systems. These structures can be equipped with an open-roof design where a portion of the roof lifts up to facilitate natural ventilation. Stand-alone greenhouses are usually oriented north to south in latitudes south of 40E, and east and west north of that line. The east–west orientation permits more wintertime light to enter the house. Because of the environmental controls in these houses, heat buildup on the south side is not an issue.

FIGURE 2.4 The polyethylene-covered, stand-along Quonset greenhouse is the most common greenhouse structure in the United States because it is inexpensive and versatile.

Gutter-Connected Greenhouses

The cost of heating a greenhouse in a given location is a function of the thermal characteristics of the covering and the amount of exposed surface area. By connecting greenhouses together, much of the exposed surface area disappears, and the structure becomes more efficient to heat. These gutter-connected houses may be Quonset style (Figure 2.5) if poly covering is used or ridge-and-furrow design if glass or rigid plastic sheeting is chosen. Gutter-connected houses are composed of a number of bays, with the width of the span between the gutters on the roof determined by the type of structure selected. The old-style ridge-and-furrow houses were high-profile units with wide individual bays. These were followed by low-profile Dutch-style, ridge-and-furrow houses with narrow bays but low roof lines that decreased the surface area of the greenhouse and kept the warm air closer to the plants. Today, the width of individual bays varies from 4 to 12.2 m (13 to 40 ft.), with most designs maintaining the low-profile roof and as wide a bay as possible. Some of the wide-span, low-profile designs employ a truss system that requires support posts for only every other gutter, thus maximizing the amount of post-free growing space.

As mechanization has become more important in greenhouses, the height of the sidewall has increased to accommodate the curtains, trolley, plumbing, lighting, and other equipment positioned above the growing space. Four-meter sidewalls are common, with many southern growers using even taller houses.

The compass arrangement of gutter-connected houses is less critical than for unheated poly tunnels. Some authorities recommend the east–west arrangement in northern areas, but there seems to be less consensus with this style house. More frequently, since these units have such a large footprint, grading considerations often

FIGURE 2.5 A gutter-connected, polyethylene-covered propagation house where cuttings are rooted and grown through their liner stage in rooting trays.

FIGURE 2.6 A low-profile, open-roofed greenhouse range covered with acrylic panels. One hundred percent of the roof opens to achieve near ambient temperature conditions, but still allowing quick protection of the crop if weather conditions change.

take precedence. Because bays within gutter-connected structures share some structural elements, the cost of construction and operation per unit area is less than for these houses than for stand-alone houses equipped in a similar fashion. Provided the environmental controls are installed in a series of zones, gutter-connected greenhouses can be maintained with different growing conditions by dropping curtains between adjacent bays.

Retractable-Roof Greenhouses

Starting about 1990, several competing firms began producing greenhouses with retractable roofs. These were primarily developed for bedding plant growers, permitting operators to open the houses during good weather and quickly cover them if bad weather threatened. Radically different flat-roofed designs were employed by some firms, whereas others kept the traditional shape of existing structures. Different manufacturers have employed a variety of methods to achieve the open-roof goal. Several companies roll up the greenhouse covering, whereas others bunch up the pleated roof covering just as a drapery is compressed as the curtains are opened. Some use a Venetian blind approach, constructing the roof as a series of heavy plastic slats that open and close. One style has ridge-and-furrow construction with the panels covered with lightweight rigid plastic sheets. The entire roof of the greenhouse opens like a crank out window (Figure 2.6). Depending on the type of design chosen, these houses cost about the same as traditional gutter-connected structures.

While retractable-roof greenhouses have not been used much in cutting propagation yet, there is little doubt that they will be in the future. They provide the same benefit to the cutting propagator as the plug grower, providing a more natural environment for plant growth. Because the plants are exposed to the elements, especially wind, fluctuating temperatures and bright sunlight, in a controlled manner, they are more hardened than the same plant grown in a completely enclosed environment. Operational costs, especially if the greenhouse is used with minimum heat, are comparable to other structures. Summertime cooling costs are eliminated by this design. Because of the increased use of electric motors, pulleys, cables, and the other assorted machinery necessary to open and close the roof, maintenance costs have been higher than with conventional greenhouses.

GREENHOUSE COVERINGS

The choices available for greenhouse covering have changed considerably since World War II when glass was the only possibility. Today, growers choose coverings based on an evaluation of what they need from the covering, effective life, initial cost, long-term cost, tax benefits, operational costs, the type of structure, and convenience. While glass is still the standard-bearer for durability—provided a hailstorm does not turn it into a truck load of shards—it is also expensive. Business decisions, especially for nursery production that may be made five to eight years before a crop is salable, are fraught with uncertainty. Many nursery managers are reluctant to face a 20-year payout for a more expensive glass structure when a less expensive, more disposable propagation structure reduces their initial capital outlay and makes it easier to amortize over a shorter period.

Polyethylene Film

Polyethylene, or "poly" in the jargon of the trade, is the dominant covering for greenhouses in most parts of the world (Figure 2.4 and Figure 2.5). While it is petroleum based, its cost is always proportionally less than the other coverings because even glass has a high energy component in its manufacture and cost. The polyethylene sheeting employed for greenhouse use has an ultraviolet (UV) inhibitor mixed with the plastic resin, permitting three to four years of use before it must be replaced. Without the UV inhibitor, poly sheeting may not last through a single winter. Polyethylene is sold according to thickness, and most growers use 0.15 mm (6 mil) thick material for the outer layer and 0.1 mm (4 mil) for the inner layer. Most growers that maintain temperatures warm enough to permit wintertime growth use two layers of poly to achieve an air-inflated insulation barrier that reduces heating costs by as much as 40% (the exception is those in warm areas, such as zones 9 and 10 in the southern United States). The insulation air layer is maintained by means of a small fan that inflates the poly sheets, maintaining a distance between layers of 50 to 70 mm. The air-filled space also cushions the outer layer of plastic, reducing abrasion, and helps extend its life. Polyethylene sheets are available in widths to accommodate all standard greenhouse designs. Poly tubes are also available, making the job of installing a double layer of poly much easier.

Designer polyethylene sheets are now available that have a variety of benefits, thanks to a new technique that permits manufacturers to extrude up to three layers at one time. All greenhouse polyethylene is manufactured for UV resistance, but in addition it may be imbued with a surfactant to reduce the drip problem during the winter, compounds to reduce electrical charge to prevent dust from clinging, or infrared (IR)-absorbing materials that reduce the amount of radiational cooling that occurs at night. Light transmission through a double layer of polyethylene film is 76%, compared to 88% for glass. The addition of IR-absorbing compounds reduces light transmission to 67%.

Other Sheet Plastics

While polyethylene dominates the plastic film market in most parts of the world, other products are available and offer some advantages. UV-resistant polyvinyl chloride available in thicknesses of 0.2 and 0.3 mm (8 and 12 mil) are the most popular in film covering for greenhouses in Japan. However, they cost about three times as much as polyethylene film while only lasting five or six years. Because they carry a static electric charge, they also tend to collect dust, which reduces light transmission.

Another product, UV-resistant ethylene tetrafluoroethylene (ETFE) film sold under brand names, such as F-Clean® and Tefzel T²®, has potential use as a greenhouse covering because it is purported to have a life expectancy of over 15 years. The product is extremely clear, with 95% light transmission, making it a popular choice for covering the surface of photoelectric cells. It also has good infrared (IR) transmission, thus more heat is radiated back outside on sunny days. It is currently over 10 times more expensive than polyethylene and not available in widths needed for commercial greenhouse applications, but if these difficulties can be resolved, it could become a useful addition for covering greenhouses.

Fiberglass-Reinforced Plastic

This corrugated, rigid plastic has been used as a greenhouse covering since the 1950s, but its popularity has declined in recent years. It has the advantage of being relatively long lived and, while new, has a light transmission coefficient of at least 88%. Because of the fiberglass particles embedded in the plastic, it tends to scatter light and gives a uniform light distribution pattern throughout the greenhouse. These corrugated sheets are strong and more resistant to hail and vandalism than other greenhouse coverings. Fiberglass-reinforced plastic (FRP) is available in a variety of widths and lengths and is flexible enough to cover a Quonset greenhouse. They tend to be tight greenhouses and slightly more efficient to heat than glass houses. The cost of FRP is intermediate between that of polyethylene and glass.

FRP does have significant drawbacks. It is combustible and tends to age poorly, turning yellow or brown over time and becoming brittle. Greenhouse grades of FRP are usually guaranteed to last for 10 to 15 years, but as the acrylic or polycarbonate covering erodes from the surface, the fiberglass particles are exposed. These capture dust and debris and further darken the surface. Resurfacing products are available, but, because of the amount of labor involved in the process, are seldom used. The shading problem is not as serious for cutting propagators as for conventional greenhouse growers because most propagators use shade cloth during at least part of the propagation cycle. Many growers used FRP to cover the ends of their greenhouses where shading is not a serious concern.

Rigid Plastic Panels

Since about 1990, a new category of greenhouse covering has gained popularity—the twin-walled rigid plastic panels. The unique twin-walled construction and large panel size makes these very tight houses, reducing heat loss by about one-half when compared to a single layer of glass. These panels are made from either acrylic or polycarbonate resins, chemicals that give the panel unique characteristics.

Acrylic panels are available in 8 or 16 mm thicknesses with widths of 120 cm (47.25 in.) and a variety of lengths. Light transmission is 83%, and that level can be maintained for at least 15 years. Some of the original greenhouses covered with rigid acrylic sheets have shown almost no loss in light transmission as they have aged. Acrylic panels are inflexible and must be used on a flat surface. They are flammable and more subject to damage by hail than polycarbonate panels. This covering is especially popular where high light issues are a concern and cost is not, for it costs more than twice as much as glass.

Polycarbonate panels have a similar appearance to acrylic panels but, unlike acrylic panels, they darken with age (Figure 2.7). New 8 mm thick panels have a 79% light transmission coefficient with about a 1% reduction expected each year. They are available in thicknesses of 4, 6, 8, 10, and 16 mm and have similar insulation advantages when compared to glass. The thinner sheets are flexible enough to be used on a Quonset-style greenhouse. Polycarbonate panels are noncombustible. They cost about twice the current cost of glass. Polycarbonate panels are more popular with growers than acrylics and are often used for side and end walls.

Glass

Greenhouse growers with sufficient capital or who can spread the cost of their facility over a number of years still prefer glass greenhouses. Unless some calamity befalls it, glass is a permanent greenhouse structure. Glass has

FIGURE 2.7 Cooling fans protrude from these gutter-connected Quonset greenhouses covered with 8 mm polycarbonate sheets. The fans are designed to replace the entire volume of air inside the greenhouse once every minute. The pipe protruding from near the ridge of the house is the exhaust port for the ceiling-mounted radiant heater used in the houses.

88% light transmissions, it is easy to maintain, and its qualities do not change over time. As greenhouse structures have become more sophisticated, the size of glass panes has gotten wider and longer. Some panes are up to a meter wide and two meters long. Most commonly, double strength 3 mm (1/8 in.) float glass is used in the United States, whereas tempered, 4 mm (5/32 in.) panes are used in Dutch houses. Because glass is heavy, the cost of the structural members is higher for glass houses than for lighter-weight rigid plastic panels. Glass greenhouses are popular for plug growers and propagators of high-value, vegetatively propagated, herbaceous liners.

BENCHING SYSTEMS

Good plant propagators have developed their own favorite way of propagating their plants. If given the choice, most would prefer to use some type of benching system in the propagation house because the plants are at a convenient height for work and inspection. Crops produced in benches usually have less incidence of disease for several reasons. Disinfestation between crops is easier, contamination by soil-borne organisms is less likely, and air circulation is better, thus reducing the incidence of foliar disease. Also, in a conventionally heated greenhouse, the floor is the coldest location in the house. Cold soil slows root development and increases the possibility of disease.

BENCH ARRANGEMENT

Benching systems can be arranged to maximize the use of floor space, with rolling benches utilizing as much as 90% of the total floor space. Peninsular benches are arranged so that each main bench extends from the greenhouse wall to the center aisle. The ends of these benches are connected by cross-benches at the walls. Peninsular bench arrangements utilize about 75% of the floor space and, while convenient, waste a lot of space. Some growers who produce large numbers of plants have modified the peninsular system to gain efficiency by using a series of inverted "T" rails at right angles to the main aisle. These rails are as far apart as the length of a carry flat. As the greenhouse is filled with cuttings, flats are slid down each row of rails, completely filling the sides of the greenhouse and leaving the central aisle for access.

Benching systems are used mostly to accommodate plants grown in individual containers nested in a carry flat, plug, or rooting tray. The size and type of benching material should be strong enough to hold the flats level and constructed of a material that can be effectively disinfested between crops. Expanded metal benching is perfect for this use, but expensive. Various kinds of wire benching systems are also effective and more economical. Bench widths of up to 2 m are acceptable, so long as the bench can be accessed from both sides. Wider benches reduce the number of aisles and increase the efficiency of the floor plan. However, the trade-off between space utilization and efficiency must be considered. If the grower cannot conveniently access the plants, any gains made in floor space utilization may be lost in overall efficiency of operation. Bench height should be about 75 cm (30 in.).

GROWING ON THE FLOOR

Although most propagators would prefer producing plants on raised benches, most choose instead to produce on the floor because the cost of the benches is eliminated and it is possible to use a higher percentage of the floor space to grow plants. The major disadvantage of growing on the floor is that it is the coldest location in the greenhouse, thus rooting is slowed. This disadvantage can be rectified by the use of ground bed heating systems, an expensive but effective way of dealing with cold soil. The ideal surface for growing on the ground is concrete because it is easy to disinfest between uses, is clean and uniform, and makes a generally agreeable working surface. It is also costly, so many growers opt for less expensive ground treatment. The most common alternative to concrete is a gravel base, sometimes with concrete aisles. Most growers cover the gravel with a woven ground cloth that serves as a weed barrier. Bare soil should never be used because of the weed and disease pressure that can be encountered. The carry trays or rooting trays are arranged to achieve the best utilization of space and to accommodate any mechanization equipment that is available. Some growers have adapted systems of mechanization that place plants onto large, mobile benches. These benches are then transported to the growing area, where they remain until their production cycle is finished, and then the bench is retrieved for shipping or for growing space.

Ground Beds

Floor production may be done in open ground beds where pine bark, sphagnum peat, sand, vermiculite, perlite, pumice, and a host of other ingredients are mixed to produce a suitable rooting medium (see Figure 2.2). Most propagators use the medium once and then disinfest it before using it again. Because of the possibility of developing wet spots in these ground beds, careful attention for surface and subsurface drainage should be considered.

HEATING SYSTEMS

The greenhouse grower and the propagator have different needs in terms of a heating system. Greenhouse growers use heat to keep their crop growing during the winter months. Propagators may or may not need to heat their crop during the winter. In some cases, it may be more advantageous to keep the crop dormant until it can be planted in spring. Before deciding on a heating system, the decision must be made on exactly how the facilities will be used.

Fuel

The array of fuels available for heating greenhouses is the same for heating a home, and the same considerations come to bear on the decision of which to use. In a greenhouse or a home, affordability, reliability, and convenience of a fuel are desirable. Greenhouses, unlike homes, are poorly insulated and lose a tremendous amount of heat. Also, greenhouses may be set up with a central heating system (a heating plant) that delivers hot water to individual houses. Some fuels, such as coal and oil, are appropriate to fire boilers, but not suited to unit heaters in individual greenhouses. Boiler systems are more popular and cost effective in colder climates than in warmer regions.

Electrical-resistance heat is an expensive way of heating and is never used for space heating in the greenhouse. If used in the greenhouse, electrically generated heat is used only to power heat mats or heating cables.

Solar heating seems like an obvious way of reducing heating costs—use the heat gained by the greenhouse during the day to heat the crop at night. Following the high energy costs of the 1970s, solar heating systems were thoroughly studied for their potential to heat greenhouses. Greenhouses trap a tremendous amount of heat during the day, so relatively little supplemental heat is needed on sunny days. The largest heat demand comes during the night or on cloudy days. Most of these heating systems that have been developed use solar energy to heat water. During periods of demand, the hot water is circulated through some type of heat exchanger to extract the heat. In colder areas, the solar collector surface area must be equal to the surface area of the greenhouse. If less collector space is used, the system cannot store sufficient heat in the water to keep up with demand. Based solely on the daily BTU output, about 42 m^2 of collector surface area must be provided to equal the heat output of 3.8 L (1 gallon) of heating oil. Because of the extensive areas of solar collector arrays necessary, most attempts at using solar heating for greenhouses were abandoned by the 1990s when fuel prices stabilized.

Natural gas is the least expensive and easiest-to-use heating fuel and, therefore, is the most popular choice for greenhouse heating. Heaters outfitted to burn natural gas are less expensive than equipment adapted to burn other fuels. Natural gas is delivered via buried supply lines, but these lines do not service every area. Access to natural gas should be one of the considerations weighed before deciding on a site for a new greenhouse location. If natural gas is not available, many growers use liquid propane (LP) gas. It is an easy-to-use and readily available compressed gas; however, it is expensive, costing about twice as much as natural gas. It is stored in aboveground tanks under pressure, where it exists as a liquid. Liquid propane prices vary considerably from summer to winter, with winter prices oftentimes double that of summer prices. Some growers find the investment in extra storage tanks is cost effective, allowing much of the fuel to be purchased during the summer when prices are low.

Heating oil is used, for example, in areas of the northeastern United States, but is less commonly available in other areas. Heating oil comes in six grades, with the higher grades being more viscous and difficult to use. Grades No. 1 and 2 can be pumped to individual unit heaters, but because they are liquid, they must be moved under pressure. The heavier grades are slightly less expensive and are used to fire central heating plants. Coal systems,

FIGURE 2.8 Plastic tubing being installed in the porous concrete floor of a greenhouse range to provide bottom heat.

while once used to fire central heating systems, are seldom seen today.

Wood heat has been used to fuel central boilers and may be cost effective if a dependable, cheap source of wood can be found. Some small growers are attracted to the low startup cost of individual fireboxes in a greenhouse, but quickly discover feeding a fire is like having a baby, always wanting to be fed when the sleeping is best. If wood heat is used, care must be taken to assure that no combustion gases enter the greenhouse. Wood smoke is a ready source of ethylene gas, which can cause serious crop injury. The most efficient wood-fired heating plants use either wood chips or 1- to 2-m-long logs that can be automatically fed into the furnace as the system requires them. The fire heats water that is then piped to the heat exchange device. Because these systems are not pressurized, they are less expensive to operate, install, and own.

Heat Delivery

In regions with cold winters and for large greenhouse ranges, central heating plants are more economical than unit heaters. These central heating plants most commonly are hot water systems that deliver the heat via either finned pipes, small fan-powered heat exchangers, or, more recently, underground pipes buried in the floor (Figure 2.8). The first two require steel pipe and have high startup costs, while the piping used to heat the floor is plastic and correspondingly less expensive.

Unit heaters are most commonly fueled with natural gas. These unit heaters heat the air and give about 80% overall fuel efficiency. The exhaust gasses must be vented outside. Some companies promote high-efficiency heaters, claiming 95% efficiency. These heaters are mounted outside the greenhouse and blow the heat and exhaust gasses directly into the greenhouses. Ethylene injury can be a major problem with these types of heaters unless fresh air is brought into the houses, negating the claimed gain in efficiency. Unit heaters can be equipped with polyethylene tubes to distribute the warmed air uniformly through the greenhouse. If propagation is being done on benches, it is possible to duct the heated air beneath the tables and warm the soil. Horizontal air flow (HAF) fans are now more commonly used than poly tubes to move the heated air around the greenhouse.

Radiant heaters have been used in some greenhouses because they claim a higher fuel efficiency than unit heaters. Radiant heaters burn natural gas inside a long pipe, which turns cherry red. Above the pipe, a stainless steel reflector deflects the infrared heat downward to the crop. Radiant heaters warm plants, soil, pots, and walks, not the air. Because the air is not heated, radiant heaters have a higher heating efficiency. The air temperature in a greenhouse heated by this method is usually 3°C to 6°C (5°F to 10°F) cooler than the air temperature inside a greenhouse with a forced air heating system. However, because they have a higher initial cost, they are not yet widely used in greenhouses.

Heated floor systems are especially well-suited for propagation purposes. Because it is more important to keep the soil warm than the air around the crop, these systems deliver the heat where it is needed most. PEX (cross-linked polyethylene), EPDM (a synthetic rubber), and polybutylene pipes have been most commonly used in the heated floor systems. The piping is usually 19 mm (3/4 in.) in diameter and placed 15 to 30 cm apart in the floor (Figure 2.8). Hot water is delivered to the floor pipe at about 50°C (122°F). A portable, tabletop heating system employing EPDM tubing is readily available and has become popular with plug growers and vegetative cutting propagators. It uses hot water heaters to provide the heat for a network of tubing that is placed beneath the plants. Some growers devise their own hot water systems to heat rooting beds (Figure 2.9).

FIGURE 2.9 Bottom heat is provided for these propagation ground beds using a homemade, circulating, hot water heating system that uses a conventional hot water heater as a source of heat.

COOLING SYSTEMS

Cooling greenhouse space becomes increasingly important as one heads toward the equator; conversely, heating is more important in northern latitudes. The greenhouse is an excellent solar collector, with temperatures inside a greenhouse often 15°C to 20°C (28°F to 36°F) warmer than the outside air temperature. In the winter months, this is not a problem, but during late spring and summer months when warm ambient conditions prevail outside, inside temperatures become too hot for most crops. Some growers in warmer climates routinely stop growing in the greenhouses during the summer and, where possible, move production to outdoor gravel beds during hot weather.

The traditional approach for greenhouse cooling in colder areas relied on natural ventilation. Roof and side vents were opened, and the natural chimney effect (convection) did the rest. However, changes in greenhouse design, especially the introduction of poly houses, saw natural ventilation systems disappear in favor of fan systems, which pull fresh air through the greenhouse. As electric energy costs have increased, passive cooling systems that rely less heavily on fans have become more practical. Higher greenhouse sidewalls are a feature of many southern greenhouses to take advantage of the fact that hot air rises in the greenhouse, thus keeping it away from the crop (Figure 2.10).

OPEN-ROOF GREENHOUSES

Several manufacturers are designing greenhouses with retractable, rollup or standup roofs that allow for efficient crop cooling. The original passively cooled greenhouses relied on roof vents of only about 10% of the roof area. These new open-roof designs have almost 100% of the roof opened. These maintain the crop temperature at or near ambient conditions and permit year-round use of

FIGURE 2.10 Sidewalls, 4 m or more in height, are common in modern greenhouse construction. Higher sidewalls keep the heat away from the crop and allow room for the various kinds of mechanization suspended above the crop.

the greenhouse space. This further increases efficiency because mechanization can be more easily justified if the space is used on a continual basis. The opening and closing of the houses occurs within a matter of minutes and is controlled by the greenhouse environmental control system. Polyethylene-covered Quonset greenhouses can now be equipped with lift-top attachments that permit lifting about half of the roof off of the structure. These hinged roofs are positioned so that the prevailing winds pass over the hinged side of the house, thus pulling the heat out as the air passes the opening. Open-roof greenhouse design seems to be the type of cooling system that will dominate in the immediate future.

Pad and Fan Cooling

Pad and fan cooling systems rely on rapid air exchange, essentially exchanging the volume in the greenhouse once every minute. The number and placement of fans is critical for proper design of these systems. Because the air continues to be heated by the sun as it is pulled through the greenhouse, the effective distance from the air intake to the exhaust fans is about 70 m (200 ft.). Simple air exchange works effectively when the outside air is cool, but when the ambient temperature is already warm, further cooling is needed (see Figure 2.7).

An additional step in cooling relies on evaporative pads to remove heat energy from the incoming air, thus introducing cooler air into the greenhouse. As air passes through the pads, water changes from the liquid to the gaseous phase, extracting heat energy from the air for the phase change. In an environment with no water vapor present, air temperature can be reduced as much as 22°C by this method. However, as the relative humidity increases, less water can be evaporated from the cooling pads, thus reducing their effectiveness. At 100% relative humidity there is no cooling effect at all. Pad and fan cooling systems work effectively in areas with low summertime humidity, but less well in areas where the humidity often exceeds 70% during the day. At 70% humidity, the incoming air temperature can only be reduced 3°C. Aspen pads were used for greenhouse cooling, but are seldom seen today. Cross-fluted cardboard pads 10 cm (4 in.) thick are the norm today for greenhouse pads.

Fog Cooling Systems

Greenhouse fog systems have been used in greenhouse cooling and rely on the same principle of evaporative cooling used in the pad and fan cooling system. However, instead of using saturated pads as the source of water, fog is pumped directly into the greenhouse space where it emerges from the fog generator as 10 to 40 micrometers (μm)-sized drops. These drops are small enough that they will remain suspended in the air. Fog cooling systems cannot be used with fan systems because the fog would simply be pulled from the house. When used, they are combined with a passive ventilation system. Fog systems can be used in propagation, so a system using the fog to prevent desiccation of the cuttings and provide cooling might be feasible.

Shading

Shading has always been a part of greenhouse cooling. Even in northern areas, sunlight is not a limiting factor during the summer, so shading is used to reduce the amount of radiant energy entering the structure. Lime- or latex-based whitewash is applied to the greenhouse roof in the summer and removed during the winter. With the advent of plastic greenhouses, painting on shading compounds was not feasible because it could not be removed in the winter, so various types of shade fabrics became available. These reduced the amount of incoming radiation from 20% to 80%, with 40% to 50% shade reduction being most commonly used (see Figure 2.4). As greenhouses became mechanized, shading was automatically added or removed as needed to maintain crop temperature. Originally shade cloths were black or green woven plastic material. Today, aluminized screens and white fabric cloths are used in shading. The newer products have several functions: they keep the crop cooler during the day, they may reduce crop water use, and they can be used as heat blankets at night to reduce nighttime heating costs.

Environmental Control Systems

Greenhouses are much more sophisticated than in the past, so most new structures combine computer-based technology with an integrated environmental control system that routinely monitors the crop environment and then responds as needed. These systems control heating, cooling, photoperiod, shading, and other environmental parameters as preset by the grower.

SUGGESTED READING

Agricultural Statistics Board. 1996. Floriculture crops: 1995 summary. USDA, National Agr. Statistics Ser., Sp. Cir. 6-1 (96). Washington, D.C.

Aldrich, R.A. and J.W. Bartok, Jr. 1994. *Greenhouse Engineering*, 3rd rev. Pub. NRAES-33. Northeast Reg. Agr. Eng. Ser., Cornell Univ., Ithaca, NY. 203 pp.

Boodley, J.W. 1997. Greenhouses, in *The Commercial Greenhouse*, 2nd ed. Delmar Publishers, Albany, NY, pp. 28–56.

Landicho, S. 2003. The Dark Ages … from black woven cloth to high-tech shade systems. *American Nurseryman*, Chicago, IL, May 15, pp. 26–28.

Nelson, P.V. 1998. Greenhouse construction, heating and cooling, in *Greenhouse Operation and Management*, 5th ed. Prentice-Hall, Inc., Upper Saddle River, NJ, pp. 35–170.

Robbins, J. and G. Klingaman. 2000. Irrigation Water for Greenhouses and Nurseries. Arkansas Coop. Ext. Factsheet FSA6061. 6 pp.

Simeonova, N. 2003. This year in greenhouse structures. *Greenhouse Products News*. Vol. 13, No. 4 (April): 40–43.

Yamada, Y. 2003. Choosing a covering. *Greenhouse Grower*, Vol. 21, No. 2 (Feb.): 38–40.

3 Holistic Thought Process for the Design of Propagation Facilities

Milton E. Tignor

CHAPTER 3 CONCEPTS

- The purpose and function of plant propagation structures is to provide an appropriate environment for plants and employees.

- Important planning considerations for propagation facilities include plant environment, market for propagation materials, and future plans for expansion.

- Plant-centric propagation facility design can save time and money by ensuring that the environment provided maximizes plant growth and development.

- Pragmatic concerns, such as cost, market, environment, labor availability, and building regulations, must be evaluated prior to construction.

The practice of plant propagation is an amalgam of science and art. In chapter 1 and chapter 2, students were introduced to the industry and the types of structures that are regularly utilized to propagate plants. Even the most skilled plant propagator needs the appropriate venue to be successful. Therefore, the design of propagation facilities is critical for both biological and economic success. This chapter will provide students with a mental framework from which design decisions for plant propagation facilities can be initiated. The central theme of the chapter is "plant-centric" design, but other important considerations are discussed. Experts have written extensively on the design and construction of plant propagation facilities. This brief chapter is not designed to provide all the details necessary to construct a propagation facility, but rather to provide the background to develop fundamental design parameters that can serve as a strategic and practical starting point for consultation with extension and construction professionals. In particular, this chapter will focus on the facility design necessary to propagate plants from cuttings and seeds.

BACKGROUND

Successful propagators are paid a premium for their efforts. Thus, there are many firms that specialize in propagation that sell plant materials to other facilities for the more straightforward task of "finishing" for final sale for wholesale or retail markets. Although the process of plant propagation can be challenging, students should realize that, with education and training, most horticultural crops are propagated relatively easily. However, there are some difficult species that require additional experience and very specific training.

Plant propagules of many different types are used in the horticultural industry. Cells, tissues, and organs can all be utilized as starting materials. Although the propagation of vegetable transplants from seed may seem strikingly different from rooting holly cuttings, many of the same processes are at work. To ensure that these processes are maximized, the plant propagation facility must be properly designed. Plant propagation can be generally thought of as a two-stage process. In stage one, seeds or cuttings are fostered in an optimum environment to promote growth. In the second stage, the new generation of plants is placed in a setting that allows hardening off. Following the second stage, the plants should be developed sufficiently to be transplanted to a final location, whether it is a window sill, a homeowner's yard, an industrial-scale landscape, or a restoration project on a stream bank. The following exercise focuses on facility design for stage one, but many of the principles are common to the design of facilities for hardening off.

Depending on the type of propagation scheme, many different processes can threaten the success of the plant propagation operation. For example, when cuttings are placed in a mist bench for rooting, they have the disadvantage of transpiring (losing water in the vapor phase primarily through the stomata) more water than can be replaced. Initially, the cuttings are without a root system to move water from the substrate to the xylem of the plant. Additionally, temperature fluctuations, pathogens, insects, and rodents can all threaten successful plant propagation. Plant environmental stress must be of primary concern to not only plant propagators, but facility designers.

In the horticultural industry, plants are propagated by either asexual (various types of cuttings) or sexual (seeds) means. In both of these types of propagation, there are many biological, physiological, and chemical processes that must take place. All methods of plant propagation require cell division, cell elongation, and cell differentiation. An ample water supply is required for all of these processes. In addition to cellular-level processes, some types of plant propagation are complicated by the presence of incomplete plant tissues and organs. For example, when rooting stem cuttings, you start with a propagule that has no root system. Genetics also plays a role. Sometimes propagation involves propagules with different genetic compositions. When grafting scions to rootstocks, the propagator starts with two different genotypes and a break in the vascular tissue that must regenerate through the cellular processes mentioned above. In these cases, the plant material is extremely susceptible to water loss that can lead to desiccation, reduced cell division, poor growth, and even death. Likewise, seedlings must go from a stage of being dependent on stored food reserves to a point where they are developmentally, physiologically, and photosynthetically competent to be established outside of the protected and optimized plant propagation environment.

Plants also require energy for these growth processes that is produced by respiration. Simply put, respiration is a process that requires both carbohydrates and oxygen to yield ATP (adenosine triphosphate), which the plant cells use to power "life processes." Since processes, such as photosynthesis, signal transduction, and cell division, all require proteins, providing optimum temperatures is critical for the propagation to occur in the most efficient manner. A plant propagation facility must be designed to meet the biological requirements of the plants first.

FACILITY DESIGN

Generally, propagators should consider the items that are listed in Table 3.1 when designing a facility. These considerations can be thought of in three hierarchal tiers of importance—primary, secondary, and tertiary. Essentially, the primary concerns must be incorporated into the design first before considering the secondary issues. Likewise, the pragmatic issues of secondary importance need to be addressed before the long-term tertiary planning issues. However, all the considerations must be addressed for success.

Unfortunately, many newcomers to the industry start with a design and then try to make the design tolerances "fit" those required by the plant species to be propagated. This strategy is destined to cost the propagator additional time, money, and stress. The most successful strategy is to make detailed decisions concerning as many design

TABLE 3.1
Initial Considerations for Successful Plant Propagation Facility Design

Consideration	Importance	Factor
Plant-centric	Primary	Environmental requirements of particular plant species
	Primary	Plant propagation technique: sexual (seed), asexual (cutting, grafting, etc.)
	Primary	Reducing plant stress
Pragmatic	Secondary	Funding available for construction of facility, costs of construction, etc.
	Secondary	Purpose of propagation: wholesale, retail, research, historical preservation, riparian restoration, etc.
	Secondary	Local environmental conditions
	Secondary	Control of pests, including insects, disease, and rodents
	Secondary	Meets governmental regulations for zoning, construction, and safety
	Secondary	Movement of people and machinery
	Secondary	Space availability
	Secondary	Need and availability of skilled labor
Long-term	Tertiary	Propagator preferences, motivation, and goals
	Tertiary	Plans for expansion
	Tertiary	Community development and maturation.

considerations as possible prior to the actual design and construction of the facility.

PLANT-CENTRIC DESIGN

Since the end result of successful propagation is the production of a healthy, vigorous marketable plant, the primary concerns that need to be addressed are the environmental requirements of the species being propagated and the technique that will be utilized. Global plant distribution is limited primarily by water availability and temperature. These two requirements are also the most important to consider in the design of a plant propagation facility. In the first stage of propagation, high levels of plant-available water, high humidity levels, and a warm root substrate are desirable for most species. Light, in addition to being an absolute requirement for plant growth, can also be a requirement for seed germination of some plant species. Other requirements that are also important for the design of plant propagation facilities are air movement and oxygen availability.

Plant-available water can be provided through many different types of irrigation including hand watering with a wand, drip irrigation, and automated misting systems. Mist bench systems use low-pressure (35 to 70 PSI [pounds per square inch]) nozzles that output a relatively high volume of water (15 to 38 L/h [4 to 10 g/h]). Typically, depending on the model, one nozzle is required for every 1.4 to 4.6 m^2 (15 to 50 ft^2) of propagation bench space (Figure 3.1). With the simple addition of a timer, the mist can be applied intermittently, typically 5 to 10 sec. every 5 to 10 min. Mist systems have the advantage of requiring relatively inexpensive pumps due to the low pressure requirement, but conversely, the high volume of water can cause increased disease pressure. The wet environment can also be uncomfortable for employees.

Alternatively, fog systems can be used to keep the temperature in the propagation area cool and raise the humidity level to reduce plant transpiration losses. The water droplets are extremely small and evaporate rapidly. Fog systems require a much lower volume of water, but higher pressure and excellent water quality are essentials that can raise long-term operational costs. However, fog systems provide an extremely uniform temperature throughout the facility and, because of the rapid evaporation rate, a comfortable work environment for employees. As root systems on cuttings develop, it becomes necessary to irrigate more regularly when using a fog system.

The total amount of water that will be needed must be considered when designing your irrigation/humidity delivery system. The maximum daily water requirements of a crop vary with species, age, propagation technique, etc. It is not unusual for a bench crop to utilize 20.4 L/m^2/day (0.5 gal/ft^2/day) when near saleable stage. When working with design professionals, building in extra irrigation capacity

FIGURE 3.1 (a) Propagation bench with mist nozzles and heating coils. Note the metal mesh that is being utilized to cover the coils, allowing the bottom heat to be distributed more evenly. (b) Mist nozzle detail. There are many different manufacturers of mist nozzles and heating coils.

is always beneficial. Determining how much extra capacity depends on many factors, such as water quality (water with high salts requires more irrigation capacity for leaching) and expansion plans (cheaper to provide capacity initially than increase capacity later).

Temperature is critical to the rapid development of a new generation of plants. Generally speaking, most species can be successfully propagated from seeds or cuttings at a temperature between 21°C and 27°C. Heat comes from the bottom up in order to warm the substrate and promote root proliferation. Commonly, steam or hot water tubing (connected to a boiler) is placed around the perimeter of the greenhouse, in concrete floors, on top of the floor, or underneath the benches. Hanging forced air unit heaters may also be used, but are less desirable because heat is provided from the top down. The sizing of boilers and heaters requires knowledge of facility size, construction, and local climate. For smaller facility needs, bottom heat can be provided using a heating mat or heating coils (see Figure 3.1). These electric devices come in several different sizes and usually have a thermostat for temperature control. For example, bearberry (*Arctostaphylos uva-ursi* (L.) Spreng.) has been rooted successfully from cuttings using intermittent mist and bottom heat at 21°C. However, students should note that there are plants that require temperatures outside of this range, and research into the species being propagated is

TABLE 3.2

Some Species-Specific Propagation Recommendations

Plant Name	Method of Propagation	
	Seed	Cuttings
African Violet[a] (*Saintpaulia ionantha* Wendl.)	30°C (86°F)	Leaf cuttings; root substrate 25°C (75°F); air temp 18°C (64°F)
Wax Begonia[a] (*Begonia schulziana* Urban & Ekman.)	28°C (82°F) for one week, then 25°C (78°F) until emergence, light required	Leaves or softwood cuttings
Ornamental Cabbage[a] (*Brassica oleracea* L.)	2 weeks at 21°C (70°F)	Not practiced commercially
Japanese Stewartia[b] (*Stewartia pseudocamellia* Maxim.)	2 months warm treatment, followed by 3 months cold, recommended treatment lengths vary	Variable success and recommendations vary
Paperbark Maple[b] (*Acer griseum* (Franch.) Pax.)	Germination rate of 1% to 2%, requires cold stratification	Juvenile material roots from stem cuttings with some success, little success with mature plants
Poinsettia[a] (*Euphorbia pulcherrima* Willd. ex Klotzsch)	Not practiced commercially	Stem cuttings with leaves using a mist bench
Sedum[a] (*Sedum acre* L.)	Day temperatures should be held at 29°C (86°F) and night temps at 21°C (70°F), light requirement	Roots so readily that cuttings are placed directly into containers
White Fringe Tree[b] (*Chionanthus virginicus* L.)	Double dormancy, requires 3 to 5 months warm for root and 5°C (41°F) cold for 1 month for shoot. If sown outdoors it takes 2 years to germinate	Extremely difficult to root cuttings, reports suggest there may only be a 1- to 2-week window when it can be rooted successfully
White Enkianthus[b] (*Enkianthus perulatus* (Miq.) Schn.)	Good germination, but seedlings extremely slow growing	Difficult, sometimes cuttings die from an undocumented disorder

[a] Reviewed in Hartmann, H. T., D. E. Kester, F. T. Davies, Jr., and R. L. Geneve. 2002. *Plant Propagation: Principles and Practices.* 7th ed. Prentice-Hall, Inc., Upper Saddle River, NJ.

[b] Dirr, M.A. 1998. *Manual of Woody Landscape Plants: Their Identification, Ornamental Characteristics, Culture, Propagation and Uses.* Stipes Publishing, Champaign, IL. Additional information provided by Adam Wheeler, personal communication.

critical for success. Some examples of specific recommendations for seed and cutting propagation are listed in Table 3.2. Several of the plant species listed in the table can be propagated with minimum difficulty. However, some of the species, including ornamental plants, such as *Acer griseum* (Franch.) Pax., *Chionanthus virginicus* L., *Enkianthus perulatus* (Miq.) C.K. Schneid., and *Stewartia Pseudocamellia* Maxim., are notoriously difficult to propagate for a number of reasons.

Although there are many technically advanced, automated, industrial-scale, propagation facilities that are successful, the principles in the previous paragraphs can be used to propagate plant material successfully on a smaller, more economical scale. In Figure 3.2, a small facility in a glass house is shown. It is capable of being adapted for propagating a variety of plant species. A mist system that may be controlled intermittently is used to raise the humidity and keep the root substrate moist. The two benches in the background have cost-effective PVC frames that can be used to support plastic barriers to further raise the humidity in the propagation area or shade cloth to reduce incident light.

PRAGMATIC DESIGN CONSIDERATIONS

Once the plant-centric environmental factors for the plant species and propagation scheme have been determined in detail, decisions can be made concerning immediate design considerations. Capital is obviously critical and will make some design decisions conclusive. Experts estimate the cost per ft^2 to construct a greenhouse in the United States ranges from $13 to $18 before any special adaptations are made for plant propagation. Material and construction costs vary from region to region. Students should also note that these figures are based on a facility that covers a minimum of 0.4 ha (1 acre). A smaller facility would have a higher cost per ft^2. Of course, if you already have the greenhouse, the cost of retrofitting a small area for plant propagation may be much less. In many cases, cheaper forms of structures are used, such as tunnel houses and shade houses, depending on the type of plant propagation and location.

Another early decision that needs to be made is to determine the impact that the purpose of the plant propagation will have on the facilities design. For example, a

FIGURE 3.2 Common small-scale propagation facility design. Note the setup of the three different propagation benches shown. Plastic can be pulled over the two framed mist benches to increase humidity.

wholesale facility does not need extra space for customers to move about, and aesthetics are much less important than in a retail facility. However, if you are planning retail sales from the same house where your propagation facilities are located, there are other important considerations, such as customer safety, freedom of movement, and handicap accessibility. Greenhouse propagation facilities for schools, prisons, retirement communities, hospitals, and other institutions have specific design challenges that require consultation with experts.

If a facility is being designed and built, selection of a greenhouse manufacturer is also a critical step in the process. There are many different companies specializing in the construction of greenhouse facilities. If you need recommendations, a good place to start is the National Greenhouse Manufacturers' Association (NGMA) (www.ngma.com). The NGMA believes in constructing structures tailored to growers' needs. NGMA has structural, component, marketing, and university member divisions. A professional designer will also help a grower work within a budget and weigh the pros and cons of design decisions, including choices of structural materials, type of glazings, type of heating and cooling systems, layout, and bench type. A great deal of thought should be given to each of these decisions. Depending on the purpose of the facility, specialized items, such as insect screens, thermal covers, or specialized irrigation systems (e.g., ebb and flow), may be needed. Automation is also on the rise, and there are companies specializing in automating greenhouses for grower needs.

These decisions also require knowledge of local environmental conditions, including incidental light levels, rainfall, temperature averages, temperature extremes, and propensity for severe weather, such as hail or hurricanes. Weather can be a local driver for municipal building codes as well as material choices. A professional firm will also know how to find information on local fire code requirements, zoning, and construction requirements that need to be addressed prior to any groundbreaking.

Labor is one item that may seem less important than some of the other factors mentioned, but the importance of labor supply cannot be underestimated. In some areas, such as the southeastern United States, labor is abundant and relatively inexpensive. However, in the northeastern United States, labor is expensive due to the high cost of living, and as a result, many growers have a perennial labor supply problem. It is interesting to note that the USDA reports that the average agricultural laborer only earns about three-fifths as much salary as the average laborer (all other industries combined) in the United States. Many agricultural workers also live below the poverty line. Design decisions should take into account worker safety and comfort when possible. Worker-friendly climate and ergonomics can make employees not only happier, but more efficient and productive.

LONG-TERM DESIGN ISSUES

Increasingly, there are entrepreneurs that establish small horticultural businesses that occupy a specific market niche. Organic production is one such successful area. Check the most current national standards from the USDA for more details on organic production in greenhouses. There are also producers who see environmental protection and sustainability as an important personal effort even when local laws and regulations do not require it. There have been many advances in energy conservation for propagation facilities. Perhaps a proprietor wishes to employ fewer people at a livable wage and wishes to accomplish this via increased automation. Alternatively, a propagator's business plan may include the desire to pass on knowledge by creating internships for college students.

These are the types of issues that can be addressed when the primary and secondary design considerations have been met.

Many times the most cost effective time to plan and build for expansion is during the initial construction phase. Evaluating long-term plans during the initial construction phase can make it much simpler to expand later on. Building orientation, water supply, electrical loads, communications, and traffic patterns are just a few examples of things that should be evaluated carefully to make future expansion smooth.

Think about the location of the current facility being planned. What will the location and surrounding area look like in the next decade? No one can forecast the future, but examining current trends can yield some valuable insights. Is urban development on the rise? Is the land use pattern fairly stable or in a period of rapid flux? Are there highways being planned that will divert customer traffic from your facility or bring more to it? Are there plans for national chain stores that may be able to sell a similar product at lower prices? Are there any local businesses that might expand to target the same market niche?

If an individual or team addresses the plant propagation design concerns mentioned in this exercise, the likelihood of a successful functioning facility is high. Additionally, the details that need to be gathered in order to address the design parameters can be incorporated into a comprehensive business plan, which is often necessary to attain capital through loans or investors. Now, explore some of the physical, economical, and biological criteria for establishing propagation greenhouses. Discuss the materials presented in Procedure 3.1.

Procedure 3.1		
Plant Propagation Facility Design Exercises		
Step	Instructions and Comments
1	Examine Table 3.2, and discuss the extremes required for germination of the different plant species listed. Which seed produces seedlings the fastest? Which seed species takes the longest to produce seedlings? How do the recommendations listed here impact the design of a plant propagation facility?
2	Choose a plant species that you would be interested in propagating. Your instructor may provide a list of choices.
3	Research the recommendations for the propagation of the species you have chosen. Some of the references at the end of the chapter may help you get started. Be sure to include the optimum environmental requirements for the propagation of the species. Be sure to include temperature/water/humidity levels. Realize that for some plant species you may need several different types of propagation areas.
4	List any unusual environmental requirements that have been recommended for successful propagation of your chosen species.
5	List the equipment you would use to meet the ideal propagation environmental specifications for your species. Assume that (a) you have an unlimited budget and (b) you are designing a minimalist structure.
6	In the future there might be a time when you might want to propagate different plant species from the one you have chosen for this exercise. What other species could be propagated in a facility providing the environmental specifications you listed in Step 2? Are there minor modifications that you could make in your design parameters that would greatly increase the number of species you could propagate without losing the ability to propagate your species of choice? Remember adaptability could enhance your chances for economic success.
7	Based on the material covered in this and other chapters, what other factors might you want to consider if you made the decision to actually build the facility?
8	Based on Steps 1 through 6, develop a comprehensive list of specific questions that you would want to ask a greenhouse manufacturer if they were going to design and build the facility for you.
9	If you were going to be the owner of this facility, are there any special modifications you would make in addition to what was covered in this chapter?
10	An old friend from the Peace Corps contacts you and wants to know how to design a plant propagation facility in Kenya for a local community. Based on what you have learned, develop a flow chart explaining the design thought process to your colleague.
11	Study Figure 3.2. List as many different components as you can find in the photograph that can be used to alter the plant environment.

REFERENCES AND SUGGESTED LITERATURE

Aldrich, R.A. and J.W. Bartok, Jr. 1994. *Greenhouse Engineering.* Natural Resource, Agriculture, and Engineering Service. NRAES-33. 212 pp.

Dirr, M.A. 1998. *Manual of Woody Landscape Plants: Their Identification, Ornamental Characteristics, Culture, Propagation and Uses.* Stipes Publishing, Champaign, IL. 1187 pp.

Hartmann, H.T., D.E. Kester, F.T. Davies, Jr., and R.L. Geneve. 2002. *Plant Propagation: Principles and Practices.* 7th ed. Prentice-Hall, Inc., Upper Saddle River, NJ. 880 pp.

Jozwik, F. 2000. *The Greenhouse and Nursery Handbook: A Complete Guide to Growing and Selling Ornamental Plants.* Andmar Press, Mills, WY. 806 pp.

Nelson, P.V. 2003. *Greenhouse Operations and Management,* 6th ed. Prentice-Hall, Inc., Upper Saddle River, NJ. 629 pp.

Northeast Regional Agricultural Engineering Service. 2001. *Energy Conservation for Commercial Greenhouses.* NRAES-3. 84 pp.

Northeast Regional Agricultural Engineering Service. 1994. *Greenhouse Systems: Automation, Culture, and Environment.* NRAES-72. 306 pp.

Wilson, S.B. (Ed.), M. Thetford, D.G. Clark, F.T. Davies, B. Dehgan, G.A. Giacomelli, M.E. Kane, K.A. Klock-Moore, S.M. Scheiber, R.K. Schoellhorn, M.E. Tignor, W.A. Vendrame, and J.F. Williamson. 2002. *Principles and Practices of Plant Propagation CD* (PLS 3221/5221), Institute of Food and Agricultural Sciences, University of Florida. For further information contact Dr. Wilson at sbwilson@mail.ifas.ufl.edu.

WEB SITES

International Plant Propagators' Society, Inc. http://www.ipps.org/.

National Greenhouse Manufacturer's Association. http://www.ngma.com.

National Organic Program. http://www.ams.usda.gov/nop/indexIE.htm.

Ohio State University Horticulture and Crop Science in Virtual Perspective. http://plantfacts.ohio-state.edu/. Searchable database; 260,000 pages of plant information.

Worldwide Greenhouse Education. http://www.uvm.edu/wge.

4 Intermittent Mist Control for Plant Propagation

David W. Burger

Intermittent mist systems provide the necessary environment for the vegetative propagation of herbaceous, softwood, and semihardwood cuttings (Figure 4.1 and Figure 4.2). Mist controls water loss from the cuttings by reducing leaf temperature via evaporative cooling and by raising the relative humidity around the cuttings.

An intermittent mist system can be constructed inside a greenhouse where many aspects of the environment (light intensity, day length, and temperature) can be controlled. However, intermittent mist systems can also be constructed outside. This is often advantageous if the plant material being propagated is adapted to higher light intensities and temperatures. Whether constructed inside or outside, intermitted mist systems require a source of high-quality water under pressure (at least 40 PSI) and electricity. The basic intermittent mist system includes a day/night time clock that turns the system on (during the day) and off (during the night). When the system is on during the day, a second clock controls the frequency and duration of misting events. This second clock is typically set so that misting occurs for 5 to 10 sec. every 5 to 10 min. However, research has shown that many plants form adventitious roots more quickly when the misting is kept to a minimum. This is difficult to do with time clocks, since they do not respond to changes in the surrounding environment. For example, on overcast or rainy days when the light intensity and temperature are lower, mist systems controlled by time clock still run as if it were a bright, hot day. Therefore, alternatives have been sought to replace time clocks with other mist-controlling mechanisms that respond to changes in the environment. These devices, often called electronic leaves, work by either depending on water to complete a circuit to activate/deactivate the mist system or on the weight of water on a simulated leaf surface (e.g., stainless steel screen). In the electric circuit completion case, while the surface of the electronic leaf is covered with a film of water, a circuit is completed, keeping the normally open solenoid valve closed. When the water evaporates from the surface of the sensor, the circuit is broken, and the valve opens, allowing water to be distributed through the mist nozzles. When mist again completely covers the sensor's surface, the circuit is completed, and the valve closes. Of course, this type of electronic leaf is not useful when using deionized water that does not conduct electricity.

In the screen balance type (Figure 4.3), as water evaporates from the surface of the screen, it slowly rises until a threshold is reached whereby the normally closed valve is opened. As mist is deposited on the screen's surface, it once again gains weight and lowers to a point

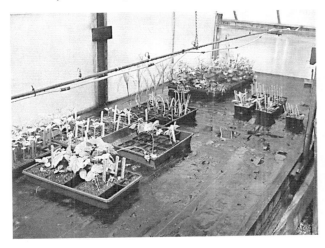

FIGURE 4.1 An example of an intermittent mist system while off.

FIGURE 4.2 An example of an intermittent mist system while on.

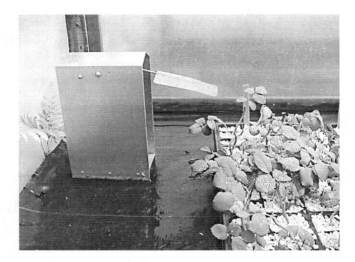

FIGURE 4.3 An example of balance-type electronic leaf.

where the valve will be turned off. Both of these electronic leaves are more responsive to changing environmental conditions and help to improve the ability of the cuttings to form adventitious roots and reduce the use of water. Another type of intermittent mist control involves photovoltaic (solar) cells (Figure 4.4). They work by converting solar energy into electrical energy and storing the energy in a capacitor until needed. During the day, as the light intensity and temperature increase, the rate of energy conversion/storage increases, and the frequency of misting increases concomitantly. Later in the day as the light intensity and temperature decrease, the reverse happens. Of course, during the night, the mist system is off. An advantage of using solar cells is that no additional electricity is required because the photovoltaic cell generates electricity that is stored in a capacitor until a threshold is reached. Once that threshold is reached, the stored electricity can be used to activate the solenoid valve.

LABORATORY EXERCISES

Two important controls the plant propagator has over the use of intermittent mist systems are the following: (1) water distribution uniformity and (2) misting frequency. The following laboratory exercises will demonstrate the effects of each on the success/failure of plants to form adventitious roots.

EXPERIMENT 1: WATER DISTRIBUTION UNIFORMITY

Uniform water distribution is a necessity during adventitious root initiation on herbaceous, softwood, and semihardwood cuttings. If water distribution from the mist system is not uniform, some plants will receive too much water and others will receive too little. This may lead to delays and/or variation in root initiation and rooted cuttings that are not uniform. A simple "catch can" test can determine how uniform the inter-

FIGURE 4.4 An example of photovoltaic mist controller.

mittent mist system is. It also serves to identify areas that receive too much/too little water, so they can be corrected or avoided, and helps determine the application rate of the mist system.

Materials

The following materials will be required for the experiment:

- Enough uniform containers (plastic cups, cat food cans, etc.) to place on each square foot (in a grid pattern) on the mist bench
- 25-mL graduated cylinder
- Metric ruler
- Tape measure for laying out the grid

Follow Procedure 4.1 to complete this experiment.

Anticipated Results

The coefficient of uniformity will be between 0 and 100. If it is less than 80, then the distribution of mist over the evaluated area is less than satisfactory. The application rate will be in units of inches/hour or centimeters/hour and can give the propagator an idea of whether they are applying too little or too much water to the cuttings. This decision will be helped by including results from the second experiment focused on misting frequency.

Questions

- What was the calculated coefficient of uniformity? If it was <80, what do you think was the cause? Are any of the mist emitters plugged or misaligned?
- How do you think air currents in/around the mist bench could affect the coefficient of uniformity?

	Procedure 4.1
	Evaluating Water Distribution Uniformity in a Mist System
Step	Instructions and Comments
1	Place the uniform containers in a grid pattern over the entire surface of the mist bench, spacing them approximately 1 ft (30.5 cm) apart. A regular pattern helps determine high/low water delivery areas after the test.
2	Activate the mist system for a known length of time. This will probably be between 5 and 15 min. The goal is to catch more than a few milliliters of water, but not more than 25 mL in the containers spaced over the bench.
3	Stop the system after a known time period and measure the volume of water in each container. Calculate the mean volume (M) in the containers by adding all the individual volumes and dividing by the number of containers (n).
4	Calculate the surface area of the container (A) by first measuring its diameter using a metric ruler. Since the radius is one half of the diameter, you can then calculate the surface area by using the following formula: surface area = $\pi \cdot (radius)^2$, where the value of π is 3.14. By knowing the surface area of the container in cm^2, and by measuring the volume of water in it in cm^3 (remember one cm^3 = one mL), you can calculate the depth of the water in cm.
5	Next, calculate the average absolute deviation from the mean (D) by taking the volume of each container and subtracting from it the mean volume (M) determined in Step 3. Take the absolute values of these deviations from the mean, add them together, and divide the sum by the number of containers (n) to get the mean deviation (D).
6	Now you can calculate the coefficient of uniformity (CU) for the mist system, which gives you a measure of how uniformly your mist system is distributing water using the following formula and values you calculated in Steps 3 and 5. CU = 100 (1 – D/M) where D is the average absolute deviation of the mean (calculated in Step 5) and M is the mean volume of the containers (calculated in Step 3).
7	The application rate for the mist system can now be calculated by dividing the average volume of water (M from Step 3) from the test (in mL or cm^3) by the surface area (A from Step 4) of the container (in cm^2) and then dividing that result by the number of minutes the mist system was on. The result is the rate of mist application (in cm per min.). Multiply this value by 60 to obtain the mist application rate per hour.

- What effect do you think water pressure would have on droplet size from the emitter and the resulting mist distribution?

EXPERIMENT 2: MISTING FREQUENCY AND ITS EFFECT ON ROOTING

In general, most plants root better when misted just enough to prevent desiccation and maintain an optimum air temperature around them. All too often, cuttings under intermittent mist systems are given too much water, either by having the mist on too long, too often (or both), or left on during the nighttime hours. Interesting results can be obtained by experimenting with different mist frequencies and times.

Materials

The following materials will be required for the completion of the experiment:

- At least two mist benches under separate timer control
- Cuttings from plants that root easily (e.g., coleus, chrysanthemum, etc.)
- Ruler

Follow Procedure 4.2 to complete this experiment.

Anticipated Results

Calculate rooting percentage by dividing the number of cuttings rooted by the number of cuttings started and multiplying the result by 100. Calculate the average root length for the cuttings of each species under each misting frequency tested. Cuttings rooted under a higher misting frequency (mist applied for longer periods of time and/or more often) will likely have a lower rooting percentage and average root length. This is most likely due to too much water in the rooting medium that interferes with oxygen diffusion to the stem/root surface. This effect is more pronounced when a rooting medium with a high water holding capacity is used.

Questions

- Were there differences in rooting percentage and root length from the species and frequencies tested? If so, what are some of the possible causes? If not, why not?

Procedure 4.2	
Misting Frequency and Its Effect on Rooting	
Step	Instructions and Comments
1	Place cuttings in flats containing a porous rooting medium, such as 2 perlite:1 peat, on mist benches of varying misting frequencies and times.
2	One such combination of treatments could be the following: one mist bench set at 5 seconds every 5 minutes and another set at 5 sec. every 10 min.
3	After 1 to 2 weeks, assess the rooting response (percentage of cuttings rooted, number of roots per cutting, and mean root length).

Part III

Plant Propagation Media and Containers

5 Media and Containers for Seed and Cutting Propagation and Transplanting

Calvin Chong

CHAPTER 5 CONCEPTS

- Most plants are propagated in soilless media unless propagation is directly in the field.

- Good aeration, moisture retention, and drainage are essential in a propagation medium.

- Small pores in the medium retain more water and large pores provide aeration.

- Propagation media may consist of ingredients or mixes, with mixes being more common. Many mixes contain sand, perlite, and/or sphagnum peat.

- Container media should be at least partially disinfested and free of pathogens, harmful pests, weeds, and toxic chemicals.

- Physical properties of media can be characterized by determining bulk density, pore spaces, and chemical properties, such as electrical conductivity and pH.

- High-density containerized systems are used increasingly for propagation.

- Many plants are propagated in individual containers or cells to increase germination rate and plant uniformity and reduce competition, damping-off, and transplant shock.

The propagation of nursery and ornamental plants is a vital part of most nurseries. During the past 25 years, much has changed in commercial propagation practices. One of these changes is that, except in special cases, such as the propagation of hardwood cuttings and seeds in field beds, soil is seldom used today. Media with soil do not provide the range of options, such as lightness in weight, ease of handling, and freedom from pathogens, pests, weed seed, and contaminants, which are required for modern propagation practices.

Nursery and ornamental plants are increasingly propagated from seeds and cuttings using soilless media in high-density containerized systems, such as plug trays or cell packs. The use of containers provides advantages, such as flexibility in time of planting, ease of handling and movement of materials, and adaptation to mechanization. Germination and rooting typically take place in modern computer-controlled greenhouses, polyhouses, or other protected environments that provide the needed moisture, temperature, and light (see chapter 3). These technologies allow more efficient, cost effective, and labor-saving ways of propagating plants.

This chapter summarizes the important functions and features of soilless media and containers used in nursery and ornamental plant propagation and transplanting practices.

FUNCTIONS OF A MEDIUM

While media for sowing seeds are generally finer in texture than those in rooting media, the functional characteristics of both are similar and can be considered together. A propagating medium provides anchorage and support and an appropriate air–moisture balance for metabolic processes.

The stem of a plant typically grows upright. The medium should be firm enough to allow and maintain adequate anchoring and support so that plants can develop into specimens suitable for transplanting or sale. Ideally, the medium should be stable and resistant to shrinkage or decomposition during production.

The ideal propagation medium should be loose or friable enough to maintain adequate amounts of air (one with an oxygen content of at least 12%) for germination and cell division at the base of the cutting and growth of new roots (Dobson, 2000). Oxygen is required for respiration to oxidize starches, fats, and other food reserves in the seed or cutting, and its utilization is proportional to the amount of metabolic activity. Good aeration in the medium also allows removal of carbon dioxide, a by-product of respiration, which may cause damage to seedlings and cuttings when present in excess (Dobson, 2000).

During propagation, it is necessary to prevent moisture loss from the medium and to maintain a humid atmosphere in the immediate surrounding air. The medium surrounding the seed or base of a cutting must remain moist. It should have a high moisture-holding capacity, yet allow excess water to drain freely so that the seeds, cuttings, or new tissues are not drowned in water. Roots may rot if the medium is poorly drained.

PROPERTIES OF MEDIA

It is often said that there are as many media as there are propagators, but there is no one best medium for all plants or all conditions. Factors, such as species, the time of year, use of growth-promoting substances, or the interaction of these and/or other variables, have a considerable influence on propagation success (Dirr and Heuser, 1987; Hartmann et al., 2002). In addition, a particular species may be more or less disposed to rooting from one year to another due to factors such as differences in vigor influenced by environmental variables.

Whether for seeds, cuttings, or transplants, the medium at the start must be at least partially disinfested and free of pathogens, harmful pests, weed seeds, and toxic chemicals. Any problems early on, such as the presence of weed seeds, will require extra labor and cost for removal later. As a rule, the medium should contain little or no nutrients. Nutrients, per se, are of little value until roots are formed and can be added with water as needed. If present in excess, nutrients can cause injury to seedlings or cuttings.

The important features of a medium can be described technically in terms of physical properties, such as bulk density and pore spaces, and chemical properties, such as electrical conductivity and pH. Knowledge of these properties provides the propagator with objective or easily measured criteria to evaluate and compare different lots or batches of the same mixture, and leads to an increased understanding of media dynamics. A comparison of the physical and chemical properties of traditional rooting medium components, such as sand, perlite, and peat, and selected waste by-products and composts is provided in Table 5.1.

PHYSICAL PROPERTIES

Methods for measuring physical properties of media are discussed in chapter 6.

BULK DENSITY

A typical propagation medium may consist of 30% by volume of solid particles and 70% space comprised of air and water. Bulk density expresses the mass of a medium (solid particles) per unit volume. Units are usually expressed in terms of grams per cubic centimeter. A high bulk density generally indicates a "tight" or slow-draining medium; a low number indicates an "open" or free-draining medium.

Measurements for physical properties, including bulk density and pore spaces should begin using dried material. This excludes the effect of water and also allows more accurate cross-comparisons of measurements between media regardless of time or space. From a practical viewpoint, however, measurements are often reported on an air-dry weight basis (Table 5.1), such as that obtained with about 12% moisture after a medium has been spread out to dry for several days on a greenhouse bench.

The bulk densities for some common materials are as follows (g/cc, air-dried basis): 0.10 to 0.15 (perlite, peat, wood chip, coir); 0.20 to 0.25 (bark, sawdust); 0.30 to 0.40 (paper mill sludges, spent mushroom compost, turkey litter compost); 0.50 to 0.60 (plant residue and municipal composts); field soil (1.25); and sand (1.50). Values up to 0.50 g/cc are used most commonly for seeds and cuttings, although mixtures or single-component materials, such as sand, with higher values have been used successfully (Maronek et al., 1985).

For liner or transplant production up to 1 q (0.95 L) sizes, additional weight is often required to prevent tipping of the containers. One way to achieve this without drastically reducing porosity is to increase the weight of the mixture by adding up to 25% by volume of sand. The added weight, however, increases handling and shipping costs.

PORE SPACES

The remaining part of a medium is referred to as the total pore space or total porosity. It is the space between particles that can potentially be filled with air or water. Values for total porosity should be at least 50% by volume and preferably in the range 65% to 70%. Heavily compacted turf soil can have total porosities below 30%, whereas some peats can have total porosities above 90%. Rock wool (mineral wool) has a total porosity of 98%.

The percent volume of a medium filled with air after it is freely drained of water is called the aeration or air-filled porosity. Aeration porosities should typically range between 15 and 30%. Higher values will not harm cuttings, but the medium may require more frequent watering. Values lower than 15% usually indicate a poorly drained or waterlogged medium. By itself, sand has aeration porosity that is close to this threshold, but drains freely.

The percent volume of water remaining in the pore space after the medium has been drained is the water

TABLE 5.1
Chemical and Physical Analysis of Traditional Rooting Medium Materials and Selected Waste By-Products and Composts

Variable	Sand	Perlite	Peat	Bark	Wood Chips	Hemp Chips	Coir	Sawdust	Paper Mill Sludge	Leaf and Yard Waste Compost	Household Waste Compost	Spent Mushroom Compost	Turkey Litter Compost
pH[a]	7.8	7.6	4.0	6.3	7.9	7.3	5.4	5.8	7.2	8.4	8.4	8.2	8.7
Salts (dS·m^{-1})[a]	0.1	<0.1	0.1	0.1	0.3	0.4	1.0	0.1	1.2	1.7	3.0	4.0	4.1
						Nutrients							
NH$_4$-N (ppm)[b]	0.5	0.02	2	0.02	0.2	2.0	0.3	0.5	37	—	4	15	103
NO$_3$-N (ppm)	5	0.02	3	0.02	0.2	50	0.1	0.01	0.02	3	0.02	89	232
P (ppm)	0.05	0.05	0.3	8	0.1	11	3	0.5	8	1	2	6	27
K (ppm)	6	0.8	0.6	42	25	420	173	54	89	733	1166	2066	2792
Ca (ppm)	30	1.3	3	34	40	33	5	9	409	114	165	871	100
Mg (ppm)	16	0.6	1	12	6	13	3	1	83	48	50	220	153
Na (ppm)	10	6	5	10	20	2	75	6	387	89	139	511	501
Cl (ppm)	14	6	7	19	8	—	203	25	136	986	848	1328	1656
SO$_4$ (ppm)	10	—	—	3	—	—	—	—	159	—	29	894	316
Fe (ppm)	0.5	0.2	0.2	0.5	0.4	1.2	0.4	0.4	1.4	0.2	1.5	1.9	11.4
Mn (ppm)	0.05	0.1	0.1	0.6	0.2	0.3	0.1	0.3	1.4	0	0.1	0.9	2
Zn (ppm)	0.05	0.1	0.1	0.05	0.2	0.3	0.1	0.2	0.5	0.1	0.1	0.4	6.3
						Physical Properties[c]							
Bulk density (g·cm^{-3})	1.50	0.12	0.11	0.20	0.15	—	0.1	0.23	0.43	—	0.55	0.39	0.31
Aeration porosity (%)	14	32	30	46	41	—	29	18	40	—	32	40	45
Water retention capacity (%)	13	24	58	32	26	—	56	53	31	—	34	31	28

[a] pH and salts measured in 1 substrate:2 water (vol/vol) extracts.
[b] Concentrations of all nutrients measured in saturated medium extraction (greenhouse) procedure.
[c] Expressed on air-dry basis.

retention capacity, commonly referred to as the water-filled porosity or water-holding capacity. Values should typically range between 25 and 35%, but are sometimes higher. Smaller pores generally function more as the water retention pores, whereas larger pores are the aeration pores. Thus, materials with excessive amounts of fine particles or dust reduce aeration porosity and should be avoided.

CHEMICAL PROPERTIES

Soluble salts are measured indirectly in terms of electrical conductivity (EC), which provides an indirect measure of the nutrient or fertility status of a material or medium (see Table 5.1 for various comparisons). Note, depending on the laboratory, salts may be extracted using different methods and/or different proportions of medium and water. Thus, the interpretative ranges for salt values will differ depending on the way it is extracted (Szmidt and Chong, 1995). In this chapter, salt levels are expressed in terms of deci-Siemens per meter ($dS \cdot m^{-1}$) measured in 1:2 (vol/vol) medium and water extracts.

SALTS AND NUTRIENTS

Soluble salts are dissolved nutrients or inorganic ions in aqueous solution. Selected waste derivatives, such as paper mill sludges and municipal composts with intermediate salt levels (EC 1.0 to 2.5 $dS \cdot m^{-1}$; Table 5.1), can be used in rooting media, if handled appropriately. In media mixed with these materials, salts leach very quickly, especially from shallow flats or plugs. Often, one or several sprinklings of water, or just leaving the media in flats under mist for a day or two, will result in sufficient leaching to lower salts to acceptable levels. High-salt-containing materials, such as spent mushroom compost and turkey litter compost (EC > 2.5 $dS \cdot m^{-1}$; Table 5.1), are not recommended. The salts may leach incompletely or inconsistently, thus posing a high risk of burning the cuttings. Always check the concentration of soluble salts of a waste-derived medium before use. If the reading is higher than 0.2 $dS \cdot m^{-1}$ (1 medium:2 water by volume extract), leach it and use only when the reading is below or very close to this value, or if you have prior knowledge that cuttings are tolerant of a higher salt level. Keep records of watering, fertilizer, and related cultural practices, and adjust these practices accordingly.

Before using any unfamiliar material on a large scale, try it on a limited scale to see how well it works. As a general rule, use no more than 30% of the material by volume in the mix. Using this smaller proportion often by itself mitigates potential harm caused by salts or other toxic substances. If the material appears to be desirable, increase its use in stages over several years.

Seedlings or cuttings are typically very sensitive to salts in the medium. Fast-rooting cuttings of woody species seem to be less susceptible to high salt levels than slow-rooting ones, and herbaceous cuttings more sensitive than woody ones. While there is some information on the relative tolerance of ornamental plants (whole) to salt levels, there is little or no similar definitive information for seedlings or cuttings during propagation (Hartmann et al., 2002; Skimina, 1980). Under intermittent mist, no negative effects on rooting of a wide assortment of woody cuttings occur in media with salt levels ≤0.2 $dS \cdot m^{-1}$. This threshold is at least five times lower than that considered ideal for use in container growing media (1.0 $dS \cdot m^{-1}$).

pH

Among factors that affect rooting, pH is still one of the least understood. For rooting of most nursery plants, pH values in the media between 5.5 and 6.5 are considered ideal, although many species will root well outside of this range. Acidophilic (acid-loving) plants, such as rhododendrons and azaleas (*Rhododendron* spp.) and other Ericaceous species, require acidic media as low as pH 3.8 to 4.5 to maintain growth, although rooting can occur at higher pH values. A wide assortment of nonericaceous, deciduous, and evergreen species has been rooted in media with pH values ranging between 4.0 and 8.4. No detrimental effects attributable to pH between 5.0 and 8.4 have been observed, but rooting of some species was suppressed at pH 4.0 using solely sphagnum peat.

This evidence suggests that many common shrubs are amphitolerant, i.e., tolerance to a wide range of pH (Maynard, 2000), and that variation in pH per se may not be as critical a deterrent for rooting of cuttings as previously thought. The primary consideration of pH typically relates to its effect on solubility or availability of nutrients. Confusion or lack of knowledge about pH tolerance may be related to inherent morphological, anatomical, and physiological differences among species, but definitive information is lacking.

BIOLOGICAL PROPERTIES

The purpose of mixing is to make media with physical and chemical properties within prescribed ranges, as discussed above. However, organic materials, such as peat, bark, composts, and by-products, such as paper mill sludges, may have biological properties that are potentially useful to seeds or cuttings during propagation. Materials such as these consist of organic matter or plant residues in various states of decomposition, but may also contain beneficial microorganisms that produce antibacterial and antifungal agents, root-promoting substances, such as auxins and cytokinins, and/or biocatalysts, such as enzymes, vitamins, and antibiotics (Atzmon et al., 1997; Hoitink et al., 1997). However, the identification of these substances in wastes and composts and their potential for promoting roots need to be elucidated. There seems to be little knowledge on compost use in propagation specifically to exploit these biological properties.

MATERIALS AND MIXTURES

Most propagation media are mixtures of various materials. While media are sometimes made with three or even four different materials, in most cases, media with equivalent properties can be accomplished using two components. These are easier to prepare and perhaps more predictable in their performance or outcome.

Many different materials have been used in propagation media. These can be classed as inorganic or organic and may be used more or less in the natural state, manufactured, or blended (Hartmann et al., 2002). Inorganic materials include the following: sand, perlite, vermiculite, polystyrene, and rock wool (mineral wool). Organic materials include peat, bark, wood by-products, and composts, although many organic or farm-derived waste materials have been used on a more limited scale. As a rule, the mineral component is used to improve drainage and aeration by increasing the proportions of large air-filled pore space.

INORGANIC MATERIALS

SAND

The use of sand has declined dramatically during the past 25 years, and it has been largely replaced by lighter or more easily-handled materials such as perlite, or by commercial premixes. Trucking costs can be high because sand is heavy, and it must be disinfested for control of pathogens and pests. A medium grade (0.02 to 0.08 in; 0.5 to 2 mm) of washed, sharp builder's sand is recommended, since finer sand reduces pore space and drainage (Dirr and Heuser, 1987). Sand contains little or no nutrients (see Table 5.1) and is used most often in mixtures with peat, perlite, and/or other organic components, and sometimes as a single-component medium. Many types of cuttings will rot from conditions created by excessive moisture in sand and peat mixtures with proportions of peat over 50% (Wells, 1985). Seeds or cuttings propagated in sand tend to have coarse root textures.

PERLITE

Originally used as an insulating material, perlite is one of the most widely used materials in propagation media, sometimes as a single component, but more often in mixtures, such as 2:1, 1:1, and 1:2, by volume ratio with sphagnum peat. Perlite is mined from crushed aluminum silicate volcanic rock that expands and extrudes when heated rapidly to 1000°C (Dickey et al., 1978; Dirr and Heuser, 1987). It is white, light (0.12 g/cc), having irregular-shaped particles and sealed internal air spaces, and cannot hold water or nutrients in its structure. Its primary uses are for increasing or maintaining pore spaces and as a lighter substitute for sand. It is sterile, does not decay, drains freely, and can be reused if redisinfested. The pH of perlite is usually close to neutral, but may vary between 6.5 and 8.4. A coarse (horticultural) grade is usually recommended for use in propagation.

VERMICULITE

Another heat-treated mineral, vermiculite is mined from a mica-like ore (aluminum-iron-magnesium silicate) (Reed, 1996). When heated to 760°C, it expands into plate-like layers that have a grey-tan color and silvery sheen. It is sterile and light like perlite and has pH usually ranging between 7.0 and 7.5, but can range from 6.0 to 9.5, depending on the processing and the source. It behaves like an organic material, having a high capacity for holding water and nutrients. It contains relatively high amounts of Ca, Mg, and K (Dirr and Heuser, 1987). It is easily compressed when moistened and does not reexpand (Dickey et al., 1978). While it is more often used with cuttings that root rapidly, such as chrysanthemums (*Dendranthema grandiflora* Tzvelev.) and coleus (*Coleus* sp.), Wells (1985) obtained successful rooting of difficult-to-root blue spruce (*Picea* sp.) in vermiculite mixtures and had failure in other types of media. Richer et al. (2004) used fine vermiculite and peat moss (1:1 by volume) to improve rooting of difficult-to-root sugar maple (*Acer saccharum* Marsh.). Vermiculite is considered to be a good "seed starter" material and used commonly in germination mixes with peat, bark, and composts. Particle sizes up to 0.25 in. (6 mm) are best for use.

POLYSTYRENE

Polystyrene flakes and beads [0.16 to 0.47 in. (4 to 12 mm) size] have been used as an inexpensive substitute for perlite (Dunn and Cole, 2000). Like perlite, it is light, near neutral in pH, and inert. It is difficult to mix well, since particles tend to float or migrate to the top of the medium.

ROCK WOOL

Another inert, sterile, heat-treated material, rock wool has a very high total pore space volume of 98% and a high aeration porosity of about 50%. It will not hold water or nutrients in its structure and drains very freely. For this reason, it is an ideal medium for rooting cuttings with succulent stems, such as coleus, begonia (*Begonia* spp.), and impatiens (*Impatiens* spp.). It is typically manufactured into various preformed shapes, such as cylinders, blocks, and wedges, and prepackaged to be used as a single-component medium. Rock wool is durable, and its properties are reliable and consistent. Its pH and soluble salts content fluctuate minimally.

SYNTHETIC FOAMS

These are plastic materials with near neutral pH and high water-holding capacities. Like rock wool, foams or synthetics are manufactured and sold primarily as preformed blocks, cubes, or wedges in ready-to-use tray plugs. They are friable, easily broken down, and must be handled carefully. Once crushed or compacted, they will not resume their original shapes.

ORGANIC MATERIALS

PEAT

Found in swampy areas, peat is the partially decomposed residue of mosses, reeds, and sedges. It occurs abundantly in cool climatic areas of Canada, Northern Europe, and Russia. Types, grades, and quality of peat vary considerably. Sphagnum peat is the type preferred for use in propagation media. It contains over 75% of *Sphagnum* sp. (acid bog moss) and a minimum of 90% organic matter (Reed, 1996). Young sphagnum peat, harvested from the top of bogs, is blond, and has very acidic pH somewhere between 3.5 and 4.0. It is light like perlite, has aeration porosity of about 30%, is relatively free of infesting organisms, and contains little or no nutrients (Table 5.1). Older, darker peats are more decomposed, usually denser (lower aeration porosity), and have higher pH values. Sphagnum peat absorbs between 10 and 20 times its weight in water, decomposes slowly, and may have fungistatic properties similar to its parent material, sphagnum moss (Hoitink et al., 1997); however, during decomposition, these properties are often lost. It is considered to be a nonrenewable or, at least, a slowly renewable resource.

BARK AND WOOD BY-PRODUCTS

A by-product of the lumber industry, pine bark and composted pine bark are used widely in container growing media and are increasingly being used in propagating media by some nurseries (Hartmann et al., 2002). It is relatively light (two to three times the weight of peat), typically low in salts (Table 5.1), and known to have disease-suppressing properties (Hoitink et al., 1997). Depending on the source, the pH of pine bark varies from 3.8 to 6.5, although sources with values between 5.0 and 5.5 are more often used. The best materials are screened through ¼ to 3/8 in. (0.6 to 1.0 cm) mesh, and typically have aeration porosity between 30% and 40% (Dirr and Heuser, 1987). It is a good substitute for peat, but has lower water-holding capacity (Table 5.1). Bark has an open structure that allows water to drain freely through it.

Related by-products, such as sawdust, wood shavings, and wood chips, have been used in the industry, but their use has been more limited. Many of these products decompose while in use and can sometimes deplete the nitrogen in the medium to the detriment of the cuttings (Hartmann et al., 2002). Sawdust, in particular, has very high water retention capacity (Table 5.1), and media mixed with it may become waterlogged.

COIR

During the past 10 years, coir, a waste remaining after coconut husks are processed for coir fiber, has made a rapid impact in the floricultural industry and is now commonly used in growing substrates. Some growers claim that it is better than peat as a growing material. It has a very open structure, yet is quite resistant to compaction and slow to decompose. It has a very high water-holding capacity, but diseases are less likely to occur with it than with peat. Depending on the source, the soluble salts content varies widely, as does the pH (Konduru et al., 1999). Excellent rooting results have been obtained using coir in proportions between 15% and 75% by volume mixed with perlite (Chong, 2002).

COMPOSTS AND ORGANIC WASTES

Attention is increasingly being placed on use of composts, leaf molds, coir, paper mill sludges, and other waste-derived materials as media (Chen et al., 2003; Chong, 2002). In the past, there has been little use of composts and related materials in North America (Hartman et al., 2002). Increasing concern for the environment and a movement toward "reduce, reuse, and recycle" may portend a growing trend for their use in propagation.

Major deterrents for use of these materials in propagation include the following: inconsistent quality, potential phytotoxicity because of high salt levels or unsatisfactory pH, differences in species response, and lack of scientific information. Furthermore, a proven paper mill by-product or compost may not always be more economical than traditional materials, such as perlite or peat, or may not be available within an affordable trucking distance.

Examination of various paper mill sludges and composts indicate that these are three to five times heavier, and their porosity and aeration characteristics are comparable to, or better than, those of peat (Table 5.1). Thus, from this standpoint, these waste-derived materials are ideal substitutes for peat in propagation media.

Paper mill sludges and municipal waste composts have successfully been used in volume proportions up to 60% and 75%, respectively, for rooting of many woody species (Chong, 1999; Chong et al., 1998). For commercial use, lower proportions are recommended. Results are generally better when these materials are mixed with perlite than with peat. Like peat, the high water retention capacity in the sludges and composts tend to make the mixes "too wet."

A paper mill sludge and perlite (1:1, vol/vol) mixture is routinely used outdoors under intermittent mist (Chong, 2002). However, it is not recommended for use in winter propagation in greenhouses with bottom heat. The sludge seems to lose its friability and integrity under these conditions. Mixtures of municipal leaf and yard waste compost and perlite (1:1 or 1:3, vol/vol) are also used. These sludge and compost mixtures are preferred because their soluble salt contents are relatively low, the salts leach easily, and these by-products are available locally.

Traditional "Mix-Your-Own" Media

Sand, perlite, and peat are perhaps the most well known mixing materials used in propagation media. These materials are predictable and consistent, and typically contain little or no nutrients (Table 5.1). As such, they are considered to be ideal raw materials for rooting media. They differ widely in physical properties and are used in mixtures primarily to modify or improve aeration, porosity, and water retention characteristics.

The porosity and water retention characteristics of a mixture may not always be the same among batches, or even among nurseries using the same mixtures. Keep detailed records to monitor the physical and chemical variations among batches or from year to year.

The solid portion of a medium is composed of particles of various sizes. The amounts of different kinds of solid particles, such as fine or coarse, determine the texture and influence the physical and chemical properties of a medium. Texture also influences the choice of cuttings and method of applying water. Fine-textured media compact more and retain more water and nutrients. These conditions may cause the survival or rooting ability of slower or harder-to-root cuttings to be reduced. When rooted sufficiently for transplanting, cuttings pull away more easily from a coarse medium, tending to leave them bare-rooted and more susceptible to "transplant shock." A coarse medium drains more freely and is more suitable when water is applied manually. A fine medium is more suitable with overhead mist where much less water is applied, or with high-pressure fog where water is applied occasionally.

Materials with different particle sizes and/or wettability, such as polystyrene and peat, are difficult to mix uniformly. Dry organic materials, such as peat and composts, are hydrophobic and difficult to wet initially or to mix with other materials. The use of a wetting agent improves wettability and makes mixing easier. Some people are allergic to dust from peat, perlite, and other materials; thus, it is recommended to wear a mask while mixing.

The actual process of mixing can dramatically affect the physical characteristics of a medium, as well as the integrity of preincorporated slow-release fertilizers, if included as a nutrient charge. Rupture of the fertilizer pellets could result in buildup of salts in the medium. Paddle-type machines blend ingredients quickly and efficiently, but have greater potential for destroying the physical structure of the materials due to overmixing (Dobson, 2000). Drum mixers require a longer time for mixing, but are more gentle on the materials. Front-end loaders with manure spreaders are usually available for mixing materials in most nurseries, but ensuring uniformity can be a problem. The use of coarser starting materials, or adding materials in batches, may mitigate degradation from mixing (Dobson, 2000). To provide uniformity without degradation, Sambrook (2000) described the following simple procedure: Spread a large sheet of thick plastic on the ground and roll the mixture back and forth. Do not store nutrient-charged media or keep in a warm and/or humid environment before use, since these conditions can cause premature release of nutrients and salt buildup.

PREMIXED MEDIA

Many commercial or brand name prepared (pre-) mixes are peat based, generally containing sphagnum peat, perlite, and/or vermiculite. These are usually referred to as peat-lite mixes. Other materials, such as compost and rock wool, are increasingly being used in premixes. Premixes may cost more than mixing your own, but they are more convenient to use, reliable, and reasonably consistent from year to year. They are usually available in formulations that vary in particle sizes, moisture, and drainage characteristics. Some may contain a nutrient charge and a wetting agent. Low levels of fertilizers containing minor nutrients are not required initially for germination or rooting. After roots have formed, however, fertilizers are required to maintain satisfactory growth and development.

CONTAINERS

Container refers to individual flats and pots or multiunit systems, such as cavity trays, inserts, plugs, and cells. Containers may be accompanied by either matching or carry trays in different counts. Blow- and injection-molded technology now used in the manufacture of plastic containers facilitates the production of many combinations of types, sizes, depth, shapes, and colors (Hall, 2004).

Preference depends on the type of plants and production system, cost and availability, or other economic reasons, such as cost for storage or disposal after use. Because of the wide choice of containers available, only selected ones are described and discussed here. These containers illustrate the type of transformations in design and manufacturing that have occurred in container technology and their relevance to propagation.

FLATS AND TRAYS

Until the late 1960s, nursery and ornamental plant propagation was done primarily in open fields or outdoor protected beds. Open, shallow flats or trays made of materials, such as wood, plastic, and styrofoam, were used to save space in cold frames and greenhouses. The use of flats facilitated easy handling and moving of plant material. When empty, plastic flats nestle in each other and are easy to store.

In modern propagation, the use of trays is an important part of the growing system rather than simply one of providing support for containers or to optimize use of space. Trays, also commonly referred to as carry trays or shuttle trays, offer more efficient shipping and effective product presentations to the consumer. They are manufactured in a variety of attractive colors and with a range of configurations to match the extensive selection and styles of containers, inserts, and cell packs. Other features may include the following: special design for automated filling machines and handling equipment, strengthened ribs for durability or use with roller conveyers and other machines, raised bottoms to allow better air–water exchange, skirted sidewalls for use with electric eyes during automated flat-filling operation, and design for easy denesting.

In flats, seeds are typically sown in shallow grooves or spread over the surface of the medium and covered with a layer of material, such as fine vermiculite or screened sphagnum peat. This ensures that the medium in contact with the seeds remains moist. In this type of propagation, microorganisms that cause diseases, such as damping-off or root rots, can spread easily (see chapter 8) and are encouraged even more by factors such as thick seeding and high humidity. Lifting the rooted cuttings or seedlings means that the intertwined root systems must be disturbed or pulled apart, which typically results in increased risk of transplant shock and delayed establishment.

POTS

Propagating small plants or liners directly in small individual pots in sizes up to 4½ in. (11 cm) filled with appropriate media is a common practice. In individual pots, competition among plants is reduced, and there is less risk of damping-off or root diseases. Germination rate is also often higher, and plants grow larger, more uniformly, and have higher quality. When plants are moved to a holding area once roots have been initiated, the same space can be reused within the traditional rooting season. With intact root systems, container-started stock are less susceptible to transplant shock and can be transplanted or lined out anytime of the season into larger containers, open or shaded beds, lath houses, or fields.

Liner pots are made of various types of materials, although ones made of plastic and compressed peat are widely used. The durability of plastic pots varies considerably. Some are thicker and durable, whereas others are thinner and fragile. Inexpensive, flexible, thin-walled ones, often joined together for easier handling, are designed to be used only once.

Pots require about three times more space than flats, but having transplants with intact root systems from pots is often more advantageous and convenient. The square types are more widely used than the round ones. Square containers are often manufactured with the upper portions joined to fit conveniently into trays designed to contain them in multiples, and are more adaptable for automatic filling and handling. Spillage of the medium between square containers is reduced compared to round containers.

Sometimes more expensive than some similarly sized plastic pots, soft-walled pots, such as those made of peat, are biodegradable and can be planted along with the plants. A major concern with peat pots is the loss of water from any portion of the pots remaining above the soil. This wicking effect can deplete water from the medium or soil immediately around the base of the transplant liner causing the plant to wither or die. Also, if plants are kept too long in peat pots, the side walls may deteriorate, or roots of neighboring plants become entwined, making them more difficult to handle. The "puck-type" peat container is simply a compressed disk of sphagnum peat surrounded by plastic netting. Sizes range from 3/4 to 2 in. (1.9 to 5.0 cm) in diameter. When water is applied, the puck expands up to seven times its volume, producing a cylinder into which a seed or cutting is inserted. Some are manufactured with a nutrient charge.

Earthenware or small clay pots are not used today, except for some nurseries that continue to use them to produce bench-grafted or other specialized stock. The porous side walls of clay pots allow moisture and air exchange, but also the buildup of salts or substances that may inhibit plant growth. These residuals can be removed by soaking in water. Clay pots can be steam disinfested, unlike plastic pots that become brittle after steam treatment. For sterilizing plastic pots, Hartmann et al. (2002) recommend dipping for three minutes in 150°F (70°C) hot water then into diluted bleach. Other products contain benzylkonium chloride, hydrogen dioxide (hydrogen peroxide), or similar ingredients.

PLUGS AND MULTICAVITY SYSTEMS

The introduction of multicavity trays in the late 1960s—first for cuttings, then seeds—revolutionized the plant-growing industry. The number of cavities per tray typically varied from about 30 to over 100. By the 1980s, plants were increasingly being produced in totally containerized, high-density systems, resulting in more efficient and rapid production.

Media and Containers for Seed and Cutting Propagation and Transplanting

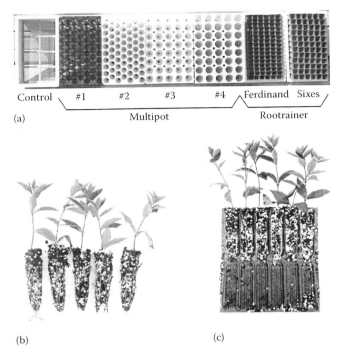

FIGURE 5.1 (a) Cylindrical-shaped and rectangular-shaped cavities of various densities; see Table 5.2 for actual dimensions. (b) Dislodged rooted cuttings from Multipot #2 plugs. (c) The Rootrainer insert opens bilaterally like a book for removal of the rooted cuttings.

Plugs are made of various types of materials, such as paper and foam (styrofoam), although most designs and types are of plastic. Foam harbors disease organisms and algae, and transplants are sometimes difficult to dislodge because of intrusion of roots into the foam. Plugs usually have a slight taper or are conical. These shapes are ideal for adapting to mechanization. Deep-cavity plugs are especially suitable for growing transplants of forestry crops and woody ornamentals and sometimes used to save labor and eliminate intermediate transplanting (Figure 5.1). Shallower plugs or cells are more suitable for rooting smaller cuttings, or for seed germination of annuals, perennials, and herbaceous plants (Figure 5.2). Some, appearing like dimpled depressions in a sheet of plastic and referred to simply as "sheets," are made of lightweight extruded plastic.

The type, size, and spacing of plugs may influence the performance of transplants. Small increases in plug volume during propagation can result in marked increases in plant size and quality. Some would argue that both depth and diameter, not just volume, also influence plant responses. To illustrate this point, I examined rooting in four cylindrical-shaped plugs (Multipot series 1, 2, 3, and 4) and two square-shaped Rootrainer types (Figure 5.1; Table 5.2). The percent rooting of Hetz juniper (*Juniperus chinensis* 'Hetzii') was moderate and not influenced by any of the plugs (Table 5.2). Andorra juniper (*J. horizontalis* 'Plumosa') rooted well in all plugs, except in the Ferdinand Rootrainer that had the smallest volume. During and after two consecutive years of growth in a container nursery, plug-rooted plants of both junipers grew significantly more than control plants rooted in an open tray (Table 5.2). There generally was a positive relationship between plug volume and percentage of rooting, and also between plug volume and subsequent growth in the nursery. In another experiment with the plugs, rooting percentage of privet (*Ligustrum vulgare*) decreased with increasing plug volume, but, as with the junipers, growth of plants in the container nursery increased with increasing plug volumes.

The size of plugs or cells also needs consideration when choosing a propagating medium. Plugs or containers with smaller volumes are more difficult to fill uniformly, particularly if the mixture is coarse. A finer mix may fill easier, but will cause the air porosity to be reduced, as also will overfilling of the medium. Containers that are wider in proportion to their depth drain less thoroughly than deeper ones.

As a general rule, once seedlings or cuttings have rooted sufficiently to enable lifting with the root system intact, they are ready for transplanting. However, note that the roots of seedlings or cuttings kept for extended periods in the initial plug or container may become potbound by forming circling or kinked roots. This condition is a primary cause of stress, decline, and death of plants, and is especially important for trees and woody species where symptoms may be delayed or become evident only years later. While the time of transplanting may vary from nursery to nursery, propagators planning to keep liners for extended periods should consider starting with a larger plug or container, or transplant earlier (Knight

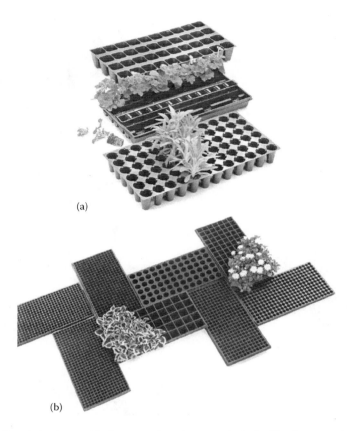

FIGURE 5.2 (a) Various shallow injection-molded propagation plug trays include 40-cell square (top) and 72-cell round (bottom) plug flats, and six-strip 60-cell sheets with matching ribbed trays to facilitate spacing the strips (center). (Photo courtesy of Dillen Products.) (b) A line of plug flats (square, 50 to 512 cells; round, 72 to 288 cells) designed to be (1) rigid enough to use without a carry tray and (2) compatible with most, if not all, automatic seeders and transplanting equipment. (Photo courtesy of ITML Horticultural Products, Inc.)

et al., 1993). Transplanting earlier is more beneficial for faster-growing than slower-growing species.

The vertical ridges along the inside wall of plugs helps to direct root growth downward (Figure 5.1b and Figure 5.1c). Ridges may reduce or delay root circling, but do not completely eliminate it. RootMaker® II is a 32-cell tray that addresses the problem of root circling in trees and woody species (Whitcomb, 2001). The original Root-Maker design was an injection-molded container 2.5 in. × 2.5 in. × 4 in. (6.4 cm × 6.4 cm × 10.2 cm) with a series of sawtooth-like ledges and openings in the sides and four bottom openings. Oak seeds planted in the RootMaker propagation containers develop roots in all directions following transplanting, not just straight down (Figure 5.3). Trees that develop large numbers of roots at the root–stem juncture and along the vertical axis of a short taproot have consistently grown faster than trees with fewer roots arising from this point.

ULTRAHIGH-DENSITY SYSTEMS

During the past decade, container technology in propagation underwent another major transformation. The greenhouse and bedding plant industry increased the use of automation and mass production, which facilitated rapid turnover of inventory. This was a response to a market environment with increasing availability of plant materials, which in turn, increased competition and resulted in marginal returns. For example, the number of plug cells per tray may vary from 288 (thimble-size) to 512 (diameter of a ball point pen) (Figure 5.2b). Seeds are sown mechanically 1 to 2 per plug, germinated, and grown at these ultrahigh densities. Depending on species, seedlings are transplanted within one to three weeks into inexpensive cell packs. The packs usually contain four or six plants that are grown-on and marketed as a unit.

Using high-quality pregerminated seeds, there is no wastage or loss of seeds. Some operations germinate seeds in special rooms or germinating chambers. These environments are usually thick walled or insulated with temperature, relative humidity, and air circulation precisely controlled. Electric light ensures early and rapid growth of the seedlings after germination and emergence from the media.

Modern computerized-control greenhouses are now designed and equipped with special fans, heaters, sen-

TABLE 5.2
Comparative Rooting and Postrooting Effects of Two Juniper Species in Various Types of Containers

Tray type	Tray Length (cm)	Tray Width (cm)	Tray Depth (cm)	Cavity Diameter (cm)	Cavity Volume (cm³)	No. per tray	Spacing Between Cuttings (cm²)	Rooting Phase Rooting (%) Andorra	Rooting Phase Rooting (%) Hetz	Growing-On Phase Shoot[d] Weight (g/plant) Andorra	Growing-On Phase Shoot[d] Weight (g/plant) Hetz
Open Rootrainer tray (control)	35.6	21.6	10.2[b]	—	109	70	12.5	90	53	589	456
Multipot #1[a]	35.5	22.2	8.8	3.1	57	67	16	77	45	676*	588*
Multipot #2	35.5	22.2	12.0	3.1	65	67	16	95	54	698*	621*
Multipot #3	60.9	35.5	12.1	3.8	99	96	25	97	63	706*	601*
Multipot #4	60.9	35.5	16.7	3.8	149	96	25	90	45	711*	623*
Ferdinand Rootrainer[c]	35.6	21.6	10.5	2.0 × 2.4	40	96	7.5	52*	46	631*	603*
Sixes Rootrainer	35.6	21.6	14.0	2.0 × 2.7	90	72	9	95	46	693*	626*

[a] Cylindrical-shaped cavities; Ropak Capilano Ltd., Mississauga, ON, Canada.
[b] Depth of open Rootrainer tray.
[c] Rectangular-shaped cavities; Spencer Lemaire Industry Ltd., Edmonton, AB, Canada.
[d] After two growing seasons in 2-gal. (6-liter) containers.
* Significantly different from the control at the 5% level of probability.

FIGURE 5.3 (a) The 32-cell RootMaker II tray. (b) Roots of oak seedlings extend in the direction they are oriented in their liner containers, a plug and RootMaker container, from left to right. (c) Seedlings were transplanted, respectively, from the plug and RootMaker container into 3-gallon (8-liter) containers, and then removed after 3 weeks to observe root development. (Photo courtesy of Dr. Carl Whitcomb, Lacebark Inc.)

sors, and irrigation/humidification systems to grow these smaller or miniature plugs successfully. Greenhouses designed with high gutters and increased air flow and volume reduce the chance for diseases and provide optimum light, water, and nutritional needs for the tiny plugs.

In many operations, the entire propagation system for bedding plants and small perennials—from sowing to market—is automated. Automation technologies, including media, flat and pot fillers, seeding and potting machines, rolling benches, etc., have increased the efficiency of many nurseries and greenhouses that are large enough to exploit the economics of scale that make these technologies cost effective (Hall, 2004). Development of seedlings or cuttings is predictable, and production is easier to manage to meet the customers' demands. The use of these technologies has all but replaced manual sowing, at least for vegetable and ornamental crops.

READY-TO-USE SYSTEMS

It is perhaps not surprising that the ready-to-use ("just add water") substrate-container systems have become quite popular in recent years. These systems offer attractive alternatives to traditional filling of trays with propagation media. There is no waste to dispose of, and no root disturbance. The substrate-container unit is planted along with the plants.

BLOCKS, CUBES, AND WEDGES

The preformed rooting substrate, commonly called "foam blocks," cubes, or synthetics, is made up of a baked polyphenolic resin and may contain wetting agents and a small nutrient charge (Figure 5.4a). They are also pH adjusted and predrilled to receive cuttings without the need for dibbling (making planting holes). In addition to benefits listed above, cubes have a high porosity, keep cuttings consistently upright after sticking, and plants rooted in cubes are easy to remove for transplanting. Until recently, the Oasis brand was the only one of its kind. Another brand, Agrifoam, was recently introduced.

Many growers have reported excellent rooting successes with cubes as well as the variant wedges or wedge-shaped types (Figure 5.4a). Compared with the cubes, the wedges seem to direct the downward growth of roots more effectively. Cuttings root out faster from wedges and are easier to remove from matching trays. Wedges are also easier to transplant because of their shape.

The Fibrgro™ predrilled rock wool cubes are manufactured in various dimensions (1.0 to 2.0 in.^{2} × 1.6 in. high; 2.5 to 5.0 cm^2 × 4.0 cm high) with between 50 and 98 per pad. The pads fit into specially designed open flats. Rock wool blocks are larger (3.0 to 8.0 in. long × 3.0 to 8.0 in. wide × 1.6 to 8.0 in. high; 7.6 to 20.3 cm long × 7.6 to 20.3 cm wide × 4.0 to 20.3 cm high). These are predrilled with one or more holes or none. Cubes, called Brownies, are also made of pressed peat [1 ½ × 1 ½ × 1 ¼ in. (3.8 × 3.8 × 3.2 cm) thick].

PLANT SLEEVES

A new ready-to-use container system, under the trade name Ellepot (Figure 5.4b), illustrates a trend toward an all-in-one substrate-container technology that is becom-

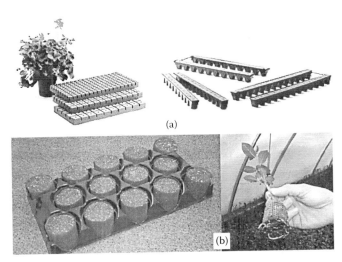

FIGURE 5.4 Ready-to-use systems. (a) Synthetic Oasis® foam blocks in various sizes (left) and plastic strips containing foam wedges in different counts (right). (Photo courtesy of Smithers-Oasis North America.) (b) Ellepot 3.1 in. (80 mm) size in 14-count tray (left); and purple-leaf sandcherry (*Prunus cerasifera*) inserted in 1.8 in. (45 mm) Ellepot from a 32-count tray can be transplanted directly in the field without removing the pot (right). (Photo courtesy of A.M.A. Plastics Ltd.)

ing more cost effective, automation friendly, and biosensitive. Small sleeves are shaped from a semitransparent paper-like, peat-based material. Sizes may vary from 0.6 to 3.9 in. (15 to 100 mm) in diameter and can be cut to any depth. There are also different choices of media for specific crops and production seasons. Each Ellepot sleeve or container is filled with medium and placed into preformed compartments within a tray. The tray is reusable, but the sleeve and medium are throwaway items. The sleeve of Fertilpot, a similar product from a different manufacturer, is made of a thin geotextile material.

The Ellepot system offers various sizes of pot and tray densities such as 1.8 in. (45 mm) in 50-plug tray; 1 in. (25 mm) in 84-plug tray; 1.2 in. (30 mm) in 3 × 26-strip tray; and 1.4 in. (35 mm) in 84-strip tray. Most plants dislodge easily from the sleeves. Fine-rooted plants, such as *Bacopa* sp. and *Bidens* sp. (both herbaceous annual plants for hanging baskets or containers), may be difficult to remove. These plants may dislodge from the sleeves when pulled for removal.

A major disadvantage of the sleeve system is the need for dedicated machinery and equipment specially designed for handling the trays and containers. Overwintering may also be a problem initially after cuttings are first inserted because of the fine texture and high peat content of the media. Mist times must be adjusted accordingly.

SUMMARY AND OUTLOOK

This chapter considered the primary features and fundamental characteristics of media and containers in relation to plant propagation and transplanting. Because there are so many variables involved in propagation, there is no one medium that is ideal or suited for all situations.

A better understanding of the technical properties of propagation media—a primary goal of this chapter—will assist the propagator to make informed judgments about the suitability of a new or novel medium and to be able to compare this with existing ones. The information will also help the propagator to achieve more consistent results over the widest range of species using as few media as possible.

In recent years, advances in materials and techniques used in manufacturing containers and ready-to-use substrate-container systems, increasing use of automation, and our precise ability to control greenhouse environments have transformed the practice of propagation to a more sophisticated or "hi-tech" level. The major advantages are increased efficiency and time and space savings.

Over the next 25 years, there will be increasing use of proprietary, compost-based, premixes and a wider array of ready-to-use systems as propagators seek more convenience and consistency. Concomitantly, with the trend to smaller transplants, there will be an increasing trend toward traditional direct sticking and propagation in larger containers. Among other reasons, direct sticking eliminates intermediate stages of propagation.

As the growing industry becomes more Earth-friendly in the years ahead, propagators need to consider the environmental impact of their practices. Organic wastes and composts will get greater consideration for use. It will be a challenge to obtain excellent and consistent results using these alternative materials in propagation.

REFERENCES

Atzmon, N., Z. Wiesman, and P. Fine. 1997. Biosolids improve rooting of bougainvillea (*Bougainvillea glabra*) cuttings. *J. Environ. Hort.* 15: 1–5.

Chen, J., D.B. McConnell, C.A. Robinson, R.D. Caldwell, and Y. Huang. 2003. Rooting foliage plant cuttings in compost-formulated substrates. *Hort Technology* 13: 110–114.

Chong, C. 1999. Rooting of deciduous woody stem cuttings in peat- and perlite-amended MSW compost media. *Compost Sci. Util.* 7: 6–14.

Chong, C. 2002. Use of wastes and composts in propagation: challenges and constraints. *Comb. Proc. Intl. Plant Prop. Soc.* 52: 410–414.

Chong, C., B. Hamersma, and K.L. Bellamy. 1998. Comparative rooting of deciduous landscape shrub cuttings in media amended with paper mill biosolids from four different sources. *Can. J. Plant Sci.* 78: 519–526.

Dickey, R.D., E.W. McElwee, C.A. Conover, and J.N. Joiner. 1978. Container growing of woody ornamental nursery plants in Florida. Univ. Florida Inst. of Food and Agr. Ser. Bull. 793. 122 pp.

Dirr, M.A. and C.W. Heuser. 1987. *The Reference Manual of Woody Plant Propagation. From Seed to Tissue Culture.* Varsity Press Inc., Athens, GA. 239 pp.

Dobson, A. 2000. A guide to propagating composts. *Comb. Proc. Intl. Plant Prop. Soc.* 50: 148–152.

Dunn, D.E. and J.C. Cole. 2000. Expanded polystyrene as a substitute for perlite in rooting media. *Comb. Intl. Plant Prop. Soc.* 50: 532–537.

Hall, C.R. 2004. Impacts of technology on the development, production and marketing of nursery crops. *Acta Hort.* 630: 103–111

Hartmann, H.T., D.E. Kester, F.T. Davies, Jr., and R.L. Geneve. 2002. *Plant Propagation: Principles and Practices,* 7th ed. Prentice-Hall, Inc., Upper Saddle River, NJ. 880 pp.

Hoitink, H.A.J., D.Y. Yan, A.G. Stone, M.S. Krause, W. Zhang, and W.A. Dick. 1997. Natural suppression. *Amer. Nurs.* 186: 90–97.

Knight, P.R., D.J. Eakes, C.H. Gilliam, and K.M. Tilt. 1993. Propagation container size and duration to transplant on growth of two *Ilex* species. *J. Environ. Hort.* 11: 160–162.

Konduru, S., M.R. Evans, and R.H. Stamps. 1999. Coconut husk and processing effects on chemical and physical properties of coconut coir dust. *Hort Science* 34: 88–90.

Maronek, D.M., D. Studebaker, and B. Oberley. 1985. Improving media aeration in liner and container production. *Comb. Proc. Intl. Plant Prop. Soc.* 35: 591–597.

Maynard, B.K. 2000. Evaluating the role of pH in the rooting of cuttings. Eastern Region, North America Research Grant Paper. *Comb. Proc. Intl. Plant Prop. Soc.* 50: 268–273.

Reed, D.W. (Ed.) 1996. *A Grower's Guide to Water, Media and Nutrition for Greenhouse Crops.* Ball Publishing. Batavia, IL.

Richer, C., J.A. Rioux, D. Tousignant, and N. Brossard. 2004. Improving vegetative propagation of sugar maple (*Acer saccharum* Marsh.). *Acta Hort.* 630: 167–175.

Sambrook, J. 2000. Experiences with simple propagation on a new nursery. *Comb. Proc. Intl. Plant Prop. Soc.* 50: 152–154.

Skimina, C.A. 1980. Salt tolerance of ornamentals. *Comb. Proc. Intl. Plant Prop. Soc.* 30: 113–118.

Szmidt, R.A.K. and C. Chong. 1995. Uniformity of spent mushroom substrate (SMS) and factors in applying recommendations for use. *Compost Sci. Util.* 3: 64–71.

Wells, J.S. 1985. *Plant Propagation Practices.* American Nurseryman, Chicago, IL. 367 pp.

Whitcomb, C. 2001. Seedling development: The critical first days. *Comb. Proc. Intl. Plant Prop. Soc.* 51: 610–614.

6 Physical Properties and Other Factors to Consider when Selecting Propagation Media

Calvin Chong, John E. Preece, Caula A. Beyl, and Robert N. Trigiano

Media used for propagation can have a profound impact on propagation by cuttings and seeds. Interestingly, a medium is not essential either to root cuttings or to germinate seeds. Cuttings will form roots in humid environments and have been rooted commercially using aeroponics. In this case, the bottoms of the cuttings are misted as the cuttings are suspended in the air and sprayed with nutrient solution. Likewise, it is easy to germinate seeds in petri dishes lined with filter paper or paper towels. Cuttings can also be rooted in a glass of water; however this is not recommended because low-quality roots form from the low oxygen levels. Bubbling air through the water enhances rooting, but is not always practical.

To be successful when germinating seeds or rooting cuttings, the propagator must furnish the proper environment. Adequate amounts of water and oxygen, proper temperature, and pH in the correct range must be provided to the propagule for successful results. A well-chosen, suitably managed propagation medium will provide a good balance of water and oxygen and an optimum pH greatly enhancing propagation results.

Light and nutrients generally are not needed from the environment for the processes of germination and rooting. Light may be necessary to overcome dormancy in some species, but has either little or a negative influence on germination of many species. Nutrients are supplied by the seed or cutting, and if stock plants were properly fertilized, additional nutrients are not required for germination, rooting, or adventitious shoot production. However, fertilization with nutrients is necessary for subsequent growth of the plants.

In experiments where different media are compared for propagation, consider these environmental factors and how they are affected by the medium. This will greatly increase understanding of propagation results. The following three experiments explore the effects of media and container type on rooting of stem cuttings, using media derived from organic wastes, and determining bulk density and pore space of the medium.

EXERCISES

EXPERIMENT 1: EFFECTS OF MEDIA AND CONTAINER TYPES ON ROOTING OF STEM CUTTINGS

This exercise is designed to evaluate and compare rooting performance of cuttings in different media and container types to determine their effects on rooting of cuttings and on subsequent growth after transplanting. Among other differences, a coarse-textured medium drains more freely than a finer one. The texture of a medium may also influence the rooting ability of cuttings and ease of handling at transplanting time. Propagating containers are available in a wide choice of designs, shapes, and capacities (depth, width, and volume). Some are sold prefilled (ready-to-use) with medium and are more convenient to use. This exercise will take seven to nine weeks to complete: three to five weeks for rooting to transplant stage with easy-rooting species, such as coralberry (*Symphoricarpus orbiculatus*), Peegee hydrangea (*Hydrangea paniculata*), spirea (*Spiraea* sp.), and weigela (*Weigela florida*), and another three to four weeks for growing-on.

Materials

Each student or team of students will need the following materials:

- Medium A: fine-textured medium, such as a commercial peat-lite mix and Medium B: coarse-textured medium, such as 100% perlite or 2:1 perlite:peat (v/v)
- Ninety unrooted cuttings per replication per species (i.e., ten replications for each of nine media/container treatments described below)
- Indolebutyric acid (IBA: 5000 mg/L)
- Media/container treatments:
  - Ten synthetic cubes (such as Oasis foam blocks)
  - Ten synthetic wedges (such as Oasis foam wedges)
  - Ten peat-filled Jiffy 7 pucks

	Procedure 6.1
	Rooting Performance of Cuttings in Various Media and Containers
Step	Instructions and Comments
1	Before use, place all media/container treatments upright in an appropriately sized tray and water to saturation.
2	Collect terminal stem cuttings. Trim cuttings of each species to uniform length (i.e., 1.6 to 2.0 in., 2.4 to 2.8 in. [4 to 5 cm, 6 to 7 cm], etc.) and remove leaves from the lower 0.6 to 0.8 in. (1.5 to 2.0 cm) portion. If leaves are large (some deciduous species), remove one-third to half of each to reduce water loss due to transpiration.
3	Dip the basal 0.4 in. (1 cm) of each cutting for 5 sec. in the hormone solution, and insert one into each individual container unit or 10 into each open tray. Root outdoor or in a greenhouse under appropriate conditions, such as intermittent mist (e.g., 8 sec./30 min at start, reducing to 8 sec./2 h after one week) and shade.
4	This experiment can be conducted with the media/container treatments arranged in a complete randomized block design with 2 or 3 replications and 10 cuttings per plot (treatment unit). Each student or group can be assigned a separate species and/or replication.
5	Between three and five weeks, most cuttings of a species will have rooted to some degree and/or be ready for transplanting. Determine the percentage of rooted cuttings in each treatment that are ready for transplanting and also the percentage that have rooted but are not yet ready. How do you know when a cutting has rooted? For the cubes, wedges, and Jiffy 7s, this would be when the roots have developed to the edge of the container. For the plugs, this would be when there is sufficient rooting to lift or dislodge the cuttings with the root system intact. For the open trays, this would be when cuttings develop a "tight" 1-in. (2.5-cm) root ball. Tightness of the ball and the amount of medium attached will vary with the number of roots and their degree of development, including secondary branching.
6	Select five rooted cuttings at random from each treatment and transplant them to larger containers (i.e., 1 qt [0.95 L] size) filled with a standard potting medium. Grow-on for several weeks using standard cultural practices, including regular fertilizer applications. Measure the height and spread of each plant and calculate the growth index, GI. GI = (height + width) / 2. Remove the tops of plants in each treatment, dry in an oven, and weigh. Express data in terms of GI and shoot dry weight per plant (mean over five plants). Tabulate and compare your results.

- Ten shallow plugs [i.e., 2 in. (5 cm) deep, cylindrical or square] filled with Medium A. Note: Most plug trays can usually be cut with a saw into sections, each having a number of cavities; for example, a tray containing 50 cavities into five 10-cavity units
- Ten shallow plugs filled with Medium B
- Ten deep plugs [i.e., 4 in. (10 cm) deep, cylindrical or square] filled with Medium A
- Ten deep plugs filled with Medium B
- One open tray [i.e., a rectangular-shaped 4 × 10 × 2 ½ in. (10 × 25 × 6 cm) deep] fiber container filled with Medium A, 10 cuttings to be inserted
- One open tray filled with Medium B, 10 cuttings to be inserted

Follow Procedure 6.1 to complete the experiment.

Anticipated Results

Most cuttings will root in almost any soilless substrate, but percentage and speed of rooting may vary with the media and/or container type. As a rule, cuttings rooted in containers with larger volumes will grow more after transplanting than those rooted in containers with smaller volumes.

Questions

- Was there any difference in rooting percentage, readiness to transplant, or subsequent growth due to the media/container treatments?
- Was there any apparent relationship between growth and any of the container dimensions?
- Was there any apparent difference between Medium A and Medium B regarding ease of filling the plugs or trays? Did one medium compact more than the other?

EXPERIMENT 2: ASSESSING ORGANIC WASTE-DERIVED MATERIALS FOR ROOTING OF STEM CUTTINGS

This exercise is designed to assess the feasibility of using organic waste-derived materials in rooting media. Many organic waste by-products, such as paper mill sludges and municipal composts, have potential for use in propagation media. The pH and salt level in these materials may be higher than desirable. Where supply is plentiful and within affordable trucking distance, use of these materials can result in significant savings. This experiment will take three to five weeks to complete for easy-to-root species, such as forsythia (*Forsythia x intermedia*), ninebark (*Physocarpus opulifolius*), and deutzia (*Deutzia gracilis*), and several weeks more for slower or harder-to-root species, such as boxwood (*Buxus sempervirens*) and purple-leaf sandcherry (*Prunus cerasifera*).

Materials

Each student or team of students will require the following materials to complete this experiment:

- Individual (unmixed) materials:
  - Sphagnum peat, perlite, and an organic waste-derived material, such as paper mill sludge, municipal solid waste compost, or leaf and yard waste compost
- Root promoting growth regulator: 5000 ppm (5000 mg/L) indolebutyric acid dissolved in windshield washer fluid, or an appropriate substitute
- Solubridge (electrical conductivity meter) and pH meter
- Ninety unrooted cuttings per replication per species
- Nine flats each about 12 × 12 × 2.5 in. deep (30 × 30 × 6 cm deep)
- Media treatments:
  - Control: 50% sphagnum peat: 50% perlite by volume
  - Perlite-based Group I (leached) media: 100% perlite, and perlite mixed with 20, 40, or 60% by volume of waste material
  - Perlite-based Group II (unleached) media: 100% perlite, and perlite mixed with 20, 40, or 60% by volume of waste material

Follow the instructions listed in Procedure 6.2 to complete this experiment.

Anticipated Results

While species responses may be variable, cuttings of easy-to-root species are expected to root well in media where the soluble salts concentration is <0.2 dS·m^{-1} at the time of cutting insertion, and some less so between 0.2 and 0.7 dS·m^{-1}. Media with salt levels above 0.8 dS·m^{-1} are expected to inhibit rooting of many, if not most, species, and may even be detrimental. Varying the proportions of material in the medium will indicate the optimal amount that is suitable.

Questions

- How quickly do the salts leach from the media? How much water is required per unit volume of media to leach to the 0.2 dS·m^{-1} salt threshold?
- How long does it take for cuttings to be affected by the salts?
- Is the waste derivative an effective alternative material for use in rooting media?
- Can the information gained from this exercise be easily reapplied or extrapolated for use of other materials?
- What conclusions can be made about the effect of pH? Are there any noticeable differences between species?

EXPERIMENT 3: BULK DENSITY AND PORE SPACE OF MEDIA

Media designed for plant growth or propagation serves several purposes including providing support, moisture, aeration, and nutrition, although the last function is normally not important for media used for propagation of cuttings. The components of the medium, their particle size, the degree of compaction, are all characteristics that affect how well the medium provides for the functions above. There must be a balance between the medium's ability to hold moisture and its capacity to stay aerated. Fine pore space will retain water and larger pore spaces provide for exchange of air supplying oxygen for aerobic respiration of roots.

Bulk density of a medium also depends upon what components have been used to make up the medium. In contrast to formulated growing media, most soils have a bulk density somewhere between 1–1.8 g/cm^3, although some very friable clays have bulk densities below that range. There is a strong inverse relationship between bulk density and pore space. Media with the greatest bulk density have less pore space and vice versa. Three common components of commercial growing media are peat, perlite, and vermiculite. Peat has a bulk density of about 0.05–0.5 g/cm^3, vermiculite 0.06–0.16 g/cm^3, and perlite about 0.24 g/cm^3. As peat degrades, more fine particles occur, increasing the bulk density and decreasing the pore space. Compare these bulk densities with that of sand, which is 1.45–1.65 g/cm^3! ProMix "PGX" (Premier Horticulture) has a bulk density of 0.13–0.16 g/cm^3, and Scott's

\multicolumn{2}{c}{**Procedure 6.2**}

Procedure 6.2
The Feasibility of Using Organic Waste-Derived Materials in Rooting Media

Step	Instructions and Comments
1	To serve as reference points, measure the electrical conductivity (EC, a measure of soluble salts concentration) and pH in 1 medium:2 water (vol/vol) extracts from each of the raw (unmixed) materials.
2	Water all media treatments to saturation. Measure the EC and pH in the control and all the Group I media. (The results should be similar in the Group II media.) If the salt level is above $0.2 dS \cdot m^{-1}$ in any of the Group I media, leach by sprinkling about 1 q (0.95 L) of water over the surface. Measure the soluble salts concentration and pH again. Repeat this procedure, if necessary, until the soluble salts are $<0.2\ dS \cdot m^{-1}$.
3	Collect terminal stem cuttings. Trim cuttings of each species to uniform length [i.e., 1.6 to 2.0 in., 2.4 to 2.8 in. (4 to 5 cm, 6 to 7 cm), etc.] and remove leaves from the lower 0.6 to 0.8 in. (1.5 to 2.0 cm) portion. If leaves are large (some deciduous species), remove one-third to half of each to reduce water loss due to transpiration. Dip the basal 0.4 in. (1 cm) of each cutting for five seconds in the hormone solution; then insert about 2.5 cm (1 in.) apart into each treatment medium. Root outdoors or in a greenhouse under appropriate conditions, such as intermittent mist (e.g., 8 sec./30 min at start, reducing to 8 sec./2 hr after 1 week) and shade.
4	This experiment can be conducted with the treatment media (flats) arranged in a complete randomized block design with 2 or 3 replications of each medium and 10 cuttings per plot (treatment unit). Within each flat, cuttings of several different species spaced in rows 0.8 to 1.2 in. (2 to 3 cm) apart can be accommodated. Each student or group can be assigned a separate species and/or replication.
5	At each of several increasing intervals during rooting (4, 8, 16, etc., days), collect samples of medium from the Group II (unleached) media and measure their soluble salts and pH.
6	Assess rooting performance in terms of percent rooting, number of roots, and length of the longest root. Count and measure only roots longer then 0.02 in. (1 mm). Note: All treatments within a species should be assessed at the same time. This is typically done when most of the cuttings in one or more treatments have rooted. Plot percent rooting responses against rates of waste material; insert percent rooting response in the control medium as a horizontal line across the graph. Repeat this procedure for root number and root length responses.

Metro Mix 200 has a bulk density of 0.21–0.24 g/cm^3, which changes to 0.75–0.85 g/cm^3 when wet.

Because of their wide usage, it is suggested that as many of the media described above in this chapter be included in this exercise. However, this laboratory is appropriate for any potting or propagation medium or blend. It is interesting and informative to see the results across several media.

Facilities

No special facilities are required for this exercise. It can be conducted in a laboratory or classroom. It is important that there be level table tops or bench tops for the balances.

Materials

These items will be sufficient for individual students or teams.

- One or more media to test
- Triple beam or electronic balance graduated in grams
- Translucent cups or other straight-sided containers that are graduated and will hold 200 mL volume
- If graduated containers are not available, paper or polystyrene cups that can be written on will suffice
- Graduated cylinder that can measure at least 200 mL
- Flexible screening cut into squares that will cover the mouth of the cup with sufficient length hanging over the sides to be secured by a rubber band; Saran greenhouse shade cloth can be cut into flexible squares for this exercise
- A rubber band

Follow the instructions in Procedure 6.3 to complete the experiment.

Anticipated Results

High bulk density, which is typical of media containing excessive amounts of fine particles (fines), results in low pore space, excessive moisture retention, and inadequate oxygen, and hence poor rooting performance. Look for these signs: poor appearance or coloration of cuttings, inconsistent or poor rooting percentages, increased incidence of browning or rotting of the cutting bases, cuttings that develop fine roots, roots that tend to be more prolific

Procedure 6.3
Bulk Density and Pore Space of Media

Step	Instructions and Comments
1	Pour 200 mL of water into the cup and mark with a line. The cup must be weighed in grams when dry to be able to subtract the weight of the cup from that of the medium. Some electronic balances will do this automatically by using the tare function.
2	Fill the cup to the 200 mL level with the chosen medium, and gently tap the bottom on a firm surface to settle the medium. Add more if necessary to fill to the 200 mL line, and then weigh in grams. Subtract the weight of the cup to determine the dry weight of the medium. You can now determine its bulk density (weight/volume) by dividing the weight of the medium by its volume (200 mL). For reference, the bulk density of water at room temperature is 1 because 1 mL of water weighs 1 g.
3	Pour 200 mL of water into the graduated cylinder and slowly pour into the medium in the cup. Stop when all of the pore spaces are filled with water or when the water begins to stand on the surface of the medium. Two problems can occur at this stage. First, some media, such as perlite and vermiculite tend to float, leaving an area with just water and, second, no medium at the bottom of the cup. This can give misleading readings. Some media, such as milled sphagnum moss and sphagnum peat, are hydrophobic and difficult to wet. This provides a valuable experience for students, but the medium will require stirring (or a wetting agent) to wet thoroughly.
4	The amount of water that was added is an accurate measure of the total amount of pore space in a medium. When growing or propagating a plant, this pore space is occupied by water and air. Both are critical to plant growth and propagation. The percentage of the original volume (200 mL) that was pore space can now be calculated by dividing the total amount of pore space calculated by the amount of water that was added by the volume of medium (200). This will give the total porosity percentage.
5	To obtain a measure of how much of the pore space of a medium is occupied by air and how much by water after irrigation and drainage, place the flexible screen over the mouth of the cup and secure it to the cup with the rubber band. Invert the cup at a 45° angle over a sink or container that can hold the water, and drain until only a few, slow drips remain. A recently watered, thoroughly drained container is said to be at "container capacity."
6	To determine how much of the pore space is now occupied by air and how much by water, reweigh the container with the moist medium and subtract the weight of the cup. Now subtract the original dry weight of the medium. The difference is the weight of the water that remains in the medium. Although this is in grams because 1 mL of water weighs 1 g, the number can be converted to mL, and this is the volume of water held in the drained medium (at container capacity).
7	To determine the amount of pore space of the wet medium that is occupied by air, simply subtract the amount of water held in the drained medium from the amount of water that was originally added to the medium to measure its total pore space. The student can then calculate the ratio of total pore space occupied by water and by air at the container capacity. This can lead to a discussion of the balance of both water and oxygen for growing and propagating plants.

near the surface of the medium where there is more exposure to oxygen and moisture is less.

Questions

- Why is oxygen in the medium so important for plant propagation and root growth and development?
- What is an ideal balance between water and air in medium pore spaces for plant propagation?
- What symptoms can be expected on seeds or cuttings being propagated if pore space occupied by air is limited in a medium?
- What are the symptoms of inadequate aeration in a medium on established plants?
- Why is medium density an important concern for propagating or growing plants?
- Why is medium density a concern when managing employees at a commercial plant propagation or horticultural facility?

7 Media for Cutting Propagation

Patricia S. Holloway

Propagation media consist of a wide variety of substances, from field soils to preformed synthetic foam blocks. The most appropriate medium is determined by a combination of factors including commercial availability, the species being propagated, the type of cutting, the season, the type of propagation facility (e.g., intermittent mist, fog, polyethylene tunnels—chapter 2), and environmental factors, such as light, air and medium temperature, and relative humidity. Traditionally, the most common commercial propagation media included combinations of *Sphagnum* sp. peat mixed with perlite, vermiculite, and/or sand. More recently, options have expanded to include a variety of regionally available composted wood products, processing by-products, such as rice hulls, and synthetic foam flakes and blocks (chapter 6). The best medium is one that physically supports the cutting and provides a moist, well-aerated, nontoxic environment for root initiation and development.

In addition to the basic medium, fertilizer is sometimes added either as a foliar mist application or incorporated into the medium to improve cutting quality following rooting. Fertilizer added to the medium is usually a slow-release formulation lasting eight months or more. These nutrients are especially beneficial during the hardening-off period following root initiation and development, but prior to transplanting. In addition to fertilizers, mycorrhizal fungi may be incorporated into the propagation medium to improve rooting and root development. These fungi associate with plant roots and aid in nutrient and water uptake by increasing the surface-absorbing area of roots. Some also produce growth-promoting substances, such as auxins, gibberellins, and vitamins. The symbiotic association (mutualism) of mycorrhizal fungi and plant roots has been shown to improve the growth of plants especially when incorporated into pasteurized media at a very early stage of development. With some plants, inoculation of the propagation or transplant medium with the fungus or a root extract may increase rooting percentages, root survival, root quantity, and plant quality and survival following transfer to a potting mix.

The purpose of these laboratory exercises is to demonstrate methods of propagating cuttings with common commercial propagation mixes as well as synthetic blocks and cubes. The effects of fertilizer and mycorrhizal fungus media amendments on rooting, root development, and transplant quality also will be explored.

SAFETY CONSIDERATIONS FOR ALL EXPERIMENTS

Propagation knives and clippers are sharp. Care must be taken to avoid cuts. Some of the propagation media, especially dry *Sphagnum* peat, vermiculite, and perlite are very dusty and should be handled in a well-ventilated room. Use a face mask suitable for nuisance dust when mixing media for propagation flats. Eye irritation is possible from dust, so wear safety goggles. Moistening the medium prior to use helps minimize dust problems.

EXERCISES

EXPERIMENT 1: ROOTING CUTTINGS IN VARIOUS MEDIA

The type of medium can influence rooting percentages, root quality and quantity, rooting time, the impact of diseases and insect pests on rooting, and the successful transplanting of rooted cuttings into growing media. This experiment will compare various rooting media and identify the best choice for a variety of plants and propagation systems. It will allow for a comparison of common media, such as perlite and vermiculite, as well as locally abundant resources, such as rice hulls and other composted materials. It will illustrate the challenges facing propagators in the choices necessary in developing a rapid, consistent, and easy-to-use propagation system for a variety of plants.

Materials

The following materials will be required to complete this experiment:

- Propagation flats
- Media—The choice of media depends on availability and may include one or a combination of: organic materials [shredded or milled *Sphagnum* or *Hypnum* peat, coconut fibers (coir), composted softwood bark (pine, fir, spruce), composted hardwood bark (oak, beech), rice hulls, composted sawdust, paper mill biosolids, composted wood fiber]; inorganic media: horticultural-grade perlite (3 to 4 mm size), horticultural-grade vermiculite (<3 mm size), water-absorbent and/

or water-repellent rock wool granules, sand (washed, pasteurized, lime free, 0.5 to 2 mm size), pumice; and synthetic media (expanded polystyrene flakes or beads, urea-formaldehyde foam resin flakes). Peat-based media attract fungus gnats whose larvae will decimate new roots. Avoid these combinations in greenhouses with fungus gnat populations

- Plant materials—Suggested greenhouse plants: Angel Wing Begonia (*Begonia corallina* Carriere and related species); Fibrous-rooted Begonia (*Begonia* Semperflorens-Cultorum Group); Rhizomatous Begonia (*Begonia* Rex-Cultorum hybrids); Christmas Cactus (*Schlumbergera bridgesii* (Lem.) Lofgr.); Thanksgiving Cactus (*Schlumbergera truncata* (Haw.) Moran); Florists' Carnation (*Dianthus caryophyllus* L.); Florists' Chrysanthemum (*Dendranthema* X *grandiflorum* Kitam. and cvs); Cigar plant (*Cuphea ignea* A. DC.); Coleus (*Coleus scutellarioides* (L.) Benth); Crown of thorns (*Euphorbia millii* Desmoul.); Fuchsia (*Fuchsia* X *hybrida* Hort. Ex Vilm.); Geranium (*Pelargonium* X *hortorum* L.H. Bailey); Scented Geranium (*Pelargonium* sp. L'Her. Ex Ait.); Heliotrope (*Heliotropium arborescens* L.); Hydrangea (*Hydrangea paniculata* Siebold); Impatiens (*Impatiens walleriana* Hook.f); Canary Island Ivy (*Hedera canariensis* Willd.); English Ivy (*Hedera helix* L. and cvs); Jade plant (*Crassula argentea* Thunb. C. argentea 'Variegata'); Flowering Maple (*Abutilon* X *hybridum* Hort.); Orchid cactus (*Epiphyllum* spp. Haw.); Poinsettia (*Euphorbia pulcherrima* Willd ex Klotzsch); Polka-dot plant (*Hypoestes phyllostachya* Bak.); Purple Heart (*Tradescantia pallida* Rose); Spider-wort (*Tradescantia albiflora* Kunth.); String of pearls (*Senicio radicans* (L.f.) Schultz-Bip.); Swedish Ivy (*Plectranthus australis* R. Br.); Tradescantia (*Tradescantia fluminensis* Vell.); Wax Plant (*Hoya carnosa* (L.f.) R.Br.); Zebrina (*Zebrina pendula* Schnizl.)
- Suggested herbaceous and woody perennials: Achillea (*Achillea ptarminca* L. *A. millefolium* L. *A. filipendula* Lam.); Boxwood (*Buxus sempervirens* L.); Buddleia (*Buddleia davidii* Franch.); Black currants (*Ribes nigrum* L. *R. hudsonianum* Richardson.); Red and white currants (*Ribes triste* Pall. *Ribes* sp. L.); Deutzia (*Deutzia gracilis* Siebold & Zucc.); Dianthus (*Dianthus superbus* L.); Redosier Dogwood (*Cornus sericea* L.); English Lavender (*Lavandula angustifolia* Mill.); Forsythia (*Forsythia* X *intermedia* Zab., *F. ovata* Nakai); Heather (*Calluna vulgaris* (L.) Hull); Hibiscus (*Hibiscus rosa-sinensis* L.); Juniper (*Juniperus horizontalis*, Moench *J. chinensis* L.); Kinnikinnick (*Arctostaphylos uva-ursi* (L.) K. Spreng.); Mock Orange (*Philadelphus coronaries* L., *P. lewisii*, Pursh.); Cottage Pink (*Dianthus plumarius* L.); Potentilla, Cinquefoil (*Potentilla fruticosa* L.); Privet (*Ligustrum japonicum*, Thunb., *L. ovalifolium* Hassk.); Rosemary (*Rosemarinus officinalis* L.); Speedwell (*Veronica incana* L. , *V. spicata* L.); Bridal wreath spiraea (*Spiraea* X *vanhouttei* (C. Briot) Zab.); False spiraea (*Sorbaria sorbifolia* (L.) A. Braun.); Pink spiraea (*Spiraea douglasii* Hook.); Tatarian honeysuckle (*Lonicera tatarica* L.); , English Thyme (*Thymus vulgaris* L.); Vinca, Periwinkle (*Vinca minor* L., *V. major* L.); Weigela (*Weigela florida* (Bunge) A. DC.)
- Propagating knife or clippers
- Dibble, stake, or pencil for making furrows
- Plastic or wooden labels and marking pens
- Rooting powder with paper towel or cup or rooting liquid
- Intermittent mist propagation system, high-humidity propagating box, or other propagation facility
- Ruler or calipers for measuring root length
- Materials for measuring bulk density, total pore space, available water holding capacity, air-filled porosity, pH, and electrical conductivity (chapter 6)

ALTERNATIVE EXPERIMENTS

The choice of medium is directly related to the type of propagation facilities, moisture availability, and environmental conditions. Experiment 1 may be repeated with additional treatments, such as the following: (1) compare rooting in media on an intermittent mist propagation bench versus a closed polyethylene propagation box (chapter 2), mist versus fog (chapter 4), etc.; (2) compare media under two or more misting frequencies or watering regimes; and (3) compare media with and without bottom heat, overhead lights, and rooting compounds.

Additional interesting comparisons for group experiments include the following: (1) organic versus inorganic and synthetic media; (2) grades (particle size) of a product, such as vermiculite, sand, peat, and coconut fiber; (3) types of peat: *Sphagnum, Hypnum,* and sedge; and (4) combinations of media commonly used by commercial greenhouses and nurseries, such as (1) *Sphagnum* peat, sand, peat/sand mix in one of the following 1:1, 3:1, or 4:1 ratios (by volume); (2) *Sphagnum* peat, perlite, peat/perlite mixes in 1:1, 2:1 or 3:1 ratios; (3) *Sphagnum* peat, vermiculite, peat/vermiculite mix in 1:1 or 2:1 ratios.

Follow the instructions outlined in Procedure 7.1 to complete the experiment.

Procedure 7.1
Rooting Cuttings in Various Media

Step	Instructions and Comments
1	Fill four flats with one of the following media: (1) sphagnum peat/sand 1:1 (by volume), (2) horticultural grade perlite, (3) horticultural grade vermiculite, (4) perlite/vermiculite 1:1 (by volume) or other combinations.
2	Measure and record for each medium: bulk density, total pore space, available water holding capacity, air-filled porosity, pH, and electrical conductivity (chapter 6).
3	Make 20, 3- to 6-in. (7.5 to 15 cm) uniform stem tip cuttings each of at least four different greenhouse plants and four herbaceous or woody plant materials.
4	Make labels for each treatment with your last name, medium type, plant name, and date.
5	Remove any flowers from each cutting, and remove leaves from the proximal (lowest) 1½ to 2 in. (4 to 5 cm) of each cutting.
6	Treat all cuttings with a rooting powder or quick dip solution appropriate for the type of cutting.
7	Beginning in the upper left hand corner of the flat, make lengthwise furrows in the medium with a dibble, stake, or pencil. Furrows should be about 3/4 the depth of medium.
8	Insert the label at the top of the furrow, and insert five cuttings at 1- to 2-in. (2.5 to 5 cm) spacing, depending on the size of the cutting. Do not push the cuttings to the bottom of the flat. Repeat with a second species, inserting a label at the head of the furrow until the row is full.
9	Firm the medium around the base of each cutting, and remove any leaves that are partially buried in the medium. Repeat with additional furrows, moving from left to right in the flat until all cuttings are stuck.
10	Place flats into the propagation box, mist bench, etc. Check cuttings weekly until first root appearance. Gently tug on cuttings to dislodge them from the medium. Once roots are observed on one cutting in each treatment, do not disturb the others. Record days to first rooting. When reinserting cuttings into the medium, avoid damage to new roots by using a dibble or stake, opening a hole in the medium, and carefully returning cuttings to the propagation medium.
11	After four weeks for greenhouse plants and six weeks for herbaceous or woody perennial plants (or at the end of the semester or quarter), harvest all cuttings, keeping treatments separate with their labels.
12	Record root color and percentage of rooting for the five cuttings in each treatment.
13	Evaluate root abundance using the scale: 0 = no roots; 1 = few roots, easy to count, no secondary branching, little medium attached to roots; 2 = moderate amount of roots, difficult to count, some secondary branching; 3 = abundant roots, nearly impossible to count, obscured by the medium, abundant secondary branching. Calculate average root abundance per treatment.
14	Measure the longest root on each cutting from point of origin on the stem to root tip. Calculate average maximum root length per treatment.
15	Select the longest root and bend it in half onto itself. Evaluate root brittleness using the following scale: 0 = root is flexible, does not break, can be bent nearly 180 degrees without breaking; 1 = root is stiff but bendable, can be bent to 90 degrees before breaking; 2 = root is brittle and easily breaks when bent less than 90 degrees.
16	Construct at least two tables and/or bar graphs: (1) medium characteristics (from Step 2) and rooting characteristics. Table or figure 2 in your laboratory report should include a comparison of media and species in relation to root quality (color, brittleness, abundance), rooting success (percent rooting), and timing (days to first rooting).

Anticipated Results

Results will vary depending on the treatments chosen as well as the formulations of the media tested. The tropical greenhouse and herbaceous species suggested generally root easily, but certain species may root better in specific media. For example, succulents, such as *Crassula argentea* (Jade Plant), may do better in well-drained media with a higher proportion of sand or perlite relative to the peat. Other species will root well in a variety of media, but the quantity and quality of the roots and the rooting pattern along the stem may differ among media. Since some of the more difficult woody plants take longer to root and thus are in the rooting medium a longer time, these species may reveal subtle differences among the rooting media that were chosen, especially the speed of rooting and the quality of roots.

Questions

- Compare and contrast media and the rooting percentages of all species. Is one medium superior to all others?
- How did root quality, quantity, and rooting speed differ among media and species?
- What physical characteristics of each medium might explain the rooting results?
- Does each medium fulfill the requirements of a good propagating medium (paragraph 1 in introduction)?

EXPERIMENT 2: COMPRESSED PELLET OR SOLID BLOCK MEDIA AND ROOTING OF CUTTINGS

Compressed pellets and solid block media are common in certain horticulture businesses, such as the mass production of chrysanthemum, geranium, and poinsettia rooted cuttings. The blocks are easy to handle and provide a uniform rooting environment, often with standard spacing. The pellet or block is transplanted with the rooted cutting into a growing medium, thus reducing transplant shock. This experiment will demonstrate the advantages and disadvantages of using pellets or blocks for mass production of rooted cuttings. It may be combined with Experiment 1 to show a variety of propagation systems available to growers.

Materials

The following materials will be needed to complete this exercise:

- Propagation flats
- At least three of the following media: compressed peat pellets or blocks with or without added fertilizer, rock wool compressed cubes, urea-formaldehyde or polyurethane foam cubes (e.g., Oasis® Rootcubes®, Wedges®, Jiffy 7, and Jiffy 9, etc.); media should be drenched, if necessary, to leach toxic substances from the blocks; saturate and expand compressed cubes with water at least 24 h before class
- Dowel, dibble, or pencil for making holes for stems if not premade
- Propagating knife or clippers
- Plastic or wooden labels and marking pens
- Rooting powder or liquid (concentration for herbaceous stem cuttings)
- Florists' chrysanthemum (*Dendranthema x grandiflorum* (Ramat.) Kitam.), carnation (*Dianthus caryophyllus* L.), poinsettia (*Euphorbia pulcherrima* Willd. ex Klotzsch), geranium (*Pelargonium x hortorum* L.H. Bailey), or other easily rooting species
- Washed quartz sand—many of the propagating blocks come with premade holes for stems; choose stems that match the diameter of the holes or use sand to fill in the hole and support the cutting
- Intermittent mist propagation system or propagating box
- Ruler or calipers for measuring root length
- Materials for measuring bulk density, total pore space, available water holding capacity, air-filled porosity, pH, and electrical conductivity (chapter 6)

Follow the instructions described in Procedure 7.2 to complete this experiment.

Anticipated Results

The media chosen have been specially formulated for use in rooting plants, and thus, all of the media should perform well. There will be differences in ease of handling, which the students will experience as they examine the rooted cuttings and collect data. If a fertilizer-enhanced medium is used and compared to a control without fertilizer, marked differences in growth may not be apparent until the cuttings are well rooted and developing. The watering and misting regime for many pellets and blocks differs from cuttings in propagating flats. Pellets, especially solid peat pellets, can hold a lot of water and cause cuttings to rot. Comparisons between the two methods should use separate misting systems.

Questions

- Compare rooting and transplant growth among cuttings in different media. Is one medium superior to all others?

Media for Cutting Propagation

Procedure 7.2
Compressed Pellet or Solid Block Media and Rooting of Cuttings

Step	Instructions and Comments
1	Prepare a flat that contains at least 10 cells of three media (compressed peat, rock wool, foam propagation blocks or others). Make sure the flat has sufficient drainage so there is no standing water.
2	Measure and record for each medium or obtain from manufacturer's published information the following: total pore space, available water holding capacity, air-filled porosity, pH, and electrical conductivity (chapter 6)
3	Make 30 cuttings of florists' chrysanthemum, geranium, poinsettia, carnation, or other easy to root species.
4	Treat all cuttings with a rooting powder or quick dip liquid with a concentration appropriate for herbaceous cuttings.
5	Some blocks have preset holes, whereas others do not. If holes do not exist, make them with a dowel, dibble, or pencil the diameter of each cutting in the top of each saturated block or pellet. Insert one cutting per hole not more than 3/4 of the total block or pellet depth. If the hole is too large to fit the cutting snugly, make a second hole close to the original or add a small amount of washed quartz sand to the hole.
6	Label the flat with your last name, plant name, and date. Place the flat in the propagation box or beneath intermittent mist. Keep media and cuttings moist.
7	Check the sides of the pellets or blocks weekly for root emergence. Record the date of first root emergence for each treatment.
8	After three to four weeks, separate all cubes or pellets and count the number of emerging roots tips. Do not try to remove the cuttings from the media. Record percent rooting for the cuttings in each type of medium, and describe the roots (color, presence of root hairs, general appearance compared to the other treatment media).
9	Rate each cutting based on the following scale: 0 = no roots; 1 = few roots visible, all less than 1 in. (2.5 cm) long, no secondary branching; 2 = moderate amount of roots visible, at least half are 1 in. (2.5 cm) long or longer, some secondary branching; 3 = abundant roots, most are 1 in. (2.5 cm) long or longer, abundant secondary roots. Calculate the average root abundance per treatment.
10	Measure the length of the longest root from the medium surface to root tip. Calculate average length of longest root for each treatment.
11	Construct a table and/or bar graph comparing media effects on root quality (color, abundance, root length), rooting success (percent rooting), and timing (days to first root appearance).
12	Plant the rooted cuttings including the propagation medium. Bury the pellet or block at least 1/4 in. (1/2 cm) into a standard greenhouse potting mix.
13	Measure the height of each cutting, and count the number of leaves immediately after planting and again four weeks later. Calculate the difference between plant height and leaf number between dates, and compare growth of rooted cuttings following transplanting
14	Construct a table or bar graph comparing plant growth following rooting among media types.

- How did root quality, quantity, and rooting speed differ among media?
- What physical characteristics of each medium might explain the rooting results?
- Does each medium fulfill the requirements of a good propagating medium (see introduction)?

Experiment 3: Rooting of Cuttings in Fertilizer-Amended Media

Growers are interested in producing high-quality rooted cuttings in the least amount of time. For some species, especially those that require long rooting times, the addtion of fertilizer, either as a liquid mist or incorporated into the medium, produces a superior rooted plant, one that will justify adding fertilizer to production costs. This experiment will demonstrate the changes in rooting and plant growth with the addition of a slow-release fertilizer to the rooting medium.

Materials

The following supplies will be required to complete this experiment:

- Propagation flats

- Peat/perlite, perlite/vermiculite (1:1 by volume) propagating medium or other combinations
- Florists' chrysanthemum (*Dendranthema x grandiflorum* (Ramat.) Kitam.), carnation (*Dianthus caryophyllus* L.), poinsettia (*Euphorbia pulcherrima* Willd. ex Klotzsch), geranium (*Pelargonium x hortorum* L.H. Bailey), or other easily rooting species
- Propagating knife or clippers
- Dibble, stake, or pencil for making furrows in the medium
- Labels and marking pens
- Rooting powder or liquid (concentration for herbaceous cuttings)
- Osmocote® 18-6-12 fertilizer or similar slow release formulation
- Intermittent mist propagation system or propagating box
- Ruler or calipers for measuring cuttings
- Peat/lite commercial potting mix and containers appropriate for the size of the rooted cuttings

Additional interesting comparisons for group experiments include the following: (1) comparisons of a variety of organic, inorganic, and synthetic media with and without fertilizer; (2) intermittent mist versus propagation boxes using flats with and without fertilizer; and (3) comparison of soluble fertilizer applied in mist or fog with media-incorporated fertilizer.

Follow the instructions in Procedure 7.3 to complete this experiment.

Anticipated Results

Addition of a slow-release fertilizer may or may not show differences in overall rooting percentages depending on the plant used, but root quantity, quality, especially total root length and branching patterns will show a response to fertilizer. The amount of time a cutting is in the propagation structure also may be shortened by addition of fertilizer. The quality of the shoots following rooting, especially changes in color, can be very different in fertilized versus nonfertilized cuttings. Following transplanting to a potting medium, differences may also be measurable in shoot growth. Rooted cuttings may also begin new shoot growth more rapidly in cuttings treated with fertilizer than for untreated cuttings.

Questions

Compare and contrast fertilizer-amended media with the control and the rooting ability of all species.

- Is there an advantage to incorporating fertilizer in the mix? How do rooting percentages differ from an unfertilized control?
- How did root quality, quantity, rooting speed, and new growth differ among treatments and species?

EXPERIMENT 4: ROOTING AND GROWTH OF KINNIKINNICK, *ARCTOSTAPHYLOS UVA-URSI*, USING MEDIUM AMENDED WITH MYCORRHIZAL FUNGI ROOT EXTRACT

Kinnikinnick is a common evergreen ornamental ground cover used in home and public landscapes. It is also one of the most studied species in terms of colonization by mycorrhizal fungi. Plant roots are easy to harvest, and those colonized by mycorrhizae are easy to distinguish from noncolonized roots with a hand lens or stereomicroscope by their thickened root tips. This experiment demonstrates a simple method of inoculating pasteurized or sterilized propagation and growing media with mycorrhizae that may significantly improve the growth and establishment of this nursery ground cover.

Materials

The following materials are needed to complete the experiment:

- Propagation flats
- Sphagnum peat, sand
- Kinnikinnick (*Arctostaphylos uva ursi* (L.) Spreng.) nursery stock or established landscape plants
- Rooted or unrooted kinnikinnick stem cuttings
- Dissecting microscope
- Blender, sieve, piece of window screen or kitchen wire strainer
- Large plastic buckets or storage containers for mixing extract
- Spray bottle or garden watering can with water breaker
- Intermittent mist propagation bench or propagation box
- Labels and marking pens

Alternative Experiments

Depending upon conditions, rooting and mycorrhizal colonization may require more time than allocated for a single semester or quarter. Use cuttings prerooted in a sterile medium and apply the mycorrhizal root drench to the transplant medium rather than the rooting medium to learn how mycorrhizae influence growth of rooted cuttings. Additional comparisons with groups include the following: treat half of cuttings in each with rooting powder or quick dip appropriate for cutting type. Compare mycorrhizal treatments among different rooting media.

Media for Cutting Propagation

	Procedure 7.3 Rooting of Cuttings in Fertilizer-Amended Media
Step	Instructions and Comments
1	Fill three flats with Sphagnum peat/sand 2:1 (by volume) or other propagating medium. Incorporate into one flat, 3.0 g/L 18-6-12 (eight to nine months) Osmocote® controlled-release fertilizer or similar product and into the second flat, 6.0 g/L Osmocote. The third flat is the no-fertilizer control.
2	Make fifteen 3- to 6-in. (7.5 to 15 cm) long, stem tip cuttings each of at least three different greenhouse plants: florists' chrysanthemum, hydrangea, English ivy, and three herbaceous or woody plant materials: potentilla, vinca, speedwell (see additional suggestions in Experiment 1).
3	Make labels for each treatment with your last name, medium type, plant name, and date.
4	Remove any flowers from each cutting, and remove leaves from the basal 1½ to 2 in. (4 to 5 cm) of each cutting.
5	Treat all cuttings with a rooting powder or quick dip solution appropriate for the type of cutting.
6	Beginning in the upper left hand corner of the flat, make lengthwise furrows in the medium with a dibble, stake, or pencil. Furrows should be about 3/4 the depth of the flat or container.
7	Insert the label at the top of the furrow, and insert five cuttings at 1- to 2-in. (2.5 to 5 cm) spacing, depending on the size of the cutting. Do not push the cuttings to the bottom of the flat. Repeat with a second species, inserting a label at the head of the furrow until the row is full.
8	Firm the medium around the base of each cutting, and remove any additional leaves that are partially buried in the medium. Repeat with additional furrows until all cuttings are stuck.
9	Place flats into the propagation box, mist bench, etc. Check cuttings weekly until first root appearance. Gently tug on cuttings to dislodge them from the medium. Once roots are observed on one cutting in each treatment, do not disturb the others. Record days to first rooting. When reinserting cuttings into the medium, avoid damage to new roots by using a dibble or stake, opening a hole in the medium, and carefully returning cuttings to the propagation medium.
10	After four weeks for greenhouse plants and six weeks for herbaceous or woody perennial plants (or at the end of the semester or quarter), harvest all cuttings, keeping treatments separate with their labels.
11	Record root color and percentage of rooting for the five cuttings in each treatment.
12	Evaluate root abundance using the scale: 0 = no roots; 1 = few roots, easy to count, no secondary branching, little medium attached to roots; 2 = moderate amount of roots, difficult to count, some secondary branching; 3 = abundant roots, nearly impossible to count, obscured by the medium, abundant secondary branching. Calculate average root abundance per treatment.
13	Measure the longest root on each cutting from point of origin on the stem to root tip. Calculate average maximum root length per treatment.
14	Measure new growth per cutting, including number of new stems and leaves. Calculate the average number of new shoots and leaves per cutting for each treatment.
15	Construct a table and/or bar graph including a comparison of fertilizer treatments and species in relation to root quality (color, abundance), rooting success (percent rooting), timing (days to first rooting), and amount of new growth (new stems and leaves).

Try other species, such as birch (*Betula papyrifera* Marsh. and other species), cottonwood (*Populus balsamifera* L. and other species), and Arbutus (*Arbutus menziesii* Pursh). Plants with ectomycorrhizal associations are the easiest to study because fungal associations may be observed under a dissecting microscope. Other types of mycorrhizae require more complex microtechniques (cell clearing and staining) to detect presence of fungi. Follow the instructions in Procedure 7.4a or 7.4b to complete this experiment.

Anticipated Results

The results of this experiment will be determined by the length of the experiment and the success in introducing mycorrhizae to the medium. Rooting percentages

Procedure 7.4a
Rooting of Kinnikinnick (*Arctostaphylos uva-ursi*) Using Media Amended with Micorrhizal Fungi Root Extract

Step	Instructions and Comments
1	Fill two propagation boxes or flats with a 2:1 (by volume) mixture of moistened Sphagnum peat/sand.
2	Collect branched stems of kinnikinnick from wild stands, ornamental plantings, or containers. Make 20 stem tip cuttings, 4 to 6 in. (10 to 15 cm) in length.
3	Remove all flowers and/or fruit from each cutting, and remove leaves from the proximal 1 to 1½ in. (2.5 to 4 cm) of the stem.
4	Stick cuttings using a dibble, pencil, or stake into the medium at 1-in. (2.5 cm) spacing.
5	Collect enough kinnikinnick roots to fill a 50-mL (¼ cup) container from wild stands or commercially propagated kinnikinnick. Remove most of the surrounding soil or potting mix, enough to locate the root tips. Examine the root tips using a stereo (dissecting) microscope to see evidence of mycorrhizal fungi associated with the roots. Mycorrhizal roots are thickened at the tips with a mantle of fungus. Roots covered with the mantle seem to be swollen, cigar-shaped, with stubby branches and whitish webbing (hyphae) surrounding the roots. Nonmycorrhizal roots are thin and hairlike, without the very obvious thickening mantle.
6	Add mycorrhizal roots to a blender half filled with water. Blend until roots are well chopped (approximately 3 min).
7	Filter the liquid through a screen (kitchen strainer, 16 mesh (1.0 mm, 0.04 in.) soil sieve, or window screen will work).
8	Mix filtrate with sufficient water to drench one-half of the propagating flats for the entire class, 500 mL per standard (20 × 10 in., 51 × 25 cm) flat. Spray the extract over the cuttings and flat to thoroughly drench the medium.
9	Label each flat with your last name, treatment, and date.
10	Place flats in a propagating box. Check cuttings weekly until first root appearance (three to four weeks). Gently tug on cuttings to dislodge them from the medium. Once roots are observed on one cutting in each treatment, do not disturb the others. Record days to first rooting. When reinserting cuttings into the medium, avoid damage to new roots by using a dibble or stake, opening a hole in the medium, and carefully returning cuttings to the propagation medium.
11	At the end of the semester or quarter, harvest all cuttings, keeping them separate by treatment.
12	Record root color and percentage of rooting for the 10 cuttings in each treatment.
13	Evaluate root abundance using the scale: 0 = no roots; 1 = one to two roots, easy to count, no secondary branching; 2 = three to five roots, some secondary branching; 3 = more than five roots, abundant secondary branching. Calculate average root abundance per treatment.
14	Evaluate micorrhizal associations in roots using the following scale 0 = no mycorrhizal fungi evident; 1 = half of root tips enlarged with a fungal mantle or hyphae; 2 = abundant evidence of mycorrhizal association in more than half of the root tips.
15	Measure the longest root on each cutting from point of origin on the stem to root tip. Calculate average maximum root length per treatment.
16	Construct a table or bar graph comparing treated and untreated media in relation to root quality (color, root abundance, fungal association scale), rooting success (percent rooting), and timing (days to first rooting).

Procedure 7.4b
Rooting of Kinnikinnick (*Arctostaphylos uva-ursi*) Using Rooting Compounds and Media Amended with Micorrhizal Fungi Root Extract

Step	Instructions and Comments
1	Fill one propagation box or flat with a 2:1 (by volume) mixture of moistened sphagnum peat/sand.
2	Collect branched stems of kinnikinnick from wild stands, ornamental plantings, or containers. Make 20 stem tip cuttings, 4 to 6 in. (10 to 15 cm) in length.
3	Remove all flowers and/or fruit from each cutting, and remove leaves from the proximal 1 to 1½ in. (2.5 to 4 cm) of the stem. Treat each cutting with rooting powder or quick dip liquid for woody plants.
4	Stick cuttings using a dibble, pencil, or stake into the medium at 1-in. (2.5 cm) spacing. Label with the name of the plant and date. Place on an intermittent mist propagation bench until cuttings are well rooted (three or more weeks).
5	When cuttings are well rooted, transplant into a sterile commercial peat-lite potting mix to which sterile sand has been added (2:1 by volume) in cell packs.
6	Collect enough kinnikinnick fine roots to fill a 50-mL (¼ cup) container from wild stands or commercially propagated kinnikinnick nursery stock. Remove most of the surrounding soil or potting mix, enough to locate the root tips. Examine the root tips using a stereo (dissecting) microscope to see evidence of mycorrhizal fungi associated with the roots. Mycorrhizal roots are thickened at the tips with a mantle of fungus. Roots covered with the mantle appear swollen, cigar-shaped, with stubby branches and whitish webbing (hyphae) surrounding the roots. Nonmycorrhizal roots are thin and hairlike, without the very obvious thickening mantle.
7	Add mycorrhizal roots to a blender two-thirds filled with water. Blend until roots are well chopped (approximately 3 min).
8	Filter the liquid through a screen (kitchen strainer, 16 mesh (1.0 mm, 0.04 in.) soil sieve, or window screen will work).
9	Add sufficient water to make 1 L of solution. Thoroughly drench half the transplants and medium with the liquid. One L will be sufficient for 48 cells in a standard flat.
10	Label each flat with your last name, treatment, and date.
11	After four weeks, measure percent survival of rooted transplants and the number of new shoots and leaves per rooted transplant. Calculate the average number of new shoots and leaves per cutting for each treatment.
12	Wash the medium from the roots and evaluate root abundance using the scale: 0 = no roots (roots killed); 1 = one to three root tips; 2 = four to six root tips; 3 = more than six root tips. Calculate average root abundance per treatment.
13	Evaluate micorrhizal associations in roots using the following scale 0 = no mycorrhizal fungi evident; 1 = half of root tips enlarged with a fungal mantle or hyphae; 2 = abundant evidence of mycorrhizal association in more than half of the root tips.
14	Construct a table or bar graph comparing treated and untreated media in relation to shoot growth and root quality (root abundance and fungal association scale).

may be increased and root quality improved (increased root branching and root length) on treated plants, but the greatest differences occur after rooting. If the time is long enough, roots on the treated plants will show colonization by the mycorrhizae (swelling of root tips), and the timing from rooting to removal from the propagation bench will be reduced. At least three months are needed if the experiment is started with unrooted cuttings and two months with rooted plants.

Questions

- How did mycorrhizal extract influence rooting percentages or rooted cutting growth?
- How did root quality, quantity, and rooting speed change with the treatment?
- What percentage of root tips showed colonization by mycorrhizae on rooted cuttings or transplants?
- If no evidence of mycorrhizae is present following rooting or transplant development, what environmental conditions may have impacted the results of this project?

SUGGESTED READINGS

Bunt, A.C. 1988. *Media and Mixes for Container-Grown Plants. A Manual on the Preparation and Use of Growing Media for Pot Plants,* 2nd ed. Unwin Hyman, London.

Dobson, A. 2000. A guide to propagation composts. *Comb. Proc. Intl. Plant Prop. Soc.* 50: 148–149.

MacDonald, B. 1986. *Practical Woody Plant Propagation for Nursery Growers.* Timber Press. Portland, OR.

Marx, D.H. 1996. Mycorrhizal associations and plant propagation. *Comb. Proc. Intl. Plant Prop. Soc.* 46: 517–521.

Oliveira, M.L. 1995. Desirable characteristics of propagation media. *Comb. Proc. Intl. Plant Prop Soc.* 45: 267–272.

Part IV

Plant Propagation Diseases and the Importance of Sanitation

8 Disease Management

Alan S. Windham

CHAPTER 8 CONCEPTS

- Common plant pathogens include fungi, bacteria, viruses, mollicutes, viroids, and nematodes.

- The most common plant diseases observed during propagation are those caused by fungi and fungi-like organisms.

- *Rhizoctonia solani*, which causes damping-off, stem rots and web blight, and *Botrytis cinerea*, which causes botrytis blight (gray mold) are two of the most common fungal pathogens observed during propagation of plants.

- The environment in which most plants are propagated is highly favorable for plant disease development.

- Plant disease management during propagation involves sanitation, disease-free stock plants, environmental and cultural controls as well as chemical controls.

- Viruses vectored by insects, such as *impatiens necrotic spot virus* (INSV), may be very difficult to manage.

- Fungicides should be used to keep healthy plants healthy, not cure diseased plants.

Disease management during plant propagation is critically important to ensure a disease-free plant for the consumer. In most production systems, there is no other time that so many immature plants are exposed to conditions so favorable for disease development. The conditions under which seedlings or cuttings are grown, essentially in monoculture in flats or beds, can account for the rapid spread of disease. Vegetatively produced cuttings often are grown at levels of high relative humidity that favor several foliar diseases. Also, in many parts of the world, cool season annual bedding plants are predisposed to some root rot diseases by exposure to high temperature. Anyone involved in plant production needs to realize that producing a healthy, saleable plant starts prior to and continues during plant propagation. At this point of plant production, you are most likely to lay the foundation for the health of the crop.

Seedlings or rooted cuttings may not die from infectious diseases, but they may be disfigured, discolored, stunted or become generally unthrifty. Many plant diseases that occur during plant propagation have the ability to reduce the quality or grade of infected plants. There are some pathogens that infect vegetatively propagated plants that are not detected in the greenhouse, nursery, or garden center. This often occurs as the signs or symptoms of disease are not recognized or are confused with the symptoms of other maladies, such as nutrient deficiencies, pesticide phytotoxicity, or injury from abiotic stresses.

Diseases of plants in propagation systems may be caused by plant pathogens, such as fungi, fungi-like organisms, bacteria, mollicutes, viruses, viroids, and nematodes. Fungi lack chlorophyll and are eukaryotic. They are generally filamentous, branched organisms that reproduce by spores and have cell walls made of chitin and other polymers. The fungi-like organisms, such as *Phytophthora* and *Pythium,* belong to a group of organisms that are responsible for causing very damaging diseases, such as damping-off, root rots, wilting rots, foliar blights, and downy mildew. This group of pathogens (Oomycetes) have filamentous growth and reproduce by spores, which is very similar to the true fungi. However, once considered fungi, these organisms are now thought to be more closely allied to the brown algae. Prokaryotic organisms, such as bacteria, are commonly found in propagation systems causing soft rots and leaf spot diseases. Bacteria are single-celled organisms that do not have a nucleus or double-membrane-bound organelles, but have a rigid cell wall. Plant pathogenic bacteria are normally rod-shaped or filamentous. Mollicutes are distinguished from bacteria by their smaller size and lack of a cell wall. They are surrounded by a plasma membrane. Since there is no rigid cell wall, these organisms may have a variety of shapes. Mollicutes are causal agents of yellows diseases of plants. Viruses and viroids are much smaller than bacteria and cannot be seen with light microscopy. Neither can reproduce alone, as each needs the host plant's replication machinery for multiplication. Viruses are nucleoproteins with nucleic acid (DNA or RNA) surrounded by a protein coat. Viroids possess many of the attributes of viruses, but are essentially naked strands of RNA without a protein coat.

VIRUS DISEASES

Plant viruses represent one of the major threats to healthy ornamental crops. There are a variety of symptoms associated with viral diseases including ringspots, oak leaf pattern, mosaic, stunting, stem cankers, leaf spots, flowering break, etc. Viruses may be spread by several different means. For example, *tomato ringspot virus* (TRSV) may be spread via infected seed; *tobacco mosaic virus* (TMV) may be spread mechanically by handling plants with tools or hands infested with the virus; *impatiens necrotic spot virus* (INSV) and the *tomato spotted wilt virus* (TSWV) are often transmitted when vegetative cuttings are taken from infected stock plants. Viruses may also be vectored by insects. Aphids often transmit *cucumber mosaic virus* (CMV), and western flower thrips (*Frankliniella occidentalis* Pergande) are common vectors of INSV and TSWV.

IMPATIENS NECROTIC SPOT VIRUS

It is extremely important to purchase vegetative cuttings from a horticultural supplier that has a virus indexing program to ensure the health of their stock plants. New Guinea impatiens cuttings should be purchased from suppliers indexing for INSV. This virus has over 450 hosts representing over 60 genera of plants. Although, infected plants rarely die, they are often severely stunted if infected as seedlings or young cuttings. Other symptoms include leaf distortions, ringspots (Figure 8.1), oak leaf patterns, and stem lesions.

To prevent damage from INSV, monitor greenhouses for western flower thrips (WFT) by placing yellow or blue sticky cards among the crop. In a very short time, WFT

FIGURE 8.1 Zonate lesion typical of impatiens necrotic spot virus on impatiens leaf.

can acquire INSV and transmit it to healthy plants. Weed management is important inside and around the perimeter of a greenhouse. Weeds may serve as asymptomatic hosts for viruses and the insects that act as vectors. If possible, isolate seedlings from vegetatively produced crops in different greenhouses. There is always danger of a virus entering the greenhouse in vegetatively produced cuttings and then spreading to seedlings via insects, such as WFT.

ROSE MOSAIC

Rose mosaic is a common disease on roses worldwide. The causal agents for rose mosaic are *apple mosaic virus* (AMV) and *Prunus necrotic ringspot virus* (PNRSV). These viruses may occur together or individually and are often spread during the vegetative propagation of roses. Common symptoms in rose include vein clearing, mosaic and line patterns, such as "oak leaf symptoms." During hot weather, infected roses may be asymptomatic. Roses infected with rose mosaic may produce fewer and smaller flowers. Rose producers may procure virus-free rootstocks and scions from propagators that have a virus indexing program.

BACTERIAL DISEASES

Bacterial diseases, such as soft rot, bacterial leaf spot, bacterial wilt, and crown gall, are commonly found in propagation systems. Soft rot caused by *Erwinia* spp. can turn cuttings under mist into a putrid, slimy mess during warm weather. Poinsettia (*Euphorbia pulcherrima* Willd ex Klotzsch) is very susceptible to bacterial soft rot during propagation. *Erwinia carotovora* and *E. chrysanthemi* are both capable of causing soft rot in plants reproduced by vegetative cuttings. Wilting and the collapse of cuttings are common symptoms. *E. chrysanthemi* may also be associated with vascular rot and discoloration of tropical foliage plants.

Several bacteria are causal agents of leaf spot diseases of ornamental crops. One of the most common is bacterial leaf spot of zinnia. *Xanthomonas campestris* pv. *zinnae* may be found on seeds, and the bacterium spreads to the leaves of seedlings just after germination. Leaf spots are angular and dark brown to black. Overhead watering spreads the bacterium to adjacent plants. This particular disease can be prevented by disinfecting seeds prior to planting.

One of the most damaging bacterial diseases is bacterial blight of geranium caused by *Xanthomonas campestris* pv. *pelargonii*. Not only can it cause leaf spots, but also v-shaped lesions, wilting, and death. Bacterial blight is very contagious and can spread rapidly among zonal and ivy geraniums. Other geranium species may serve as hosts. Zonal cuttings should be purchased from suppliers that index their stock plants for this disease. Isolate

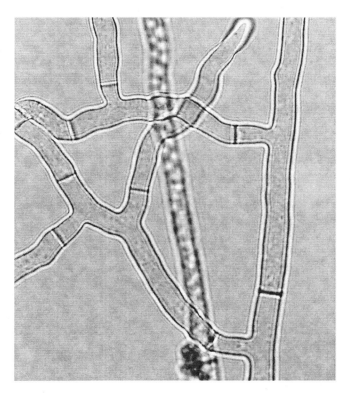

FIGURE 8.2 Right angle branching typical of *Rhizoctonia solani* hyphae.

seed geraniums from those vegetatively propagated. Also, geraniums should not be carried over from one season to the next.

Another bacterial wilt disease of geranium is caused by *Ralstonia solanacearum*. The symptoms of this disease look very much like those of bacterial blight, with the exception of leaf spots. Infected plants become chlorotic, wilt, and die. In recent years, *R. solanacearum* race 3 biovar 2 has been detected in geranium cuttings from major propagators. This pathogen has been imported into the United States in geranium cuttings from other countries. This pathogen poses a serious threat to not only ornamental plants, but to important horticultural crops in the solanaceae family that are grown for food, such as potatoes and tomatoes.

FUNGAL DISEASES

Fungal diseases are the most common diseases of ornamental plants. Included in this group are powdery mildew, downy mildew, rust, fungal leaf spots, rhizoctonia stem rot, pythium root rot, phytophthora root and crown rot, black root rot, fungal leaf spot, and damping-off of seedling crops.

DAMPING-OFF AND WEB BLIGHTS

Damping-off and stem rot are often caused by the fungus, *Rhizoctonia solani* (Figure 8.2). This fungus is ubiquitous and may be found anywhere plants are grown. Young seedlings often fall over after their stems are girdled (damping-off) (Figure 8.3), and older plants may be stunted and suffer stem breakage. This fungus is a soil inhabitant and may be spread in contaminated pots, flats, media, and tools. As it rarely sporulates in nature, *Rhizoctonia solani* spreads by mycelial growth, killing seedlings as it grows. In mist propagation or in polyhouses where temperatures and humidity are often high, "web blight" may develop at the base of cuttings or on the foliage. Growth resembling spider webs is often easily visible on infected plants.

FIGURE 8.3 Damping-off on magnolia seedlings caused by *Rhizoctonia solani*.

FIGURE 8.4 Pythium root rot on rooted cutting of garden mum. (From Trigiano, R.N., M.T. Windham, and A.S. Windham, *Plant Pathology Concepts and Laboratory Exercises,* 2nd ed., CRC Press, Boca Raton, FL, 2008. With permission.)

Botrytis Blight

Botrytis blight, also called gray mold, is probably the most prevalent of all ornamental diseases. The causal agent, *Botrytis cinerea,* is favored by cool, wet conditions, and leaf wetness is important for infection. Older leaves or flowers are most frequently damaged, but under ideal conditions, such as stagnant air and high humidity, almost any plant part is at risk of damage. Botrytis blight can be particularly damaging on geranium, poinsettia, exacum, and begonia. *Botrytis cinerea* may exist in most greenhouses as a saprophyte on dead plant tissue. It sporulates readily on plant tissue, and almost any activity in the greenhouse may initiate the release of spores. Spores must come into contact with free water to germinate. Humidity reduction, increased air movement, and good sanitation are important management strategies.

Root Rot Diseases

Pythium root rot is one of the more common root rot diseases of floral crops. It can be found on chrysanthemums, poinsettia, and many bedding plants. Wilting is one of the first signs of a problem (Figure 8.4). Upon closer examination, diseased roots are dark brown and decayed. Infected plants may wilt rapidly and die. *Pythium* is a "water mold," and may be spread like *Rhizoctonia* or spread in irrigation water. High soluble salt levels in media favor damage from pythium root rot. Phytophthora crown or root rot is not as common as pythium root rot. Like *Pythium, Phytophthora* is a water mold and can be spread in irrigation water as well as from plant to plant in splashed rain or irrigation water. These fungi-like organisms can also be spread by contaminated media that have been improperly stored or become contaminated with soil. *Phytophthora* affected stems and leaves may be dark brown to black. Symptoms include a root rot that extends into the stems of infected plants. Saturated, poorly drained media and overwatering favor both pythium and phytophthora root rots. *Phytophthora ramorum* is thought to be an exotic pathogen in the United States that has been associated with the dieback and death of native oaks in coastal California, called Sudden Oak Death (SOD). Unlike many *Phytophthora* species that cause root rots, this fungus causes canker diseases, foliar blights, and leaf spots, depending on the plant species infected. The host range is growing larger as more is learned about this pathogen. Common hosts include rhododendron, viburnum, mountain laurel, pieris, and camellia.

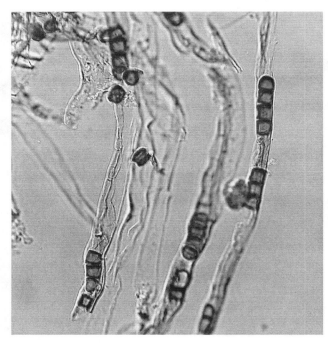

FIGURE 8.5 *Thielaviopsis basicola* aleuriospores in pansy root hairs.

Black root rot (*Thielaviopsis basicola*) is a problem on pansy and may be found on plants as diverse as fuchsia, verbena, vinca, petunia, and million bells. Infected plants generally look unthrifty with poor color and slow growth. Infected root systems may range in color from a dirty off-white, to brown and black (Figure 8.5). This disease may spread very rapidly and infect whole crops. Dark aleuriospores of *T. basicola* are often easily spotted inside infected roots (Figure 8.6). Media pH values that are near 7 or alkaline favor infection. Heat stress may also play a role in the infection process for cool-season crops.

Fungal leaf spots are not common in greenhouses, but may appear on rooted cuttings and pot plants. Poinsettia scab is an example that was first noticed in 2000. Fungal leaf spots may be zonate, irregularly shaped, or angular. Cercospora leaf spot can be particularly damaging on pansy plugs.

Powdery Mildew and Downy Mildew

Powdery mildew is easily recognized on most crops by the white, powdery, fungal growth on leaves. These fungi

Disease Management

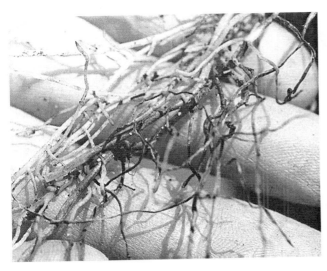

FIGURE 8.6 Blackened roots of Japanese holly infected with *Thielaviopsis basicola*.

FIGURE 8.7 Powdery mildew on flowering dogwood caused by *Erisyphe pulchra*.

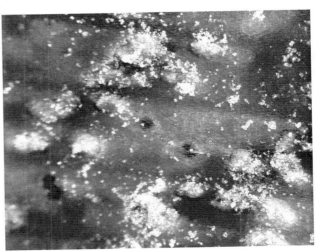

FIGURE 8.8 Rust lesions on daylily bibs caused by *Puccinia hemerocallidis*.

may disfigure and stunt the leaves and flowers. The fungi that cause powdery mildew, such as *Erysiphe*, *Microsphaera*, and *Podosphaera*, are rarely found attacking plants in mist propagation systems, as free water inhibits spore germination. Rather, they may be a problem on stock plants (Figure 8.7). While powdery mildews do not need free moisture for infection, high relative humidity is necessary for growth and development of the pathogen. Downy mildew is similar to powdery mildew in only one way, its name. The fungi-like organisms that cause downy mildew are actually more closely related to *Pythium* and *Phytophthora*, organisms that cause stem rots, root rots, and foliar blights, than to the fungi that cause powdery mildew. Downy mildew is becoming increasingly common on a variety of plants including rose, impatiens, salvia, snapdragon, and pansy. Infection is favored by cool, wet weather. On herbaceous plants, symptoms include stunting, foliar chlorosis, and leaf necrosis. On woody plants, such as rose, severe defoliation may occur. In the early stages, angular brown, black, or maroon leaf spots may appear on foliage. Also, infected shoots may become systemically infected and wilt rapidly. Downy mildews are often visible on the undersides of leaves. Off-white to gray fungal growth is often visible directly under leaf spots that are visible on the upper surface of leaves.

RUST

Rust diseases may appear on snapdragon and geranium and perennials, such as potentilla and daylily. By the time yellow leaf spots are noticed, the fungus is usually sporulating and producing millions of yellow-to-orange or brown spores in pustules that make identification of the fungus easy. Symptoms include angular leaf spots and yellowed foliage, and fungal growth may be spotted on the undersides of leaves. In 2000, daylily rust (Figure 8.8) caused by *Puccinia hemerocallidis* was first detected in the United States. This pathogen of daylily is now endemic in many areas of the southern United States and is endemic to Asia. The propagation and sale of daylily by hobbyists make regulation of this disease nearly impossible.

NEMATODE DISEASES

Nematodes are microscopic, worm-like animals, and most are harmless to plants and animals. Plant parasitic nematodes have a mouthpart, called a stylet, which is similar to a hypodermic syringe. It is used to rupture cell walls, inject enzymes into the cell that aid in digestion, and reabsorb the contents of the cell. Root knot nematodes (*Meloidogyne* spp.) may be found in floral crops where soil has contaminated media. Infected roots are swol-

FIGURE 8.9 Angular lesions caused by an *Aphelenchoides* sp. (a foliar nematode) on hosta.

len and knotted. A larger nematode problem is the foliar nematode (*Aphelenchoides* spp.). It feeds on the foliage of several floral crops, such as African violet and begonia. Many shade perennials, such as hosta (Figure 8.9) and Lenten rose, may be infected with foliar nematode. It is usually spread during propagation of infected plants. Infected plants should be discarded and not propagated. Hot water treatment can be used to clean up plants, but it is time consuming, difficult, and there is a risk that the plant material will be killed.

DISEASE MANAGEMENT TACTICS

- Sanitation: Disinfectants that contain ammonium chloride, such as Physan, Consan, and Green Shield are useful for killing bacteria and fungi on benches, pots, flats, tools, and walkways. Each greenhouse should have a trash can with a lid to hold plant debris until it can be destroyed. Rogueing, the removal of infected plants, can help slow the spread of diseases, such as INSV and bacterial blight.
- Media smarts: Soilless media are not sterile; therefore, you may occasionally recover trace amounts of plant pathogens, such as *Rhizoctonia, Thielaviopsis,* and *Pythium.* Media should be protected from soil contamination by storing and mixing on a clean bench or concrete pad. Keep hose nozzles off the ground to prevent the introduction of plant pathogens into clean media.
- Disease-free plant material: Take cuttings from healthy plants free of obvious disease. If available, take cuttings or budwood for fruit trees from a block of trees certified free of common plant viruses. Buy cuttings from horticultural suppliers that index their stock plants for INSV (impatiens, gloxinia, etc.) and bacterial blight (geranium). The added expense is worth the cost of not having to deal with these diseases. Check all incoming plants for symptoms or signs for insect and disease. Remember to check the root systems of incoming plants as well as the foliage. Black root rot can be easily detected if you take the time to look at the roots of a few plants. If there is a problem with incoming plant material, contact the plant inspector with the state Department of Agriculture to document the damage. This can make things go more smoothly when dealing with your supplier.
- Environmental controls: Humidity control is important for managing diseases, such as powder mildew, downy mildew, and botrytis blight. Vent moisture-laden air in late afternoon and heat incoming air to reduce humidity levels. This can be crucial in preventing condensation from forming on leaf surfaces at night when infection from *Botrytis* or downy mildew is likely to occur. Horizontal airflow fans can be used to reduce leaf wetness and condensation on plastic greenhouse coverings. Mount fans just above the crop and circulate air in a racetrack orientation. If possible, avoid overhead watering, especially on crops that are susceptible to botrytis blight. Sub or drip irrigation can reduce bacterial and botrytis blight. Overwatering and saturated media favor the

- water molds, *Pythium* and *Phytophthora*, as well as insects, such as fungus gnats and shore flies. Space plants to allow for air movement and light penetration of the plants' canopies. Closely spaced plants lose lower leaves that are soon colonized by *Botrytis*.
- Nutrient monitoring: Monitoring the nutritional status of your crop, along with media pH and soluble salts, allows you to make adjustments to your fertility program before a major problem arises. Pythium root rot is favored by high soluble salt levels. At least 50% of the floral crop problems diagnosed by plant disease clinics are related to the lack of nutrient monitoring. Make sure that stock plants are well maintained and receive optimum amounts of nutrients. Fertilizer is not necessary for rooting cuttings or seed germination, but will be needed shortly after.
- Scouting: At least one person should be designated as a scout for disease, insects, and early signs of poor plant growth. Spotting diseases early and rogueing infected plants is a sound management strategy. Scouts should note the plant species affected, make a tentative diagnosis, and note the location of the diseased plants.
- Insect management: Insects can vector plant pathogens that attack floral crops. Western flower thrip can transmit INSV; high thrips populations are sometimes responsible for total losses of crops susceptible to INSV. Fungus gnats and shore flies can spread fungi, such as *Pythium*, *Thielaviopsis*, and *Botrytis*. White flies can spread bacterial blight of geranium. Sticky cards that are monitored and changed frequently can give you important information about the insect populations in your greenhouse.
- Weed control: Weed-free greenhouse floors and a clean perimeter around the greenhouse can cut down on western flower thrips and virus diseases, such as INSV and TSWV. Weeds are often controlled by herbicides, hand weeding, or with a flame.
- Fungicide drenches and sprays: Most fungicides are not specifically labeled for use during plant propagation. If you are going to use a fungicide on young seedlings or unrooted cuttings, test the fungicide on a small group of plants and observe for obvious signs of phytotoxicity. Also, check to see if the fungicide has negatively affected root initiation, root number, or root mass on vegetatively produced cuttings.
- Fungicides can be an important tool in the prevention (not curing) of fungal diseases (Table 8.1). You should be aware of which fungicides have efficacy against the target pest in question. Also, consider the use rate and interval prior to choosing a fungicide. The least expensive product is not always the product that is cheapest per pound. Many crops should be drenched with fungicides to prevent stem and root rots when rooted cuttings are potted. Fungicide sprays can be used to protect the foliage of plants susceptible to powdery mildew and botrytis blight. Rotate fungicides used for botrytis blight. Recent surveys have shown that in many greenhouses, *Botrytis* isolates resistant to benzimidazole and dicarboximide fungicides are quite common.
- Don't delay in asking for assistance. Acting quickly to get a pest or disease identified on the front end of an outbreak can save plants and your profits. Too often, growers wait too late in an epidemic to ask for assistance in identifying plant problems. Plant disease clinics at land grant universities and private laboratories can be important sources of information for solving disease problems.

REFERENCES AND SUGGESTED READINGS

Albajes, R., M.L. Gullino, J.C. Van Lenteren, and Y. Elad. 1999. *Integrated Pest and Disease Management in Greenhouse Crops.* Kluwer Academic Publishers, Dordrecht, Netherlands. 545 pp.

Daughtrey, M.L., R.L. Wick, and J.L. Peterson. 1995. *Compendium of Flowering Potted Plant Diseases.* APS Press, St. Paul, MN. 90 pp.

Horst, R.K. and P.E. Nelson, 1997. *Compendium of Chrysanthemum Diseases.* APS Press, St. Paul, MN. 62 pp.

Jarvis, W.R. 1992. *Managing Diseases in Greenhouse Crops.* APS Press, St. Paul, MN. 288 pp.

Jones, R.K. and D.M. Benson. 2001. *Diseases of Woody Ornamentals and Trees in Nurseries.* APS Press, St. Paul, MN. 482 pp.

Trigiano, R.N., M.T. Windham, and A.S. Windham. 2008. *Plant Pathology Concepts and Laboratory Exercises,* 2nd ed. CRC Press, Boca Raton, FL. 558 p.

TABLE 8.1
Chemical Control of Diseases of Seedlings, Cuttings, and Liners of Ornamental Plants

Disease	Susceptible Plants	Chemical Control
Stem/Crown Rot		
Southern blight—Usually occurs in gardens, perennial borders, and nurseries during hot weather, near midsummer. Symptoms include wilting, leaf scorch, followed by plant death. Signs of disease include white mycelium on the stem of infected plants and tan to reddish-brown round, spherical resting structures of the fungus (sclerotia) on the stem and soil surface.	*Southern blight*—Ajuga, apple, clematis, crabapple, forsythia, hosta, many annual and perennial flowers	*Southern blight*—azoxystrobin, fludioxonil, flutolanil, fluoxinil, PCNB
Sclerotinia crown rot—Unlike southern blight, this disease usually appears during midspring to early summer when conditions are cool and moist. Affected plants usually wilt and die. White mycelium may be visible on stems near the soil surface. Black, oblong sclerotia may be present on the outer surface of woody plants or in the stem pith of herbaceous plants. Diseased stems should be split lengthwise and examined for signs of sclerotia.	*Sclerotinia stem rot*—Campanula, euonymus, several herbaceous flowers	*Sclerotinia crown rot*—thiophanate methyl
Rhizoctonia stem rot/damping-off—This disease is often the cause of damping-off (stem rot) of seedling plants. Seedling annual or perennial flowers or woody ornamentals may be killed by this fungus after it attacks the stem near the soil surface. Diseased seedlings often fall over and die. In the field, the fungus may move short distances down the row killing several adjacent plants. In propagation beds or flats, diseased plants may be killed in circular areas as the fungus moves outward.	*Rhizoctonia stem rot*—many herbaceous plants and seedlings of woody plants and conifers	*Rhizoctonia diseases*—azoxystrobin, fludioxonil, iprodione, thiophanate methyl, triflumizole
Gray Mold (Botrytis Blight)		
Gray mold may be found on herbaceous and woody ornamentals, usually during cloudy, cool, moist weather. Stems, leaves, and flowers may be attacked. Woody ornamentals in overwintering structures may become infected. Symptoms of infection are blighting of flowers, tan to brown leaf spots, shoot blights, and stem rot. A sign of disease is gray-brown mold on diseased plant parts.	Almost any herbaceous or woody plant. Geraniums are particularly susceptible to gray mold.	Chlorothalonil, fenhexamid, fludioxonil, iprodione, mancozeb, thiophanate methyl, triflumizole, vinclozolin

Disease Management

Powdery Mildew

Powdery mildew is easily identified by the presence of white to gray mycelium on affected leaves and/or flowers. The first sign of disease is usually isolated colonies of white fungal growth. With time, whole leaves may be totally covered with fungal growth. On some plants, such as pin oak, mildew may be present only on the undersides of leaves. On dogwood, crape myrtle, and nandina, infected leaves may be curled, twisted, or otherwise distorted. Leaves may be abnormally red with little mycelium visible; on sedum, lesions are scabby and brown.

Amelanchier, apple, azalea, begonia, columbine, crabapple, crape myrtle, dogwood, euonymus, hydrangea, lilac, nandina, phlox, rhododendron, rose, sedum, tulip tree, zinnia

Azoxystrobin, chlorothalonil, kresoxim methyl, myclobutanil, piperalin, propiconazole, thiophanate methyl, triadimefon, trifloxystrobin, triflumizole

Downy Mildew

Although this sounds similar to powdery mildew, the diseases are very different and caused by fungi from entirely different taxonomic classes. The fungi-like organisms that cause downy mildew are more closely related to fungi-like organisms that cause phytophthora and pythium root rots than the fungi that cause powdery mildew. Symptoms of downy mildew can range from leaf spots and defoliation to rapid blighting of diseased shoots. Angular leaf spots on rose may range from red to brown to black. Signs to look for include gray tufts of mycelium on the undersides of leaves directly below chlorotic lesions. Look for mycelium early in the morning while the leaves are still wet.

Alyssum, brambles, grape, pansy, rose, salvia, snapdragon, tobacco, viburnum

Dimethomorph, fosetyl Al, potassium salt of phosphoric acid

Root Rot Diseases

Plants affected with fungal root rots may be stunted, wilted, look generally unthrifty (mimic nutrient deficiency), and eventually die. Discolored decayed roots are sure symptoms of root rot diseases. Poor drainage, standing water, improperly constructed landscape beds, planting infected plants, and excessive irrigation favor phytophthora and/or pythium root rots.

Black root rot–Japanese holly, blue holly, vinca, pansy, petunia

Phytophthora root rot–azalea, dogwood, forsythia, fir, holly, juniper pieris, rhododendron, yew

Pythium root rot–herbaceous ornamentals

Black Root Rot–fludioxonil, thiophanate methyl, triflumizole

Phytophthora and Pythium root rot–etridiazole, fosetyl Al, mefenoxam, potassium salt of phosrphoric acid, propamocarb

Rust (Leaf, Stem, Needle)

Signs include bright yellow, orange, reddish-brown, or chocolate-brown raised pustules that are visible usually on the undersides of leaves. Gelatinous tendrils of rust spores are produced from galls each spring on eastern red cedar infected with cedar-apple rust. Pine needle rust produces pustules on pine during spring. Early symptoms on leaves are yellow leaf spots. Rust galls may appear on stems of pine, cedar, and hawthorn. Twig rust may cause branch dieback on plants as diverse as hawthorn and hemlock.

Amelanchier, apple, aster, azalea, buckeye, cedar, crabapple, daylily, fuchsia, geranium, grasses, hawthorn, hemlock, hollyhock, iris, jack-in-the-pulpit, juniper, mayapple, oak, pear, pine, potentilla, quince, snapdragon, sunflower

Azoxystrobin, chlorothalonil, flutolanil, kresoxim methyl, mancozeb, myclobutanil, propiconazole, triadimefon, trifloxystrobin, triflumizole

Continued

TABLE 8.1 (Continued)
Chemical Control of Diseases of Seedlings, Cuttings, and Liners of Ornamental Plants

Disease	Susceptible Plants	Chemical Control
Leaf Spot Diseases		
Leaf spot diseases are usually caused by fungi, but a few may be caused by bacteria. These are among the most common plant diseases. Symptoms vary depending on the pathogen and host. Some common symptoms include the following: frogeye or bull's eye spot marked with concentric rings; irregular, round tan spots with small black fruiting bodies; angular tan or black spots; black or tan spots surrounded by a yellow "halo"; oval-shaped leaf spots; and tan to gray spots with red or purple margins. Fungal leaf spot diseases are usually favored by wet seasons, high humidity, and/or frequent overhead irrigation. Many leaf spots cause premature defoliation.	*Alternaria LS*—aucuba, impatien, marigold, zinnia *Bull's eye LS*—magnolia, maple *Cercospora LS*—buckeye, crape myrtle, leucothoe, laurel, red bud, rose *Entomosporium LS*—hawthorne, pear, photinia *Phyllosticta LS*—holly, magnolia, maple, witch hazel	Azoxystrobin, chlorothalonil, copper hydroxide, flutolanil, fludioxonil, iprodione, kresoxim methyl, mancozeb, myclobutanil, propiconazole, thiophanate methyl, triadimefon, trifloxystrobin, triflumizole
Shot-Hole Diseases		
Some plants shed diseased leaf tissue in response to fungal or bacterial infections. Infected leaves are covered with circular, "shot" holes where diseased tissue has fallen out. Infected leaves may become chlorotic and drop prematurely. Shot-hole diseases may be caused by fungi or bacteria. Damage from shot-hole disease may be confused with insect feeding.	Almond, apricot, cherry, cherry-laurel, peach, plum (plants in the genus *Prunus*)	***Bacterial shot-hole***—copper hydroxide ***Fungal shot-hole***—myclobutanil, propiconazole
Anthracnose Diseases		
Anthracnose refers to diseases that cause leaf, stem, and/or fruit lesions. These diseases may appear as irregular leaf spots/lesions along leaf margins and across or between veins. Anthracnose may kill entire leaves, young shoots, and twigs, plus cause premature defoliation. Diseased leaf tissue may fall out of leaf lesions. Stem cankers may form at the base of succulent shoots.	Ash, dogwood, euonymus, hosta, maple, oak, sycamore	Chlorothalonil, mancozeb, myclobutanil, propiconazole, thiophanate methyl

Disease Management

Crown Gall

Rough-surfaced, hard or soft, spongy, swollen tumors or galls up to several inches in diameter may form on stems or roots. Galls are induced by infection with *Agrobacterium tumefaciens*, a bacterium. Galls may be flesh colored, greenish, or dark. Galls are usually found near or below the soil line. Galls may form at wounds made during propagation. As galls continue to develop and enlarge, surface layers may become brown, woody, and roughened. Plants with crown gall usually become unthrifty and possibly stunted. Plant death may eventually occur.

Euonymus, holly, maple, peach, plum, rhododendron, rose, willow

No chemical controls. Use rogueing, crop rotation, resistance. Clean pruning tools often or between plants.

Nematode Diseases

Millions of nematodes may live in a square meter of soil; however, only a few are parasites of plants. Most plant parasitic nematodes attack plant roots, some attack foliage. Nematode damage can be difficult to diagnose, as most of the damage occurs below ground. Plants damaged by nematodes may appear stunted, unthrifty, discolored and have discolored roots with lesions or galls. One sure way to identify nematode problems is to submit a soil and/or root sample for analysis at a plant diagnostic laboratory; submit symptomatic foliage where foliar nematode is suspected.

Root knot nematode—Abelia, aucuba, boxwood, dogwood, gardenia, holly, hydrangea, impatien, ligustrum, nandina, photinia, rose

Foliar nematode—African violet, begonia, hosta, many perennials

Lesion nematode—boxwood, juniper

Foliar nematode—chlorfenapyr (greenhouse use only)

Note: For each fungicide, consult the pesticide label for the following information: registered for legal use in your locale, target disease, target host, cropping system (nursery, greenhouse, commercial, or residential landscape), rates, use intervals, compatibility with other pesticides, information on resistance management, mammalian toxicity and phytotoxicity, and other valuable information.

9 Disinfestation of Soil and Planting Media

James J. Stapleton

Infestation of nursery soil and planting media with pests, including pathogens, nematodes, insects, and weed propagules, can be extremely destructive to crops. The pests may be disseminated widely via plant shipments and cause "exotic" pest outbreaks in their new locations. This, in turn, may trigger financially ruinous regulatory quarantines, lawsuits, and/or customer boycotts. To protect nursery industries, national and regional governments usually require specific soil treatments and handling procedures to ensure against infestation by certain soilborne pests of nursery stock grown in the field, containers, flats, or frames for commercial planting. Ornamentals and plants destined for noncommercial uses, such as in homes and gardens, are sometimes exempt from these regulations. Chemical fumigation and heat treatments have been widely used for many years to disinfest soil and planting media for nursery production.

For the purpose of demonstrating soil disinfestation to students, the use of dangerously toxic chemical fumigants is not recommended. The same eradicative effects may be obtained with heat, using the relationship of temperature × exposure time to achieve lethal dosage, rather than active ingredient concentration × exposure time, which governs lethal dosage of chemical toxicants. In addition to the soil or planting media, nursery plant containers for reuse also should be disinfested.

THERMAL DISINFESTATION

Exposure of soil or planting media to excessively high temperature [greater than 80°C (176°F)] may release phytotoxic compounds via thermal decomposition of the organic constituents. For that reason, use of aerated steam at lower temperature [e.g., 70°C (158°F)] rather than live steam or autoclaving may be preferable. The introduction of moisture to heated soil or planting media will allow more uniform heating and also will hydrate resistant structures of pests to increase their metabolic activity and render them more susceptible to treatment effects. Certain solarization techniques that heat soil and planting media to 60°C to 70°C (140°F to 176°F) also may be useful.

FACILITIES

This experiment may be conducted in the laboratory, greenhouse, or a growth chamber.

EXERCISE

EXPERIMENT 1: EFFECT OF HIGH TEMPERATURE DOSAGES ON SOILBORNE PEST SURVIVAL

This procedure describes the use of weed propagules as test organisms for determination of soil disinfestation because they are the most ubiquitous. However, other pest organisms, such as fungal and bacterial plant pathogens, phytoparasitic nematodes, and soil-dwelling arthropods, may be used if present in the experimental soil or planting media. In this case, appropriate materials and procedures for determination of numbers should be devised by the instructor. Also, portions of the treated or untreated soil may be used to conduct bioassays of pest survival and activity on susceptible plants. Achievement of lethal (temperature × exposure time) dosage will provide 100% control of unwanted pest propagules. Sublethal dosages will allow survival of varying levels of pest organisms. Students should use appropriate caution when working with hot water to avoid spills or burns.

Materials

The following items will be needed for individual students or teams:

- Sufficient soil or planting media infested with known pest organisms (pathogens, nematode parasites, insects, and/or weed propagules). Seeds (50 to 100/tube) from weedy plants important to local horticulture can be used from previous collection of wild sources. Alternatively, check with weed scientists for seed supplies and suggestions. The same principles can be illustrated substituting crop or turfgrass seeds for weed seeds (Figure 9.1)

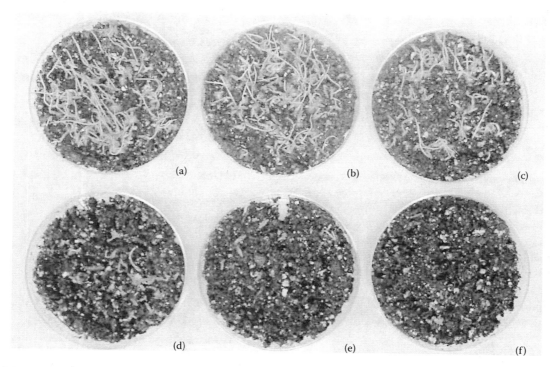

FIGURE 9.1 Heat (60°C) disinfestation of soilless medium using grass seeds as a substitute for weed seeds. (a) Soilless medium not heated; (b) soilless medium heated for 10 min; (c) soilless medium heated for 20 min; (d) soilless medium heated for 30 min; (e) soilless medium heated for 40 min; and (f) soilless medium heated for 60 min. Note: The numbers of grass seeds that germinated and developed into seedlings were dramatically reduced by heat treatments of 20 min or greater. (Photo courtesy of R. N. Trigiano, University of Tennessee.)

- Soil or planting media should be uniformly moistened to approximate field capacity prior to experiment initiation
- Three large glass or plastic culture tubes (e.g., 25 × 150 mm) with fitted lids or other closures
- Two tube racks
- Three 100 × 15 mm petri dishes for each treatment
- Three discs of 90 mm #1 filter paper for each treatment
- One or more large plastic vegetable crispers
- One squirt bottle for distilled water
- Two constant-temperature waterbaths
- Thermometer or electronic temperature logger

Follow the instructions provided in Procedure 9.1 to complete this experiment.

Anticipated Results

Numbers of weed propagules and other pest organisms will be maximal in soil or planting media from the nonheated (ambient) treatment. Reduced survival will be noted in the treatment heated for 10 minutes, whereas complete disinfestation will be observed in the treatment heated for 60 minutes (Figure 9.1).

Questions

- What kinds of organisms are soilborne pests?
- Why is pest-free soil or planting media so important to nursery operations?
- What are the parameters for calculating lethal dosage for chemical disinfestation? For thermal disinfestation?
- Why is heat treatment of moist soil more effective than of dry soil?
- What is the risk to nursery plants of disinfesting soil at temperatures greater than 80°C (176°F)?

Disinfestation of Soil and Planting Media

	Procedure 9.1 Effect of High Temperature Dosages on Soilborne Pest Survival
Step	Instructions or Comments
1	Treat soil or planting medium to be disinfested at a constant temperature of 60°C (140°F). Place sufficient untreated moist soil or planting medium to fill approximately half the volume each of the tubes when the soil is packed by gently tapping the bottom of the tube on a solid surface. Prepare three tubes for each treatment.
2	Cover tube openings and placed in racks that are then immersed in water baths for treatment. The surface of the water bath should reach to at least 25 mm above the level of the soil in the tube. Control (unheated) tubes are maintained in a water bath at room temperature [e.g., 18°C to 24°C (64°F to 75°F)]. Use appropriate care with hot water to avoid spills or burns.
3	Remove one tube from the 60°C water bath after 10 min. of exposure; remove the other tube after 60 min. exposure time (more tubes may be prepared for other intervals of time, e.g., 20, 30, 40, and 50 min if desired). After the tubes are lifted from the water baths, soil is removed by tapping the inverted tubes onto moistened filter paper discs in three labeled 100 × 15-mm petri dishes containing moistened discs of #1 filter paper. If the soil will not tap out of the tubes, scraping tools may be used, taking care not to contaminate treated soil with less-treated soil.
4	Spread soil over filter paper surfaces and cover with lids.
5	Place the petri dishes in vegetable crisper to retain moisture and then incubate in growth chamber, greenhouse bench, or other location with a light/dark and temperature cycle appropriate for growth of weed and/or bioassay plants.
6	Numbers of germinated weed propagules should be determined for each dish at least once per week. Germinated weeds may be removed after recording data from each time to avoid overgrowth, but be sure to keep a running total of germinated seeds for each of the treatments. Deionized water can be added to the petri dishes or crispers as needed to maintain the original moisture level during the weed emergence period. After the final weed emergence determination, the cumulative percent weed emergence from each heated tube is compared to that of the control (unheated) tube. If additional time samples of the soil high temperature treatment are used (Step 3), then plot the means of each treatment, and calculate the regression line of the mean number of germinating seeds versus time in the heating water bath.
7	The number of emergent weeds from each treatment should be determined at least once each week for five weeks. If soilborne nematodes or insects, or fungal or bacterial pathogens are present and recovery or bioassay of these are desired, the appropriate procedures should be done once at the most suitable time for the organism. This may be immediately after heat treatment or after a period of plant growth in the experimental soil, which will allow for sufficient symptom development or pest reproduction.

10 Botrytis and Other Propagation Pathogens

Mark P. McQuilken

There is a great risk of attack by a range of pathogens during plant propagation (chapter 8). The environmental conditions are often conducive to serious outbreaks of Botrytis gray mold on rooting stem cuttings and damping-off during seed germination and seedling growth. Both of these major diseases are the subjects of this experimental chapter.

Botrytis gray mold, caused by the fungus *Botrytis cinerea*, remains one of the most damaging diseases affecting stem cuttings of ornamental shrubs, such as heather (*Calluna vulgaris* (L.) Hull), sunrose (*Helianthemum nummularium* (L.) Mill.), lavender (*Lavandula* spp.), *Rhododendron* spp., and *Hebe* spp. It also affects herbaceous plants, such as the common geranium (*Pelargonium* x *hortorum* L.H. Bailey), *Cyclamen* spp., primroses (*Primula* spp.), and *Fuchsia* spp., as well as vines such as English ivy (*Hedera helix* L.) during rooting (Figure 10.1). It typically starts from the cut end or on the foliage tips, and symptoms include dieback and severe browning followed by poor rooting and cutting death. Characteristic furry, gray-brown masses of spores often develop on affected cuttings. The main sources of gray mold are diseased plants and crop debris. Infections can often remain in a latent state for several weeks, and cuttings taken from apparently healthy stock plants may be infected. Spores produced on diseased plants and on dead and decaying crop debris spread to other plants on air currents and by water splash. Spores landing on susceptible plant tissues either can germinate and invade tissue immediately or remain dormant for up to three weeks and infect when conditions become more favorable for infection.

Damping-off (Figure 10.2 and Figure 10.3) is by far the most common disease during seed germination and

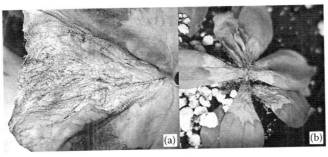

FIGURE 10.1 Foliar lesions caused by *Botrytis cinerea*. (a) Botrytis blight of geranium. Note necrotic lesion with gray "fuzz," which is sporulation of the pathogen. (b) Lesions on lisianthus leaves. (Photos courtesy of Alan S. Windham, University of Tennessee.)

FIGURE 10.2 Diseases caused by *Rhizoctonia*. (a) Damping-off of impatiens. (b) Stem rot of oakleaf hydrangea. Note the discolored (red to brown) lesions. (Photos courtesy of Alan S. Windham, University of Tennessee.)

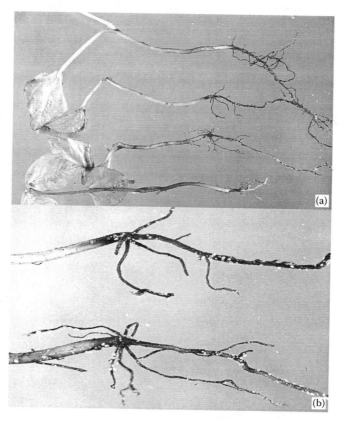

FIGURE 10.3 Damping-off. (a) Damping-off of Boston ivy. (b) Enlargement of roots and stems shown in (a). Note that the lateral roots are also affected by the pathogen. (Photos courtesy of Alan S. Windham, University of Tennessee.)

seedling growth. Fungal species of *Rhizoctonia, Fusarium,* and even *B. cinerea,* as well as fungal-like pathogens, such as species of *Pythium* and *Phytophthora,* may be the cause of economically important diseases, resulting in serious loss of seeds, seedlings, and young plants during propagation. These diseases occur at various stages of production. Seeds or seedlings can be attacked before emergence, resulting in decay and rotting (preemergence damping-off). Seedlings that have already emerged are attacked at the roots and stems at or below the surface of the propagation medium. The invaded tissue becomes water soaked and discolored, and the seedling collapses and dies (postemergence damping-off). Sometimes, the seedling may remain alive and standing, but the stem becomes girdled and stunted, and the plant eventually dies (wire stem). In larger plants, the damping-off pathogens may attack and kill the rootlets. Such attacks cause plants to become stunted and eventually die (root rot). Species susceptible to damping-off include the following: pansy (*Viola* spp.), *Lobelia* spp., alyssum (*Lobularia maritima* (L.) Desv.), *Salvia* spp., and other common plants, such as pea (*Pisum sativum* L.), beet (*Beta vulgaris* L.), tomato (*Lycopersicon esculentum* Mill.), and cucumber (*Cucumis sativus* L.).

The following three experiments will show the typical symptoms of gray mold and damping-off and demonstrate how these two important diseases can be controlled and managed during propagation. Teams of three to five students are recommended for each exercise.

GENERAL CONSIDERATIONS

Growth of Stock Plants and Cutting Preparation

Stock plants of *C. vulgaris* are maintained to provide a source of cuttings for Experiments 1 and 2. Eight to ten stock plants will provide enough material for a class of 10 to 15 students to conduct both experiments. The cultivars Arran Gold, Dark Beauty, Flamingo, Robert Chapman, Silver Queen, Sun Rise, or Velvet Fascination are recommended, as they are very susceptible to gray mold. Grow stock plants in 2-L pots containing sphagnum peat medium amended with dolomitic limestone (1.8 kg/m^3), superphosphate (0.75 kg/m^3), fritted trace elements (0.3 kg/m^3), and controlled-release fertilizer (1.0 to 1.5 kg/m^3, either 8 to 9 or 12 to 14 month formulation). Addition of a wetting agent will aid water absorption by the peat. It is a good idea to spread pine bark on the top of the medium to reduce moss growth. Ideally, grow the stock plants under a shaded polyethylene tunnel, on well-spaced capillary sand beds to keep the foliage dry and reduce the risk of

Botrytis and *Rhizoctonia*. Repot plants every year and replace every third year. Once cuttings are taken, prune plants to produce even growth.

Small branches of semi-ripe material should be cut with a pair of scissors from March/April through to September. Cuttings are prepared by breaking 15 to 20 mm of stem from the branch and drawing fingers down the stem to remove lower leaves. For both experiments, fill 15 plastic trays (approximately 23 × 18 × 6 cm) with a 1 peat:1 fine bark mixture (by volume) propagation medium without any lime or fertilizers. Insert 30 cuttings into each tray with pre-made holes. For rooting, place trays of cuttings on capillary sandbeds, under low polyethylene propagation tunnels within a glass-covered greenhouse. Alternatively, cuttings can be rooted in the moist atmosphere created under a polyethylene sheet supported just above the cuttings. A polyethylene sheet laid directly over the cuttings should be avoided because its removal lifts out the cuttings from the medium. The objective is to provide a high-humidity, stagnant-air environment to favor *Botrytis* growth; therefore, a clear plastic bag that is sealed at the top should also be sufficient. Suitable alternative cuttings to use include lavender, geranium, *Cyclamen* spp., primroses, *Hebe* spp., and *Fuchsia* spp., as well as vines, such as English ivy.

Botrytis cinerea Inoculum

Propagation beds and glasshouses that are used regularly for producing plants should contain sufficient amounts of *B. cinerea* inoculum (spores) to initiate infection of cuttings. Natural inoculum can be supplemented by placing sporulating cultures (lids of petri dishes removed) of *B. cinerea* within the experimental area. Using aseptic technique, sporulating cultures are produced by centrally inoculating plates of Difco 3% (w/v) malt extract agar (MEA, which should be prepared according to the manufacturer's instructions printed on the product label) in petri dishes (9 cm diameter) with a 0.5 cm^2 agar plug of *B. cinerea* and growing the pathogen for 7 to 10 days under fluorescent lights (16 h photoperiod) at 20°C to 22°C. Alternatively, commercial preparations of nutrient agar (NA) or Potato Dextrose Agar (Fisher Scientific, Atlanta, GA) are suitable for growing inoculum.

Disease Assessment

Assess each cutting for incidence of *Botrytis* (dieback, browning, and/or presence of a furry, gray-brown mass of spores) and disease severity, four and eight weeks after the start of each experiment. Disease severity is rated on a 0 to 5 scale (% proportion of cuttings with browning), where: 0 = no visible symptoms/healthy cutting; 1 = up to 25%; 2 = 26% to 50%; 3 = 51% to 75%; 4 = >75%; 5 = cutting totally brown and dead.

Experimental Design and Statistical Analyses (See Chapter 13)

Arrange Experiments 1 and 3 in a randomized block design with a minimum of three replicates for each treatment. Use a split-level design for Experiment 2, with each of two propagation beds being a treatment. Data from Experiments 1 and 3 are analyzed using an analysis of variance (ANOVA) after angular transformation of percentage data where required. Compare treatments using the least significant difference test (LSD) at a probability of 5% ($P = 0.05$). Use a Student's *t*-test to analyze data from Experiment 2.

EXERCISES

Experiment 1: Development and Chemical Control of Gray Mold on *Calluna vulgaris* Cuttings

Gray mold is difficult to control on cuttings because the conventional environment for rooting is conducive to serious outbreaks of the disease. The damp and humid environment established for optimum propagation promotes spore germination of the pathogen and disease spread. A single control measure is unlikely to be effective, but integrated control is more likely to provide effective and durable disease management. Fungicide sprays in combination with cultural control measures, such as practicing good crop hygiene and reducing humidity, are recommended for effective control. Preventative spray programs as part of an integrated control program are generally the most effective. Once gray mold is established, spraying often provides very poor disease control. The aim of this experiment is to compare the efficacy of two foliar-applied fungicides, iprodione (or thiram) and azoxystrobin (or pyrimethanil or tolylfluanid), against gray mold on *C. vulgaris* cuttings. Iprodione is slightly phytotoxic at recommended rates, and products that contain it are labeled either "general use pesticides," or under some circumstances, "restricted use pesticides." Azoxystrobin is a "reduced risk" fungicide. The experiment will require about 8 to 10 h in the course of eight weeks to complete.

Materials

The following items are needed for each group of three to five students:

- Access to healthy, disease-free stock plants of *C. vulgaris*
- Two to three pairs of scissors
- Peat bark propagation medium
- Nine plastic trays (approximately 23 × 18 × 6 cm)
- Labels and permanent marker

The following items are needed for each class of students:

- Fungicide A (iprodione; Rovral and other trade names, such as Chipco 26019, DOP 500F, Kidan, LFA 2043, NRC 910, and Verisan)
- Fungicide B (azoxystrobin; Amistar and other trade names, such as Abound, Bankit, Heritage, and Quadris)
- Azo precision sprayer fitted with a medium flat fan nozzle
- Personal protective clothing (overall, boots, gloves, face shield and mask)
- Propagation beds with subirrigation

CAUTION: It is recommended that fungicide sprays are prepared and applied by a trained member of the staff with proper pesticide applicator's licensing and wearing the appropriate personal protective clothing.

Follow the protocols outlined in Procedures 10.1 to complete this experiment.

Anticipated Results

After two to three weeks, cuttings in the untreated trays will show typical symptoms of infection by *Botrytis*. Symptoms should first appear on the leaf tips or the cut ends and will include dieback, browning, and cutting death. Sporulating *B. cinerea* may also be seen as characteristic furry, gray-brown masses of spores on affected cutting parts. After four and eight weeks, disease incidence and severity should be greatest on the untreated cuttings. Number of rooted cuttings is also likely to be reduced. Disease incidence and severity should be less on fungicide-treated cuttings. Of the two fungicides (iprodione and azoxystrobin) tested, disease incidence and severity should be less on cuttings treated with azoxystrobin. A reduction in size may be observed in some cultivars of cuttings treated with azoxystrobin, demonstrating slight phytotoxicity of the fungicide.

Questions

- Several propagators and growers have reported poor control of *Botrytis* using iprodione. Why?
- What strategies should a propagator/grower adopt to minimize the development of fungicide resistance?
- Describe the mode of action of azoxystrobin against *B. cinerea*.
- Why are fungicides rarely incorporated into propagation medium when rooting cuttings?

	Procedure 10.1
	Development and Chemical Control of Gray Mold on *Calluna vulgaris* Cuttings
Step	Instructions and Comments
1	Fill nine plastic trays (approximately 6 × 18 × 23 cm) with peat bark propagation medium. Take cuttings from stock plants, prepare, and insert into each tray (30 per tray) as described under stock plants and cutting preparation.
2	Label three replicate trays "Control—water," and spray with tap water.
3	Label three replicate trays "Fungicide A—iprodione," and the remaining three "Fungicide B—azoxystrobin." Place the trays in a designated spraying area for spraying by a trained and licensed member of the staff.
4	*Preparation and application of fungicides by trained staff only.* Wearing appropriate personal protective clothing, prepare the following rates of fungicide products per liter of water: iprodione (as Rovral WP; 50% a.i. w.p.), 4 g; azoxystrobin (as Amistar, 25% a.i. s.c.), 4 mL. Adjust rates if different products of these fungicides are used. Apply fungicides to appropriate labeled trays of cuttings (approximately 10 to 15 mL/tray) using an Azo precision sprayer (first spray).
5	After 24 h post-spraying, arrange the trays in a randomized block design under a low polythene tunnel on a sand bed with subirrigation and drainage or other suitable high-humidity, stagnant-air environment. Introduce inoculum of *B. cinerea* if necessary as described under inoculum production.
6	After two weeks, remove the trays from the high humidity environment of the low polyethylene tunnels and place in a designated spraying area for respraying (second spray) by a trained member of staff.
7	Wearing gloves, replace the sprayed cuttings into the high-humidity environment.
8	Assess each cutting for incidence and severity of *Botrytis* after four and eight weeks as described under disease assessments. The class should share and analyze the data using an ANOVA after angular transformation of percentage disease incidence data. **CAUTION:** Always wash your hands after handling trays of treated cuttings.

	Procedure 10.2
	Effect of Ventilating Propagation Beds on Development and Control of Gray Mold on *Calluna vulgaris* Cuttings
Step	Instructions and Comments
1	Fill six plastic trays (approximately 1 × 6 × 23 cm) with peat bark propagation medium. Take cuttings from stock plants, prepare, and insert into each tray (30 per tray) as described under stock plants and cutting preparation.
2	Fill six plastic trays (approximately 1 × 6 × 23 cm) with peat bark propagation medium. Take cuttings from stock plants, prepare, and insert into each tray (30 per tray) as described under stock plants and cutting preparation.
3	Label three replicate trays "Unventilated," and the remaining three as "Ventilated."
4	Place the three replicate "Unventilated" trays under a low polyethylene tunnel on a sand bed with subirrigation and drainage, together with the trays of other groups of the class. Label the propagation bed "Unventilated," and do not remove the polythene during the course of the experiment. Polyethylene tents over greenhouse benches or tables near windows may also be used provided that there is subirrigation.
5	Place the three replicate "Ventilated" trays under a second low polyethylene tunnel on a sand bed with subirrigation and drainage, together with the trays of other groups of the class. Label the propagation bed "Ventilated." During the course of the experiment, ventilate the bed by opening the tunnel on alternate mornings for four hours.
6	Assess each cutting for incidence and severity of *Botrytis* after four and eight weeks as described under disease assessments. The class should share and analyze the data using a Student's *t*-test.

Experiment 2: Effect of Ventilating Propagation Beds on Development and Control of Gray Mold on *Calluna vulgaris* Cuttings

Prolonged high humidity and/or leaf-surface wetness favors infection by *B. cinerea* and subsequent development of gray mold. Therefore, crop management practices such as ventilation, which reduces prolonged periods of high humidity, could affect disease development (see chapter 8). Cuttings grown under reduced humidity should be less at risk to infection. This experiment is designed to demonstrate the effect of ventilating propagation beds, by opening low polythene tunnels periodically, on the development of gray mold on *C. vulgaris* cuttings. The experiment will require about 8 to 10 h in the course of eight weeks to complete.

Materials

The following items are needed for each group of three to five students:

- Access to healthy, disease-free stock plants of *C. vulgaris*
- Two to three pairs of scissors
- Peat bark propagation medium
- Nine plastic trays (approximately 23 × 18 × 6 cm)
- Labels and permanent marker

The class requires two propagation beds with subirrigation: one ventilated and one unventilated. (See chapter 2 for details on how to construct an inexpensive polyethylene propagation tent.)

Follow the protocols described in Procedures 10.2 to complete this experiment.

Anticipated Results

Cuttings will show typical symptoms of *Botrytis* after two to three weeks as in Experiment 1. After four and eight weeks, disease incidence and severity should be greatest on the "Unventilated" and least on the "Ventilated" cuttings.

Questions

- How may periodic ventilating of propagation beds affect the rate of rooting?
- When is prolonged high humidity in a greenhouse likely to occur?
- Explain why it is important to maintain stock plants free of gray mold and not to take cuttings from diseased plants.

Experiment 3: Biological Control of *Pythium* Damping-Off during Seed Germination

Controlling disease during seed germination and seedling establishment is one of the most important tasks of the propagator. Traditional methods of control have included the integration of cultural practices, such as the manipulation of the environment, with the application of fungicides to the growing medium. However, fungicide application is not the most desirable method of controlling damping-off for several reasons. For example, fungicides are expensive, and their registration and use varies among countries. Pathogenic strains of *Pythium* have also developed resistance, or considerable tolerance, to the commonly used fungicides, etridiazole, furalaxyl, and propamocarb hydrochloride. Furthermore, there are often problems with them causing stunting and chlorosis of young seedlings. The difficulty in managing damping-off has increased the need for alternative or supplementary control methods, such as biological control through the use of antagonistic fungi. These fungi are antagonistic toward the causal pathogens via either antibiosis or mycoparasitism.

Prospects are good for control of *Pythium* damping-off by treating seed, roots, or propagating medium with spores or mycelia of suitable antagonistic fungi. For example, *P. oligandrum* is a vigorous, necrotrophic (refers to organisms that feeds on dead tissue) mycoparasite, which is active against several pathogenic species. In particular, it is very aggressive against *P. ultimum*, a major cause of seed decay and pre- and post-emergence damping-off during propagation. The aims of this experiment are to apply oospores of *P. oligandrum* to cress (*Lepidium sativum* L.) seed and assess the ability of this treatment to control pre- and post-emergence damping-off. Cress seed is used in this experiment because it is relatively cheap compared to other seed (e.g., herbaceous bedding plant seed) and is an ideal size for ease of coating and sowing. Suitable alternative seed to use include pansy, *Lobelia* spp., alyssum, *Salvia* spp., pea, beet, tomato, and cucumber. The experiment will require about 12 h in the course of eight weeks to complete.

Materials

The following items are needed for each group of three to five students:

- Cornmeal agar (CMA; Fisher Scientific, Atlanta, GA)
- Pathogenic culture of *P. ultimum*
- Culture of *P. oligandrum* antagonistic against *P. ultimum*
- Cane molasses
- Carboxymethylcellulose (CMC)
- Eight 250-mL Erlenmeyer flasks
- Two to three L sterile distilled water
- Buchner flask with funnel and Whatman No. 1 filter paper
- Cheesecloth
- Campbell's V8 tomato juice
- Ten 9-cm diameter petri dishes
- Vermiculite (grade DSF)
- 5.25% sodium hypochlorite (commercial liquid chorine bleach contains 5.25% NaOCl)
- 3.5 g cress seed (approximately 350 seed/g)
- Large plastic bags
- Peat-based propagating medium (without sand) designed for seed sowing and general purpose propagation
- Twenty plastic seed trays (approximately 20 × 16 × 5 cm)
- Labels and permanent marker pen
- Three to five pairs of small blunted forceps

Follow the protocols outlined in Procedures 10.3 and 10.4 to complete the experiment.

Anticipated Results

Damping-off should not occur in trays not infested with *P. ultimum*, whereas extensive damping-off will occur in trays infested with *P. ultimum* and sown with uncoated and CMC-coated seed. Preemergence damping-off (i.e., seeds rotted in growing medium or seedlings failed to emerge) should be evident within three to five days. Postemergence damping-off (seedlings emerged, but collapsed and died) should be evident within 7 to 10 days and will increase to reach a maximum by 21 days. There should be less damping-off in trays infested with *Pythium* and sown with P. *oligandrum* oospores + CMC-coated seed.

Questions

- Outline how you would isolate *Pythium* from a diseased seedling and undertake Koch's postulates (Mullen, 2008).
- How may the effectiveness of *P. oligandrum* be improved? Identify commercial biocontrol products available to the propagator.
- How may the propagator spread *Pythium*? State the main factors that encourage damping-off.

Procedure 10.3
Bulk Production of Oospores of *Pythium oligandrum* in Liquid Culture and Application to Seeds

Step	Instructions and Comments
1	Using aseptic technique, maintain *P. ultimum* and *P. oligandrum* on dishes of cornmeal agar (CMA). Prepare the CMA according to the manufacturer's instructions on the product label.
2	Add 45 mL of liquid cane molasses (30 g/L distilled water) to five 250-mL Erlenmeyer flasks, autoclave at 120°C and 103 kPa for 15 min, and leave to cool.
3	Inoculate flasks aseptically, each with two 20-mm diameter CMA discs cut from a 3-day culture of *P. oligandrum*, and incubate at 25°C in darkness for 18 to 21 days.
4	Remove oospore biomass by vacuum filtration (Büchner flask with funnel) onto filter paper, wash in two changes of sterile distilled water, and air-dry overnight in a laminar flow cabinet at room temperature. Each flask produces approximately 70 mg (dry weight) of oospore biomass containing 2.0×10^5 to 2.5×10^5 oospores/mg.
5	Resuspend oospore biomass (approximately 300 mg) in 50 mL sterile distilled water, and blend in a blender for 2 min. Filter homogenate through two or three layers of cheesecloth to give a dense oospore suspension.
6	Add equal volumes of 3% (v/v) carboxymethylcellulose (CMC) (10 mL) to oospore suspension, and mix thoroughly for 15 min on a mechanical shaker.
7	Surface sterilize seed in 5.25% (v/v) sodium hypochlorite (commercial bleach) for 10 min and rinse in three changes of sterile distilled water.
8	Add approximately 250 (0.7 to 0.8 g) seed to 20 mL CMC-oospore suspension and mix as described before.
9	Spread seeds sparsely in sterile, open petri dishes and dry overnight in a laminar flow cabinet.
10	Prepare batches of seed using fungus-free CMC coating as described before.

Procedure 10.4
Biological Control of *Pythium* Damping-Off During Seed Germination

Step	Instructions and Comments
1	Produce *P. ultimum* inoculum by placing 10-g samples of vermiculite in each of three 250-mL Erlenmeyer flasks, and moisten with 35 mL V8 tomato juice diluted 1:3 v/v with distilled water. After autoclaving, inoculate each flask with two 0.5 cm diameter discs cut from a three-day-old culture of *P. ultimum* and incubate for seven days at 25°C. Shake flasks vigorously by hand each day to ensure mixing of nutrients and organism.
2	Fill six plastic seed trays (5 × 16 × 20 cm) with peat-based propagating medium (approximately 1.5 L per tray). Label three replicate trays "Uninfested control—uncoated," and sow each tray with 60 uncoated seeds (6 rows of 10 seeds each, with each seed being approximately 2 cm from its neighbor and 5 mm deep). Label three replicate trays "Healthy control—CMC-coated," and sow each with 60 CMC-coated seed.
3	To infest the propagating medium, add 20 g of *P. ultimum* vermiculite culture to 20 L of propagating medium (equivalent to 2 g per 1.5 L), and mix thoroughly in a large plastic bag.
4	Fill nine seed trays with the *P. ultimum*-infested propagating medium. Label three replicate trays "Pathogen control—uncoated," and sow each tray with 60 uncoated seeds. Label three replicate trays "Pathogen control—CMC-coated," and sow each with 60 CMC-coated seeds. Label three replicate trays "Pathogen—oospore + CMC-coated," and sow each with 60 oospores + CMC-coated seeds. Sprinkle vermiculite over the surface of the propagation medium in each tray.
5	Arrange the trays in a randomized block design in a controlled environment growth room or glasshouse maintained at 18°C to 22°C, with alternating periods of 16 h light and 8 h dark. Water trays regularly with a fine spray to keep propagating medium moist, but not saturated.
6	Record diseased seedlings every 3 to 5 days for 21 days. Calculate the total percentage of mortality (both pre- and post-emergence damping-off) after 21 days for each replicate. The class should share and analyze the data using an ANOVA after angular transformation of percentage data.

LITERATURE CITED AND SUGGESTED READINGS

McQuilken, M.P. 2000. Evaluation of novel fungicides and irrigation methods for grey mould control on *Calluna vulgaris*. *Comb. Proc. Intl. Plant Prop. Soc.* 50: 137–141.

McQuilken, M.P. 2002. Control of grey mould (*Botrytis*) in container-grown ornamentals: unheated greenhouse crops. HDC Factsheet 23/02.

McQuilken, M.P., Whipps, J.M. and Cooke, R.C. 1990. Use of oospore formulations of *Pythium oligandrum* for biological control of *Pythium* damping-off in cress. *J. Phytopathol.* 135: 125–134.

McQuilken, M.P., Whipps, J.M. and Cooke, R.C. 1990. Control of damping-off in cress and sugar-beet by commercial seed-coating with *Pythium oligandrum*. *Plant Pathol.* 39: 452–462.

Mullen, J.M. 2008. Plant disease diagnosis, in *Plant Pathology Concepts and Laboratory Exercises,* 2nd ed., Trigiano, R.N., M.T. Windham, and A.S. Windham (Eds.). CRC Press, Boca Raton, FL, 2008, pp. 447–463.

O'Neill, T.M. and McQuilken, M.P. 2000. Influence of irrigation method on development of grey mould (*Botrytis cinerea*) in glasshouse crops of calluna, cyclamen and primula. *Proc. BCPC 2000.* 267–272.

11 Crop Certification Programs and the Production of Specific Pathogen-Free Plants

Danielle J. Donnelly and Adam Dale

CHAPTER 11 CONCEPTS

- Certified plants are produced under all guideline requirements of the certification program. These plants are tagged accordingly with the name of the cultivar and the name and address of the certified grower. For potato only, seed tubers designated certified include additional information on the farm and field in which they have been grown.

- The nuclear level refers to a group of plants held in a germplasm repository as micropropagated plantlets or in a greenhouse or screen house that are true-to-cultivar, physiologically healthy, and free of specific disease organisms and pests known to affect the industry. The nuclear level is the first (entry) level in the certification program. It is cuttings from these plants that are designated elite level. In the potato industry only, nuclear-level plantings may include up to two generations of potato plants grown under protected cultivation in a greenhouse or screen house. These propagules are used to plant a pre-elite generation in the field.

- The shoot tip is the growing point of a stem consisting of an apical dome and several pairs of attached leaf primordial with an overall length of ≥0.5 mm. Shoot tips are used as explants for Stage I micropropagation.

- Specific pathogen-tested (SPT) plants have repeatedly tested negative for specific pathogens (viral, fungal, bacterial, etc.) that are known to affect the industry. Applies to nuclear plants held in germplasm repositories as source material and elite cuttings taken from them for the certification programs.

- Elite plants are grown from cuttings (elite cuttings) taken from nuclear-level plants. Elite cuttings are planted into an elite nursery source block by certified growers. Cuttings can be taken from these elite plants and sold as certified plants (elite level) or used to produce a foundation nursery increase block. In the potato industry only, pre-elite potato plants are derived from nuclear-level seed tubers and are grown under a system of rigorous inspection in the field. Elite 1 describes the potato plants and seed tubers grown in the field during the next year, and these seed tubers, in turn, are used to produce successive seed generations described as elite 2, 3, or 4.

- Foundation stock plants are grown from cuttings (elite cuttings) removed from elite plants in an elite nursery source block and used to produce a much larger propagation block called a foundation nursery increase block. Cuttings can be taken from these foundation plants and sold as certified plants (foundation level). In the potato industry only, foundation level seed tubers are derived from elite seed tubers in the field, increased over successive generations for from one to four years. Tolerances for disease and pests are less stringent at the foundation level compared with elite-level seed tubers.

- Isolation distance is the planting distance between blocks. These distances are described in the certification guidelines for each species and are designed to minimize cultivar mixing and disease transmission between each individual nursery source and increase block.

- A meristem tip is an apical dome and one pair of attached leaf primordia with an overall length of <0.5 mm. Meristem tip culture is used following thermotherapy for specific virus elimination from virus-infected clones.

- A meristem tip source clone includes one meristem tip explant and all micropropagated propagules derived from it. One, or a limited number, of meristem tip or shoot tip source clones may represent the elite source material for certified propagation of a cultivar.

- Rogueing is the process of removing and destroying poor quality plants. This is practiced during the certification process to eliminate plants that appear to have symptoms of disease, physiological problems of any kind, or are mixed in with other cultivars.

CROP CERTIFICATION PROGRAMS

The term certification describes both a strategy and a process used to restrict or eliminate crop pests and diseases. Through the certification process, a crop is endorsed if it meets specified standards involving pest and disease occurrence, physical appearance, and genetic off-types. These standards are described in published guidelines that are prepared by agricultural authorities and may be supported by legislation. Propagators who participate in the certification process (certified propagators) must follow these guidelines to produce plants that may be distributed both nationally and internationally as "certified plants."

THE IMPORTANCE OF CROP CERTIFICATION

Crop certification originated because scientists and growers recognized that, when propagules are moved (vegetative propagation materials and true seed), they can spread plant pests and diseases. In agricultural industries that use seeds, bulb, corms, and other relatively hardy structures, diseases can be controlled if the propagules are heated or treated with chemicals to eliminate pathogens or pests. However, this cannot be done easily when herbaceous cuttings or woody plants are the major propagule. For this reason, a system of practices has evolved to ensure that a supply of specific pathogen-free planting material is supplied by various governmental and nongovernmental agencies to a select group of propagators (certified propagators) who, in turn, supply planting material to commercial growers. In Canada, as in many other countries around the world, certification programs are in place to limit the spread of diseases of vegetatively propagated plant species of economic importance. The crops for which certification programs are most developed in Canada and elsewhere, and where governmental agencies are often involved, include potato (*Solanum tuberosum* L.), red raspberry (*Rubus idaeus* L.), strawberry (*Fragaria x ananassa* Duch.), and a limited number of other species, such as currant and gooseberry (*Ribes* spp.). Within the ornamentals industry, similar programs are used for select species, but with less involvement from government agencies.

REGULATORY AGENCIES

The United Nations Food and Agriculture Organization's International Plant Protection Convention of 1951 was signed by Canada, the United States, and many other countries that recognized the need to regulate the international movement of plants and plant products. Under this agreement, countries can request that imported plant material be accompanied by a phytosanitary certificate and an import permit. They can also prohibit the importation of certain plant species. A phytosanitary certificate is a document attesting that a consignment has been grown and inspected according to the import requirements of the destination country. Under the authority of two parliamentary acts, the Seeds Act and the Plant Protection Act, the Canadian federal government regulates the movement of plants and plant products, both internationally and between provinces, conducts national certification programs, and attempts to control the entry, spread, and export of diseases and pests. In the United States, this authority is granted under the Federal Seed Act and coordinated by the Association of Official Seed Certifying Agencies.

IMPORT AND EXPORT OF PLANTS

To import or export plants, growers and plant propagators may have to fill certain conditions. Growers often import plants across national borders, and propagators import and export them. First, they need to know whether they can legally ship the plants to or from the country concerned. Many crop species are prohibited from entering a particular country, and the main regulatory agencies, the Canadian Food Inspection Agency (CFIA) in Canada, and the Animal and Plant Health Inspection Service (APHIS) in the United States, keep detailed records of these plants. Permission to import plants may require an import permit. This is issued by the national regulatory agency, may have conditions attached to it, and may require an inspection, quarantine period, or testing after the plants have entered the country concerned. Once plants can be shipped legally, a propagator might find that his/her sales territory for a certain variety is restricted by a Plant Patent or Plant Variety Right. Often, these commercial agreements come with restrictions; for example, a sales territory may be restricted to North America. This means that the propagator can only sell plants in Canada and the United States and is prohibited from selling plants to the rest of the world. It also means that a grower may be unable to buy European cultivars to grow in North America. Finally, the propagator will need to provide a phytosanitary certificate from the relevant regulatory agency, the CFIA in Canada or State Plant or Animal Inspection Services in the United States; APHIS delegates the export inspection to state regulatory agencies.

CERTIFICATION PROGRAMS

Many agencies cooperate in certification programs, including federal (CFIA and APHIS), provincial, and nongovernmental agencies, such as universities and grower's associations. These authorities produce guidelines that describe the certification procedures needed for each species, and these are published. The guidelines are designed to direct the agencies that produce specific pathogen-tested plants and the certified propagators who produce high-quality planting stock. They detail the conditions needed to minimize virus and mycoplasma infection and control fungal diseases, insect, mite, and nematode pests. The examples used in this chapter are taken primarily from the guidelines distributed by the Ontario Plant Propagation Programs for strawberry, red raspberry, currant, and gooseberry and the New Brunswick Plant Propagation Program for potato.

SPECIFIC PATHOGEN TESTING

When new cultivars enter the program, they are tested for pest and disease presence. Usually, an organization that enters the cultivar into the program provides material that has been tested and determined to be specific virus-negative. This comes either as cultured plant material from a laboratory (micropropagated shoots or plantlets *in vitro*) or field plants. The material is then examined critically by a pathologist trained to recognize all of the major pests (insects, mites, nematodes, etc.) and disease organisms (bacteria, fungi, viruses, etc.) known to affect the crop. The type of testing done goes well beyond standard microbiological tests for bacteria and fungi. It includes serological assays, such as enzyme-linked immunosorbent assay (ELISA), DNA probe technology, and tests where plant parts are grafted or sap is inoculated onto indicator plants. Most pests and disease organisms are readily eliminated through a combination of routine propagation by cuttings from noninfected parts of the plant or, if these do not exist, the selective use of chemicals, such as systemic fungicides. However, if virus is detected, a complex system of virus elimination is necessary. This is done through a combination of heat treatment (thermotherapy) at a relatively high temperature (37°C to 38°C) for a prolonged interval (two to eight weeks, sometimes longer) followed by meristem tip culture. Meristem tip explants are tiny shoot apices less than 0.5 mm in length that are composed of the meristematic dome and one pair of leaf primordia. The buds or shoot tips are first disinfested (cleaned on all their outer surfaces); then the meristem tips are removed under a dissecting microscope using sterile technique and placed into tissue culture. This procedure is described in Procedure 11.1 for meristem tip isolation. Within the medium, which may be different for each plant species, antiviral agents such as ribavirin may be included. The preparation of isolation stage (Stage I) culture media for meristem tips is described in Procedure 11.2 for red raspberry, strawberry, and potato.

Each meristem tip and its clonally derived propagule are known as a meristem tip source clone. When sufficient plant material is available, these meristem tip source clones are tested and retested for the virus(es) that were present in the source plant. Testing is done both on plantlets *in vitro* and after cultured plantlets are transferred to a greenhouse or screen house. In the greenhouse or screen house, plants of each clone are kept separate under conditions that are designed to favor virus multiplication within the plants. Under these conditions, the virus, if present, will reach tissue levels that are detectable by serological, molecular, or plant assays. If these meristem tip source clones repeatedly test free of the specific virus in question (specific virus-free), they will be used to increase nuclear stocks of this cultivar. Each cultivar may be represented by a limited number of specific virus-free meristem tip source clones (one to three is common). Negative test results for a specific disease agent, such as a viral pathogen, does not guarantee freedom from that virus because this is limited by the sensitivity of the test used. Also, other virus(es) may be present for which there is no test. For this reason, the terms specific pathogen-tested or virus-tested are legally and scientifically preferred to the older terms pathogen-free or virus-free.

TRUE-TO-CULTIVAR

Within each certification program, economics dictates the number of meristem tip source clones or mother plants that are used to initiate the propagation sequence that can last for several years before new source material is used. A great many propagules will be generated from a relatively small amount of source material. During this process, cultivars can be mixed and genetic mutations can occur, so confirmation is necessary that source material is true-to-cultivar (true-to-type). Until recently, the only way this could be done accurately was by a grow-out test in which commercially important characteristics are assessed through to a commercial harvest; for berry crops these include foliage, flower, and fruit characteristics and in the case of potato, foliage, and stem tuber characteristics. More recently, cultivar mixtures can be detected using various types of DNA analysis, including RFLP (restriction fragment length polymorphism), PCR (polymerase chain reaction), or similar technology.

BERRY CERTIFICATION GUIDELINES (STRAWBERRY, RED RASPBERRY, CURRANT, AND GOOSEBERRY)

Stock plants of the berry cultivars grown in Canada are maintained by AAFC Research Stations at Agassiz (BC), Kentville (NS), or the University of Guelph (ON). These stock plants are referred to as nuclear-level stock plants because they have been tested and determined to be negative for specific viruses and other disease organisms (specific pathogen-tested). They also have been examined by experts who have confirmed that they are true-to-type and physiologically healthy. These nuclear-level plants may be micropropagated plantlets *in vitro* or greenhouse- or screen house-grown plants.

Annually, nuclear stock plants are used to make cuttings that are made available to a small group of select propagators (certified plant propagators or certified propagators) on a contractual basis. These cuttings or plants are referred to as elite plants to show they are derived from nuclear-level stock plants. Certified propagators must request elite cuttings from the clonal repository through a formal contractual application process. This is done up to one year prior to planting, depending on the province. Many cultivars that are requested by the certified propagators are protected under Canadian Plant Breeder's Rights. When this is the case, additional agreements and costs are the responsibility of the certified propagator who transfers these costs to the fruit grower. The price for elite-level material varies with the province; these costs were subsidized for many years, but are now more likely to cover the real cost of production. In some provinces, when the application for nuclear material is made, an application must also be made to register the growing sites where cuttings will be planted.

Once they have received the elite material, the certified propagators must follow a strict set of guidelines that varies for each berry species and, to some extent, between provinces. These elite cuttings have tested negative for all organisms that are known to be detrimental to the industry, including viruses and other serious pathogens, insects, mites, and nematodes. The elite cuttings are used by the certified growers to plant an elite nursery source block for each cultivar received. This elite nursery source block is used to produce saleable plant material or plant material for a second generation of nursery plants.

The elite plants that are not sold to growers are used by certified propagators to establish another larger planting or nursery increase block. This is called the foundation level, to show it is derived from the elite level, and the increase block is known as a foundation nursery increase block. Plants from this nursery block may be less healthy than the elite block as they have already been exposed to potential infection in the field for one year. This foundation nursery increase block may be used to produce saleable material for a limited number of years. The life span of this planting is determined by the industry and specified in the guidelines. This can sometimes be extended for an additional year, upon request, if inspection shows that the plants have remained relatively free of disease organisms.

Under less than ideal circumstances, elite cuttings or plants may not be available for distribution to certified propagators. In this case, administrative authorities of the program may allow foundation stock derived from elite cuttings to be used to establish foundation nursery increase blocks. The guidelines are similar, but the duration of the nursery increase block will be limited (for example, to two years).

Different systems of propagation are used for the different berry crops. With strawberry, nuclear stock is usually grown in a screen house for one year, followed by one year of elite and foundation crops. Raspberry may be propagated from the elite block for up to four years, or the elite block is dug and replanted as a foundation crop. For currant and gooseberry, an elite planting referred to as a stool-bed is used for up to 10 years to take cuttings that are either sold as unrooted shoots or grown for one year and sold as rooted plants.

PLANTING AND CULTURAL PRACTICES

Elite and foundation plantings are determinate plantings, in the sense that they may be used as a source of propagules for the commercial industry for a limited number of years (typically one to four years) before they must be destroyed. Prior to planting, the land should not have been used for at least five years (this varies with the province and is sometimes reduced if the land is fumigated) to grow related species or species that are alternate hosts for diseases important to the industry. Soil fumigation for nematode control may be advised, following soil testing for nematode numbers. Each nursery block must be planted respecting minimum isolation distances from wild plants and those that are not being grown under the propagation program, for example, fields where a commercial crop of the same species is being grown. Plants that may host pests and diseases of the crop being propagated may be tested if they are growing near the nursery block. These isolation distances are maintained and plant testing done to prevent pests and pathogens from being introduced. Also, minimum distances between nursery blocks of different cultivars are specified to prevent cultivars from being mixed.

A cultural control program is specified in the guidelines to control insects, diseases, and weeds. A spray program to control insects must be maintained and fully documented. For example, log books are copied and included in inspection reports. Weed control is essential. Weeds compete very successfully with the nursery crops and can reduce propagation rates substantially. They can also harbor pests and pathogens. Blossom removal is usually advised. This is done to control pollen-borne viruses and to prevent seed production, which can lead to off-types if the seeds germinate and grow. Any fruit that is produced cannot be consumed because of pesticide residues related to strict pest control. Inspection is carried out at the discretion of the agricultural authorities and is done following planting and at intervals (usually annually) throughout the duration of the planting. The plants or propagules are usually stored before sale; stored stock may also be subject to inspection and testing.

Inspectors are usually, but not always, government agents. They are trained in the identification of pests, plant diseases, and in cultivar recognition. They visually inspect the crop and review planting plans, test reports from soil and disease-testing agencies, and grower log reports for cultural practices such as fertilizer and spray applications. It is their responsibility to work with the certified grower to follow the certification guidelines. It is up to them to determine if the crop falls within the maximum tolerances permitted by the guidelines for disease and pest incidence, cultivar mixing, and off-types.

MAXIMUM TOLERANCES

Viral, bacterial, and fungal disease tolerances may permit up to five mother plants or their progeny with visible symptoms per 1000 mother plants. All mother plants with visible symptoms and all plants derived from them must be destroyed. For diseases subject to quarantine, a zero tolerance may be set. In this case, if any visible symptoms are found, the whole nursery block of that cultivar will be rejected. Nematode numbers per kg of soil are recorded and must not exceed the guidelines for certain problem-causing nematodes. Threshold levels are designated for all potentially problematic organisms. The tolerance for cultivar mixing is zero. Cultivar mixtures and plants from an area that has become mixed must be removed. Any off-type plants or diseased plants spotted in the field are immediately rogued (removed from the field and destroyed). The size (height, stem diameter, number of roots, and their aggregate length) and condition of the plants (dormant) at the time of sale are clearly designated in the guidelines.

CERTIFICATION TAG

At the time of sale, certified plants carry a tag to designate that they were produced under all guideline requirements. On the tag is written the name of the cultivar and the name and address of the certified propagator. Advertisements are carefully worded to avoid confusion regarding the virus status of plants. For example, terms such as specific virus-tested or specific pathogen-tested are used.

SEED POTATO CERTIFICATION GUIDELINES

Potato nuclear stock is held *in vitro* in a limited number of facilities in Canada; one is federal (the "node" for potato is at the Agriculture and Agri-Food Canada (AAFC) Research Station in Fredericton, NB), several are provincial, and some are private tissue culture laboratories. All nuclear stock is comprehensively tested by the CFIA and multiplied by tissue culture, greenhouse, or screen house production. When new cultivars enter the program, they are disease-tested. If virus is detected, these plants must undergo thermotherapy and meristem tip culture (see Procedure 11.1 for meristem tip isolation). These explants are micropropagated to yield plantlets that can be retested *in vitro* and at intervals after transfer to a greenhouse or screen house. If these meristem tip source clones test free of the specific virus in question, they will be used to increase nuclear stock of this cultivar.

Within the potato industry, terminology varies slightly from that used in the berry industry. This is related, in part, to the greater influence of nuclear propagules derived from micropropagation and microtuberization. These plantlets and microtubers (pea-sized tubers formed from plantlets in culture) may be planted under protected cultivation in a greenhouse or screen house to produce a crop of minitubers (golf ball-sized tubers that develop on plants derived from plantlets or microtubers) and may be described as nuclear level over the course of one to two generations within these protected structures.

Starting with nuclear stock, a field generation (pre-elite) is produced. The number of field generations for propagule increase is limited to no more than seven generations from nuclear stock. These seed tubers advance at least one class every year after this, including elite 1, 2, 3, 4, foundation, and certified. Fields of each class of seed are inspected from two to three times annually. Rogueing is used to eliminate diseased or off-type plants. Tolerance for cultivar mixing is zero until elite 4 and is less than 0.1% for certified seed tubers. Disease tolerance is zero for certain severe disease-causing organisms, such as bacterial ring rot. Disease tolerance for other pathogens remains zero until elite 2 and may increase to 2% for certified seed tubers. Inspection is done at harvest, during storage, and at packing and shipping points. Certification tags are used to identify the cultivar and seed tuber class, province, farm, and field of production.

ECONOMIC AND LEGAL LIABILITIES

Clonally propagated plant material is traded between commercial growers, retail nurseries, propagators, and governmental and nongovernmental agencies. Some plants are traded across international borders. This trade is controlled by commercial law and phytosanitary regulations in the countries and legal jurisdictions involved. For the most part, this trade functions smoothly, but problems can occur, and disputes need to be settled. Since different countries have different trade and commercial laws, dispute resolution can be difficult.

In Canada, the trade in clonally propagated plant material is based on three principles: (1) the buyer is expected to pay for the plants that are received, (2) the propagator is expected to provide material fit for the purpose for which it is designed, and (3) the liability for negligence cannot be limited.

The buyer is expected to pay fair value for the plants, plus any royalties, and applicable taxes. Propagators will usually vary the price that they charge depending on the quantity of plants bought; larger quantities command a lower price. Often groups of growers will join together to obtain a lower price for their plants. Propagators have to deal each year with growers defaulting on their payments. In response, propagators demand at least part of the purchase price with the plant order and may not extend credit on the balance of the payment.

Despite the propagators' and the agencies' best efforts, problems do occur in the propagation programs. These usually involve cultivar mixtures, off-types, or the latent (hidden) transmission of economically devastating pathogens. The effects of these problems can be minor, but can lead to a shortage of plant material for several years, and can affect large numbers of growers. The propagators and the agencies usually insert clauses into their sales agreements and guidelines to limit their liabilities. The legal effect of these clauses is uncertain. Propagators routinely prominently display on their order forms and invoices statements that liability will be limited to the replacement costs of the plants sold. Governmental and nongovernmental agencies are increasingly inserting clauses that do not guarantee the disease-free or true-to-type status of the plants.

Usually, most cases are settled by negotiation, and only occasionally do cases reach the courts, but when they do, the consequences can be far-reaching. In the only case of its kind in Canada, Blue Mountain Nurseries successfully sued the Province of Ontario over "Latham" red raspberry plants, which were not true-to-type. As a direct result of this case, the Province abruptly cancelled the Ontario Strawberry and Raspberry Plant Certification Act. This curtailed the supply of specific pathogen-tested strawberry and raspberry plants to Ontario and eastern Canadian growers. It took the concerted efforts of the whole strawberry and raspberry industry, several years of negotiations, and the formation of the Ontario Berry Growers Association, to reinstate the program as the Ontario Strawberry and Raspberry Plant Propagation Program, which is now backed by legislation.

THE CONTRIBUTION OF CERTIFICATION PROGRAMS

Certification programs serve the agricultural community by providing a carefully monitored procedure to restrict the spread of crop pathogens and ensure the distribution of the highest-quality planting material to commercial growers. The guidelines are prepared by agricultural authorities and may be supported by legal means. These guidelines outline a set of standards and describe cultural processes to limit pests, diseases, and genetic off-types. Certified propagators who follow these guidelines are able to produce certified plants that meet both national and international standards.

EXERCISES

EXPERIMENT 11.1: MERISTEM TIP ISOLATION

The purpose of meristem tip isolation is to obtain specific virus-free plants from infected source plants. Often, the source plants have been subjected to thermotherapy. In thermotherapy of temperate species, plants are exposed to elevated temperatures (37°C to 38°C) for extended intervals (two to eight weeks is typical). Heat treatment is stressful to the plant and often results in elongated, thin (spindly) branches, and sometimes plants die. Meristem tip isolation occurs when experience suggests that the virus in question will have greatly reduced titer; by this time the plant may be severely heat stressed. Procedure 11.1 can be carried out using heat-treated red raspberry or other *Rubus* spp. or simulated on healthy greenhouse-grown plants of a wide range of other species. An experienced horticultural technician can isolate one meristem tip per minute. This technician might dissect 10 to 20 meristem tips per day, at intervals of several days, once source plants have been subjected to the recommended heat treatment. The laminar air flow cabinet is set up in the conventional way, except that a dissecting microscope is cleaned (70% alcohol) and located centrally within the unit.

Materials

Students will need the following items to complete this experiment:

- Heat-treated source plants (can simulate this)
- 1-L Erlenmeyer flask
- Square of plastic or wire netting or cheesecloth
- Elastic band to secure netting or cloth
- Plastic cover or square of Parafilm
- Single-sided razor blade or scissors
- 10% Javex bleach (contains a wetting agent)
- 1 L sterile, distilled water (refrigerated)
- 70% ethanol (for cleaning microscope, tools)
- Kimwipes or paper towel (for cleaning)

Procedure 11.1
Meristem Tip Isolation from Heat-Treated Source Plants of Red Raspberry or Its Relatives (*Rubus* spp.)[a]

Step	Instructions and Comments

The first five dissection steps are illustrated in Figure 11.1 through Figure 11.5.

1. Sever sections of stem (potato, *Rubus* spp.) from potted, heat-treated plants.
2. Remove leaves, leaving a short length (1 to 2 cm) of petiole attached.
3. Shorten stems to sections (two to three nodes long) that will fit into the 1-L container you will use for disinfestation.
4. Wash stem sections for 30 min. under a gentle stream of cool tap water. Use a square of plastic netting held in place with an elastic band to prevent loss of stem sections from the container.
5. Disinfest these stem sections in 10% liquid chlorine bleach containing a wetting agent, such as Tween 20 (1 mL/L) for 15 min, with agitation at intervals, and rinse three times with sterile distilled water. Leave a small amount of sterile water in the disinfestation container, and place the container of disinfested shoots toward the back of the clean laminar air flow cabinet.
6. Remove one short length of stem with three to four nodes from the container of disinfested shoots, and lay it horizontally on sterile paper toweling on the dissecting platform of the microscope.

The next four dissection steps are illustrated in Figure 11.6 through Figure 11.9.

7. Hold the shoot down with one pair of forceps, pull the petiole down firmly with another pair of forceps; pull until the petiole strips away from the stem to reveal the axillary bud.
8. Hold the shoot firmly and work quickly, to avoid the shoot tip drying. Use the microscope (10× to 50×) to help you to remove the outer pairs of leaf primordia. Do this using the back of the scalpel tip; slide the scalpel down to the base of each leaf primordium, and apply outward pressure to break it off at the base or to cut it off. The meristem tip is usually surrounded by from two to four pairs of leaf primordia that must be removed.
9. Once exposed, sever the meristem tip (meristematic dome plus one pair of leaf primordial) at the base, using a fresh scalpel.
10. Steady the scalpel, holding the tip close to the paper, and measure the length of the meristem tip using your calibrated eyepiece (make a note of this measurement).
11. Quickly transfer the meristem tip to a test tube containing Stage I medium by touching it to the top of the wick; the surface tension will pull on the meristem tip and permit you to orient it vertically using the scalpel tip. Cap the tube quickly and set it into the rack.
12. Pull back the petiole at the next node of your shoot, exposing the axillary bud, and repeat the explantation procedure. Use fresh scalpels each time; one or more to remove the leaf primordia, and another to cut off the meristem tip.
13. Once all of the meristem tips have been dissected from the shoot, discard the plant remnants and used paper, clean the dissecting platform using 70% ethanol, and place fresh paper toweling to receive the next shoot.
14. Test tubes should be placed on a lighted bench (16 h cool white fluorescent light of approx. 36 $\mu mol \cdot m^{-2} \cdot sec^{-1}$) at 23±2°C for at least one month.
15. Following one month in culture, your meristem tip should have developed into a small shoot (Figure 11.10).

Note: Steps in Procedure 11.1 are illustrated in Figure 11.1 through Figure 11.10.

[a] Potato (*Solanum tuberosum* L.), currant, or gooseberry (*Ribes* L. spp.) can be used instead of red raspberry (*Rubus idaeus* L.) or other *Rubus* species with no change in the dissection protocol. Strawberry (*Fragaria x ananassa* Duch.) can also be used. However, in preparing strawberry runners for disinfestation and dissection, nodes are usually handled individually.

FIGURE 11.1 Primocanes of red raspberry cut from potted, heat-treated plants.

FIGURE 11.2 These 2-cm long petiole sections from red raspberry primocanes are retained when leaves are cut from the canes.

FIGURE 11.3 Stems of red raspberry primocanes are cut into two- to three-node lengths and inserted into 1-L Erlenmeyer flasks.

Crop Certification Programs and the Production of Specific Pathogen-Free Plants

FIGURE 11.4 Stem sections of red raspberry primocanes are washed under cool tap water for 30 min, retained by plastic netting held in place with an elastic band.

FIGURE 11.5 Stem sections of red raspberry primocanes are disinfested for 15 min in 10% household bleach, followed by three rinses with sterile water.

FIGURE 11.6 The axillary bud of red raspberry is located just above the scar left from removal of the petiole.

FIGURE 11.7 The meristem tip of red raspberry is exposed once several pairs of leaf primordial have been removed.

FIGURE 11.8 The meristem tip of red raspberry is removed using a sterile scalpel.

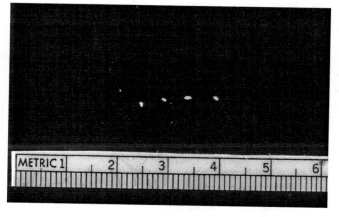

FIGURE 11.9 Meristem tip explants of red raspberry are shown lined up against a metric ruler so that their size can be put into perspective.

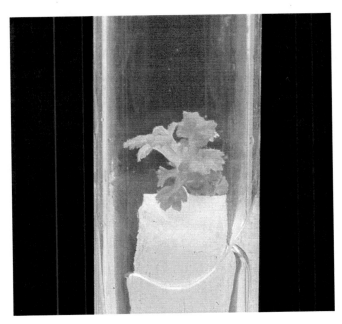

FIGURE 11.10 The meristem tip of red raspberry has grown into a small shoot (approx. 1-cm long) after one month in culture.

- Sterile pack of paper towels or alternate materials (for sterile dissection surface)
- Sterile pack containing several pairs of long- or short-handled forceps
- Sterile pack containing several pairs of scalpels (#7 scalpel handles, #11 blades)
- Dissecting microscope with a calibrated, reticulated eyepiece
- Prepared medium (described in Procedure 11.2.)

Follow the instructions listed in Procedure 11.1 to complete this experiment.

Anticipated Results

The dissection procedure may seem challenging at first. This is especially true for students who do not have much experience at dissecting shoot tips or other explants. However, these difficulties can be quickly overcome by practicing under a dissecting microscope placed on the open bench. Practice in pairs, with one student acting as "spotter" while the other student works. A brief discussion on tactful, helpful criticism may be in order. All manipulations should be practiced: dissection and the transfer of the isolated meristem tip onto the wick in the test tube. Be sure to touch the meristem tip to the wick without bringing fingers too close to the lip of the test tube. There are various ways to safely position the test tube cap when you insert the explants. These can be discussed and practiced. When dissection efficiency and test tube manipulations have improved, the procedure can be carried out in the laminar air flow cabinet with full attention to aseptic practice.

Once the meristem tip is positioned upright anywhere near the top of the wick, the tube is capped and carefully set into the test tube rack. Meristem tips that are displaced from the wick in the first few days of incubation rarely survive. These "sinkers" should be aseptically rescued and replaced onto their wick.

The meristem tips, which have little color at first, should start to become green and enlarge over the first two to three days. These can be measured at intervals and their growth rate plotted. A lag period may precede growth for the tiniest meristem tips, and some may not survive. Growth rate tends to be faster for the larger meristem tips, which may not exhibit an initial lag period. Contaminated tubes may indicate poor surface disinfestation (microorganisms associated with the explant) or poor laminar air flow technique (microorganisms randomly distributed within the test tube). After they have become established, shoots or plantlets can be readily subcultured to fresh, solidified medium. This is usually done one month following excision.

Questions

- Why do specific pathogen assays precede thermotherapy treatment?
- What is a meristem tip source clone?
- What is the relationship between meristem tip size, survival, and virus elimination?
- Why is rapid dissection necessary for meristem tips?
- How could you investigate possible sources for contamination of meristem tip cultures?

	Procedure 11.2
	Media Preparation for Meristem Tip Explants of Red Raspberry, Strawberry, and Potato (Stage I)[a]
Step	Instructions and Comments
1	To a 1-L Erlenmeyer flask add 700 mL of double-distilled water.
2	Prepare Murashige and Skoog (1962) basal salt medium formulation using your stock solutions or commercial prepared powdered medium.
3	Dissolve 30 g sucrose and 100 mg myo-inositol in the medium.
4	Add 1.0 mg (4.44 µM) BAP, 0.1 mg (0.49 µM) IBA (red raspberry); 0.25 mg (1.11 µM) BAP, 0.025 mg (12.30 µM) IBA, 0.025 mg (72.17 µM) GA_3 (strawberry); or no growth regulators (potato).
5	Make to 1 L volume by adding double-distilled water.
6	Adjust pH to 5.7.
7	Dispense into test tubes into which you have inserted a filter wick. The wick should be made of Whatman number 2 or 3 filter paper, cut chromatographic paper, or fiber cloth, such as Miracloth. The wick should project 0.5 to 1.0 cm above the medium meniscus.
8	Cap your tubes and autoclave for 15 min. at 15 psi.
9	Cool autoclaved medium in a laminar air flow cabinet.
10	Use fresh or store in the fridge in plastic bags for up to one month.

[a] If Stage II culture will follow, growth regulators may have to be adjusted, and medium is usually solidified with agar.

EXPERIMENT 11.2: MEDIA PREPARATION FOR MERISTEM TIP EXPLANTS

Meristem tip explants are usually placed onto a wick, made of one of several possible materials. Wicks can be cut from #2 or #3 filter paper, chromatography paper, or Miracloth. The wicks are folded in half and inserted into small test tubes. The medium is usually Stage I medium, but in liquid form, as more explants survive in this medium. It is carefully dispensed to the side of the wick to avoid collapsing the wick. The wick should project just above the level of the medium. A small test tube is preferred, so that the scalpel on which the meristem tip is balanced can be safely inserted into the tube without touching fingers to the lip of the container.

Materials

Students will need the following items to complete this experiment:

- Small (<20 mL) test tubes
- Wicks cut to fit tubes and folded in half, inserted into tubes
- 1-L Erlenmeyer flask
- Double-distilled water
- Stock solutions of basal salts or prepared, packaged salts for 1 L medium
- 30 g sucrose, 100 mg myo-inositol
- Stock solutions for growth regulators (usually 10 mg growth regulator per 100 mL water)
- Medium dispenser
- pH apparatus and buffers
- Stirring platform
- Autoclave

Follow the instructions provided in Procedure 11.2 to complete the experiment.

Anticipated Results

Choose test tubes carefully. When the test tubes are too large, explant insertion is harder, so you get more contamination and the wicks tend to collapse against the sides of the tube. Wicks should fit snugly and remain erect and intact during medium dispensing and autoclaving.

Questions

- What are the characteristics of an ideal wick, and what are these used for?
- Why is liquid medium preferred over agar-solidified medium for meristem tip explants?
- Why is double-distilled water preferred over tap water for medium preparation?
- Why should autoclaved medium be cooled in a laminar air flow hood?
- Why is fresh medium preferred over medium that has been stored for several months?

SUGGESTED READINGS

Animal and Plant Health Inspection Service. http://www.aphis.usda.gov.

Buonassisi, A.J., H.A. Daubeny, and B. Peters. 1989. The B.C. raspberry certification program. *Acta Hort.* 262: 175–185.

Canadian Food Inspection Agency. http://www.inspection.gc.ca.

Dale, A. 1999. Guidelines Ontario raspberry plant production program. University of Guelph, Dept. of Plant Agriculture, Simcoe, ON. 11 pp.

Dale, A. 1999. Guidelines Ontario *Ribes* plant production program. University of Guelph, Dept. of Plant Agriculture, Simcoe, ON. 10 pp.

Dale, A. 1999. Guidelines Ontario strawberry plant production program. University of Guelph, Dept. of Plant Agriculture, Simcoe, ON. 10 pp.

De Boer, S.H. 1994. The role of plant pathology research in the Canadian potato industry. *Can. J. Plant Pathol.* 16: 150–155.

De Boer, S.H., S.A. Slack, G.W. van den Bovenkamp, and I. Mastenbroek. 1996. A role for pathogen indexing procedures in potato certification, in *Advances in Botanical Research, vol. 23, Incorporating Advances in Plant Pathology*. Academic Press Ltd., New York, pp. 217–242.

Hartmann, H.T., D.E. Kester, F.T. Davies, Jr., and R.L. Geneve. 2002. *Plant Propagation: Principles and Practices*. 7th ed., Prentice Hall, Inc., Upper Saddle River, NJ. 880 pp.

McDonald, J.G. 1995. Disease control through crop certification: herbaceous crops. *Can. J. Plant Pathol.* 17: 267–273.

Murashige, T. and F. Skoog. 1962. A revised medium for rapid growth and bioassays with tobacco tissue cultures. *Physiol. Plant.* 15: 473–497.

12 Integrated Pest Management for Plant Propagation Systems

William E. Klingeman

CHAPTER 12 CONCEPTS

- Integrated Pest Management (IPM) enables growers to incorporate multiple practices into decision-making strategies that restrict pest populations below levels causing economic or aesthetic plant loss.

- The first step in successful IPM is to identify the pest.

- Scouting, or routine monitoring of propagated plants, is the key to successful IPM.

- Propagators have several options for managing pests in production systems including host plant resistance, cultural, physical, mechanical, biological and chemical control, as well as doing nothing.

- Records of IPM actions and outcomes can be used to train employees and scouts. Well-kept records may limit grower liability in wrongful injury claims.

- IPM programs are ineffective unless observations and recommended management actions are recorded.

- IPM is incomplete until the effects of management actions have been assessed.

In sum, the objectives of Integrated Pest Management (IPM) in plant propagation are to restrict pest entry into production systems and manage pest populations below levels causing economic or aesthetic crop losses using the best available biologically based control strategies. Once understood, the principles of IPM can be readily applied to nursery production fields, container-grown ornamentals, and within plant propagation structures. A grower determines which IPM tactics to use based upon their direct knowledge and expectation of pest-related injury to ornamental plants. An experienced grower will also consider historical observations of pest populations at their production facility. To be successful, this knowledge is combined with an awareness of alternative management actions and their consequences.

Some producers might argue that species of propagation stock are too diverse and rotation cycles are too short in commercial propagation for IPM to be practical. In fact, IPM principles can be applied in propagation operations as readily as they can be incorporated into greenhouse, nursery, and landscape management, and although they may not know it, most experienced plant propagators are already practicing aspects of IPM. Regardless, proactive growers who understand and apply IPM for propagation systems are at a competitive advantage when they can balance use of restricted chemical pesticides, worker re-entry intervals, employee safety, and pesticide resistance to produce high-quality plants with attention to both economic and environmental sustainability.

From an economic perspective, rooted cuttings or tissue-cultured plantlets can make substantial profits for high-quality growers. But it is important to remember that early in propagation, individual plants account for nearly the minimum initial investment. Infested or poor quality plants can be thrown away with the least economic loss. As steps in production continue, however, plants must be transplanted into a larger container, surrounded by amended soilless media, and then plants receive regular fertilizer and irrigation applications. As plants mature, they must also be pruned and spaced, repotted and re-fertilized, lifted, stored, and finally, shipped to market and sold. Because each step adds costs, competent plant propagators demand superior plant quality and performance from the onset of their efforts. Many of the very best producers are willing to throw away inferior plant products knowing that they will recover those costs through customer satisfaction, loyalty, and an earned reputation for growing quality liners and plants.

FACILITIES AND THE PROPAGATION ENVIRONMENT

The hobbyist can readily propagate many species and cultivars of plants with very little investment in structural facilities. For example, English ivy (*Hedera helix* L.), pothos, and many willow (*Salix* sp.) species produce functional roots if cut stems are immersed in just water. On the scale of commercial production, where heavy volume and a diverse plant mix are commonplace, plant propagators require greenhouses or polyethylene film (sheet-plastic) covered houses to protect and acclimatize their plant material. Conditions in propagation structures are well suited to rapid development of both pest populations and food resources like fungi and algae, on which pests depend. In turn, environmental conditions surrounding plants help direct the management strategies for pests that a propagator is likely to encounter. Synthetic woven fabric in different shade densities is often used to reduce the level of direct light that plants receive. Cuttings and media are kept moist with mist systems, spray stakes, or overhead irrigation, and humidity in the greenhouse or under plastic can frequently exceed 90%. Temperatures are often high in combination with reduced airflow and high humidity. Rooted cuttings and container-grown liners are stored pot-tight during winter. In spring, temperatures in plastic-covered poly-houses can rise quickly. Pest and disease populations can rapidly damage overwintered stock left under cover for too long in spring.

TOOLS FOR THE IPM SCOUT

Material investments required to monitor pests are relatively inexpensive. A scout should purchase a small 10× to 15× hand magnifier with a lens not less than 20 mm (0.75 in.) across. The diameter of the lens should be large enough that it is comfortable to use for extended durations. A pocketknife and small, sealable plastic bags can be used to collect and preserve root or leaf tissues until they can be more thoroughly examined. Production rows, borders, and blocks of stock plants can be sampled quickly using a sturdy canvas sweep net. Delicate branches and foliage can be shaken over a square "beat" cloth or a light-colored pan. Alternatively, a clipboard or field notebook with a light surface can be used as a backdrop to count mites or thrips shaken from plant foliage. A light surface is important to allow contrast with darker colors and shadows of small and slow-moving pests.

A scout should also invest in a small aspirator. An aspirator is essentially a miniature vacuum that enables small arthropods to be drawn into a collection vial. The scout sucks a small volume of air into a screened intake tube. The opening of a second flexible tube is held close to the insect and the vacuum pulls the insect into the vial.

A small volume of 70% isopropyl (rubbing) alcohol preservative can be added to collection vials, which should be labeled and the specimen returned to the office for subsequent identification. Logically, a scout's portable kit should include 10- to 20-mL (0.4- to 0.8-fl. oz.) watertight vials and a permanent labeling pen or pencil.

Digital cameras are invaluable tools, particularly if they can be mounted onto a dissecting microscope. With digital images and e-mail, professionals who are miles away from a grower or problem can quickly identify the causal agent. Many universities have established "Distance Diagnostics" Extension resources that allow digital images of pests, pest injury, disease symptoms and signs to be submitted online to specialist experts. Correct identification of pest and disease organisms is often impossible without clear images of small but diagnostic characteristics. This technology also lets experts rapidly recommend action or request additional information that may initially have been overlooked in the field.

SCOUTING AND MONITORING TECHNIQUES

Consistent plant monitoring is key to successful integrated pest management for plant propagation. Monitoring allows the scout to identify pest outbreaks before economic and aesthetic losses occur. Production areas should be divided into logical units and mapped prior to scouting. Maps may reflect individual propagation houses and production areas that can be subdivided by similar crop types, irrigation type or schedule, or crop age. Plugs and easily rooted cuttings are short-cycle crops that may need to be scouted every two to four days. Slower-to-root cuttings and seedling liners can be monitored at weekly or biweekly intervals. Pests of field-grown tree whips and container-grown liners of woody ornamentals can be detected by scouting biweekly throughout the ornamental growing season.

Scouts do not need to inspect every plant within a production block to determine if pest populations are not approaching economically or aesthetically damaging levels. Many pest populations develop in patchy distributions, rather than evenly throughout production areas. A scout should inspect more plants in a production block to detect pests in patchy or aggregated populations. Fewer plants can be scouted to detect pests that are distributed evenly throughout the block.

To maximize scouting efforts, mark the point of entry into a production block with a colored flag. Vary the point of entry into production blocks and travel in a routine pattern through stock blocks and propagation beds. At regular intervals, select and inspect a plant for signs and symptoms of pest or disease presence. It is appropriate to be "biased," particularly in commercial operations, in selecting which plants are inspected. Chlorotic, wilted, and damaged foliage are often indicators of diseases or arthropod pest activity. Plants with characteristic signs and symptoms should not be ignored. On return visits to

Integrated Pest Management for Plant Propagation Systems

the production block, start above or below the last entry point and inspect different plants.

Within about 100 square meters (1075 square feet) of production area, inspect 20 or more plants. It should take an experienced scout less than 15 min. to scan the upper and lower leaf surfaces of both new and old foliage, flowers, leaf axils, bark on the main trunk and stems, and roots of 20 plants showing evidence of pests or disease. Inspect whole individual plugs or cuttings. On larger specimens, examine three to five, 10 cm (4 in.) long, terminal stem sections per plant. Inspect stock blocks for cuttings at low and high positions on each cardinal direction around the surface of individual trees or shrubs. In propagation and greenhouses, start inspections near the entrance. Pay attention to bench space adjacent to vents and horizontal airflow (HAF) fans. Do not ignore bench middles, but concentrate on the edges and ends of benches where pest populations usually develop first. Scout beneath benches for pests persisting on weeds. Do not forget to inspect hanging baskets.

While most pests develop in patchy distributions, the probable location of outbreaks is seldom *un*predictable. "Hot-spots" are problem areas that, more or less regularly, contribute to seasonal population outbreaks. These often include shaded corners, low areas that collect standing water, drip-zones around greenhouse cooling pads and beneath greenhouse purlins and mist irrigation lines. Spider mites prefer warm, dry areas near greenhouse vents, sidewalls, and gravel roadways. When pests are encountered, inspect a greater number of plants to provide a more accurate estimate of pest population size. Insert a colored flag into the pot or flat to mark the outbreak location. Make a rough count of the ratio of affected to unaffected plants within the production unit.

STICKY CARDS AND SENTINEL PLANTS

Propagation pests can be attracted to yellow (e.g., whiteflies, aphids, fungus gnats, leafminer flies, and shore and crane flies) or blue (esp. thrips) sticky cards. Sticky cards are used to monitor pest populations. Cards generally do not work to control pest populations once they become established. Cards should be mounted on bamboo or metal stakes. Suspend cards at canopy height to intercept flying pests. For best effect, sticky cards should be counted weekly. Use a grid layout with at least one card per 100 square meters (~1075 square feet).

Plants that are susceptible to pests can be exploited within a Propagation IPM program. For example, rye (*Secale* sp.) and wheat (*Triticum* sp.) seedlings can be grown beside ornamentals as trap crops for fungus gnats. Garden beans (*Phaseolus vulgaris* L.) and roses (*Rosa* sp.) can be used to monitor spider mite and broad mite populations. Tomatoes, lantana, and poinsettia can become trap crops for whiteflies, and aphid outbreaks can be detected on chrysanthemum. Because pest populations can quickly build to plant-threatening levels on untended sentinel and trap plants, once installed, sentinel and trap plants must be observed regularly. Infested sentinel plants should be removed and destroyed to avoid contaminating the ornamental crop.

ARTHROPOD PROFILES: A ROGUE'S GALLERY OF PROPAGATION PESTS

Ornamental plants have high aesthetic thresholds—plants must look good to sell. As a result, economic loss to propagated plants can occur when relatively few pests are found. Competent growers should be able to recognize the propagation pests that threaten ornamental crops. An efficient propagator will also train employees to recognize pests and diseases that are common production problems. Once an arthropod is discovered, it must be identified and confirmed as a pest. Next, the scout should thoroughly sample production blocks to determine if pest numbers will translate to plant damage and economic loss.

COMMON PROPAGATION PESTS

FUNGUS GNATS

The cosmopolitan distribution of *Bradysia* sp. and *Sciara* sp. fungus gnats has established them as the most familiar pests of commercial and hobbyist propagators. Fungus gnats develop quickly during mild winters and in spring (Figure 12.1a). Fungus gnats persist in propagation systems that provide too much irrigation. Seedlings, stem cuttings, and young plants grown in rich, organic rooting media are particularly susceptible. Individual adult females may live as long as seven days, during which they can deposit 100 to 150 eggs. Adult fungus gnats are readily observed on moist media surfaces where they deposit eggs. Larval fungus gnats can be found in groups feeding on foliage, fine roots, and root hairs on or near the surface of the medium, and at the basal wound on cuttings. Larval feeding injury provides entry points for plant pathogenic fungi. Plant pathogens that cause root and crown rots, including *Pythium, Verticillium,* and *Fusarium,* are also transferred on the bodies and feet of adult fungus gnats. Subsequent plant wilting may be attributed as much to plant disease as to direct root loss from larval feeding.

Correctly identifying fungus gnats is important. Fungus gnats are often confused with shore flies, which are generally a nuisance pest. In contrast to shore fly larvae that have a light-colored head that is largely soft tissue, fungus gnat larvae have a hardened, black head capsule. Adult fungus gnats are about 2.5 mm (0.1 in.) long and have long legs and long antennae. Shore fly adults (Figure 12.1b) are smaller than fungus gnats (about 2.0 mm, or 0.08 in. long), and have shorter legs and stubby antennae.

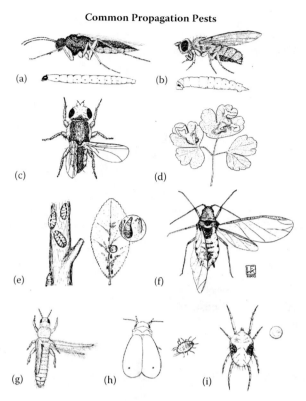

FIGURE 12.1 Common propagation pests. Growers, propagators, and employees should learn to identify the most common arthropod pests in propagation systems. These include fungus gnat adults (a, top) and larvae (a, bottom), shore fly adults (b, top) and larvae (b, bottom), serpentine leafminers (c) and leafminer damage on columbine leaflets (d), soft-scales (e, left) and armored "oystershell" scales (e, right), wingless and winged (alate) adult aphids (f), thrips (g), whitefly adults (h, left) and a pupa (h, right), and twospotted spider mite adults and an egg (i). (Drawings by W.E. Klingeman.)

Adult fungus gnats can be attracted to sentinel pots in which rye, wheat, or other small grains are grown. Eggs hatch within four days, larvae mature in about 14 days, and pupation requires about three days at 22°C (72°F). Once infested, sentinel pots should be removed and destroyed and sentinel pots replaced at two-week intervals during propagation cycles. Adult fungus gnats are also attracted to yellow colors and can be monitored with commercially available paper or plastic sticky traps. Presence of fungus gnat larvae can be monitored by using small potato wedges about 25 mm (1 in.) in diameter. The potato wedge should be placed flat on the media surface and scouted regularly for presence of larvae. Insecticides applied as drenches control larval fungus gnats, rather than adults. Results of control efforts may not be immediately apparent.

Shore Flies

Shore flies (*Scatella* sp.) are also common inhabitants of propagation houses (Figure 12.1b). Unless propagation areas are adjacent to retail sales areas, shore flies are generally of very little economic concern. While adult shore flies are primarily a nuisance pest, large populations of flies can spread plant pathogens. Populations can build quickly where moisture and humidity have allowed algal food resources to develop. Shore fly larvae also feed on decaying organic matter. Prompt sanitation to reduce these food resources is an important strategy for effective cultural control.

Leafminers

Larvae of several fly and beetle species develop in tunnels cut beneath leaf epidermal cells or within the palisade or spongy mesophyll cell layers of leaves. Chrysanthemum leafminers (*Chromatomyia* [= *Phytomyza*] sp.) and serpentine leafminers (*Liriomyza* sp.) (Figure 12.1c) are pests of many different greenhouse crops. Annual and perennial crops, including sunflowers, zinnia, daisies, columbine (Figure 12.1d), and fruits and vegetables (e.g., tomatoes and onions) are susceptible. Sanitation is essential in infested propagation areas. When humidity is high, chrysanthemum leafminer larvae can successfully develop and pupate within cut or aborted foliage that drops onto greenhouse benches and floors. Adult chrysanthemum leafminers, which are about 2.5 mm (0.1 in.) long, first appear in April and early May in the mid-Atlantic states. Three to four generations can be completed within a year, but flies can persist yearlong in climate-controlled greenhouses.

Scale Insects and Mealybugs

Some 6000 species of scale insects occur worldwide—about 1000 are found in North America. Among the most commonly encountered are mealybugs and armored and soft scales. Adult female mealybugs are small (1 to 5 mm long), wingless, soft-bodied insects. Tufts of cottony white wax exude from specialized cells and often highlight body segmentation of these pests. Around the margins of many mealybug species, wax is exuded into short filaments. In species like the longtailed mealybug (*Pseudococcus longispinus*), these waxy secretions extend from the body in filaments up to 4 mm long. Waxy secretions also hide the tiny, yellow eggs of mealybugs, forming a protective "ovisac." Soft scales are generally larger (2 to 6 mm long) than armored scales (1 to 3 mm long). Soft scales do not produce the hardened wax covering that armored scales exude to protect the insect and its eggs. Soft scales like *Pulvinaria* sp. (Figure 12.1e, left) can infest a broad range of host plants. Tea scale (*Fiorinia theae* Green) is an armored "oystershell" scale that can develop on many ornamental hosts, including hollies, dogwood, euonymus, and camellia (Figure 12.1e, right).

Feeding activity of scale insects and mealybugs can occur on leaf undersides, along shoots and stems, on roots, and in shoot axils. Throughout their lives, mealybugs remain mobile, moving slowly throughout host plants and inserting piercing-sucking mouthparts to feed on plant sap. Many species of mealybugs inject a salivary toxin into plants while feeding. These toxins can result in plant stunting, contorted leaves and branches, stem dieback, and plant death. Plants infested with scales may look generally unhealthy. Infested leaves may become chlorotic, particularly on the upper surface of leaves where scales are feeding. Leaves may wilt and drop prematurely.

In spring and early summer, newly hatched scale nymphs, called "crawlers," migrate away from the protection of their mother's dead body and water-repellent wax testa to colonize new plant tissue. Unlike adults, the crawler stage is readily controlled with insecticides, horticultural oils, and insecticidal soaps. Emergence of crawlers can be monitored by placing a small section of double-sided sticky tape around a small stem just above a female scale. Many scales are capable of multiple generations a season and require continuous monitoring. Scale populations may develop quickly following broad-spectrum insecticide applications that kill parasitic wasps and other natural enemies of scales.

Blocks of ornamental stock plants, cuttings, and bare-rooted liners should be inspected for scale and mealybug populations before they are introduced to propagation areas. Ants tend scale insects and mealybugs for the sugary secretions scales exude and can be used to locate pest populations.

Aphids

Several species of aphids, including green peach aphid [*Myzus persicae* (Sulzer)] and melon (= cotton) aphid (*Aphis gossypii* Glover), are common pests in propagation and production systems. Individual female nymphs are about 3.0 mm (0.12 in.) long. Female aphids reproduce for 20 to 30 days, during which they can give live birth to 60 to 100 female nymphs. Newborn nymphs become reproductive in 7 to 10 days, depending on temperature. If aphids become crowded or plant nutritional quality declines, winged females (alates) develop that migrate to new hosts (Figure 12.1f). Like many aphid species, green peach aphid produces pink and green color morphs. Color dimorphism in aphid species may be controlled by genetic, temperature, and plant nutritional factors.

Feeding aphids extract plant sap through piercing stylet mouthparts that penetrate phloem tissues. Excess ingested fluids are excreted as "honeydew," a sugar-rich solution. Honeydew may be forcibly ejected by the aphid and will thinly coat adjacent foliage. Ants quickly find insects that produce honeydew and protect them from natural enemies. Ants can be followed to locate aphid, mealybug, and scale insect outbreaks. In time, a layer of dark fungal hyphae called sooty mold will grow on honeydew. While sooty mold is aesthetically unpleasant, it also reduces plant photosynthetic potential and subsequent plant growth. Because sooty mold needs time to develop, it is found more commonly among blocks of propagation stock plants than among rooted cuttings or container-grown liners.

The wooly apple aphid [*Eriosoma lanigerum* (Hausmann)] is another pest in propagation. Wooly apple aphids use elms (*Ulmus* spp.) as their primary host, but will colonize roots of rosaceous plants as secondary hosts. Wooly apple aphids are frequently overlooked on rootstocks of apples and crabapples (*Malus* sp.), pyracantha, pear (*Pyrus* sp.), and quince (*Chaenomeles* sp.). As a result of aphid feeding, wooly apple aphids cause bark cracks and galling of stem and root tissues. Fibrous roots are reduced on heavily infested plants. Affected trees can become stunted and die.

Thrips

There are numerous species of thrips common in ornamental production and landscape systems (Figure 12.1g). Many different thrip species, including *Echinothrips americana* (Morgan), western flower thrips [*Frankliniella occidentalis* (Pergande)], greenhouse thrips [*Heliothrips haemorrhoidalis* (Bouché)], and banded greenhouse thrips [*Hercinothrips femoralis* (O. M. Reuter)] have an extensive host range among ornamental plants. Adult thrips are about 1.5 mm (0.06 in.) long. Thrips have rasping-sucking mouthparts used to pierce plant cells and ingest cell contents. Feeding injury from thrips may range in severity from silvery stipples and streaks of chlorosis to contorted flowers and leaves. Affected leaves fail to develop fully and may prematurely senesce. Thrips also vector tomato spotted wilt virus (TSWV) and plant diseases.

Like many insects, thrips reproductive rates are accelerated at higher temperatures. Thrips can complete 12 to 15 generations during a growing season, but persist yearlong in greenhouse environments. Adult and larval thrips are found in partly opened flower buds, at the base of expanded flowers, and on leaf undersides. Slow-moving and light-colored larval thrips can be encouraged to move (and thus be detected) by exhaling warm air onto flowers and buds. Adult thrips are attracted to yellow and blue cards coated with sticky adhesive and placed at plant canopy height.

Whiteflies

Silverleaf whitefly (*Bemisia argentifolii* Bellows and Perring), bandedwinged whitefly [*Trialeurodes abutilonea* (Haldeman)], and greenhouse whitefly [*Trialeurodes vaporariorum* (Westwood)] are common propagation pests. Each of these species has an extensive host range among ornamental plants. Adult whiteflies are active fliers when disturbed. Species of whiteflies are difficult to identify from the adults, which look like small (about 2 mm long) white moths (Figure 12.1h, left). Pupae or pupal cases are examined to identify species correctly (Figure 12.1h, right). Whiteflies colonize and feed on leaf undersides, reducing plant vigor. Whiteflies also vector plant viruses and diseases. Like aphids and scale insects, whiteflies produce honeydew. Aesthetic loss from sooty mold will reduce the economic value of infested propagation stock.

MITES

Several species of mites challenge commercial and hobbyist propagators. Until population levels become high, mites generally feed on leaf undersides. Feeding injury results in chlorotic stippling of leaves followed by premature leaf drop. Mites are often first active within the protected centers of shrubs, where they can be overlooked. Twospotted spider mites (*Tetranychus urticae* Koch) are widespread pests of hundreds of plant species including ornamental hosts and weed species utilized as refuges (Figure 12.1i). Twospotted spider mite populations develop quickly in hot, dry environments. Populations develop quickly in the drought-shadow behind entrances to propagation houses. Production areas adjacent to fan vents and dirt or gravel roads are common "hot spots" for spider mite outbreaks. Individual adult female spider mites, which are about 0.35 mm (0.015 in.) long, lay 100 to 300 eggs during their two- to four-week lifespan. As populations build, twospotted spider mites produce silk webbing that can be seen, especially on the tips of new, young shoots.

Other mite pests that feed on a broad range of hosts include European red mites (*Panonychus ulmi*) and southern red mites (*Oligonychus ilicis*). Like European and southern red mites, spruce spider mites (*Oligonychus ungunuis*) are cool-weather pests. Populations of these mites develop rapidly in spring and fall.

Broad mites [*Polyphagotarsonemus latus* (Banks)] and cyclamen mites [*Steneotarsonemus pallidus* (Banks)] feed on a wide range of ornamental and weedy host plants. Broad and cyclamen mite populations can be found in high numbers on buds and small adjacent leaf initials. As broad and cyclamen mites feed on plants, newly emerging leaves fail to expand fully and become contorted and brittle. Studies indicate that salivary toxins cause plant injury when they are excreted into plant tissues as mites feed. Broad mites have been collected from legs of whiteflies and other insects. This behavior, called phoresy, allows broad mites to spread rapidly through production areas. Broad and cyclamen mites can also become passengers on humans and are readily spread on infested cuttings and discarded plant tissues.

SEED PARASITES

Propagators who rely on seeds to generate commercial stock may encounter insect species that parasitize plant seeds. Insect larvae slowly consume seeds while they develop inside the protective seed coat. Rose chalcids and several torymid wasps (e.g., Figure 12.2a) parasitize seeds of rosaceous plants. Seeds of *Rosa*, hawthorn (*Crataegus* sp.), and mountain ash (*Sorbus* sp.) are often infested. Bruchid beetles (Figure 12.2b) parasitize seeds of many legumes including ornamental scotchbroom (*Cytissus* sp.), honeylocust (*Gleditsia triacanthos* L.), Kentucky coffeetree (*Gymnocladus dioicus* (L.) K. Koch) and eastern redbud (*Cercis canadensis* L.). Acorns of oaks (*Quercus* sp.) can be infested with a wide range of beetle, wasp, fly, and moth larvae. Infested seeds of many species can be detected using a simple float test, in which seeds are submerged in tap water. A drop of soap can be added to break the surface tension surrounding small seeds. Infested

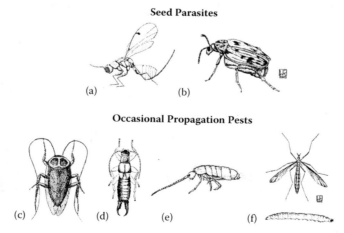

FIGURE 12.2 Seed parasites and occasional propagation pests. Rose chalcids and torymid wasps (a) and bruchid weevils (b) feed on the endosperm of seeds. During development, the larvae of these seed parasites are protected within the hard seed coat. Cockroaches (c), earwigs (d), springtails or collembola (e), and larval crane flies (f, bottom) are occasionally pests in propagation facilities. An adult crane fly is also illustrated (f, top). Large populations of these pests can develop under optimum climatic conditions. (Drawings by W.E. Klingeman.)

seeds will float and can be discarded or cut open to confirm presence of pest larvae.

OCCASIONAL PROPAGATION PESTS

Cockroaches (Figure 12.2c) and earwigs (Figure 12.2d) inhabit tight, protected spaces. These pests can be transported to greenhouses and nurseries in cardboard boxes and supplies used for shipping and growing ornamental liners. Once established, these pests are difficult to manage. As populations of these pests expand, they shift from dead or dying plant tissue to consume root hairs, roots, and young foliage.

Collembola, or springtails, are also occasionally pests of propagation stock (Figure 12.2e). Large springtail populations develop in damp environments and may move into propagation and production areas during drought. Collembola are opportunists and will consume fine roots in addition to decaying organic matter. Similarly, larvae of winter crane flies and other crane fly species (Figure 12.2f) develop in shaded, wet greenhouse environments. Like springtails, crane fly larvae will feed on plant roots as well as dead plant tissue.

MANAGEMENT STRATEGIES

Pest management is expensive. Costs include people-hour labor associated with scouting for pests, mixing and applying pesticides, and cleaning pesticide equipment. Pesticides themselves are expensive. Time and money a grower invests in liner stock are lost when plants are discarded. For grafted liners of named cultivars ready for sale, losses may exceed several dollars per plant. Earning a reputation for shipping poor quality plants or infested stock is not only the most costly loss associated with production pests, it is a recipe for failure. Preventing pest populations before they outbreak is not only logical … it makes "cents."

After identifying the pest and determining the size and location of pest populations, the scout must make a recommendation for action. If propagated liners are ready to be sold, pest presence demands immediate control. Other management recommendations may be based on prior experience with the pest and expectations of plant injury. An informed grower has several management options to choose from, including cultural, physical, mechanical, biological, and chemical controls as well as incorporating host plant resistance. In addition, the grower may elect to do nothing. Natural enemies and plant vigor can out-compete many pest population cycles.

CULTURAL CONTROL

SANITATION

In propagation and nursery production systems, sanitation is consistently the most neglected option for cultural control that limits pest population introduction and outbreaks. Dead and dying plant tissues, including discarded cutting material, should be regularly removed from propagation areas and destroyed. Nursery cull piles are decomposing mounds of substandard-quality, dead, and disease- or pest-damaged plants. Plants in nursery cull piles should be buried, composted, burned, or hauled off-site. Too frequently, unsalable plant rejects are carted to a bordering waste area and tossed onto the pile. These slow-dying, drought-stressed plants accelerate pest development or become attractants to opportunistic pests that leave cull piles and move into production areas once their life cycles are completed and their food resources decompose.

PASTEURIZATION

Disease and pest populations may develop in soilless and soil-based propagation media mixes exposed to the environment. Media can be pasteurized by injecting steam beneath the surface of media covered with a tarp. Media temperatures at the center of the pile should reach 60°C to 82°C (140°F to 180°F) and be kept stable for 30 minutes. At this range, temperatures will kill arthropod pests, weed seeds, nematodes, and most harmful bacteria and fungi. Fewer beneficial organisms are killed at lower temperatures of the range. Media pasteurized at the lower temperature will contain populations of surviving beneficial organisms that out-compete pathogens in the event of recontamination.

IRRIGATION

Unrooted cuttings and young liners in containers and field beds need different volumes of water to thrive. Too much water washes nutrients and minerals from the plant root zone. High humidity and standing water enable fungal and mold growth, and fungus gnats follow. Providing plants with too little water causes plant stress and subjects plants to outbreaks of moisture-intolerant mite pests. An efficient propagator recognizes plant needs and provides the appropriate irrigation volume needed to maintain plant health and sustain optimal plant growth. It is preferable to irrigate in the morning and early afternoon, allowing foliage to dry before nightfall, thus limiting fungal development.

WEED MANAGEMENT

Many weeds and grasses supply pollen and nectar food resources for arthropod natural enemies and refuges for both predatory and prey insects. Yet, well-managed ornamental crops perform the same function. In general, plant propagators and nursery managers cannot tolerate the competitive and aesthetic losses attributed to weed populations. Weedy plants should be excluded from production rows and greenhouses. Aggressive weeds and weeds that produce abundant seeds should be closely controlled throughout production areas, including nursery borders.

Grassy weeds and grass borders near propagation areas should be regularly mowed to reduce pest populations.

Weeds that border propagation and production areas are also overlooked as reservoirs of pests and diseases. For example, common ragweed (*Ambrosia artemisiifolia* L.) and spiny sowthistle [*Sonchus asper* (L.) Hill] support banded-wing, greenhouse, and silverleaf whitefly populations. Twospotted spidermites can persist on woodsorrel (*Oxalis* sp.) and sowthistles (*Sonchus* sp.). Broad mites overwinter on hairy galinsoga [*Galinsoga ciliata* (Raf.) Blake], prickly sida (*Sida spinosa* L.), and beggar's tick (*Bidens* sp.). Western flower thrips overwinter on spiny sowthistle and common chickweed (*Stellaria media* (L.) Vill.) and find seasonal refuge on Japanese honeysuckle (*Lonicera japonica* Thunb.), multiflora rose (*Rosa multiflora* Thunb.), and dandelion (*Taraxacum officianale* L.). The American serpentine leafminer can be found on spiny sowthistle and hairy galinsoga.

QUARANTINE AND STAGING AREAS

A principal goal of the propagator or nursery manager is to limit introduction of new pest populations on imported plant materials. Many well-planned plant propagation facilities are designed to provide a quarantine or staging area for incoming plant stock. All new plant material, including cuttings and budwood, is thoroughly inspected in a staging area before it is allowed into greenhouses, poly-houses, or other production areas. Infested plants or plant parts can be returned to the supplier, treated, or discarded before they contaminate the production area.

Hot Water Disinfestations

Cuttings, rooted liners, bulbs or tubers, and even entire flats of plants can be immersed in heated water for brief periods to successfully control all life stages of scale insects, mites, mealybugs, and nematodes. Propagators may be challenged to make this control option practical, however. Not all plants respond similarly to hot water immersion. Plant tissues of many species can be easily injured. Plants should be tested individually before entire blocks of plants are treated. Water-bath temperatures should be closely observed and held constant during submersion. Plants can be submerged in heated water kept at 38°C to 43°C (100°F to 110°F) for durations ranging from 5 to 20 minutes, depending on subsequent plant performance. Many plants respond poorly unless preconditioned in 26°C (78°F) water, and most benefit by being hosed or rinsed off with cool water following submersion.

Employee Work Patterns and Clothing Colors

Routine and consistent monitoring efforts keep nursery managers and propagators aware of pest "hot-spots" and problem areas. Once identified, plants in these production areas should be the last ones visited during daily work routines. Pest populations can be factored into efficient task scheduling to keep employees from spreading pests to uninfested areas. In general, choose neutral tan and beige-colored uniforms, and restrict clothing colors that attract pests, like red, yellow, light blue, and green.

PHYSICAL CONTROL

Screening and Fabrics

Commercially available insect screens and spun bonded fabric barriers exclude arthropod pests from propagation areas. Screen mesh openings as small as 0.15 mm (0.0059 in.) prevent entry of thrips into greenhouses. As mesh openings get smaller, however, airflow is restricted. Powerful (and expensive) fans are needed to move the same volume of air through production areas. Solutions to this challenge include pleated screens with larger surface areas and screened annexes that are built externally around air intake vents. Spun-bonded and woven fabrics are also available that are lightweight and reflective. Fabrics shelter field-grown propagation blocks from pests as well as protect plants from light frost damage.

Barriers and Bands

Several pests can be restricted from greenhouse propagation benches by inverting and securing plastic or aluminum pans where the support post meets the bench. The container interior is coated with Pestik or Tanglefoot insect glue. These barriers are an effective against flightless pests. Snails, slugs, some beetles, aphid-tending ants, and juvenile cockroaches and earwigs, which are flightless when young, are kept below the barrier.

MECHANICAL CONTROL

Thrips, aphid, and mite populations decline following some mechanical actions like regular plant handling and brushing or shaking foliage. Because brushed plants are typically smaller than unbrushed plants, mechanical brushing has also been used as a nonchemical alternative to plant growth regulators. Brushing and handling activity physically disrupt pest feeding activity. Mechanical actions also induce plant stress, which stimulate chemical plant defenses. In limited greenhouse trials, mechanically induced stress has been used as a pest management tactic on some seedling species and rooted cuttings.

BIOLOGICAL CONTROL

Predators and parasites are often efficient foragers for prey arthropods in open environmental systems. If they

TABLE 12.1

Arthropod Pest	Common Name	Scientific Name
Fungus gnats	Bt	*Bacillus thuringiensis* ssp. *israelensis*
	predatory mites	*Hypoaspis miles*
	nematodes	*Steinernema feltiae*
Shore flies	Bt	*Bacillus thuringiensis* ssp. *israelensis*
	predatory mites	*Hypoaspis miles*
	nematodes	*Steinernema feltiae* and spp., *Heterorhabditis* spp.
Leafminers (Diptera)	Parasitic wasps	*Dacnusa* spp., *Diglyphus* spp.
Scales	Predatory ladybeetles	*Chilocorus sp.*, *Rhyzobius* spp.
	Parasitic wasps	*Aphytis* spp., *Encarsia* spp., *Metaphycus* spp.
Mealybugs	Lacewing predators	*Chrysopa* and *Chrysoperla* sp. larvae
	Parasitic wasp	*Leptomastix dactylopii*
Aphids	Parasitic fly	*Aphidoletes aphidimyza*
	Parasitic wasps	*Aphelinus semiflavus*, *Aphidius matricariae*, *Lysiphlebus testaceipes*
	Lacewing predators	*Chrysopa* and *Chrysoperla* sp. larvae
	Predatory bug	*Orius insidiosus*
Thrips	Lacewing predators	*Chrysopa* and *Chrysoperla* sp. larvae
	Minute pirate bug	*Orius insidiosus*
	Predatory bug	*Eusieus* spp.
	Predatory mite	*Macrotracheliella* sp.
	Predatory thrips	*Franklinothrips vespiformis*, *Scolothrips maculatus*
Whiteflies	Predatory ladybeetle	*Delphastus pusillus*
	Parasitic wasps	*Encarsia* spp., *Eretmocerus* spp.
Spider mites	Lacewing predator	*Chrysopa* sp.
	Predatory bug	*Orius insidiosus*
	Predatory mites	*Amblyseius californicus*, *Phytoseiulus persimilis*, and *P. longipes*
	Predatory thrips	*Franklinothrips vespiformis*, *Scolothrips maculates*
	Spider mite destroyer ladybeetle	*Stethorus punctillium*

Note: Many of the most effective natural enemies of propagation pests are mass-reared or cultured for commercial sale. In general, natural enemies sold for pest management have specific prey preferences. Correct pest identification in a propagation system allows IPM practitioners to pair the appropriate parasite, predator, or pathogen with its preferred prey or host.

can be conserved or released into production areas, populations of natural enemies and entomopathogens are also valuable components of propagation IPM. Many of the most effective natural enemies are mass-reared for commercial sale (Table 12.1). Once established, survival of biological control agents is dependent on their successful integration with other management strategies. Long-residual, systemic, and broad-spectrum (general) insecticide and miticide applications should be avoided before and after releasing natural enemies. Fungicidal sprays are toxic to entomopathogenic fungi like *Beauveria bassiana* (Balsamo) Vuillemin and *Metarhizium anisopliae* (Metsch.) Sorokin. Many natural enemies must be supplied with alternative pests, pollen, or nectar resources to survive when their preferred food resources decline.

CHEMICAL CONTROL

Chemical pesticides are a practical component of successful IPM programs. Within an IPM framework, a pesticide is chosen after considering its expected ability to control the pest, the action of the pesticide on predatory and parasitic natural enemies, and the mode of action of its chemical constituents. Emphasis is placed on making spot treatments to centralized pest populations. Spot sprays are used to manage localized populations rather than relying on calendar-based or cover sprays. Spot treatments are seldom effective against general outbreaks that require "curative" action. Retreatment intervals are timed to catch pests that emerge from pesticide-resistant eggs and pupae. The reentry interval, during which unprotected workers are excluded from treated areas following a pesticide application, is also an important consideration in propagation systems. Growers prefer to use pesticides that allow workers to handle plants as quickly as possible after treatment.

MANAGEMENT OF PEST RESURGENCE AND PESTICIDE RESISTANCE

Pest resurgence occurs when a broad-spectrum (nonselective) insecticide is used to control a pest. When natural enemies of a pest are also eliminated, pests that survive treatment or that recolonize the crop develop without pressure from parasites and predators. Subsequent pest populations may be larger than the initial pest population. Efficient pesticide use is accomplished by closely monitoring populations of identified pests. Pesticides that are selective for that pest are applied to the most susceptible life stages of the pest while pest populations are still growing.

Mode of action describes the insect physiological or functional metabolic pathways interrupted by the chemical pesticide. Disrupting these pathways leads to pest mortality. Mode of action is often related to the chemical class that categorizes a pesticide active ingredient. For example, Sevin is a carbamate class insecticide that, like other carbamates, disrupts the nervous system by allowing continuous axon excitation when arthropod pests contact or ingest the active ingredient carbaryl.

In general, pests that develop rapidly and have multiple generations within a season more readily develop resistance to pesticides. Resistance develops through genetic mutations or physiological shifts that allow recessive alleles to be expressed. Resistance is more likely to occur when the same pesticide or mode of action is applied continually to control pests. Pesticide resistance can also be limited by rotating modes of action after several treatments have been applied. Alternatively, strips or groups of untreated plants allow populations of susceptible gene-dominant pests to survive and interbreed with resistant, gene-recessive populations.

PHYTOTOXICITY

Plants in propagation systems are often susceptible to phytotoxicity, or damage from pesticide volatilization, drift, and direct contact with pesticides, which would not have affected the parent plant. Rootlets and leaflets on tissue-propagated microplants are not fully functional until they have been acclimated. Microplants and unrooted cuttings are more sensitive to pesticides than parent plants. Exposure to direct sunlight and temperatures in excess of 30°C (86°F) also increase the severity of plant phytotoxic response. Common symptoms of phytotoxicity include loss of the protective waxy bloom (particularly among conifers treated with horticultural oils and insecticidal soaps), premature anthocyanin production (chlorosis and reddening of foliage), spot chlorosis and necrosis on foliage and flowers, leaf curl, and leaf drop.

TISSUE CULTURE

Tissue culture presents unique challenges for managing pests on plantlets. By volume, very little pesticide product would be purchased by tissue culture propagators. Few, if any, chemical pesticides are labeled for use in tissue culture systems. Thus, propagators have very few chemical options for controlling arthropod pests, mite, thrips, and aphid populations that develop in tissue culture. These pests are most readily managed with sterile techniques and use of only pest-free stock plant material.

HOST PLANT RESISTANCE

Many chemical, morphological, and physiological resistance mechanisms have evolved in plants. Susceptibility to pests and diseases often varies within genera and among cultivars of ornamental plant species. For example, resistance to pests may be attributed to the density or repellency of trichomes on leaves and stems. Pests may simply be unable to reach the leaf surface of some hairy species or cultivars. Plants may also have glandular trichomes that exude chemical byproducts of cellular metabolism. Terpenoids, essential oils, and tannins in plant tissues offer a measure of protection against many sap-feeding and defoliating pests.

Disease- and pest-resistant plants offer advantages to growers and landscape managers. Fewer chemical, labor, and maintenance expenses are needed to produce these crops. Lower-maintenance plants can be developed as a commercial niche and marketed to homeowners. Landscapes designed using pest- and disease-resistant plants are environmentally sustainable. Yet, even resistant plants can be overcome by pest pressure when subjected to water, nutritional, or other stress. The "right plant" must be grown in the "right place," whether in the landscape or the production nursery.

RECORDKEEPING

There are several incentives for growers to maintain thorough and accurate IPM and scouting records. Good records can limit a grower's liability in workplace-related lawsuits. A scout's IPM records provide an invaluable, site-specific training tool for new employees in the event that an IPM scout quits or has his/her employment terminated. Records will provide an easy method of tracking the type, rate, and volume of pesticides applied. Perhaps most importantly, a scout's records will reveal the location of pest and disease "hot spots." Hot spots are characteristic of even the best-run operations. Once identified,

TABLE 12.2

Date: _____ Name of Scout: _____
Stock Block (or area of nursery): _____
Time In/Time Out: _____ / _____
Host Plant (with cultivar name): _____ Plant Size: _____
Plant Stage (dormant, budding, flowering, leafing out, fruiting, mature): _____

Type of pest/natural enemy	Stage(egg/larvae/nymph/adult)	Number observed	Location	Comment

Damaged Area (Percent of stem, bark, flowers, leaves): _____
Number of Damaged Plants: _____
Natural Enemies (None, Few, Common, Abundant): _____
General Comments: _____

Recommended Management Action: _____

Pesticide/Date/Rate/Volume Applied: _____
Date of Follow-Up Evaluation: _____
Success of Treatment/Comments: _____

Note: A sample sheet from a scout's notebook should provide space for thorough records. List the pests and natural enemies that were observed, the number and type of host plants (and what parts) were affected, and what management actions are recommended. Specify the pesticide and date applied, and the rate that was used, and the volume of material applied. Additional space should be provided to allow follow-up observations and future recommendations for managing the pest.

however, scouts benefit by using these areas as monitoring sites that announce early pest outbreaks.

In the field, scouting reports should be entered into a database or made in indelible ink or pencil. The report log can consist of individual pages that are designated by date or production area. Pages should be maintained and filed in a logical order with the expectation that they will become a useful management resource (Table 12.2).

ASSESSMENT OF MANAGEMENT ACTIONS

A propagation IPM program is incomplete without the evaluation of management tactics that were implemented. Each decision should be reevaluated, based on aesthetic and economic satisfaction with management results. The final assessment should be recorded and filed, along with any associated observations (e.g., weather, new plant/pest introductions, changes in staff, new equipment, etc.) that were expected to have influenced the success or failure of the management action. Invariably, these notes will become a more valuable resource as the IPM program matures. Importantly, maintenance of accurate and thorough records is prudent insurance against the loss or relocation of employees who have been responsible for implementing the program. A newly hired scout or IPM manager should use past records as a practical training guide.

SUGGESTED READING

Ali, A.D. 1987. Integrated arthropod management in plant tissue culture production. *Comb. Proc. Intl. Plant Prop. Soc.* 37: 104–106.

Daar, S., H. Olkowski, and W. Olkowski. 1992. *IPM Training Manual for Wholesale Nursery Growers.* Bio-Integral Resource Center, Berkeley, CA, and Wholesale Nursery Growers of America, Washington, D.C. 84 pp.

Dreistadt, S.H. 2001. *Integrated Pest Management for Floriculture and Nurseries.* Publ. 3402. University of California Press, Berkeley. 422 pp.

Flint, M.L. 1992. Resources for establishing an IPM program for ornamental plants. *Comb. Proc. Intl. Plant Prop. Soc.* 42: 251–255.

Flint, M.L. and S.H. Dreistadt. 1998. *Natural Enemies Handbook.* University of California Press, Berkeley. 154 pp.

James, R.L., R.K. Dumroese, and D.L. Wenny. 1995. *Botrytis cinerea* carried by adult fungus gnats (Diptera: Sciaridae) in container nurseries. *Tree Planters' Notes* 46: 48–53.

Johnson, W.T. and H.H. Lyon. 1991. *Insects That Feed on Trees and Shrubs,* 2nd ed. Cornell University Press, Ithaca, NY. 560 pp.

Latimer, J.G. and R.D. Oetting. 1994. Brushing reduces thrips and aphid populations on some greenhouse-grown vegetable transplants. *HortScience* 29: 1279–1281.

Parella, M.P. 1992. On overview of integrated pest management for plant propagation. *Comb. Proc. Intl. Plant Prop. Soc.* 42: 242–245.

Rosetta, R. 1995. How to use biological control to manage propagation pests. *Comb. Proc. Intl. Plant Prop. Soc.* 45: 272–275.

Schoonhoven, L.M., T. Jermy, and J.J.A. van Loon. 1998. *Insect-Plant Biology: From Physiology to Evolution.* Chapman and Hall, London. 409 pp.

Spooner-Hart, R.N. 1988. Integrated pest management with reference to plant propagation. *Comb. Proc. Intl. Plant Prop. Soc.* 38: 119–125.

Zagory, E.M. and R. Rosetta. 1992. Making the change to integrated pest management. *Comb. Proc. Intl. Plant Prop. Soc.* 42: 246–250.

Part V

Evaluation of Propagated Plants

13 Evaluation of Data from Propagation Experiments

Michael E. Compton

CHAPTER 13 CONCEPTS

- The most commonly used form of hypothesis for plant propagation research is the "null hypothesis." This popular hypothesis states that all treatments will elicit a similar response.

- An experimental unit (EU) is the smallest unit that receives a single treatment. The observational unit (OU) is the object being observed or measured. However, a cutting may be considered as both the EU and OU in cases in which only one cutting is stuck per pot.

- The completely randomized, randomized complete block, incomplete block, and split plot designs are most commonly used for plant propagation research. These designs differ in their randomization schemes and experimental requirements.

- Many researchers find it beneficial to culture multiple explants in a culture vessel. This conserves resources and time, but leads to a situation in which multiple measurements are made for each replicate (e.g., culture vessel). Statistically, this is referred to as subsampling or repeated measures.

- The nature of the data influences the type of statistical procedures that should be used. Most observations made in plant propagation experiments result in data that can be classified as continuous (shoot length, root length, etc.), counts [number of organs per explant or callus (shoots, embryos or protocorm-like bodies)], binomial [response or no response (e.g., percentages)], or multinomial (shoots, roots, leaves, or no response). Therefore, the statistical methods used to evaluate the treatment effects vary for each type of data observation.

Efficient and reliable propagation practices are necessary to supply the millions of plants required annually by the green industry. Practices used by nurseries and greenhouses were developed through research projects aimed at pursuing ideas that optimized propagation and/or solved specific problems associated with propagating plants. From these ideas, planned experiments were developed in which treatments were devised and tested in controlled conditions under which plants were typically propagated. Data gathered from observations made during experimentation were evaluated using statistical analysis to ascertain the effectiveness of each treatment at improving plant propagation rates or correcting specific problems associated with propagation.

The purpose of this chapter is to introduce you to the basic elements of plant propagation research and data analysis. After reading this chapter you should understand the basics of the scientific approach and be able to select and apply experimental designs and statistical methods to analyze, interpret, and present plant propagation data.

METHODOLOGIES ASSOCIATED WITH EVALUATING DATA

Organization is an important part of conducting research. Thoughts and materials must be organized into a logical progression of events that can be applied in experimental conditions. This section focuses on the factors and events important for conducting research and provides examples of how these items can be organized to improve research effectiveness.

ESTABLISHING A HYPOTHESIS AND EXPERIMENTAL OBJECTIVES

Ideas aimed at optimizing plant propagation need to be formulated into a hypothesis that can be tested. When formulating a hypothesis, it is important to examine published literature to learn from peers that have conducted similar research. Your hypothesis can be refined, if necessary, and a set of experimental objectives developed.

The most commonly used hypothesis for propagation research is the "null hypothesis." This popular hypothesis states that all treatments will elicit a similar response among all propagules. Application of the null hypothesis to plant propagation research can be illustrated by using an example in which one concentration of four rooting factors [indole-3-butyric acid (IBA), K-IBA, naphthaleneacetic acid (NAA), and a solvent control with no rooting factor] were tested to evaluate their ability to stimulate adventitious root formation in softwood cuttings of Koreanspice viburnum (*Viburnum carlesii* Hemsl.). In this case, the null hypothesis would be stated as follows—softwood cuttings of Koreanspice viburnum will root similarly when treated with equal concentrations of IBA, K-IBA, or NAA, or a solvent (50% ethanol) control. In this case, the control is added to ensure that the results obtained were due to the rooting factor treatments and not the solvent or an internal factor.

Selecting the Components of the Experiment

The treatments and supportive materials required to conduct the experiment are selected following the development of a hypothesis and experimental objectives. Treatments are usually referred to as any condition(s) that characterize the population of interest and whose effect on the population can be measured and tested (Lentner and Bishop, 1986). In asexual propagation experiments, treatments may be rooting compounds, as demonstrated in the previous paragraph, or include environmental factors, such as misting frequency and duration, light level and photoperiod, temperature and humidity, and rooting media or physiological factors, such as stage of development (softwood versus semihardwood versus hardwood), developmental phase (mature versus juvenile), etiolation (banding), degree of wounding, or any other treatment that influences stock plant or cutting development. Researchers may choose to examine one treatment factor at a time or test several factors simultaneously [e.g., stock plant genotype and plant growth regulator (PGR) concentration]. Regardless of the number of treatments selected, it is important that they be chosen according to the experimental objectives and evoke a wide range of responses so that an optimum level can be easily identified.

Supportive materials are not treatments, but are items found under the conditions in which the problem is observed. In asexual propagation studies, these materials may be mist irrigation, rooting substrate, growth regulator type and concentration, growth environment (light level and photoperiod), or any item important for satisfactory development of cuttings. Supportive materials should not interfere with the treatment effects.

The experimental and observational units are chosen once the treatments and supportive materials are selected. By definition, an experimental unit (EU), or replicate, is the smallest unit that receives a single treatment, whereas an observational unit (OU) is the object being observed or measured (Lentner and Bishop, 1986). A replicate may be a container housing several cuttings to which one treatment is applied, or individual cuttings stuck in a container to which all treatments are applied. In the latter situation, each propagule would be considered as both the EU and OU. During an experiment each OU is observed and its reaction to the experimental treatment(s) recorded.

Replication occurs when more than one EU is evaluated for each treatment. Replication is necessary in research so that an accurate estimate of treatment effects can be obtained. The number of replicates required per treatment is based on the amount of variability expected to occur in the dataset and the degree to which the researcher wishes to detect treatment effects (Zar, 1984). At least three replicates are required per treatment to calculate means and measure error. Most researchers rely on prior experience, review of literature, and speculation to determine the number of replicates to be used in an experiment. However, insufficient replication can reduce the ability of statistical procedures to detect treatment differences (Lentner and Bishop, 1986). To provide assistance in determining replicate numbers, Kempthorne (1973) provides a formula that uses information from previous studies to estimate the number of replicates required per treatment.

Experimental designs are used to arrange replicates and OUs in a specific randomization scheme to measure variation associated with experimental treatments and supportive materials as well as nontreatment factors that might influence the outcome of the experiment. Randomization also reduces experimenter bias by assigning treatments in a planned random, fair, and unbiased manner. It is important to examine the supporting materials and population of interest for possible variation. The experimental design should be chosen after the EUs and OUs have been selected and should be simple to employ, efficient in regard to measuring treatment effects and residuals, and should match the experimental objectives (Compton and Mize, 1999). Typical experimental designs include the completely randomized (CR) and randomized complete block (RCB) designs. These designs differ in their randomization schemes and experimental requirements.

The completely randomized design is most commonly used for propagation research because it is easy to use and does not use a patterned randomization scheme. This allows researchers to maximize the number of replicates used and employ equal or unequal treatment replication (Little and Hills, 1978). This is important because unequal replicate numbers often occur in propagation studies due to death of propagules from nontreatment factors, such as mist failure or inaccurate mist coverage, insufficient number of available cuttings, or other similar factors not associated with the treatment.

Evaluation of Data from Propagation Experiments

Completely Randomized

①	②	④	②	③	④	④	①	③	④
④	③	②	①	④	①	②	③	①	②
②	①	①	③	④	②	③	④	③	③
③	④	①	②	②	③	①	②	④	①

(a)

Randomized Complete Block

1	2	3	4	5	6	7	8	9	10
①	②	④	③	③	②	①	④	③	②
②	④	③	①	①	④	④	③	①	③
③	①	②	④	②	③	②	①	④	④
④	③	①	②	④	①	③	②	②	①

(b)

FIGURE 13.1 Randomization schemes for a completely randomized (a) and randomized complete block (b) designs with four treatments. Each treatment was replicated 10 times in each design.

Randomization of replicates in the CR design can be illustrated using an example in which the effectiveness of four rooting factors on adventitious root formation on softwood cuttings of Koreanspice viburnum was examined. Cuttings were dipped in a 48 mM aqueous solution of one of three rooting factors (IBA, K-IBA, or NAA) for 30 sec. before being placed in a mist bed containing sand. Control cuttings were dipped in 50% ethanol (solvent for the auxins) for the same duration. Cuttings were misted with reverse osmosis water for 10 sec. duration on 10 min. intervals. Application of mist was controlled by a Gemini6 control unit with a photoelectric cell. The randomization of replicates for this experiment would appear as shown in Figure 13.1a.

Several techniques can be used to assign propagules to treatments. It is usually best to prepare and sort cuttings into groups of similar size and quality before assigning them to treatments. Sorted cuttings from each size group are selected at random and the treatment applied. A plan for randomization of treated cuttings in the propagation area should be established beforehand by drawing numbers representing treatment replicates from a hat and diagramming the results on paper. However, most statistical analysis software packages can be used to establish the randomization scheme.

From a statistical perspective, the CR design is most efficient because it allows researchers to maximize degrees of freedom (df) for error, which is important for detecting treatment differences during statistical analysis (Little and Hills, 1978). However, for the CR design to be effective, all replicates should be as homogeneous as possible. Situations in which EUs or OUs are highly heterogeneous increase variability in the experiment and decrease the likelihood that treatment differences will be detected (Little and Hills, 1978).

Designs that employ blocking are used when a high degree of heterogeneity exists among the EUs. Unevenness or inconsistency among environmental factors in the greenhouse, differences among stock plants from which propagules are obtained, differences among seasons in which cuttings are harvested, variability among technicians, as well as differences among batches of cuttings, are a few nontreatment factors that introduce heterogeneity.

When using the randomized complete block design, replicates of each treatment are assembled into blocks. This creates a high level of uniformity among treatments within a block and reduces variation not associated with a treatment (Little and Hills, 1978). When using blocking, it is important that only measurable factors are blocked and that all propagules in a block are as uniform as possible. For example, propagules from the same stock plant often have similar rhizogenic competence and can be used for treatments that appear in the same block.

The viburnum study could be used to illustrate the RCB design by using propagules from one stock plant to establish replicates of each treatment in a block. In this case, 10 stock plants would be required to produce enough blocks to meet the requirement of 10 replicates per treatment (Figure 13.1b). Using the RCB in this situation would maximize uniformity within blocks by utilizing the homogeneity present within each stock plant.

It is impossible to be absolutely sure if blocking will be required when designing experiments. The need for blocking can be determined by testing for blocking when conducting the statistical analysis. Kempthorne (1973) outlines a simple method for determining if blocking is necessary.

Determining How Many Times an Experiment Should Be Conducted

Most statisticians and researchers believe that an experiment must be repeated to validate the results (Lentner and Bishop, 1986). However, repeating experiments does not guarantee a similar outcome even if the same stock plants are used, as rhizogenesis is often influenced by the stock plant environment as well as endogenous physiological factors. Repeating experiments may introduce time as a nontreatment factor if changes in stock plant physiology have occurred with increased stock plant age or time in the growing season. This can lead to differences in experimental results. Researchers generally have two options when differences between runs in an experiment are obtained. The entire experiment can be repeated, which can be costly, or each run may be examined as a block or main plot. The best option would be to first look at the data from the two runs and determine if similar trends exist. If the trends are similar, the experiment should not be repeated. However, the circumstance(s) that caused the difference(s) should be examined and explained. If the results obtained in the two runs contrast, the experiment should be repeated. If differences among runs persist, the possibility of seasonal effects on the results should be examined experimentally. A decision should be made during the planning phase regarding how many times the experiment should be conducted and what will be done if differences between runs of the experiment are observed.

Choosing the Best Data and Planning the Statistical Analysis

The data recorded should help evaluate treatment effects in accordance to the experimental objectives. Recording the wrong data or failing to record important parameters will reduce the likelihood that the researcher will be able to properly assess propagule response. Most researchers review relevant literature and combine the information learned with previous experience when deciding which observations to measure in their experiments. In the viburnum illustrations used throughout this chapter, the number and percentage of propagules producing roots, the number of roots per propagule, root dry weight, and root length are observations that most researchers would use to evaluate propagule competence. However, to get the most information from the project, the researcher may wish to make additional observations, such as a rating of cutting vigor, shoot weight, and the number of and percentage of propagules that produced callus. It is usually a good idea to record as much relevant data as possible and chose the best parameters when presenting results.

CONDUCTING THE EXPERIMENT

Experimentation may begin once planning is completed and thoroughly reviewed. Personal bias must be avoided during all phases of the experiment to avoid swaying the outcome. This can be accomplished by using the structure of the experimental design to set up the experiment (Mize et al., 1999). Do not prepare all propagules for one treatment at a time. Instead, you should prepare the cuttings and sort them into groups of like sizes or vigor ratings before applying the treatments.

It is also important to avoid circumstances, such as fatigue, that introduce error into the experiment. The time required for each phase of the experiment should be estimated during planning and tasks that are taxing should be staggered to avoid worker strain. Mistakes may also occur when propagules are stuck into the rooting substrate or labeled. Be careful to label each cutting or replicate with the correct information. If using codes, be sure to record keys to your codes in your laboratory notebook. There are many avenues for introducing operator error into your experiment, so it is important to take the precautions necessary to encourage accurate and safe experimentation.

Recording Data

Observations made during an experiment are recorded on data sheets. This activity can lead to mistakes if the sheets are not well organized. Designing data sheets in the manner that each datum will be entered into the computer helps to avoid mistakes during transcription. Many statistical software packages will produce data sheets for you. If not, data sheets should be designed during planning to ensure that the researcher is considering observations that measure treatment effects.

As when starting an experiment, you should use the structure of the experiment when recording observations. Observations of replicates in one block should be recorded before moving to the next block. This helps to avoid personal bias. Be meticulous and take your time. Sloppy technique introduces errors and inconsistency. Take breaks to avoid fatigue when recording data. When entering data on the computer, it is important to save your work frequently. In addition, be sure to save several electronic copies of your files and distribute them into several secure locations to avoid losing valuable information. One year, a colleague's laboratory was destroyed in a fire, and over 26 years of valuable data were lost.

Data should be checked for errors after they are recorded. Erroneous data entries lead to incorrect treatment evaluation and interpretation (Mize et al., 1999). Make sure that data of each treatment fall within the expected boundaries. Data outside these boundaries are called outliers and may result from errors in reading, recording, or transcribing data, or may be values obtained from plant tissues with inherent problems or genetic mutations. Outliers are often discarded. However, researchers should exercise caution when deleting data, as personal bias may be unintentionally introduced and distort the results. Several statistical procedures can be used to detect outliers and are helpful to reduce bias when editing (Barnett and Lewis, 1984). Treatment codes should be examined while checking data. Do not assume that codes were entered correctly. In the viburnum example, the four treatments were identified by assigning them numbers 1 through 4 (e.g., IBA was assigned treatment number 1, K-IBA the number 2, NAA was number 3, and the control treatment assigned the number 4). Errors associated with entering codes can be identified by calculating the mean for the coded variables (Mize et al., 1999). In this example, obtaining a mean of 2.5 for the rooting factor treatment identification codes indicates that the numbers were entered correctly. Editing data can be facilitated by printing data sheets and thoroughly reviewing them. This can be a time-consuming task but is worth the effort.

DATA ANALYSIS AND INTERPRETATION

Data must be analyzed statistically and the results interpreted as planned. The nature of the data influences the type of statistical procedure that should be used. Most observations recorded in propagation experiments result in data that can be classified as counts (number of roots per cutting), binomial (response or no response, e.g., percentages), multinomial (root, callus, or no response), or continuous (root length) (Mize et al., 1999). The statistical methods used to evaluate treatment effects vary for each of these types of data observations. Outlining the sources of variation and degrees of freedom during planning can help identify statistical procedures most appropriate for the type of data recorded as well as the experimental objectives. In analyzing data, a general analysis is conducted first to determine if there are any differences between the treatment levels. Further analysis of treatment means is conducted if differences are detected by the general analysis (Compton, 1994; Mize et al., 1999).

The previously described example in which rooting of *Viburnum carlesii* softwood cuttings was examined following treatment of cuttings with three rooting factors will be used to discuss the common types of general analyses. To understand how each statistical analysis applies to the experiment, it is important to know the methods and materials used. Softwood cuttings (about 15 cm long) containing five to six nodes were collected on two separate dates, two weeks apart, from field-grown plants in July. Each collection date represented a replication of the experiment in time. Since the stock plants themselves were all the same clone, it was decided that cuttings from all plants could be pooled together. All cuttings were sorted and only the most vigorous propagules used for the experiment. The stems of all remaining cuttings were trimmed to just below the fourth node from the apex, with leaves from the lower two nodes removed before being dipped into a solution containing 48 mM concentration of one of three rooting factors (IBA, K-IBA, or NAA dissolved in 50% ethanol) or a 50% ethanol control for 30 sec. Following treatment, cuttings were placed in 5 cm (high) × 27.5 cm (wide) × 55 cm (long) containers filled with coarse sand and incubated in the greenhouse in diffused light (maximum 250 μM m^{-2}·sec^{-1}) and natural photoperiod. Cuttings were arranged within the containers in a completely randomized design with 10 propagules per treatment. Cuttings were irrigated using mist applied through flora-mist nozzles (15 L/h) at a duration of 8 sec. every 10 min. Misting commenced one hour before sunrise and concluded one hour after sunset. Rooting of treated cuttings was observed for eight weeks, after which time the number of cuttings with roots (binomial), the number of roots per cutting (count), and root length (continuous) were recorded.

ANALYZING RESPONSE/PERCENTAGE DATA

Response data, the ability of propagules to survive and produce roots, are important to evaluate the influence of treatments on rhizogenesis. These data have a binomial distribution, which causes treatment variances to be dependent on cutting performance (Lentner and Bishop, 1986; Mize et al., 1999). Percentages calculated from observations are often considered important because researchers are interested in identifying treatments that optimize propagule response and propagation efficiency. This can translate directly to increased profits. Logistic regression is best suited for analyzing response data because the procedure calculates a special coefficient and standard error value that is not dependent on cutting performance (Mize et al., 1999).

In the viburnum example, data on the percentage of cuttings that produced roots were binomially distributed, with datum being recorded as a zero (no roots/response) or 1 (roots observed). Analyzing these data using logistic regression revealed that the percentage of propagules that produced shoots was influenced by the rooting factor used (Table 13.1). The p value (level of significance) was obtained by comparing the calculated statistic with a known set of values. Your computer software will calculate the logistic regression coefficients and generate a p value, and p values equal to or less than 0.05 are considered significant. This level is chosen by most researchers because 0.05 indicates that a similar result would be obtained 95% of the time that the experiment is conducted, signifying a high level of confidence for the outcome. Another reason for selecting 0.05 is that most journals require researchers to test at this level of significance. In our experiment, a significant p value of 0.0001 for rooting factor was obtained, indicating that there were significant differences among the treatments, and there is a 99.99% probability that we would observe a similar response the next time we conduct the experiment. This is important when considering that there may be nurseries or greenhouses throughout the world that will be interested in using our technique.

Before analyzing response data, it is important to determine if there are treatments in which the response did not vary. This usually occurs when all or none of the

TABLE 13.1
Logistic Regression Analysis Table Demonstrating the Influence of Three Rooting Factors and a Control on the Percentage of *Viburnum carlesii* Softwood Cuttings That Produced Roots[a]

Predictor Variables	Logistic Regression Coefficient (A)	Standard Error (B)	Logistic Regression Statistic (A/B)	*p* Value
Constant	−2.579	0.6876	−3.75	0.0002
Rooting factor	1.032	0.2551	4.04	0.0001

[a] Data are hypothetical.

propagules responded. This occurred in our control treatment in which none of the propagules produced roots. Researchers have two options when this occurs. Data in treatments in which no response or a 100% response was observed can be altered before analysis (Mize et al., 1999). Zeros may be changed to a slightly higher value (0.000001) and 100% values reduced slightly (0.999999). On the other hand, data in treatments with an all-or-nothing response may be discarded. It is important to decide during planning how these values will be handled if this situation occurs. In addition, you should also indicate if treatments were dropped or values altered when writing reports and manuscripts. In this example, none of the control cuttings responded, and it was decided to alter the zero values obtained in the control treatment to 0.000001 before analyzing the data.

Because of its popularity and ease of use, many researchers wish to use analysis of variance (ANOVA) for binary data. This practice may lead to erroneous interpretations because percentage data, or proportions, generally are binomially distributed (Compton, 1994; Fernandez, 1992; Finney, 1989; Zar, 1984). In this situation, treatment variances are dependent upon the success of the treatment in stimulating a response (Mize et al., 1999). Treatments with poor success percentages (0% to 30%) often have small variances, whereas those that stimulate a high success rate (70% to 100%) have large variances. Treatments in which the response rate is between 30% and 70% generally have more normal variances compared to treatments with greater or smaller success rates. Because ANOVA produces separate estimates for treatments and experimental error and because the standard errors for treatment parameters are a function of success rate, use of ANOVA for datasets containing treatment parameters with poor or high success rates often leads to a poor estimate of treatment differences by ANOVA (Mize et al., 1999).

However, ANOVA can be used for analyzing response data, provided that the data are transformed before analysis (Compton, 1994). The arcsin transformation is generally used before analyzing response data with ANOVA. The equation $P' = \arcsin(P)^{0.5}$ (P equals the original percentage value) is commonly used to transform response data (Bartlett, 1937; Finney, 1989; Zar, 1984). However, when values of 0% or 100% occur, the formula $\arcsin(1/4n)$ and $\arcsin[100 - (1/4n)]$ should be substituted, respectively. This transformation helps to correct some of the distribution problems associated with response data, allowing ANOVA to work more accurately. The arcsin transformation is not required when treatment values fall between 30% and 70%, as their distribution is usually more normal (Fernandez, 1992). It has been observed that results of ANOVA and logistic regression are similar in the latter situation (Mize et al., 1999).

Data that have been transformed should be expressed in the original units when presented in tables and graphs, and the results of the statistical analyses applied to the original data. This is done because means of transformed data provide little information to the reader (Compton, 1994). Zar (1984) instructs that data transformed by arcsin can be converted back to the original scale by using the formula $P' = (\sin P')2$, where P' equals the arcsin value and P' is the new data value. In either case, it is important to explain to the readers, in your methods description, how your data were transformed, analyzed, and converted back to the original scale.

Count Data

Count data are not normally distributed because treatment variances are equal to the average response of the treatments (Zar, 1984). Because of their distribution, count data (number of roots per cutting) should be analyzed using Poisson regression since the procedure uses a logarithmic value of the mean counts, making the data more normalized (Mize et al., 1999). Poisson regression calculates a coefficient that is divided by the standard error to determine if the model is significant. In the viburnum example, observations on the number of roots per cutting is considered a count variable. Because none of the control cuttings responded, there are no data values for the number of roots per cutting among propagules in this treatment. It

TABLE 13.2
Poisson Regression Summary Table for the Number of Roots per *Viburnum carlesii* Softwood Cuttings Treated with Three Rooting Factors[a]

Predictor Variables	Poisson Regression Coefficient (A)	Standard Error (B)	Poisson Regression Statistic (A/B)	p Value
Constant	0.1247	0.293	0.43	0.6703
Rooting factor	0.3527	0.088	4.00	0.0001

[a] Data are hypothetical.

is not correct to place zero values for the number of roots per cutting because doing so would skew the dataset and influence our interpretation. Therefore, only data for cuttings treated with the rooting factor that responded were analyzed. In our case, Poisson regression indicated that the rooting factors influenced the number of roots per cutting ($p = 0.0001$; Table 13.2).

As stated in the section discussing response data, many researchers often wish to use ANOVA to analyze count data. Mize et al. (1999) stated that ANOVA is unreliable when count values are less than 10. In this case, Poisson regression yields more accurate results than ANOVA. However, ANOVA may be used as long as the data are transformed using a square-root transformation prior to analysis (Compton, 1994).

ANALYSIS OF CONTINUOUS DATA

Continuous data are more normally distributed than binomial and count data, having treatments with similar variances that are well suited for statistical analysis using ANOVA (Mize et al., 1999). During ANOVA, the value of each observation is subtracted from the overall mean, and the differences between each datum and the overall mean (considered random error or residual) are used in the analysis (Mize et al., 1999). A model statement is created and tested that identifies the dependent variables (observations recorded) and independent variables (treatments, treatment interactions) based on the experimental design used. A summary table is generated during ANOVA that provides the results of the model tested (Lentner and Bishop, 1986). Information in the summary table includes the degrees of freedom (df), sources of variation (SS and MS), mean square error (MSE), F-statistic (F) and p value. In ANOVA, the F-statistic is obtained by dividing the MS for treatment by the MSE, and the resulting value is used to determine if the hypothesis is rejected or accepted (Lentner and Bishop, 1986).

In the viburnum example, the root length variable is considered continuous. Since a CR design was used in this study, rooting factor and experimental error were identified as sources of variation (Table 13.3). Because cuttings receiving the control treatment did not produce roots, and entering zero values would skew the dataset, values for this treatment were not included in the analysis. Likewise, values for cuttings that failed to root in the rooting factor treatments were not analyzed. According to the results, root length was influenced by the type of rooting factor used ($p = 0.0001$; Table 13.3). This was determined by looking at the F and p values for the rooting factor variable. In our ANOVA, the F-statistic (181.46) was calculated by dividing the MS for rooting factor (1530.7) by the MSE (8.43561). The MS value for rooting factor reflects the degree of variation associated with the source. The MSE measures variation from any other nontreatment sources. The p value (level of significance for the F) can be obtained by comparing the calculated F against standardized values for ANOVA found in a statistical textbook.

TABLE 13.3
Analysis of Variance Summary Table for the Number of Roots Formed per *Viburnum carlesii* Softwood Cuttings Treated with 48 mM IBA, K-IBA, or NAA

Source	DF	SS	MS	F	p
Rooting factor	2	3061.4	1530.7	181.46	0.0001
Experimental error	38	320.553	8.43561		
Total	40	3381.95			
Contrast					
NAA versus IBA derivatives	1	4040.6	4040.6	478.99	0.00001
IBA versus K-IBA	1	37.604	37.604	4.458	0.08

Note: Propagules (softwood cuttings) were prepared from plants grown in the nursery. Excised softwood cuttings were trimmed to 15 cm and their basal 3 cm dipped for 30 sec. in a 50% ethanol solution containing 48 mM IBA, K-IBA, or NAA, or a 50% ethanol control. There were 10 propagules per treatment, and the experiment was completed twice. Cuttings were arranged in a completely randomized design. Data for the controls as well as data for nonresponding cuttings in the auxin treatments were deleted before analysis. Data for runs were combined, since they were not significantly different. Data are hypothetical.

ANOVA is preferred by most researchers because it is easy to perform, can be used to evaluate data obtained from virtually all experimental designs, and generates a mean squared error (MSE) value that is considered to be the best estimate of experimental error (Mize et al., 1999). ANOVA provides a good overall analysis because it accurately measures how treatments relate to each other when used in the proper circumstances.

Methods for Comparing Treatment Means

One of the main objectives of propagation research is to identify treatments that optimize propagule response. To fulfill this objective, it is necessary for researchers to apply a method(s) to determine significant differences among the treatments applied. In experiments in which only two treatments are tested, the general analysis provides enough information about treatment differences. A significant p value indicates that the two treatments differ, and the researcher can calculate the mean for each treatment and declare the treatment with the greater mean to be the best. However, most researchers design experiments that compare more than two treatments, and in this situation the general analysis does not provide enough specific information about how the treatments differ or which treatment(s) is(are) best, requiring further analysis of treatment means. The easiest way to compare treatment means is to rank them in ascending or descending order and pick the best treatment. However, this method does not consider within-treatment variation and does not account for variation in how the propagules responded to the treatments. Failure to consider variation among cuttings within a treatment and between treatments increases the likelihood that the researcher will make a mistake when interpreting the results (Mize et al, 1999). There are many mean separation procedures that account for within-treatment variation that can be used for evaluating data from propagation experiments.

Standard Error of the Mean

The standard error (SE) of the mean is commonly used by researchers to compare treatment means. An SE value is easy to calculate and is derived using the sample standard deviation (Zar, 1984). Most researchers use SE values for mean separation purposes by presenting the treatment means and respective SE together and the difference between paired values calculated. Treatments are declared different if their collective values do not overlap. Problems occur when standard error terms are used to compare the means of treatments ranked far apart. Treatments with small values have a reduced SE when compared to treatments with larger values, violating the assumption of ANOVA that treatment variances are equal. This often leads to interpretations that do not accurately reflect population variance. To yield more realistic results when using SE values to compare treatment means, one SE value should be calculated from the ANOVA MSE or SE values obtained from Poisson or logistic regression and used to compare adjacent means (Mize et al., 1999). Most statisticians believe that use of SE to compare treatment means yields poor information compared to mean comparison procedures that use the sample variance when calculating the mean comparison statistic.

Multiple Comparison and Multiple Range Tests

Multiple comparison and multiple range tests are statistical procedures that use the population variance to calculate a numerical value for comparing treatment means. Means are ranked, and the difference between the compared means is calculated. The calculated difference between means is compared with the critical value computed by the mean comparison statistic. If the difference between the compared means exceeds the computed statistical value, the treatments are considered statistically different. However, if the difference between treatment means is equal to or less than the computed statistical value, the treatments are considered similar (Mize et al., 1999). When presenting means in a table or graph, means designated as different are assigned different letters, whereas treatments declared similar are assigned the same letter (Table 13.4).

In the viburnum example, Bonferroni's, Tukey's honestly significant difference (Tukey's HSD), and least significant difference (LSD) were used to compare the effect of different rooting factors on rhizogenesis (Table 13.4).

TABLE 13.4

Comparison of Results Obtained from Analyzing Treatment Means Using Least Significant Difference (LSD) and Tukey's Honestly Significant Difference (Tukey's HSD) Mean Comparison Tests[a]

Auxin Type	Root Length (mm)	Results of Mean Comparison Tests		
		Bonferroni[b]	Tukey's[c]	LSD[d]
IBA	27.5	a	a	a
K-IBA	25.3	a	a	b
NAA	6.9	b	b	c

[a] Data were analyzed following ANOVA for a single-factor completely randomized design. Data are hypothetical.
[b] Means with the same letter are not significantly different according to Bonferroni's test at 0.05.
[c] Means with the same letter are not significantly different according to Tukey's HSD at 0.05.
[d] Means with the same letter are not significantly different according to LSD test at 0.05.

Evaluation of Data from Propagation Experiments

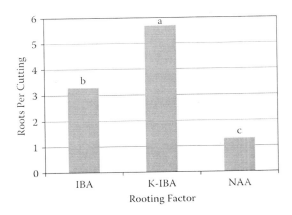

FIGURE 13.2 Bar graph demonstrating the use of lower case letters to illustrate differences among rooting factor treatments as assigned by Bonferroni's multiple comparison test. Bars with different letters are considered statistically different at the 0.05 level.

Normally, only one comparison test is used to compare treatment means. However, three are used here to demonstrate differences due to the formulae used to calculate statistical values. Means of each treatment were ranked from highest to lowest and compared. Different results were obtained from the three tests. Results of LSD suggested that the cuttings responded differently to each of the three rooting factor treatments. However, Bonferroni's and Tukey's HSD differed from LSD, indicating that root length of cuttings treated with IBA and K-IBA was similar and collectively greater that cuttings treated with NAA. When looking at the treatment means, a difference of 1.8 mm in root length between the K-IBA and IBA treatments is probably not practically different and probably would not impact establishment of rooted cuttings in containers or the field. It is the opinion of many statisticians that use of LSD for mean comparisons should be limited and not used unless treatments are considered different in a general test (ANOVA, etc.). LSD is considered by many statisticians to be the most liberal mean comparison procedure, i.e., most likely to declare means different. This often leads researchers to declare means different that are similar, which occurred in this example. Bonferroni's and Tukey's HSD are considered moderately conservative tests and yield more accurate results. Results of mean comparison tests should be presented in either tables or bar graphs. If using bar graphs, letters assigned to specific treatments should be positioned above each bar (Figure 13.2).

Multiple comparison and multiple range tests should only be used when treatments are unrelated (Lentner and Bishop, 1986). In propagation studies, these are treatments in which different species, substrate media, or unrelated rooting factors are tested. These mean comparison procedures should not be used when treatments are related, e.g., different concentrations of the same rooting factor

(Compton, 1994; Mize et al., 1999). Multiple comparisons and multiple range tests can be used for count and binomial data, but the data must be normalized first using square root (count data) or arcsin (binomial data) transformation procedures (Compton, 1994). Once results of transformed data are calculated, the data are converted back to the original scale for presentation (Zarr, 1984) and letters designations obtained using transformed data assigned to the original treatment means.

ORTHOGONAL CONTRASTS

Orthogonal contrasts are used to compare treatments, or groups of treatments, with similar characteristics (Compton, 1994; Mize et al., 1999). In propagation studies, these may be plant growth regulators or rooting factors with similar activity. Contrast statements are usually performed as part of ANOVA, but the number of comparisons is restricted to the number of df for the treatment variable (Lentner and Bishop, 1986). A contrast statement is written that specifies the treatments or group of treatments to be compared, and ANOVA calculates the df, SS, and MS for the comparison. An F-statistic is calculated by dividing the MS for the contrasts by the MSE and the level of significance (p value) determined. One df is used for each comparison.

In the viburnum example, two contrasts of interest were planned before the experiment—(1) IBA and K-IBA versus NAA and (2) IBA versus K-IBA—for the root length variable. Results of orthogonal contrasts are usually presented in a table containing ANOVA computations (see Table 13.3). Interpreting the results of the contrasts for root length indicated that cuttings that rooted following treatment with 48 mM IBA or K-IBA produced longer roots than those that rooted following treatment with a similar concentration of NAA ($p = 0.0001$). However, no difference in root length was observed among cuttings treated with IBA or K-IBA ($p = 0.08$).

Orthogonal contrasts can be used to analyze binomial or count data, given that the data are transformed before analysis (see previous sections for these types of data). Information gained from orthogonal contrasts is often much more useful than outcomes of multiple comparison or multiple range tests. See Lentner and Bishop (1986), Little and Hills (1978), or Zar (1984) for more information regarding the use of orthogonal contrasts.

TREND ANALYSIS

Trend analysis is the most effective method for analyzing data of treatments consisting of various levels of a single factor (different concentrations of the same growth regulator) or increments in time (Kleinbaum et al., 1988). The primary objective in these types of experiments is to identify a concentration, or time period, that simulates optimum

TABLE 13.5
Results Obtained from Simple Linear Regression Analysis of the Percentage of *Viburnum carlesii* Softwood Cuttings That Produced Roots Following Treatment with 0, 12, 24, 36, or 48 mM K-IBA[a]

Predictor Variables	Coefficient	Std Error	Student's t	p
Constant	−0.22	0.06913	−3.18	0.0017
Treatment	0.0223	0.00210	10.62	0.0001
R^2	0.3627	Residual MS	0.159	
Adj. R^2	0.3595	Std Deviation	0.399	

Source	DF	MS	F	p
Linear trend	1	17.956	115.16	0.0000
Quadratic trend	1	0.5000	3.21	0.0748
Cubic trend	1	0.4840	3.10	0.0797
Residual	196	0.15931		
Total	199			

[a] Data for 0 K-IBA were deleted because none of the cuttings produced roots. Data are hypothetical.

propagule response. Models identifying specific trends (linear, quadratic, and cubic) are tested in a stepwise fashion from simplest (linear) to most complex (cubic in most cases) until a nonsignificant trend is identified. At this time the last significant trend is considered to best describe the data (Kleinbaum et al., 1988). Trend analysis uses SS, t, and R^2 values to indicate significant trends. Trends may be tested through regression analysis or polynomial contrasts statements in ANOVA (Compton, 1994).

Trend analysis can be illustrated by using an example in which the effect of five concentrations (0, 12, 24, 36, and 48 mM) of K-IBA on rooting of *V. carlesii* softwood cuttings was examined. In this example, softwood cuttings were prepared as described previously and dipped in one of the K-IBA solutions for 30 sec. Data recorded included the percentage of cuttings with roots, the number of roots per cutting, and root length. Since percentage values are binomial in nature, all data were transformed before analysis using the arcsin transformation. Also, since none of the cuttings receiving the control treatment (0 K-IBA) produced roots, all data for this treatment were excluded from the analysis. Results from trend analysis indicated that rooting performance (percentage of cuttings with roots) was dependent on K-IBA concentration (Table 13.5), with the linear model (y = −22 + 2.23 × K-IBA) giving the best fit ($p < 0.00001$; $R^2 = 0.36$). Higher-order models, such as the quadratic model ($p = 0.07$; $R^2 = 0.389$), did not provide a better explanation of how the cuttings responded to the application of K-IBA. Because the highest K-IBA level yielded the best results, it would be wise to conduct another experiment comparing at least the two highest levels used in this experiment (36 and 48 mM) with even higher doses.

Means of dosage treatments should be presented in graphs displaying the regression equation and fitted line, R^2, individual data values, and confidence intervals (Figure 13.3). Similar to mean comparison tests, confidence intervals of treatments that overlap are considered similar. In the viburnum example, none of the confidence intervals for the treatments overlapped. Therefore, the 48 mM treatment would be considered best for promoting rooting of softwood cuttings.

Trend analysis and polynomial contrasts may be used for evaluating optimum treatment levels, even if those levels were not directly tested but lie within treatment boundaries (Compton, 1994). Predicted values can be obtained using information generated from the analyses, allowing researchers to estimate the effects of treatments that were not evaluated.

MULTIFACTOR EXPERIMENTS

In the previous examples, single-factor experiments were emphasized (e.g., experiments in which one treatment factor, such as auxin type, is examined during experimentation). However, it is common for researchers to look at the effects of several factors simultaneously (multifactor treatment designs). In some experiments, multifactor treatment designs are established in such a way that all treatment factors are present in each treatment (complete factorial treatment designs). These experiments are popular among experimenters because the effects of each factor individually as well as interactions among them can be examined in one experiment

Evaluation of Data from Propagation Experiments

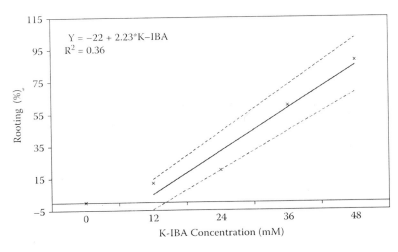

FIGURE 13.3 Line graph demonstrating the use of simple linear regression to illustrate differences in rooting percentage among softwood cuttings treated with 0 to 48 mM K-IBA. Treatment means (x), calculated regression line (—), and 95% confidence intervals (- -) are shown.

rather than two (Lentner and Bishop, 1986). Another advantage of multifactor experiments is that they use resources most efficiently (Compton, 1994). It is not a requirement that treatments for multifactor experiments be arranged in a complete factorial. For example, augmented factorial treatment designs are set up so that some treatments do not consist of all treatment factors. The main criterion is that all treatments make good practical and scientific sense.

The split plot (SP) design is often used for multifactor experiments. In this design, levels of one factor are placed into groups (blocks) called main plots, which contain all levels of the second factor. Separate error terms are used to test for significance of each factor. For example, the interaction between the main plot treatment and block is used as the error for the main plot treatment (Table 13.6). In contrast, the block by subplot treatment interaction was used to test for significance of the subplot treatment, and the main plot treatment by block by subplot treatment term was used as error to compare the main plot by subplot treatment interaction.

TABLE 13.6
Analysis of Variance Summary Table for the Split Plot Design in Which *Trichoderma harzianum* Concentration (0, 5, and 25 mg/kg of Rooting Medium) Was Main Plot and Viburnum Species Were Subplots[a]

		Dependent Variables								
		Root Dry Weight			Rooting Percentage			Stem Dry Weight		
Source	DF	MS	F	p	MS	F	p	MS	F	p
T. harzianum conc. (T)	2	732.040	290.58	0.00001	0.0875	0.89	0.4464	1396.39	506.74	0.00001
Block (B)	4	5.69179	2.26	0.1518	0.4729	4.83	0.0282	4.68366	1.70	0.2423
T*B	8	2.51924			0.0979			2.75567		
Viburnum species (V)	3	36.2634	9.67	0.0016	0.4486	1.07	0.3965	48.7371	26.59	0.00001
B*V	12	3.74831			0.4174			1.83292		
T*V	6	2.72972	0.65	0.6917	0.0819	0.42	0.8604	5.43448	1.90	0.1223
T*B*V	24	4.21614			0.1965	2.86139				
Residual	898	0.84473			0.06375	0.39606				
Total	957									
T. harzianum contrasts										
0 vs. 5 and 25 mg	1	988.66	196.22	0.0005	NA	NA	NA	1890.5	343.02	0.0011
5 vs. 25 mg	1	475.43	94.36	0.0005	NA	NA	NA	902.29	163.72	0.0005

[a] Dependent variables examined were root dry weight, rooting percentage, and shoot weight. Data are hypothetical.

Trichoderma harzianum	1_4	3_1	2_5	2_2	1_4	3_4	....	1_2	2_1	3_5	3_3	2_4	1_3
	①	④	②	③	③	②		④	②	①	①	②	③
	③	①	④	②	①	①		③	①	④	③	④	①
	④	②	③	①	④	③		①	③	②	④	①	④
Viburnum species	②	③	①	④	②	④		②	④	③	②	③	②

FIGURE 13.4 Randomization scheme for a split-plot design in which main plots (*Trichoderma harzianum* concentration) are randomized into blocks and subplots (*Viburnum* species) were randomized within each main plant. *T. harzianum* concentrations (0, 5, and 25 mg/k) were replicated five times. *Viburnum* species were assigned to subplots [*V. carlesii* (①), *V. dentatum* (②), *V. lantana* (③), *V. trilobum* "Compactum" (④)].

To illustrate use of the split plot experimental design in factorial treatment experiments, an example in which the influence of two concentrations (5 or 25 g/kg of rooting medium) of *Trichoderma harzianum* isolate T-22 on rooting of four viburnum species will be examined. Three of the viburnum species selected were considered easy to root—arrowwood viburnum (*V. dentatum* L.), dwarf American cranberry viburnum (*V. trilobum* Marsh 'Compactum'), wayfaringtree viburnum (*V. lantana* L.), whereas Koreanspice viburnum (*V. carlesii*) was considered difficult to root (Dirr, 1998). Softwood cuttings of each species were given a 30-sec. dip in a 48 mM solution of K-IBA before transfer to rooting medium treated with either 0, 5, or 25 g of *T. harzianum* (isolate T-22) per kg medium. Rooting medium consisted of a mixture containing equal parts (v/v) of sphagnum peat and perlite. The prescribed amount of *T. harzianum* (isolate T-22) was added to moistened medium before being dispensed into 24-cell 26.25 wide × 52.5 long × 7.5 deep cm plastic trays. Cuttings were arranged in a split plot design with *T. harzianum* concentration as main plots and viburnum species as subplots. Each tray was considered a main plot and only possessed medium containing one concentration of *T. harzianum* to avoid cross contamination. There were five main plot blocks per *T. harzianum* concentration with eight replicates of each species per block, resulting in 40 replicates of each species per treatment (Figure 13.4). Individual trays (blocks) were randomized on greenhouse benches with bottom heat set to maintain a minimum root zone temperature of 25°C. Cuttings were irrigated with mist for 10 sec. on 10-min intervals. Cutting growth was measured by recording stem and root weight as well as percentage of rooting eight weeks after initiating the experiment. The experiment was conducted twice.

Examining the results of ANOVA, significant differences in root dry weight were observed for *Trichoderma* concentration (0.0001) and viburnum species ($p = 0.0016$; Table 13.6). However, there was no interaction between species and *Trichoderma* concentration for root dry weight ($p = 0.6917$), indicating that the four species reacted simi-

TABLE 13.7

Use of Orthogonal Contrast for Determining Significance in Rooting among Viburnum Softwood Cuttings Treated with Three Concentrations of *Trichoderma harzianum*[a]

T. harzianum Inoculation Density[b]	Number of Cuttings	Root Dry Weight (kg)	Rooting Percentage	Shoot Dry Weight (kg)
0	320	1.6	91	2.3
5	320	2.9	91	4.1
25	318	4.6	94	6.5
Contrasts[c]				
0 vs. 5 and 25		0.0005	NS	0.0011
5 vs. 25		0.0005	NS	0.0005

[a] Data are hypothetical.
[b] *Trichoderma harzianum* inoculation density in mg per kg of rooting medium.
[c] *p* values as determined by ANOVA at 1 and 8 df as calculated using a split plot design with *Trichoderm harzianum* inoculation density as main plots.

TABLE 13.8
Root and Shoot Dry Weight Plus Rooting Percentage of Softwood Cuttings Collected from Four Viburnum Species[a]

Viburnum Species	Number of Cuttings	Root Dry Weight[b]	Rooting Percentage[c]	Shoot Dry Weight
V. dentatum	240	3.5 a	95	4.8 a
V. lantana	238	3.2 ab	95	4.4 a
V. trilobum 'Compactum'	240	2.8 bc	92	3.9 b
V. carlesii	240	2.7 c	86	3.9 b

[a] Data are hypothetical.
[b] Means within columns with the same letter are not different according to Bonferroni at 3 and 12 df.
[c] Means not significant according to analysis of variance.

larly to the *Trichoderma* treatments. Two general contrasts (0 *Trichoderma* versus 5 and 25 mg *Trichoderma* per kg of rooting medium, and 5 versus 25 mg *Trichoderma* per kg rooting medium) were conducted to examine the influence of *Trichoderma* on root growth. Examining the contrasts (Table 13.6) and treatment means (Table 13.7) revealed that adding *Trichoderma* to the rooting medium improved root growth of viburnum softwood cuttings ($p = 0.0005$). In addition, the second contrast and treatment means indicate that adding 25 mg of *Trichoderma* per kg of rooting medium at the time of sticking cuttings improved root growth more than adding 5 mg/kg ($p = 0.0005$). Examination of rooting percentages among species revealed that all viburnum species tested rooted similarly when using 48 mM K-IBA ($p = 0.3965$) and that *Trichoderma* acted similarly among the species ($p = 0.4464$). Addition of *Trichoderma* did improve stem growth of all viburnum cuttings ($p = 0.0001$). However, it is unclear if increased stem growth occurred directly as a result of *Trichoderma* treatment or was stimulated by improved root growth. Stem growth also varied among viburnum species, with cuttings of *V. dentatum* and *V. lantana* displaying more growth than those of *V. carlesii* and *V. trilobum* 'Compactum,' likely because the former displayed better root growth (Table 13.8).

In this experiment, the SP was the most efficient experimental design to use because it allowed the researchers to minimize cross contamination of *Trichoderma* between treatments. However, the SP design may also be used in situations in which two treatment factors are expected to have highly different variances. In any case, the experimental setup, data recording, as well as data analysis and interpretation are more complicated for the SP design than the CR and RCB designs.

CONCLUSIONS

This chapter has outlined the steps of experimentation and many of the procedures used in the analysis of plant propagation data. It is important to remember that well-conceived research projects are planned carefully before implementation, which helps to ensure that the experiment is conducted properly and will answer the desired questions. Make sure to use the most appropriate experimental design for your conditions and experimental objectives. Doing so can enable you to maximize experimental precision and minimize the influence of nontreatment factors. Be sure to minimize bias, fatigue, and errors during the course of conducting an experiment. These factors lead to untrue results that mislead researchers. Use the experimental design while conducting the experiment and recording and analyzing data. This helps to provide an objective, nonbiased way to accurately evaluate treatment differences. Used properly, statistics is a valuable tool to the scientific researcher, allowing investigators to declare experimental results with confidence.

LITERATURE CITED

Barnett, V. and T. Lewis. 1984. *Outliers in Statistical Data*. John Wiley & Sons, Chichester, U.K.

Bartlett, M.S. 1937. Some examples of statistical methods of research in agriculture and applied biology. *J. Royal Statist. Soc. Suppl.* 4: 137–170.

Compton, M.E. 1994. Statistical methods suitable for the analysis of plant tissue culture data. *Plant Cell Tissue Organ Cult.* 37: 217–242.

Compton, M.E. and C.W. Mize. 1999. Statistical considerations for in vitro research: I—Birth of an idea to collecting data. *In Vitro Cell. Dev. Biol. Plant* 35: 115–121.

Dirr, M.A. 1998. *Manual of Woody Landscape Plants*. Stipes Publishing L.L.C. Champaign, IL.

Fernandez, G.C.J. 1992. Residual analysis and data transformations: Important tools in statistical analysis. *HortScience* 27: 297–300.

Finney, D.J. 1989. Was this in your statistical textbook? V. Transformation of data. *Expt. Agr.* 25: 165–175.

Kempthorne, O. 1973. *The Design and Analysis of Experiments*. Robert E. Krieger Publishing Co., Malabar, FL.

Kleinbaum, D.G., L.L. Kupper, and K.E. Muller. 1988. *Applied Regression Analysis and Other Multivariable Methods,* 2nd ed. PWS-Kent Publishing Co., Boston.

Lentner, M. and T. Bishop. 1986. *Experimental Design and Analysis.* Valley Book Company, Blacksburg, VA.

Little, T.M. and F.J. Hills. 1978. *Agricultural Experimentation: Design and Analysis.* John Wiley & Sons, New York.

Mize, C.W., K.J. Koehler, and M.E. Compton. 1999. Statistical considerations for in vitro research: II—Data to presentation. *In Vitro Cell. Dev. Biol. Plant* 35: 122–126.

Zar, J.H. 1984. *Biostatistical Analysis,* 2nd ed. Prentice-Hall, Inc., Englewood Cliffs, NJ.

Part VI

General Concepts for Successful Vegetative Propagation

14 Plant Growth Regulators Used in Propagation

Zong-Ming Cheng, Yi Li, and Zhen Zhang

CHAPTER 14 CONCEPTS

- Plant growth regulators (PGRs) are organic molecules that affect plant physiological processes in very low concentrations. PGRs may be naturally occurring or manmade (synthetic).

- Auxins, cytokinins, gibberellins, abscisic acid, and ethylene are considered the traditional classes of PGRs.

- More recently, polyamines, jasmonates, salicylic acid, and brassinosteroids have been added as classes of PGRs.

- PGRs produced in one tissue or part of the plant may be transported to other parts of the plant.

- Auxins, such as NAA and IBA, are used in the propagation industry to induce adventitious root formation on stem cuttings; the other classes of PGRs are typically not used in conventional plant propagation.

- Synthetic auxins, such as 2,4-D and dicamba, are typically not used to promote adventitious rooting, but are used as broad-leaf herbicides. These auxins are often employed in tissue culture protocols to induce somatic embryogenesis.

- Cytokinins and auxins, and to a lesser extent some of the other classes of PGRs, are extensively used in micropropagation (tissue culture) of plants. Cytokinins induce multiple shoot formation from preformed meristems. They also can be used to generate adventitious shoots from somatic cells, such as those found in leaves and stems.

Plant growth regulators (PGRs) are a group of organic substances that are naturally occurring or manmade (synthetic), which affect physiological processes of plants at very low concentrations, usually in nanomolar (10^{-9} M) to micromolar (10^{-6} M) amounts. Another term, plant hormones, is used usually to designate those growth regulators that are naturally produced by plants. Plant hormones and PGRs are indistinguishable when considering the application of these substances to achieve specific horticultural results. In this chapter, plant hormones primarily will be discussed because synthetic PGRs generally have similar effects as plant hormones.

Auxins, cytokinins, gibberellins (GAs), abscisic acid (ABA), and ethylene are the five classes of compounds into which plant hormones have historically been classified. In addition to these five classical plant hormone classes, polyamines, jasmonates, salicylic acid, and brassinosteroids all have been listed as plant hormones, although it is unclear whether these compounds have universal effects or act in just a few special growth and developmental events (Davies, 2004a,b). Another group of synthetic compounds, plant growth retardants, are PGRs, but not plant hormones,

and have been used widely in greenhouse crop production. Since they have a relatively minor role in conventional plant propagation, this group of PGRs will not be discussed in this chapter. Each of the nine main groups of plant regulators will be discussed generally and specifically as they relate to plant propagation.

AUXINS

Auxin was the first class of plant hormones discovered. Charles Darwin first described the effects of light on movement of canary grass (*Phalaris canariensis*) coleoptiles in his book, *The Power of Movement in Plants*, presented in 1880. The bending of coleoptiles toward light was eventually attributed to the effects of indoleacetic acid (IAA), an auxin (Normanly et al., 2004). The most common naturally occurring auxin is IAA and was discovered in 1935. Indolebutyric acid (IBA) also occurs naturally (Ludwig-Muller and Epstein, 1994; Figure 14.1). Both auxins are synthesized from tryptophan and occur primarily in leaf primordial, young leaf tissues, and developing seeds (Normanly et al., 2004). Compounds serving

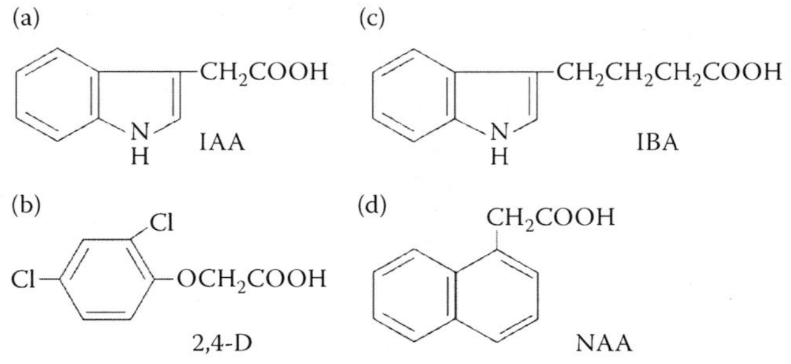

FIGURE 14.1 Chemical structures of some auxin-type PRGs. (a) Indolyl-3-acetic acid (IAA); (b) 2,4-dichlorophenoxyacetic acid (2,4-D); (c) indolyl-3-butyric acid (IBA); (d) 1-naphthalenacetic acid (NAA). (From Gaba, V. P. 2005. In *Plant Development and Biotechnology*. Trigiano, R.N. and D.J. Gray (Eds.), CRC Press, Boca Raton, FL, pp. 87–99. With permission.)

as IAA precursors, such as indoleacetaldehyde, may also exhibit some auxin activity. Auxin that is synthesized in leaves is transported through cell-to-cell movement into the stem. Transport in stems to roots may also involve the phloem tissues (Davies, 2004a,b).

Since auxins play many roles in plant growth and development, the concentrations of these compounds are critical and highly regulated by the plant. One way that plants regulate the amount of IAA in tissues is by controlling its biosynthesis. Another regulatory mechanism involves the conjugation of auxins with sugars and amino acids into inactive forms. The formation of auxin conjugates permits plants to store and transport the auxins without expressing the growth regulator effect. Conjugated auxins can be activated rapidly by environmental stimuli. The third mechanism for regulating auxin levels is by oxidation of IAA, and presumably IBA, by the enzyme, IAA oxidase (Normanly et al., 2004). Conjugates of IAA and synthetic auxins, such as 2,4-D, cannot be destroyed by the activity of IAA oxidase.

Shortly after natural auxin was discovered, auxin was synthesized and became available to plant biologists. Much of our earlier knowledge on functions of auxin was derived from its exogenous applications to experimental plant systems. In the last 20 years, characterization of many auxin-related mutants (Reid and Howell, 1995) and genetic engineering with auxin biosynthesis genes from *Agrobacterium tumefaciens* (Phillips, 2004; Cheng et al., 2006) have also contributed to our current understanding of auxin biosynthesis, mode of action and signal transduction pathways, and interactions with other hormones.

Auxins are essential for numerous aspects of plant growth and development (Davies, 2004a). The most relevant role of auxins in plant propagation is that they stimulate root initiation on stem and leaf cuttings and the development of branch roots (chapter 17). This function of auxins is essential for cutting propagation of many plant species used in horticultural and forestry industries. Auxins have been used as powerful agents to induce adventitious root formation in many species that are naturally difficult to root (Dirr, 1998). Because of this, the horticultural industry has become more efficient in production and has become more flexible in producing plants all year around. Auxins are used commonly in tissue culture for inducing adventitious root formation on microcuttings. Use of auxins for inducing adventitious roots in stem cuttings is described in chapter 17 through chapter 20 and using them in tissue culture is described in chapter 31.

Although auxins are required or helpful for inducing root primordial, high concentrations of auxins often inhibit root primordial growth and elongation of roots from stem cuttings or microcuttings. This has been demonstrated by high-level expression of the *iaa*M gene from *Agrobacterium tumefaciens* (Klee et al., 1987; Sitbon et al., 1992). In those transgenic plants, abundant root primordial sites exist, but few developed into normal roots.

Since IAA is broken down rapidly to inactive products by light and microorganisms, it is not widely used in plant propagation. Since IBA and naphthalene acetic acid (NAA) are more resistant to degradation by microbes and plants, they appear to have a more pronounced and longer effect than IAA and, therefore, are more widely used in the horticultural industry for plant propagation. However, responses of individual plant species to different auxins are not universal. Therefore, trials with different auxins and various concentrations and carriers are necessary for each species before successful rooting procedures can be developed and used for propagation. Several synthetic, phenoxy-type herbicides, such as 2,4-D and 2,4,5-T (Figure 14.1), used at low concentrations, have auxin-like effects in promoting callus formation either alone or in combination with cytokinins. However, they are rarely used for inducing adventitious roots from cuttings because of the propensity to form callus on the bases of cuttings, which is undesirable during rooting. These synthetic compounds are widely used in inducing somatic embryos in tissue culture of many species including cereal crops and turfgrasses.

In addition to promotion of adventitious root formation, auxins synthesized in the apical buds and young leaf tissues suppress the growth of lateral buds, therefore, maintaining the apical dominance. This auxin effect has been confirmed by transformation. The plants transformed

Plant Growth Regulators Used in Propagation

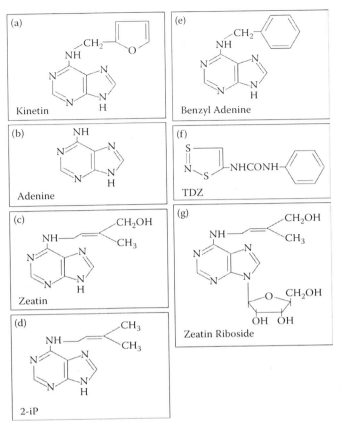

FIGURE 14.2 Chemical structures of some cytokinin-type PGRs. (a) Kinetin; (b) adenine; (c) zeatin; (d) N^6-(2-isopentyl) adenine (2iP); (e) 6-benzyladenine (BA); (f) thidiazuron (TDZ); and (g) zeatin riboside. (From Gaba, V.P. 2005. In *Plant Development and Biotechnology*. Trigiano, R.N. and D.J. Gray (Eds.), CRC Press, Boca Raton, FL, pp. 87–99. With permission.)

with auxin biosynthesis gene from *Agrobacterium* maintain strong apical dominance with relatively few branches. This effect of auxins can be overcome by pinching the terminal shoot tip, applying exogenous cytokinin, or by transforming plants with a cytokinin biosynthesis gene.

Auxins also stimulate cell division in the cambium and vascular tissue differentiation, including xylem and phloem (Aloni, 2004). In combination with cytokinin, they also induce cell division and organ differentiation (Schwarz et al., 2004), such as shoots and somatic embryos (Gray, 2004). They also stimulate root differentiation from somatic tissues in tissue culture (chapter 31). Other effects induced by auxins include mediating tropistic movements (such as bending toward or away from a stimulus, responses of shoots to light and of roots and shoots to gravity), delaying leaf senescence, stimulating parthenocarpic fruits (fruits setting without fertilization) and seedlessness, delaying fruit ripening, and promoting femaleness in dioecious flowers via stimulation of ethylene production (Davies, 2004).

CYTOKININS

Cytokinins are a class of plant hormones that are essential to plant development because of their role in promoting cell division. Cytokinins were discovered by Dr. Skoog's laboratory when they were searching for chemicals that could induce cell division in combination with an auxin in tobacco pith cultures. They found some complex and undefined materials, such as coconut milk, vascular tissue extracts, autoclaved DNA, and yeast extracts restored cell division in the presence of an auxin (Sakakibara, 2004). The cytokinin zeatin, or 6-(4-hydroxy-3-methylbut-2-enyl-amino) purine, was first purified from immature kernels of *Zea mays* (Letham, 1963). Cytokinins are derivatives of adenine, a component of both DNA and RNA, and are present in ribosides and ribotides forms (Figure 14.2). Cytokinins are synthesized in root tips and developing seeds and transported from roots to shoots via xylem tissue.

Cytokinins have multiple functions in plant growth and development (Davies, 2004a; Werner et al., 2001). The interaction of cytokinin with auxin in promoting cell division is one of the primary functions related to plant propagation. Exogenous applications of cytokinins induce cell division and morphogenesis in tissue culture in the presence of auxin. A high cytokinin/auxin ratio favors shoot formation, and a low ratio favors rooting. A combination of high concentration of both auxin and cytokinin generally induces mass callus formation. The promotion of cell division has also been confirmed in the crown galls induced by *Agrobacterium tumefaciens,* which transfers the cytokinin biosysnthesis gene, *ipt,* and auxin biosynthesis genes, *iaa*M and *iaa*H, into plant cells, and in plant cells that are transformed with the isolated *ipt* gene (Li et al., 2003).

Cytokinins are present in relatively high concentrations (mM) in developing seeds, but decrease quickly as the seeds mature. Cytokinins also interact with gibberellins and abscisic acid in controlling seed dormancy. Cytokinins also counteract the effects of ABA and have synergistic effect with GA in breaking seed and bud dormancy, but do not seem to have a direct role in seed germination (chapter 41). Cytokinins also cause release of lateral buds from apical dominance, which is one of the main functions of auxins. Application of exogenous cytokinins or expression of the *ipt* gene in transgenic plants results in the "bushy" or the excessively "branchy" phenotype. Another major role of cytokinins is to delay senescence of leaves and detached flowers. This delay may be related to the effect of cytokinins in retarding degradation of chlorophyll and cellular proteins. The antisenescence effects of cytokinins are critical to the cut-flower industry.

ABSCISIC ACID (ABA)

ABA (Figure 14.3), a ubiquitous weak acid in lower and higher plants, was first identified in the early 1960s as a growth inhibitor and is synthesized indirectly from carotenoids (Walton and Li, 1995). ABA is mostly uncharged when present in the relatively acidic apoplastic (fluid-occupied space between cells) compartment of plants and can easily cross the plasma membrane into cells. Recent genetic studies, especially in *Arabidopsis,* have identified many gene loci involved in ABA synthesis and responses (Gonzalez-Guzman et al., 2004; Koiwai et al., 2004).

ABA is involved in the regulation of many aspects of plant growth and development, including embryo maturation, seed dormancy, germination, root development, stomata function, cell division, and elongation, and responses to such environmental stresses as flood, salinity, cold, drought, pathogen attack, and UV radiation (Davies, 2004a,b; Gonzalez-Guzman et al., 2004; Koiwai et al., 2004). However, ABA, unlike the name indicated, does not appear to control directly organ (e.g., leaf) abscission, but does promotes senescence and/or stress responses, the processes directly preceding abscission. ABA has historically been considered as a growth inhibitor; however, some young, actively growing tissues have relatively high levels of the PGR. In contrast, ABA-deficient mutant plants are severely stunted; therefore, the name of ABA does not match the roles it plays in plant growth and development.

The role of ABA in plant propagation is primarily associated with the following two aspects of growth and physiology of plants: seed and bud dormancy and somatic embryo (chapter 36) maturation in tissue culture. ABA appears to play a key role in preventing "viviparous germination" of developing embryos and inducing primary dormancy. ABA has been isolated from seed coats of dormant peach, plum, apple, rose, and walnut. Another indication that ABA may be involved in seed dormancy is that the concentration usually decreases during seed stratification. ABA's role in preventing viviparous germination and inducing dormancy has also been observed similarly in bud dormancy/dormancy release. Application of ABA can inhibit germination of nondormant seeds and counteract the effect of applied GA, which promotes germination. In somatic embryogenesis in tissue culture, ABA tends to promote somatic embryo maturation (Gray, 2004).

GIBBERELLINS (GAs)

The gibberellins (GAs) are a group of tetracyclic, diterpenoid acids with a common *ent*-gibberellane ring structure

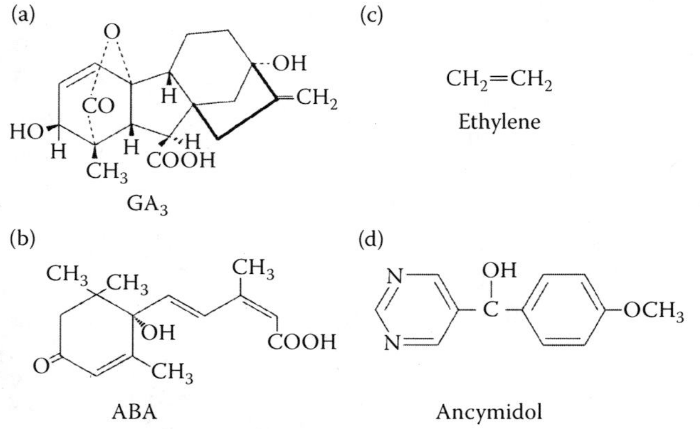

FIGURE 14.3 Chemical structures of PGRs of varied actions. (a) Giberellic acid (GA_3); (b) abscisic acid (ABA); (c) ethylene; and (d) ancymidol. (From Gaba, V.P. 2005. In *Plant Development and Biotechnology.* Trigiano, R.N. and D.J. Gray (Eds.), CRC Press, Boca Raton, FL, pp. 87–99. With permission.)

(Figure 14.3). Gibberellins were first isolated from the fungus *Gibberella fukikuori*, which causes abnormally tall rice plants (foolish rice disease). GAs are ubiquitous in both angiosperms (flowering plants) and gymnosperms (conifers, cycads, gingko, Gnetales) as well as in ferns. They have also been isolated from lower plants, such as mosses and algae, and some fungal and bacterial species. All gibberellins are assigned numbers as gibberellin A_{1-x}, irrespective of their origin. To date, there have been 126 GAs isolated, many of which are most likely nonessential to the plant. Instead, these forms are probably inactive precursors or breakdown products of active gibberellins (Sponsel and Hedden, 2004).

The most widely available GA is gibberellic acid or GA_3, which is a fungal secondary product. GA_1 is the primary gibberellin responsible for stem elongation in plants (Sun, 2004). GAs are synthesized in young tissues of shoots, although the exact location is uncertain, and in developing seeds. It is also uncertain whether or not GA is synthesized in roots. GAs are probably transported both in xylem and phloem tissues (Davies, 2004b).

GAs plays many roles in plant growth and development. One of the major roles is to promote stem elongation by stimulating both cell elongation and cell division, resulting in tall plants. One of the seven traits, dwarf–tall, that Mendel studied in peas to establish the Mendelian inheritance law, is now known to involve GA biosynthesis. The dwarf plant has the recessive gene, *le*, which encodes a defective enzyme to synthesize GA. The tall plant possesses the dominant gene, *Le*, which encodes a functioning enzyme for GA biosynthesis. This can be confirmed by the fact that the dwarf phenotype can be reversed to the tall phenotype by treatment with GA_3. Similar results have been obtained with dwarf mutants of other plant species. The mode of action of GAs in promoting stem elongation has been reviewed in great detail (Sun, 2004).

GAs cause stem elongation of plants in response to long days, resulting in bolting of long-day plants in the spring. In a number of long-day plants that are cultivated under short-day conditions, or vernalization-requiring plants, such as *Hyoscyamus, Daucus, Crepis, Silene,* without a cold stimulus, flower formation does not take place. Treatment with GA_3 induces flower development even without the necessary external signals. This is attributed to elongation of the stem axis that is a prerequisite for flower formation and bolting. Gibberellins do not affect flower formation in short-day plants and most other long-day plants in which the stem axis does not elongate immediately before flower formation. Exogenously applied GA also promotes fruit set and growth without pollination in certain species, such as grape.

The major roles of GA in plant propagation are inducing seed germination for some seeds that normally require cold treatment (stratification) or light exposure, and releasing buds from dormancy. In seeds that require light to germinate (e.g., lettuce), gibberellins are able to substitute for P_{FR} (far red light). Gibberellins also regulates the formation and secretion of hydrolytic enzymes, such as alpha-amylase, a protease, and a ribonuclease, in the aleuron layer (a part of the endosperm tissues of cereal seeds) of cereal grains during the early stages of embryo germination. These enzymes help mobilize endosperm storage compounds supplying the embryo with nutrients during germination.

ETHYLENE

Ethylene ($H_2C=CH_2$) (Figure 14.3) is a gaseous hormone with a simple structure and is the only member of its class. Endogenous synthesis of ethylene by plants was first demonstrated in 1934 by Gane and was proposed as a plant hormone in 1935 by Pech et al. (2004). It is ubiquitous to all higher plants and is usually associated with fruit ripening (Reid, 1995). Ethylene is produced from methionine in essentially all tissues in all higher plants. Methionine is first converted to S-adenosyl methionine (SAM), which is then converted to 1-amino-cyclopropane-1-carboxylic acid (ACC) by ACC-synthase. ACC is then oxidized to produce ethylene. ACC synthase is the rate-limiting step for ethylene production, and it is down-regulated to delay fruit ripening in the "flavor saver" tomatoes (Phillips, 2004).

Although ethylene is a simple compound, as a plant hormone, it affects many plant processes (Bleecker and Kende, 2000; Davies, 2004a), including fruit ripening and responses to stresses, such as flooding, drought, wounding, and senescence. Ethylene has been shown either to promote or inhibit adventitious root formation. For example, the ethylene releaser, ethephon, inhibited adventitious root formation in mung bean (*Vigna radiata*) (Geneve and Heuser, 1983), while in derooted sunflower seedlings when cotyledons and apical buds were present, ethylene or ethephon promoted adventitious root formation (Liu and Reid, 1992). Its role in root induction may be related to how it interacts with auxin because ethylene and auxin biosyntheses share some pathways, or may be related to the stress environment. For example, in a wetland plant, *Rumex palustris,* flooding induced ethylene production; plants in flooded soil exhibited increased sensitivity of the root-forming tissues to endogenous IAA, thus inducing adventitious roots (Visser et al., 1996). However, since high concentrations of auxins induce ethylene synthesis, it is difficult to differentiate the role of auxins and ethylene. Clark et al. (1999) conducted an elegant experiment using ethylene-insensitive tomato (*Lycopersicon esculentum*) and petunia (*Petunia × hybrida*) plants to determine if normal or adventitious root formation is affected by ethylene insensitivity. They found that application of auxin (IBA) increased adventitious root formation on stem cuttings of wild-type plants, but had little or no effect on rooting of ethylene-insensitive (NR) mutant plants and in ethylene-insensitive transgenic petunia plants. Application of ACC increased adventitious root formation on vegetative stem cuttings from NR and wild-type plants, but NR cuttings produced fewer adventitious roots than wild-type cuttings. They concluded that the promotive effect of auxin on adventitious rooting may be mediated by ethylene.

POLYAMINES (PAs)

Polyamines in plants are represented by diamine putrescine (Put), triamine spermidine (Spd), and tetramine spermine (Spm) (Galston and Kaur-Sawhney, 1995). They appear to be essential to growth and development because mutants lacking the ability to synthesize PAs are unable to grow and develop normally. Furthermore, the addition of PAs to these mutants generally restores normal growth and development. Abnormal growth results from applications of inhibitors of the PA biosynthetic enzymes further confirm the indispensability of PAs to plants, although some controversy exists as to whether or not polyamines should be classified as plant hormones (Davies, 2004a). PAs have a number of functions, including stabilizing membranes and nucleic acids and controlling protein structure and enzyme activity. The compounds also have been linked to cell division, vascular differentiation, somatic embryo formation in tissue culture, root initiation, adventitious shoot formation, flower initiation and development, and control of fruit ripening and senescence (Galston and Kaur-Sawhney, 1995). PAs can enhance rooting in stem cuttings when applied with auxin, but PAs alone seem to have limited effect in adventitious root formation, at least are not considered as essential. Recent work on genetic engineering of polyamine biosynthesis has shown that the PAs enhance somatic embryogenesis in carrots cultured in hormone-free medium, and even in medium with inhibitory 2,4-D-containing concentrations (Bastola and Minocha, 1995).

JASMONATES

Both jasmonic acid (JA) and its fragrant methyl ester, methyl jasmonate (meJA), are widely present in species of the plant kingdom. MeJA is a fragrant oil that contributes to the distinctive aroma of certain fruits and flowers, such as jasmine, which is how the name jasmonic acid was derived. Since both compounds display biological activity, they are collectively called jasmonates. Jasmonates are biosynthesized from α-linolenic acid (Howe, 2004) and inhibit many plant growth and developmental processes including seed germination. They also promote senescence, abscission, fruit ripening, pigment formation, and tendril coiling. One of the important roles of jasmonates appears to be in plant defense by inducing the synthesis of proteinase inhibitors that deter insect feeding (Davies, 2004a). Jamonates also promote potato tuber formation. The role of jasmonates in plant propagation has not been well established. Some research has shown that they promote adventitious root formation in mung bean derooted cuttings as well as in potato *in vitro* shoots (Zhang and Cheng, 1996).

SALICYLIC ACID (SA)

Salicylic acid is found in the bark of willow (*Salix* species) trees where the name was derived, and was only recently recognized as a plant growth hormone. Salicylic acid is found ubiquitously in plants species and is biosynthesized from the amino acid phenylalanine. Pathways of SA biosynthesis and metabolism in tobacco have been recently identified. SA, an endogenous regulator of disease resistance, is a product of phenylpropanoid metabolism formed via decarboxylation of trans-cinnamic acid to benzoic acid and its subsequent 2-hydroxylation to SA. SA, and its close analog, aspirin, have a general role in resistance to pathogens by inducing pathogenesis-related proteins. SA delays cut flower senescence, possibly by inhibiting ethylene biosynthesis. It also elicits a plethora of responses, including blocking wounding responses, and other physiological processes. SA has been shown to inhibit seed germination, and when applied with IAA, it stimulates adventitious rooting in mung beans (Delaney, 2004). Overall, SA plays a very limited role in plant propagation (Davies, 2004a).

BRASSINOSTEROIDS (BR$_X$)

Brassinosteroids are a group of over 60 different steroidal compounds represented by the compound brassinolide (BR$_1$) that was first isolated from *Brassica napus* pollen. They are widely distributed in plants. BR$_x$ promote stem elongation, ethylene synthesis, and epinasty and inhibits root growth and development. BR$_x$ do not seem to have a direct role in plant propagation (Choe, 2004; Davies, 2004a), unless it would be interactions with GAs or auxins.

CONCLUSION

This chapter has introduced the basic concept of plant hormones and plant growth regulators, with a focus on their roles related to plant propagation, particularly cuttings, tissue culture, and seed and bud dormancies. Each class of growth regulators plays key, multiple roles in the process of plant growth and development. Some roles or effects are more obvious and can be easily characterized; others are less clear. All plant hormones interact with each other in regulating plant growth and development. When applying PGRs, one needs always to consider the types, target plant tissues, and their interactions with other types of hormones or PGRs, either endogenously or exogenously. Details of hormonal functions and roles in each type of plant propagation strategy will be discussed in relevant chapters and laboratory practices.

LITERATURE CITED AND SUGGESTED READINGS

Aloni, R. 2004. The induction of vascular tissues by auxin, in *Plant Hormones, Biosynthesis, Signal Transduction, Action!* 3rd ed. Davies, P.J. (Ed.), Kluwer Academic Publishers. Dordrecht, The Netherlands, pp. 471–492.

Bastola, D.R. and S.C. Minocha. 1995. Increased putrescine biosynthesis through transfer of mouse ornithine decarboxylase cDNA in carrot promotes somatic embryogenesis. *Plant Physiol.* 109: 63–71.

Bleecker, A. and H. Kende. 2000. Ethylene: A gaseous signal molecule in plants. *Annu. Rev. Cell. Dev. Biol.* 16: 1–18.

Cheng, Z.-M., W. Dai, M.J. Bosela, and L.D. Osburn. 2006. Genetic engineering approach to enhance adventitious root formation of hardwood cuttings. *J. Crop Improvement* 17: 211–225.

Choe, S. 2004. Brassinosteroids biosynthesis and metabolism, in *Plant Hormones, Biosynthesis, Signal Transduction, Action!* 3rd ed. Davies, P.J. (Ed.), Kluwer Academic Publishers. Dordrecht, The Netherlands, pp. 156–178.

Clark, D.G., E.K. Gubrium, J.E. Barrett, T.A. Nell, and H.J. Klee. 1999. Root formation in ethylene-insensitive plants. *Plant Physiol.* 121: 53–60.

Davies, P.J. 2004a. The plant hormones: Their nature, occurrence, and functions, in *Plant Hormones, Biosynthesis, Signal Transduction, Action!* 3rd ed. Davies, P.J. (Ed.), Kluwer Academic Publishers. Dordrecht, The Netherlands, pp. 1–15.

Davies, P.J. 2004b. Regulatory factors in hormone action: Level, location and signal transduction, in *Plant Hormones, Biosynthesis, Signal Transduction, Action!* 3rd ed. Davies, P.J. (Ed.), Kluwer Academic Publishers. Dordrecht, The Netherlands, pp. 16–35.

Delaney, T.P. 2004. Salicylic acid, in *Plant Hormones, Biosynthesis, Signal Transduction, Action!* 3rd ed. Davies, P.J. (Ed.), Kluwer Academic Publishers. Dordrecht, The Netherlands, pp. 635–653.

Dirr, M.A. 1998. *Manual of Woody Landscape Plants*, 5th ed. Stipes Publications. Champaign, IL.

Gaba, V.P. 2005. Plant growth regulators in plant tissue culture and development, in *Plant Development and Biotechnology*. Trigiano, R.N. and D.J. Gray (Eds.), CRC Press, Boca Raton, FL, pp. 87–99.

Galston, A.W. and R. Kaur-Sawhney, 1995. Polyamines as endogenous growth regulators, in *Plant Hormones, Biosynthesis, Signal Transduction, Action!* 2nd ed. Davies, P.J. (Ed.), Kluwer Academic Publishers. Dordrecht, The Netherlands, pp. 158–178.

Geneve, R.L. and C.W. Heuser. 1983. The relationship between ethephon and auxin on adventitious root initiation in cuttings of *Vigna radiata* (L.) R. Wilcz. *J. Amer. Soc. Hort. Sci.* 108: 330–333.

Gonzalez-Guzman, M., D. Abia, J. Salinas, R. Serrano, and P.L. Rodriguez. 2004. Two new alleles of the abscisic aldehyde oxidase 3 gene reveal its role in abscisic acid biosynthesis in seeds. *Plant Physiol.* 135: 325–333.

Gray, D. 2004. Propagation from nonmeristematic tissues: Nonzygotic embryogenesis, in *Plant Development and Biotechnology*. Trigiano, R.N. and D.J. Gray (Eds.), CRC Press, Boca Raton, FL, pp. 187–200.

Howe, G.A. 2004. Jasmonates, in *Plant Hormones, Biosynthesis, Signal Transduction, Action!* 3rd ed. Davies, P.J. (Ed.), Kluwer Academic Publishers. Dordrecht, The Netherlands, pp. 610–634.

Kepczynski, J. and E. Kepczynska. 1997. Ethylene in seed dormancy and germination. *Physiol. Plant.* 101: 720–726.

Klee, H.J., R.B. Hoesch, M.A. Hinchee, M.B. Hein, and N.L. Haffmann. 1987. The effects of overproduction of two *Agrobacterium tumefaciens* TDNA auxin biosynthetic gene products in transgenic petunia plants. *Genes Dev.* 1: 8696.

Koiwai, H, K. Nakaminami, M. Seo, W. Mitsuhashi, T. Toyomasu, and T. Koshiba. 2004. Tissue-specific localization of an abscisic acid biosynthetic enzyme, AAO3, in Arabidopsis. *Plant Physiol.* 134: 1697–1707.

Letham, D.S. 1963. Zeatin, a factor inducing cell division from *Zea mays*. *Life Sci.* 8: 569–573.

Li Y., H. Duan, Y.H. Wu, R.J. McAvoy, Y. Pei, D. Zhao, J. Wurst, Q. Li, and K. Luo. 2003. Transgenics of plant hormones and their potential application in horticultural crops, in *Genetically Modified Crops, Their Development, Uses, and Risks*. Liang, G.H. and D.Z. Skinner. (Eds.), Haworth Press, New York, pp. 101–112.

Liu, J.-H. and D.M. Reid. 1992. Auxin and ethylene-stimulated adventitious rooting in relation to tissue sensitivity to auxin and ethylene production in sunflower hypocotyls. *J. Exp. Bot.* 43: 1191–1198.

Ludwig-Muller, J. and E. Epstein. 1994. Indole-3-butyric acid in Arabidopsis thaliana III. *In vitro* biosynthesis. *J. Plant Growth Regul.* 14: 7–14.

Normanly, J., J.P. Slovin, and J.D. Cohen. 2004 Auxin biosynthesis and metabolism, in *Plant Hormones, Biosynthesis, Signal Transduction, Action!* 3rd ed. Davies, P.J. (Ed.), Kluwer Academic Publishers. Dordrecht, The Netherlands, pp. 36–62.

Pech, J.-C., M. Bouzayen, and A. Latché. 2004. Ethylene biosynthesis, in *Plant Hormones, Biosynthesis, Signal Transduction, Action!* 3rd ed. Davies, P.J. (Ed.), Kluwer Academic Publishers. Dordrecht, The Netherlands, pp. 115–136.

Phillips A.L. 2004. Genetic and transgenic approaches to improving crop performance via hormones, in *Plant Hormones, Biosynthesis, Signal Transduction, Action!* 3rd ed. Davies, P.J. (Ed.), Kluwer Academic Publishers. Dordrecht, The Netherlands, pp. 582–609.

Reid, J.B. and S.H. Howell, 1995. Hormone mutants and plant development, in *Plant Hormones, Biosynthesis, Signal Transduction, Action!* 3rd ed. Davies, P.J. (Ed.), Kluwer Academic Publishers. Dordrecht, The Netherlands, p. 448–485.

Sakakibara, H. 2004. Cytokinin biosynthesis and metabolism, in *Plant Hormones, Biosynthesis, Signal Transduction, Action!* 3rd ed. Davies, P.J. (Ed.), Kluwer Academic Publishers. Dordrecht, The Netherlands, p. 95–114.

Schwartz, S.H. and J.A.D. Zeevaart. 2004. Abscisic acid biosynthesis and metabolism, in *Plant Hormones, Biosynthesis, Signal Transduction, Action!* 3rd ed. Davies, P.J. (Ed.), Kluwer Academic Publishers. Dordrecht, The Netherlands, pp. 137–155.

Schwarz, O.J., A.R. Sharma, and R.M. Beaty. 2004. Propagation from nonmeristematic tissues: Organogenesis, in *Plant Development and Biotechnology*. Trigiano, R.N. and D.J. Gray (Eds.), CRC Press, Boca Raton, FL, pp. 159–172.

Sitbon, F., S. Hennion, B. Sundberg, C.H.A. Little, O. Olsson, and G. Sandberg. 1992. Transgenic tobacco plants coexpressing the *Agrobacterium tumefaciens iaa*M and *iaa*H genes display altered growth and indoleacetic acid metabolism. *Plant Physiol.* 99: 1062–1069.

Sponsel, V.M. and P. Hedden. 2004. Gibberellin biosynthesis and metabolism, in *Plant Hormones, Biosynthesis, Signal Transduction, Action!* 3rd ed. Davies, P.J. (Ed.), Kluwer Academic Publishers. Dordrecht, The Netherlands, pp. 63–94.

Sun, T.-P. 2004. Gibberellin signal transduction in stem elongation and leaf growth, in *Plant Hormones, Biosynthesis, Signal Transduction, Action!* 3rd ed. Davies, P.J. (Ed.), Kluwer Academic Publishers. Dordrecht, The Netherlands, pp. 304–320.

Visser, E, J.D. Cohen, G. Barendse, C. Blom, and L. Voesenek. 1996. An ethylene-mediated increase in sensitivity to auxin induces adventitious root formation in flooded *Rumex palustris* Sm. *Plant Physiol.* 112: 1687–1692.

Walton, D.C. and Y. Li. 1995. Abscisic acid biosynthesis and metabolism, in *Plant Hormones, Biosynthesis, Signal Transduction, Action!* 2nd ed. Davies, P.J. (Ed.), Kluwer Academic Publishers. Dordrecht, The Netherlands, pp. 140–157.

Werner, T., V. Motyka, M. Strnad, T. Schmülling. 2001. Regulation of plant growth by cytokinin. *Proc. Natl. Acad. Sci. U.S.A.* 98: 10478–10492.

Zhang, Z. and Z.-M. Cheng. 1996. The effect of jasmonic acid on *in vitro* nodal culture of three potato cultivars. *HortScience* 31: 631.

15 Juvenility and Its Effect on Macro- and Micropropagation

Caula A. Beyl

CHAPTER 15 CONCEPTS

- As plants transition from juvenile to mature, they may exhibit changes in leaf shape, pigmentation, stem characteristics, and rate of growth. The most important change for propagators is that, as a plant matures, it becomes increasingly more difficult to induce adventitious roots.

- The area of the tree containing tissue that is the most juvenile is called the "cone of juvenility" and encompasses a vaguely cone-shaped region broadest at the crown and diminishing as the distance from the crown increases for both the aboveground portion and the root system.

- Various techniques have been used to rejuvenate mature plants so that they are more amenable to clonal propagation including hedging, serial propagation, serial grafting, shoot forcing, etiolation and blanching, and tissue culture.

- Shoots at the top of the tree are the youngest chronologically, but from an ontogenetic standpoint, are the most mature. Those at the base of the tree or closest to the crown are the oldest chronologically, but the most juvenile ontogenetically.

- The transition from juvenile to adult is believed to be epigenetic in nature, with internal or external cues changing gene expression, but not the genes themselves, in a stable and persistent manner.

- Not only is juvenile tissue the most valuable from a macropropagation point of view, but juvenile tissue is also more amenable to establishment in tissue culture (micropropagation), typically exhibits better shoot proliferation rates, and the *in vitro* shoots are more amenable to rooting than for mature tissue.

As woody and herbaceous plants grow from seedlings to fully mature flowering plants, a process occurs that is much akin to the changes that we go through as we transition through childhood, adolescence, and adulthood. Maturity is attained when the apical meristem changes from vegetative to flowering in response to some environmental stimulus to flower, usually day length or chilling. In its juvenile phase, a plant is incapable of flowering or sexual reproduction even if the proper environmental cues to flower are given. This process is of profound impact to plant propagators because one of the characteristics of the mature phase is the difficulty encountered with vegetative propagation either by traditional means or tissue culture. Often the true value of an ornamental, forest, or fruit selection cannot be determined until it is fully mature and has exhibited the flower, form, or fruit characteristics that make it unique and prized. However, this is also the time when it is the most difficult to propagate either by conventional cuttings or *in vitro*.

Plant scientists often use terms, such as chronological age, physiological age, and ontogenetic age, to describe plants as they go through this "change of phase." Chronological age of a tree refers to how many years it has been growing. Chronological age of a tissue refers to how long it has been in existence. As time passes, the chronological age increases. A new shoot at the top of the tree, which has developed in the current growing season, is chronologically young. Plant scientists also talk about biological or physiological age, which refers to the degree of maturation caused by hormonal or environmental and nutritional changes. The tissues formed when the tree was a seedling or very young are the most juvenile and these are located in the part of the tree that is closest to the crown (the point where the trunk meets the roots). This also means that the new shoot at the top of the tree, which is "young" chronologically, is the most "mature" physiologically. Another term important in describing the processes involved in maturation phenomena is "ontogenetic aging," which refers to the relatively irreversible changes in gene expression resulting in maturation.

Typically, the transition to the mature form is not an abrupt one. Individual traits that are characteristic of the juvenile phase may disappear at different rates as the plant ages, and new traits characteristic of the mature phase, such as a different leaf form, flowering, and loss of rooting ability, may appear at different times during ontogenetic development. The change of phase from juvenile to mature is not genetic, but believed to be epigenetic. Epigenetic changes are caused by differential expression of genes, or genes being turned on and off, resulting in cells and tissues with distinct attributes. Some features of the juvenile phase are characteristics, such as a tendency toward rapid growth, different forms for leaves and stems, thorniness, leaf retention, and, most importantly to propagators, easier adventitious rooting than in the mature phase. In some plants, the ability to root easily is compromised very quickly, and in others, the difference in the ability to root between the juvenile and mature forms may not be as extreme. Maturity and the loss of rooting competence of woody shrubs and trees is a daunting challenge to propagators, and for some species, such as fringetree (*Chionanthus virginicus* L.) and pawpaw (*Asimina triloba* (L.) Dunal), inducing adventitious roots may continue to be a challenge for the next generation of propagators.

EPIGENETIC NATURE OF THE JUVENILE-TO-MATURE TRANSITION

The nature of new cells, tissues, and organs is determined by the master architects, the apical meristems. The process of ontogenetic aging is dependent upon the meristems or, in actuality, the number of cell divisions behind each meristem. If you look closely at the juvenile apical meristems, you will find that they are smaller than their mature counterparts and contain cells with smaller nuclei. Architects direct construction of new buildings based on instructions, and in the apical meristem, these instructions reside in the genes and the way that they are expressed. The juvenile-to-mature transition is said to be epigenetic in nature because factors, such as the environment, influence how a cell behaves as a result of its genetic programming (genotype), not by changing the DNA directly but by changing how it is expressed. This gene expression is stable and tends to persist with each successive cell division occurring in the meristem, thus resulting in juvenile or mature tissue in propagules. An architect must sometimes alter the plans when the environment changes. So too does the apical meristem change when gene expression is altered in response to internal factors (ontogenetic or hormonal) or external factors (photoperiod or chilling), ultimately resulting in either the mature phenotype or a transitional form in between juvenile and mature. These changes are described as epigenetic and, once induced, they are perpetuated in the clones.

JUVENILE TRAITS

Many woody plants have different leaf forms while they are juvenile than they exhibit when they are mature. Sometimes the change is expressed as a difference in the degree of lobing and, in others, juvenile leaves may be simple, whereas the mature leaves may be compound. One of the plants most often cited when the differences between juvenile and mature plants are related is English ivy (*Hedera helix* L.). Juvenile leaves are lobed (Figure 15.1a), whereas the mature leaves (Figure 15.1b) are entire (lack lobing). Differences between the two forms can also be seen in the stem which, when juvenile, supports many adventitious roots. Shoot growth of the juvenile form is vine-like, whereas the more mature form is more like an upright shrub. Juvenile passion flower (*Passiflora*) has cordate (heart-shaped) leaves, but the mature passion flower has very divided leaves, as does *Pothos*, a tropical plant often used for interior landscaping. For conifers, such as Japanese cedar (*Cryptomeria japonica* D. Don), juvenile foliage is awl-shaped, and the more mature foliage is scale-like. Juvenile leaves of many species are pubescent or hairy, such as the leaves of juvenile pecan trees (*Carya illinoinensis* (Wang) K. Koch.), which have a great deal of pubescence particularly on the lower surfaces. Mature pecan leaves are shiny and lustrous. The arrangement of leaves on the stem or "phyllotaxy" may also be different between juvenile and mature stems; for example, juvenile grape (*Vitis vinifera* L.) stems have a 2/5 phyllotaxy (five leaves arranged in two spirals around the stem before a leaf is directly above the leaf used as the starting point) and no tendrils. Mature grape stems have a 1/2 phyllotaxy and tendrils.

Just as traits like leaf shape and pubescence may vary with the ontogenetic phase, so may pigmentation. For some species, such as pecan, young leaves and even shoots of juvenile plants have a reddish pigmentation in the spring. Leaf retention is another characteristic of juvenile trees. This characteristic can be seen most vividly in beech (*Fagus* spp.), oaks (*Quercus* spp.), and some maples (*Acer* spp.). On older trees, the pattern of leaf retention is a reflection of the portions of the tree that are the oldest chronologically, but the most juvenile. Those areas nearest the crown and the center of the tree are retained the longest (Figure 15.2), and this also gives an idea of the shape of the tree when it was younger. Leaves of juvenile trees are also able to tolerate shade, whereas leaves of mature forms may not. Not only leaf characteristics can be different between juvenile and mature forms, but also stem characteristics. Juvenile seedling trees of honey locust (*Gleditsia triacanthos* L.) and *Citrus* spp. have an abundance of thorns relative to their mature counterparts.

FIGURE 15.1 *Hedera helix* in its (a) juvenile form is a vine bearing lobed leaves. The (b) mature form has a more compact and shrub-like growth habit than the juvenile form, as well as a more rounded leaf shape with no pronounced lobing.

As the age of the plant increases, it begins to lose juvenility, which has an impact on clonal propagation success in terms of root induction on stem cuttings, shoot formation on root cuttings, and either organogenesis or embryogenesis in tissue culture. In traditional propagation systems, this loss of juvenility is expressed in terms of a lower percentage of cuttings that form roots or the length and number of roots formed. Sometimes, it may even be expressed in terms of how well the cuttings survive and grow on after forming roots.

JUVENILITY AND ITS EFFECT ON MACROPROPAGATION SUCCESS

Loss of rooting competence presents a difficult problem for horticulturists and foresters, as they normally select superior clones based on mature characteristics, such as a vivid flower color; fruit color, shape, or taste; shoot architecture or form; and wood characteristics. However, once the candidate for cloning has attained maturity and displayed its characteristics, it has now also attained the dubious distinction of losing the ability to form adventitious roots and shoots easily.

This loss of ability for adventitious root formation can be quite dramatic; for example, the cotyledonary node of eucalyptus (*Eucalyptus grandis* W. Hill ex Maiden) has good rooting potential, but by the time the fifteenth node has developed, the rooting potential is completely lost. Many plant scientists have demonstrated a loss of the ability to root as the stock plant ages, expressed as a low percentage of rooting and/or longer rooting times. In some cases, higher concentrations of IBA (indolebutyric acid) are needed to induce roots from mature cuttings than those needed for juvenile ones.

Taking cuttings from the more juvenile portions of a mature tree, often referred to as the "cone of juvenility" (Figure 15.3), increases the chances of obtaining adventitious roots. This has been demonstrated with many species, such as olive (*Olea europaea* L.), Norway spruce (*Picea abies* (L.) Karst), and apple (*Malus domestica* Borkh.).

FIGURE 15.2 Leaf retention is a juvenile characteristic in deciduous trees. The pattern of leaf retention in this oak tree vividly illustrates the branches within the cone of juvenility.

These juvenile tissues are those that were formed when the tree was young and include the crown of the tree where the shoot joins the root, the lower trunk and the interior portion of lower limbs. In a similar pattern, the most juvenile portions of the root are those nearest the crown. The closer the apical meristem is to the roots of the plant, the more juvenile it is likely to be. This feature is exploited by techniques such as hedging or stool bedding that employ severe pruning to decrease the distance between the new growth and the root system, thus acting to rejuvenate the plant and benefit from the ease of rooting that is characteristic of the juvenile phase. The outside of the mature tree or the area farthest away from the crown and root system is also the most mature or adult and, hence, the hardest to propagate via cuttings. Root suckers (vigorous shoots arising from the roots), because of their proximity to the roots, are often very easy to root adventitiously, a trait put to use by propagators of hard-to-root species.

JUVENILITY AND ITS EFFECT ON MICROPROPAGATION SUCCESS

The mature or adult form also represents a challenge to successful micropropagation as well as macropropagation. The first difficulty is often disinfesting the explant. Explants taken from mature material and disinfested with bleach or hydrogen peroxide solutions are prone to necrosis (death of cells and tissues), whereas juvenile material is less sensitive. With *in vitro* propagation, just as in conventional macropropagation, the juvenility of the plant that is the source of the explant tissue profoundly influences how successful the tissue culture propagation will be, whether an organogenesis route or an embryogenesis route is intended. Mature explants also have a greater tendency to produce phenolic compounds, which leach into the medium, giving it a dark appearance and often becoming toxic to the explant. This is a serious limitation with establishment of some woody species, such as black walnut (*Juglans nigra* L.). Even the callus derived from explants taken from juvenile plants may be different with respect to cell size, rate of growth, and ability to form roots.

TECHNIQUES TO CIRCUMVENT MATURITY-RELATED LOSS OF ROOTING COMPETENCE

Propagators faced with recalcitrant species that are difficult to root can experiment with various techniques that reinvigorate or partially rejuvenate the plant, thereby making it easier to root adventitiously. Some of these techniques have become mainstays of the commercial nursery industry for propagation of certain species. These include hedging and stool bedding, etiolating and blanching, serial propagation or serial grafting, application of growth regulators, establishment and culture *in vitro*, and the use of specialized structures such as lignotubers.

Juvenility and Its Effect on Macro- and Micropropagation

FIGURE 15.3 A stylized tree showing regions both above ground and below ground that contain the most juvenile tissues, those tissues nearest to the crown. This is referred to as the "cone of juvenility." These juvenile areas of the tree are shaded darkly and the transition to more mature tissue is indicated by the gradual loss of shading. (Drawing courtesy of Collier Collins, Portland, OR.)

Hedging and Stool Bedding

Severe pruning can be used to rejuvenate hard-to-root species and make them more amenable to cutting propagation. Repeated shearing or hedging has been a useful technique to maintain juvenility of Monterey pine (*Pinus radiata* D. Don) for up to six years and yellow cedar (*Chamaecyparis nootkatensis* (D. Don) Spach) for up to ten years. When a stock plant is hedged, terminal buds on the periphery of the tree, which are also the most mature, are removed. The new shoots that form from dormant buds nearer to the crown are more juvenile. With stool bedding, the plants are pruned back to the ground in the dormant season, and the shoots that form in the spring have juvenile characteristics and are called "juvenile reversion shoots." Stool bedding or stool bed layering is a common practice for the production of rootstocks of apple. The same severe pruning techniques can be applied to other difficult-to-root species. If poten-

tially valuable selections are divided into two groups, one group can be maintained in the juvenile state as a source of cutting material, and the other group can be allowed to mature normally for evaluation of adult characteristics.

Etiolating and Blanching

Exclusion of light from green growing shoots using a light barrier results in etiolation and, for many plants, promotes greater adventitious root formation. The practice of applying a band of black tape or Velcro around a green shoot to exclude light is called "blanching." Both practices of light exclusion have been useful in promoting greater rooting frequency (see chapter 22 and chapter 42). When McIntosh apple (*Malus domestica*) shoots were blanched until they had reached 7.5 cm and the basal portion of the blanched shoot was kept in darkness with tape, cuttings taken from these shoots rooted at very high percentages. The effect of etiolation in promoting the rooting of apple shoots tends to persist even if the etiolated shoots are allowed to become green for several weeks, and even up to nine months later, an enhancement of rooting can be seen. Avocado (*Mangifera indica* L.) shoots lose the etiolation-induced enhancement of rooting success with light exposure gradually over a week's period of time, but if the base is kept in the dark longer for up to five weeks, even though they have been exposed to light for seven days prior to taking the cuttings, they root with a high success rate. This same promotive effect of light exclusion on adventitious root formation has been documented in many species and across family lines. Insertion of cuttings into a rooting medium also has the effect of excluding light to the basal portion of the cutting. Although exclusion of light affects auxin and other rooting cofactors, that effect is transitory and does not explain the persistence of the light exclusion effect. The changes that are induced that enhance rooting may be epigenetic, acting directly on the meristem and the cells derived from it and resulting in altered expression of the genes that regulate the adventitious root formation process.

Exclusion of light is a technique with value for micropropagation as well. Shoot explants taken from sweet chestnut shoots that had been stripped of leaves and then wrapped in foil to exclude light for four months had not only better establishment in culture (79% relative to 22%), but also had enhanced shoot proliferation.

Serial Propagation and Grafting

Serial propagation by taking cuttings at intervals, rooting them, and using them to replace the original stock plant helps to maintain juvenility or, if the plant is already mature, helps to rejuvenate it. The ontogenetic maturity in this case is influenced by the proximity of the meristem to the new adventitious root system or perhaps some physiological influence related to the newly generated roots. Juvenility can be maintained for many years using serial propagation. Grafting a scion onto juvenile rootstocks also acts to partially reverse ontogenetic aging, an effect that is compounded when the grafting is repeated several times. It has been proposed that the endogenous growth regulators and concentrations being produced by the roots are responsible. The rejuvenation effect is also enhanced by minimizing the size of the mature scion and also allowing leaves to remain on the juvenile rootstock. The presence of leaves on the adult scion can inhibit the process.

Many of the same treatments that have been developed to circumvent maturity-related loss of rooting competence for cuttings also benefit tissues to be cultured *in vitro*. For example, when plants of sweet chestnut (*Castanea sativa* Mill.) were serially grafted four times and then compared to adult trees, not only did explants of those that had been serially grafted have better establishment and excellent shoot proliferation, but also the microshoots derived from them rooted better than their counterparts from adult explants.

HIGH TEMPERATURE EXPOSURE

When mature plant material is exposed to high temperature, growth of juvenile shoots results. If potted plants are exposed to bottom heat, juvenile shoots tend to form at the bases. This technique was applied to plants of Japanese witch hazel (*Hamamelis japonica* Sieb. & Zucc.), saucer magnolia (*Magnolia soulangiana* Soul.-Bod.), and star magnolia (*M. stellata* Maxim.) to obtain explants that were amenable to *in vitro* culture. In another example, temperatures of 27°C enhanced the juvenile reversion effect of grafting mature Algerian ivy (*Hedera canariensis* Willd.) onto juvenile rootstock.

GROWTH REGULATOR TREATMENTS

The role of various growth regulators in promotion of rooting is well recognized. Application of exogenous auxins, particularly IBA or NAA (naphthalene acid), to the bases of cuttings is a well-accepted horticultural practice that promotes rooting. This clear-cut benefit of localized application of a growth regulator to promote rooting is in contrast to the much less clear effects of applying growth regulators to plants to rejuvenate them and thus promoting rooting. Treating mature shoots of English ivy with gibberellins can cause them to develop juvenile characteristics again. For other plants, either spraying the entire plant or injecting it with cytokinins provides the stimulus for a reversion to a juvenile-like state. For some conifers, production of female and male cones, the signature characteristic of maturity, can be induced with auxin/gibberellin combinations or gibberellin (GA) alone. For some woody plants, spraying with combinations of a cytokinin and a gibberellin can result in vigorous new juvenile growth that could serve as a source of cuttings. When propagators cut dormant shoots and bring them indoors to force bud break and growth, if they including a cytokinin in the forcing solution, the forcing can be more successful. For narrowleaf ash (*Fraxinus angustifolia* Vahl), the presence of the cytokinin, benzyladenine (BA) in the forcing solution promoted more bud sprouting than inclusion of GA_3 or the auxin, NAA.

Application of growth regulators can sometimes aid in producing shoots that have more juvenile characteristics and serve as a better source of explant material, but if gibberellins are used, they can result in poor establishment, interference with root formation, and inhibition of shoot formation.

TISSUE CULTURE

The tissue culture environment can serve as a means of reinstating the juvenile state or a pseudo-juvenile state. Because the mature influence is not erased completely, as evidenced by the fact that plantlets can sometimes be induced to flower *in vitro*, or plants which had been developed from *in vitro* material seem to be able to flower sooner than if they had been seedling material, some scientists prefer the terms "apparent rejuvenation" or "re-invigoration" rather than rejuvenation. If mature tissues are established *in vitro*, over time, they begin to exhibit more juvenile characteristics, such as faster growth rate of shoots, smaller leaves on shoots, and easier induction of adventitious roots. The advantages of going through a tissue culture cycle also extend to when the plantlet has been acclimatized and established. When tissue cultured plants are used as stock plants, the shoots taken from them root more easily and at higher rates than those taken from conventional stock plants. This technique of using tissue culture-derived stock plants has worked well with common lilac (*Syringa vulgaris* L.) and 'Kwanzan' flowering cherry (*Prunus serrulata* Lindl.).

Rooting may still be challenging in the *in vitro* environment if mature explants are used to establish the cultures. With mature cultures of *Eucalyptus grandis*, microcuttings taken from juvenile shoots rooted with 60% success, as opposed to those from mature shoots that only rooted with 35% success. In other studies with several *Eucalyptus* species, roots could not be induced on nodes from mature trees, although they could be induced on those from one-year-old seedling trees. Not all species exhibit the same difference, as evidenced by rooting of microshoots from both mature and juvenile silver maple (*Acer saccharinum* L.) and shoots from mature and juvenile apple rootstock.

LIGNOTUBERS AND BURLS

Some trees and woody shrubs form specialized woody swollen structures near the root crown called "lignotu-

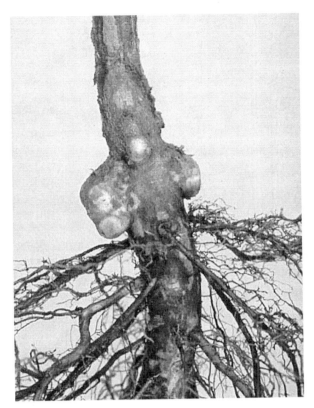

FIGURE 15.4 The large swellings just beneath the ground are lignotubers of Eucalyptus. These are thought to be a means of ensuring survival, since new shoots sprout from the fire resistant lignotuber that survives fire events. (Image courtesy of Dr. Ed Barnard, Gainesville, FL.)

bers." Sometimes these are referred to as "burls," but the term burl can refer to any swelling that has swirled woody grain or is the result of injury or infections, and technically these are not the same as lignotubers. Burls form higher on the plant than lignotubers. What makes lignotubers different and important from a propagation standpoint is that ontogenetically they are juvenile. Lignotubers contain buds and meristematic tissue that developed early in the life of the seedling in the axils of the cotyledons or the first few seedling leaves and retain their juvenile nature. When a lignotuber is induced to form shoots, the shoots exhibit juvenile characteristics including the ability to root easily. Lignotubers evolved as a mechanism for survival. If the original shrub or tree is damaged, the dormant buds become active generating suckers, one of which usually ends up becoming dominant. The Mediterranean shrub tree heath (*Erica arborea* L.) regenerates from the lignotuber after fires; in fact, the fire-resistant lignotubers are used to make pipes. Examples of other species that form lignotubers are *Eucalyptus* spp. (Figure 15.4), many of the Ericaceae family, such as mountain laurel (*Kalmia latifolia* L.) and *Rhododendron* spp., creosote bush (*Larrea tridentata* (D. C.) Colville), and redwood (*Sequoia sempervirens* D. Don) Endl.). Aerial lignotubers of maidenhair tree (*Ginkgo biloba* L.) are abundant on older trees. In Japan, these swellings on the trunk are called "chichi" meaning mothers' breasts because of their pendulous appearance (Figure 15.5). The *chichi* can grow into the ground, root, and form shoots. Instant bonsai can be created by severing the lignotuber of Ginkgo in the dormant season, and placing it upside down in the growing medium. New shoots quickly develop from the lignotuber, resulting in the "stalactite trunk" bonsai style.

SUMMARY

As you have discovered, the age and state of maturation of a plant has a profound effect on how easily it can be propagated vegetatively. Through observation and experimentation, propagators have developed various techniques

FIGURE 15.5 Venerable ancients like the "Ginkgo of Kitakanegasawa" in Aomori Prefecture, the oldest and biggest Ginkgo in Japan, bear numerous "chichi" on their trunks. These pendulous formations can grow into the ground, root, and form shoots. Bonsai trees can be developed from the severed, inverted chichi. (Photo courtesy of Tatsuo Komori, Yokohama, Japan.)

and practices that allow them to optimize the chances of rooting cuttings successfully. These include etiolation and blanching, hedging and stool bedding, serial grafting, high temperatures, growth regulator treatments, *in vitro* culture, and use of specialized structures. Successful plant propagators have found some techniques more successful with some species than others. This is why patience, experience, and education are all the hallmarks of a good propagator.

EXERCISES

EXPERIMENT 1: DIFFERENCES IN ROOTING RESPONSE OF LEAF PETIOLE CUTTINGS OF JUVENILE AND MATURE FORMS OF *HEDERA HELIX*

A classic example of the difference between juvenile and mature forms of the same species exists for *Hedera helix* (English ivy). Not only is leaf shape and growth habit different for the two phases, but also the phyllotaxy (the way leaves are arranged on the stem) changes from alternate in juvenile stems to spiral in adult stems. Stem pigmentation and whether or not the stem has adventitious roots also varies with the maturity. This species represents an excellent system for examining the effects of maturation on the ease of rooting because English ivy is readily available, and the differential response of the juvenile versus the mature phase is so well documented. Juvenile English ivy cuttings root easily with extremely high success rates, but mature English ivy roots at much lower percentages. The manner in which adventitious roots are initiated also differs in the juvenile versus the mature form. For juvenile cuttings, adventitious roots arising from the phloem and ray parenchyma of the stem appear at about two weeks. For mature cuttings, wound callus forms from cells in the pith, secondary xylem, phloem, cambium, and cortex first. It takes about four weeks from the time the cutting is taken before roots begin to emerge from the callus at the base of the cutting. If leaf petiole cuttings and some form of auxin are used to initiate roots, a similar pattern is seen. Juvenile leaf petiole cuttings root easily, but mature cuttings root more infrequently and slowly, since the roots form from tissue that must dedifferentiate into callus and then redifferentiate again into root primordia. This adds to the time required for a rooting response to be seen.

Materials

The following items are needed for each student or group of students:

- Ten leaf petiole cuttings from juvenile and 10 leaf petiole cuttings from mature English ivy. A leaf petiole cutting consists of the leaf with most of the petiole still attached. Each student or group may act as one replication of the treatment.
- 0.1% (1000 ppm) IBA. Commercial formulations in talc can be used for convenience.
- Weighing boats or petri dishes to hold the rooting powder.
- Sharp knife or pruning shears for severing the petioles and blade from the stem.
- Labels and markers.
- Flats containing 1 peat: 3 perlite (by volume), one per student or group. Dry peat is sometimes resistant to wetting. Adding a little wetting agent (Tween, Triton X, or dishwashing detergent) to the water helps.
- Intermittent mist system.

Follow the experimental protocols outlined in Procedure 15.1 to complete this experiment.

Anticipated Results

Petioles of the juvenile leaf cuttings will form roots more rapidly in response to auxin treatment because they contain preexisting competent cells that are able to respond to the auxin and then form roots. Petioles of the adult leaf will form callus in response to auxin, but then some of the callus cells divide to form root primordia. This process takes a longer time.

Questions

- What morphological differences did you observe between the juvenile and mature English ivy specimens from which you took the leaf petiole cuttings?
- Would results have differed if leaf bud cuttings had been used? Stem cuttings?
- In this experiment, you measured the percentage of cuttings forming roots, the time required until emergence of the first roots, and the number of roots formed per cutting. What other parameters could you have observed to determine the response to the IBA treatment as a function of maturity?
- What variations of this experiment could have been conducted to explore more fully the concept of change of rooting competence?

EXPERIMENT 2: FORCING AND ROOTING (1) EPICORMIC SHOOTS FROM DETACHED STEM SEGMENTS OR (2) SUCKERS FROM ROOT SEGMENTS

On mature trees, the easiest material to take for cuttings is on the periphery of the tree, but that propagation material is more mature than the tissues of the main trunk and interior portions of the lower branches. The area within the cone of juvenility has the best potential for rooting, but may not offer a large supply of suitable shoots. Fortunately,

Procedure 15.1
Comparison of Rooting Response of Leaf Petiole Cuttings from Juvenile and Mature Forms of English Ivy
(*Hedera helix*)

Step	Instructions and Comments
1	Enough shoots of both the mature and the juvenile forms of English ivy should be obtained to yield at least 10 leaf petiole cuttings of each type for each student or group of students.
2	Before the shoots are used to obtain cuttings, students should note the characteristics that indicate the state of maturity in each of the forms.
3	Leaves that are healthy and have undamaged blades should be used. Using a knife or pruning shears, students should sever the petioles of chosen leaves at a position on the petiole just above where it connects to the stem.
4	About 1 cm of the base of the leaf petiole should be dipped into the 0.1% IBA in talc. When using IBA in talc, always use a clean scoop to remove some from the original container and place the powder in a container such as a plastic weighing boat or a petri dish. That way, when cuttings are dipped into the powder, the material in the original container is not contaminated. The petiole should be tapped to remove the excess. When done, do not put the unused IBA in talc back into the container.
5	Place the cuttings in rows in the plastic flat containing 1 peat: 3 perlite (by volume), and then label the rows with the name of the treatment. Label each flat with the name of the student or group that has performed the treatment.
6	After two weeks, students should begin periodically checking the cuttings for root formation. Carefully record the date that the first roots emerge, so that elapsed time to first root emergence can be compared between the mature and juvenile cuttings.
7	At the end of the six-week period, terminate the experiment and collect data on percentage of cuttings rooted and number of roots induced.
8	The two treatments can be statistically compared (chapter 3) if cuttings prepared by each student or group are treated as replications. Alternately, comparisons can be made within each group if each leaf petiole cutting is considered one experimental unit or replication.

epicormic shoots can be forced from branch segments that are cut, laid horizontally in a well-drained substrate like perlite, and placed under mist. These shoots can then be used for shoot cuttings and rooted relatively easily. With species that sucker readily, below ground, within the cone of juvenility, large roots can be severed near the crown, cut into lengths, and suckers or shoots can be forced as well. These also serve as a very good source of cuttings and can be rooted with IBA treatment and intermittent mist.

Materials

The following items are needed for each student or group of students:

- Pieces of the main stem or branches, approximately 40 to 50 cm long, harvested from within the cone of juvenility of a mature silver maple (*Acer saccharinum* L.). Other species may prove suitable for this exercise, such as *Eucalyptus* spp., *Quercus* spp., *Betula* spp., and *Pyrus* spp. Ideally, the branches should be about 5 to 10 cm (~2 to 4 in.) in diameter, but larger diameters will work as well. One piece may be used for each student or group, representing one replication of the treatment
- Pieces approximately 20 cm (~8 in.) long, harvested from thick horizontal roots near the crown of trees or shrubs that are known to sucker can be used. Suggested species that are suitable for forcing shoots are aspens (*Populus* spp.), oak leaf hydrangea (*Hydrangea quercifolia* Bartr.), yaupon holly (*Ilex vomitoria* Ait.), and lacebark elm (*Ulmus parvifolia* Jacq.), although any species that naturally produces abundant suckers will work well
- Standard flats filled with coarse perlite, for forcing stem and root segments, one for each student or group of students
- Standard flats filled with 1 peat:3 perlite (by volume) for rooting of the forced shoots
- Intermittent mist
- Metric ruler

Follow the experimental protocols outlined in Procedure 15.2 to complete this experiment.

	Procedure 15.2
	Forcing and Rooting: (1) Epicormic Shoots from Detached Stem Segments or (2) Suckers from Root Segments
Step	Instructions and Comments
1	Each student or group of students should prepare one stem segment that is 40 to 50 cm (~20 in.) long. The stem segment should be embedded horizontally into the flat containing perlite so that only about half of its diameter is exposed. The student or group of students should then prepare a root segment that is 20 cm long and embed the piece into the flat in the same way as was done with the stem segment.
2	The flat should be placed into a greenhouse at 25°C and watered well. The flats should be misted frequently. A cycle of 6 sec. every 6 min can be used. If necessary, a broad-spectrum fungicide/algaecide like Zerotol can be applied as a drench to discourage fungal and algal growth in the flats.
3	Begin examining the stem and root segments for shoot induction weekly, and record the dates when the shoots first emerged.
4	Four to six weeks after forcing the stem and root segments, shoots that are large enough to be used as cuttings (>5 cm long) may be measured for shoot length, harvested, and then used in the next phase of the experiment.
5	Immediately after collection, shoots should be trimmed to uniform lengths, their basal ends dipped into 0.1% (1000 ppm) IBA in talc and tapped to remove the excess, and then inserted into the flats containing rooting medium, 1 peat: 3 perlite (by volume).
6	The flats should be watered thoroughly and then placed under intermittent mist on a schedule of about 4 sec. every 6 min. This is also a good opportunity to acquaint students with how intermittent mist timers are set to obtain various schedules of misting. Mist is only necessary during daylight hours.
7	After six weeks, the cuttings can be examined for rooting. Collect data on percentage of cuttings rooted, number of roots, and length of longest root. What other kinds of data can be collected?
8	The two treatments can be statistically compared if cuttings prepared by each student or group are treated as replications.
9	Variations of this experiment can include comparisons of different cultivars within a species, comparisons among several species, use of different lengths of stem or root segment, and using segments of stems farther up the tree or roots farther away from the crown.

Anticipated Results

Within 28 days, shoots should have developed from the stem segments. Depending on the species chosen for the exercise, root segments may begin yielding root suckers at about the same time or a little bit later. Both sources of shoots should provide cuttings that root more easily than cuttings taken from the mature portions of the tree, although, depending again on the species chosen, shoots derived from root cuttings may exhibit a better rooting response.

Questions

- What problems could occur with the protocol used above if the shoots being forced were intended to be used as a source of explant material for *in vitro* culture?
- How can the diameter and the length of either the stem or the root segment being used affect the initiation and growth of shoots?
- Would the response in terms of number of shoots forced be different if the stem segments were taken from farther up into the tree or if the root segments were taken from farther away from the crown?
- Will all cultivars of the same species respond in the same way to the forcing and root inducing conditions?

SUGGESTED READINGS

Abo El-Nil, M. 1982. Method for asexual reproduction of coniferous trees. U.S. Patent No. 4353184.

Ballester, A., M.C. Sanchez, and A.M. Vieitez. 1989. Etiolation as a pretreatment for *in vitro* establishment and multiplication of sweet chestnut. *Physiol. Plant.* 77: 395–400.

Debergh, P. and L. Maene. 1985. Some aspects of stock plant preparation for tissue culture propagation. *Acta Hort.* 166: 21–23.

de Fossard, R.A., P.K. Barker, and R.A. Bourne. 1977. The organ cultures of nodes of four species of Eucalyptus. *Acta Hort.* 78: 157–165.

Giovannelli, A. and R. Giannini. 1999. Effect of serial grafting on micropropagation of chestnut (*Castanea sativa* Mill.). *Acta Hort.* 494: 243–245.

Hackett, W.P. 1987. Juvenility and maturity, in *Cell and Tissue Culture in Forestry*, Vol. 1. Bonga, J.M. and D.J. Durzan (Eds.), Martinus Nijhoff, Dordrecht, The Netherlands, pp. 216–231.

Hackett, W.P. and J.R. Murray. 1997. Approaches to understanding maturation or phase change, in *Biotechnology of Ornamental Plants*. Genever, R.L., J.E. Preece, and S.A. Merkle (Eds.), CAB International, Wallingord, UK, pp. 73–86.

Henry, P.H. and J.E. Preece. 1997. Production and rooting of shoots generated from dormant stem sections of maple species. *HortScience* 32: 1274–1275.

Libby, W.J. and J.V. Hood. 1976. Juvenility in hedged radiata pine. *Acta Hort.* 56: 91–98.

Perez-Parron, M.A., M.E. Gonzalez-Benito, and C. Perez. 1994. Micropropagation of *Fraxinus angustifolia* from mature and juvenile plant material. *Plant Cell Tiss. Organ Cult.* 37: 297–302.

Pharis, R.P. and R.W. King. 1985. Gibberellins and reproductive development in seed plants. *Annu. Rev. Plant Physiol.* 36: 517–568.

Plietzsch, A. and H.H. Jesch. 1998. Using *in vitro* propagation to rejuvenate difficult-to-root woody plants. *Comb. Proc. Intl. Plant Prop. Soc.* 48: 171–176.

Preece, J.E., C.A. Huetteman, W.C. Ashby, and P.L. Roth. 1991. Micro- and cutting propagation of silver maple. I. Results with adult and juvenile propagules. *J. Amer. Soc. Hort. Sci.* 116: 142–148.

Sankara Rao, K. and R. Venkateswara. 1985. Tissue culture of forest trees: Clonal multiplication of *Eucalyptus grandis*. *Plant Sci.* 40: 51–55.

Stoutmeyer, V.T. and O.K. Britt. 1961. The behaviour of tissue cultures from English and Algerian ivy in different growth phases. *Am. J. Bot.* 52: 805–810.

16 Chimeras

Robert M. Skirvin and Margaret A. Norton

CHAPTER 16 CONCEPTS

- A chimera is a plant with two or more genetically dissimilar tissues growing side-by-side.

- Higher plants have layered meristems that originate from a few cells in the central zone of the shoot apical meristem. The outer layers maintain their integrity because they divide anticlinally.

- Chimeras arise from genetic changes in one or more of the layers in the apical meristem.

- There are three types of chimeras: mericlinal, in which the genetically different tissue is found in part of a single meristem layer; sectorial, in which the genetically different tissue is found in part of all meristem layers; and periclinal, in which the genetically different tissue makes up one entire meristem layer.

- Chimeras can be reliably propagated only from axillary buds.

- Mericlinal and sectorial chimeras are unstable in propagation. Only periclinal chimeras can be reliably propagated.

- Mericlinal and sectorial chimeras can be stabilized as periclinal chimeras by selection of axillary shoots.

Plant chimeras have intrigued scientists and plantsmen for centuries. Their sometimes odd appearance and unpredictable behavior during propagation have made them the object of speculation and experimentation. Chimeras have been the source of numerous attractive and desirable ornamental plants, and they have given rise to valuable new fruit and vegetable cultivars. Chimeras are surprisingly common. A visit to the produce department at your local grocery store will likely turn up one or two chimeras in the citrus bins. Others may lurk in among the apples and pears. A walk through a residential neighborhood will take you by many more (Figure 16.1).

What is a chimera? A chimera is a plant with two (or more) genetically dissimilar tissues growing side-by-side. The term chimera was borrowed from mythology: The Chimaera was a mythical beast with a head like a lion, a body like a goat, and a serpent for a tail. In this chapter, we will discuss the origin of plant chimeras, their importance in horticulture, and methods used both to maintain and stabilize chimeras.

HISTORY

Some of the first known chimeras arose as a consequence of grafting. Probably the oldest chimera on record is the bizzarria orange, found in a garden in Florence, Italy, in 1644. It is believed to have originated from an unsuccessful graft of a sour orange scion (*Citrus aurantium* Linn.) onto a citron rootstock (*C. medica* Linn.). The scion died, but a shoot grew from an adventitious (not preformed) bud at the graft union. The resulting branch produced different kinds of fruit: some resembled a smooth-skinned orange, others a rough-skinned lemon, and occasionally a fruit that seemed to be part orange and part lemon. Similar plants caught the attention of botanists: a shoot that seemed to be part *Cytisus* and part *Laburnum*, arising from a graft of *Cytisus purpureus* Scop. onto *Laburnum vulgare* Berch. & J. Presl; a branch from the graft union of medlar (*Mespilus germanica* L.) on hawthorn (*Crataegus monogyna* Jacq.) that had characteristics intermediate between the two genera; and a shoot from a pear (*Pyrus communis* L.) on quince rootstock (*Cydonia oblonga* Mill.) that seemed to be a blend of both plants.

Originally, botanists thought these plants were graft hybrids that had formed when cells from the rootstock fused with cells from the scion to form new cells combining the genetic traits of both plants. Hans Winkler, a scientist, attempted to prove that graft hybridization was possible. He grafted tomato (*Lycopersicum esculentum* Mill.) scions onto black nightshade (*Solanum nigrum* L.) stocks and vice versa, allowed the union to heal, and then cut the stem across the graft union. Callus cells from both

FIGURE 16.1 Chimeras. (a) Chimeral apple with light sector. (b) Chimeral orange with light sector. (c) Chimeral orange with thick rind sector. (d) Chimeral *Hoya* with albino variegation. (e) Chimeral rubber plant with albino variegation. (f) Chimeral *Ficus* with albino variegation.

stock and scion formed over the cut surface, and eventually adventitious shoots formed from this callus. Winkler chose to use tomato and nightshade because they are related, but have distinctive leaf and fruit characteristics. Tomato has a lobed, compound leaf, whereas nightshade has an entire leaf. The tomato Winkler used had a plum-shaped red fruit; the nightshade had a small, round, dark purple fruit.

One of the adventitious shoots that formed resembled tomato on one side and nightshade on the other. Winkler called this shoot a "chimera" after the mythical beast. Eventually, Winkler produced a series of shoots that he thought were graft hybrids. The leaves of Winkler's graft hybrids were intermediate in morphology between tomato and nightshade. Some resembled a tomato leaf with less dissection, while others resembled a nightshade leaf with lobes (Figure 16.2a).

Another botanist, Bauer, developed a different hypothesis. Working with a variegated zonal geranium (*Pelargonium zonale* L. 'albomarginata'), he concluded that the leaf pattern, green center surrounded by an albino margin, was the result of genetically different tissues growing adjacent to each other in layers. Instead of fusing, the cells maintained their genetic integrity. The outer layers of albino tissue were wrapped completely around a green central core in what Bauer termed a periclinal arrangement. He linked this layering in mature tissues to the layering found in plant meristems. In his chimera hypothesis, he concluded that the epidermis was formed from the outer cell layer of the apical meristem, which he termed the LI; another layer of cells formed from the second layer of the meristem, which he called the LII; and the central part of the meristem, the LIII, formed the core of the plant. He suggested that Winkler's graft hybrids were actually graft chimeras, with an epidermis formed from tissue of one plant and a core from tissues of the other (Figure 16.2b). Bauer's concept of chimeras laid the groundwork for our understanding of chimeras.

PLANT SHOOT TIP MERISTEMS

To understand chimeras thoroughly, we need to take a close look at how a plant's shoot tip meristem is structured and how it behaves when it divides. This will help us choose the best methods to propagate chimeras.

Chimeras

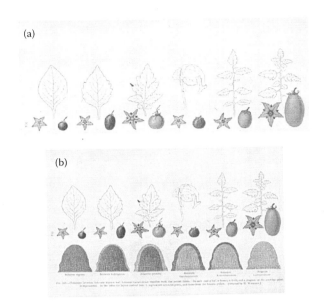

FIGURE 16.2 (a) Winkler's "graft hybrids": leaf, flower, and fruit morphologies. (b) Graft chimera explanation for the "graft hybrids." (From Fitting, H. et al. 1921. *Strasburger's Text-Book of Botany*, 5th English ed. Macmillan and Co., New York, p. 302.)

Meristem Structure

A meristem is an area of actively dividing cells. The shoot apical meristem is a region located at the topmost growing point of a plant's shoot. The apical dome is a small group of cells located above the first leaf primordium, a swelling on the flanks of the meristem that will eventually become a leaf (Figure 16.3)

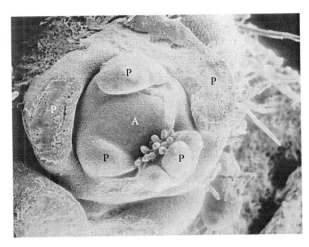

FIGURE 16.3 Scanning electron micrograph of a shoot tip meristem showing the apical dome (A) and leaf primordia (P).

In longitudinal section, the most striking characteristic of a shoot meristem, at least in higher plants, is layering. Ferns and lower plants do not have layered meristems; their tissues arise from one or more apical initials, which can contribute to any part of the plant (Figure 16.4a). Gymnosperm and angiosperm meristems, however, exhibit distinct layers. The pattern of layering differs among gymnosperms and angiosperms. Gymnosperms generally have a group of apical initials that form both the surface layer of the meristem (tunica) as well as the central core (corpus) (Figure 16.4b). Angiosperms have the most complex meristem structures. They have at least one tunica layer and often have many more (Figure 16.4c).

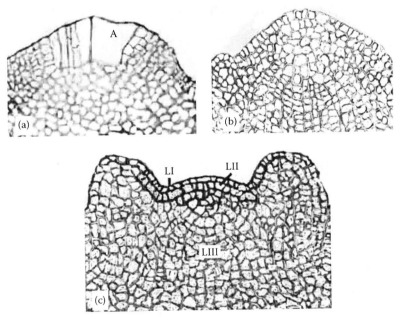

FIGURE 16.4 (a) Diagram of a longitudinal section through a fern shoot apical meristem showing the apical initial (A). (b) Diagram of a longitudinal section through a gymnosperm shoot apical meristem showing the apical initials and layering. (c) Diagram of a longitudinal section through an angiosperm shoot apical meristem showing layering.

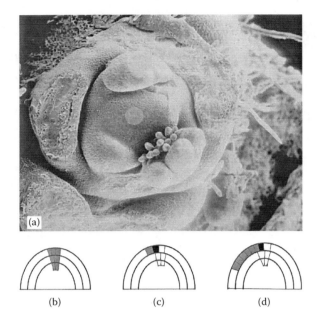

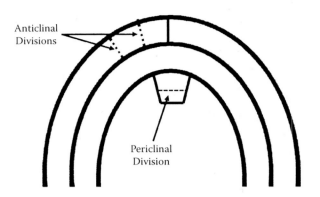

FIGURE 16.6 Diagram of longitudinal section of shoot tip meristem showing anticlinal cell division and periclinal cell division.

FIGURE 16.5 (a) Shoot apical meristem showing central zone (in circle). (b) Diagram of a longitudinal section through the shoot apical dome showing central zone cells (in gray). (c) Diagram of a longitudinal section through the shoot apical dome showing an apical initial (black) and daughter cells (gray). (d) Diagram of a longitudinal section through the shoot apical dome showing an apical initial (black) with daughter cells and their derivatives (gray).

Unlike gymnosperms, each angiosperm tunica layer has its own set of apical initials. As a result, the tunica layers are maintained distinct from one another.

The outermost layer in the meristem is designated LI, and the next layer LII (Figure 16.4c). Numbering continues, with the corpus being given the highest number. Most angiosperms have an LI, LII, and LIII, with the LIII being the corpus.

Cell Division in the Shoot Tip Meristem

All the cells that make up a plant's shoot come from a few cells in the meristem. These cells are found in the central zone of the apical dome (Figure 16.5a,b). Cells in the central zone divide slowly. One cell will remain in the central zone as an initial, while the other is pushed away from the central zone (Figure 16.5c). This second cell (daughter cell) will continue to divide for some time, producing a patch of derivative cells (Figure 16.5d). Eventually these derivative cells cease to divide. All cells of a plant's body can be traced back to one particular central zone cell.

Because the cells in the central zone divide slowly, they undergo fewer divisions than the other cells in the apical dome. As a result, there is less opportunity for an error to occur during cell division, making the central zone cells a kind of reservoir to preserve the genetic integrity of the plant. Occasionally a daughter cell will displace a central zone cell instead of being pushed outside the central zone. In this case, the daughter cell will take over the function of the displaced central zone cell, and the displaced cell will act as a daughter cell.

Cells can divide in many directions, but in the meristem, they generally divide in two planes: anticlinal (perpendicular to the central axis of the plant) and periclinal (parallel to the central axis of the plant) (Figure 16.6). Anticlinal divisions result in a sheet of cells one layer thick, whereas periclinal divisions result in an increase in plant girth.

Tunica cells in the angiosperm meristem divide almost exclusively by anticlinal divisions. It is this pattern of cell division that causes the layering in meristems. As a result of these anticlinal divisions, each layer is maintained one cell thick and remains distinct from the adjacent layer. The layers in gymnosperm meristems are less distinct because the tunica occasionally divides periclinally, and its cells mingle with cells in the corpus. Cells in the corpus divide both anticlinally and periclinally, resulting in an increase in stem length and girth.

The Ultimate Fate of Cells in the Shoot Tip Meristem

The kind of mature, differentiated cell that a daughter cell becomes depends on the meristem layer from which it originates. Cells from each layer in the meristem tend to contribute to certain tissues because of their relative position in the maturing plant. For example, in plants with a three-layered meristem, LI cells will form the epidermis of the shoot, LII cells will form tissue just beneath the epidermis, including much of the leaf lamina and the gametes in the flower, and LIII cells will form the bulk of the structure of the plant, including adventitious roots, which nearly always form from cells in or around the vascular tissue. In a plant with a two-layered meristem (LI and LII), the LI will become the epidermis (possibly many layers thick) and part of the leaf lamina, and the LII will become the remainder of the plant, including the gametes. The

FIGURE 16.7 Rex begonia (*Begonia* Rex-cultorum) is a variegated plant that has a heritable pattern in its leaves. It is not a chimera because its pattern arises from differential gene expression, not mutation.

precise contribution a particular meristem layer makes to the mature plant varies greatly among species.

The relationship of the meristem layers to tissues in the mature plant was determined using a type of chimera called a cytochimera. Cytochimeras have meristem layers with different ploidy or numbers of chromosomes. When a cell doubles its chromosome number (spontaneously or following treatment with a chemical called colchicine), the volume of its nucleus also doubles. This is easy to visualize because nuclei can be stained dark. The layers of a periclinal cytochimera that vary in ploidy level can be distinguished from each other under a microscope. A plant with a polyploid LI, for example, would appear to have an LI with large nuclei and LII and LIII cells with normal-sized nuclei; such a chimera could be labeled a 4-2-2 cytochimera (tetraploid LI, diploid LII, and diploid LIII). By examining the tissues in a mature plant derived from a 4-2-2 cytochimera, we learn that all epidermal tissues of the shoot are tetraploid and all internal tissues are diploid, and we can deduce that the epidermis is derived from the LI. Similarly, analysis of a 2-4-2 cytochimera reveals that much of the internal body of leaves and fruits as well as all gametes (male and female) are derived from the LII. Analysis of a 2-2-4 cytochimera would show that the most of the internal tissue of the plant and all roots are tetraploid and thus derived from the LIII.

HOW CHIMERAS FORM

Most people assume that all of a plant's cells are genetically identical. After all, most plants and animals begin their lives as a single cell derived from the union of male and female gametes, so theoretically all cells of an organism should be identical. However, genetic changes occur all the time in plants. Cells mutate spontaneously following exposure to cosmic rays, ultraviolet light, or intentional mutagenic treatments. These mutant cells can be maintained by ordinary cell division. Most of these cells are never observed because they occur away from the central zone of the meristem, in a region of the plant that will eventually become nonmeristematic. A mutant cell outside the central zone may divide to form an isolated patch of clonal cells, but as the plant ages, this patch of mutant cells may be shed as a leaf dies or a layer of bark falls off, or it may be buried within the plant as its girth increases. Some scientists call a plant such as this a mosaic rather than a chimera, preferring to reserve the term chimera for plants with mutations in the central zone cells, which give rise to more or less permanent streaks or layers of genetically different tissue.

Chimeras arise from spontaneous mutations as well as from grafting. Chimeras are most likely to be observed if they affect a color or morphologic characteristic. For this reason, the best known chimeras are those that express variegation in leaf color. This is most often seen as green (chlorophyll) and albino tissues growing together (see Figure 16.1d,e,f). However, just because a plant is variegated does not mean it is a chimera. Some plants exhibit genetic patterns that are passed on through the sexual cycle (Figure 16.7). Other chimeras can include observable changes in growth habit, thorniness, leaf shape, and fruit color (see Figure 16.1a,b,c), or less apparent changes in metabolic products, enzyme levels, or other biochemical processes.

KINDS OF CHIMERAS

Chimeras can be classified into three forms. These include mericlinal chimeras and sectorial chimeras, which are marked by instability, and the more stable periclinal chimeras, in which one tissue completely encloses another like a hand in a glove.

Mericlinal Chimera

Mericlinal chimeras are chimeras in which genetically different tissue occurs in part of one meristem layer.

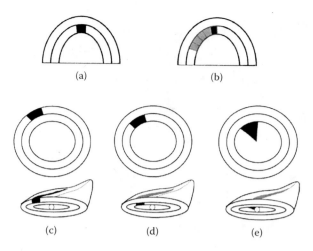

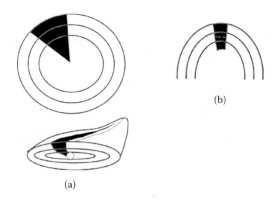

FIGURE 16.8 (a) Diagram of longitudinal section of shoot tip meristem showing mutant central zone cell in LII. (b) Diagram of longitudinal section of shoot tip meristem showing mutant central zone cell in LII, with its derivatives. (c) Diagram of longitudinal section of shoot tip meristem showing mutant central zone cell and its derivatives in the LI, extending into the leaf primordium and mature leaf. (d) Diagram of longitudinal section of shoot tip meristem showing mutant central zone cell and its derivatives in the LII, extending into the leaf primordium and mature leaf. (e) Diagram of longitudinal section of shoot tip meristem showing mutant central zone cell and its derivatives in the LIII, extending into the leaf primordium and mature leaf.

Mericlinal chimeras are common, arising when a central zone cell in one layer of the meristem mutates (Figure 16.8a). It divides to form a daughter cell, which in turn divides to form derivatives (Figure 16.8b). As the mutant cell in the central zone continues to form more daughter cells, the continual production of cells forms a streak of mutant tissue extending the length of the shoot. If the mutant central zone cell is in the LI (Figure 16.8c), its progeny will appear as a streak of mutant epidermis in the mature shoot. If the mutant central zone cell is in the LII, its progeny will lie just beneath the epidermis in the shoot and in the lamina of the leaf (Figure 16.8d). If the mutant central zone cell is in the LIII, its progeny will form a wedge of tissue in the central part of the stem and may also be seen around the midrib of the leaf (Figure 16.8e).

Sectorial Chimera

A sectorial chimera is similar to a mericlinal chimera, except that the mutation occurs in part of each layer of the meristem (Figure 16.9a). The result is a wedge of mutated tissue in the mature stem that extends from the axis of the plant to the epidermis. Sectorial chimeras often occur in seedlings early in embryo development before meristem layering has developed, but they may also arise in mature plants when daughter cells from a single mutant central zone cell divide periclinally and displace adja-

FIGURE 16.9 (a) Diagram of longitudinal section of shoot tip meristem showing mutant central zone cells in LI, LII, and LIII, with their derivatives, extending into the mature leaf. (b) Diagram of longitudinal section of shoot tip meristem showing mutant central zone cell in LII dividing periclinally to replace central zone cells in LI and LIII.

cent central zone cells in the other layers of the meristem (Figure 16.9b).

Periclinal Chimera

The most stable form of chimera is one in which all central zone cells in a single layer are mutated (Figure 16.10a). This is called a periclinal or "hand-in-glove" chimera. In this situation, cells of one genotype completely surround those of another. Periclinal chimeras most often occur when an axillary bud forms within the mutant streak of tissue in a mericlinal chimera (Figure 16.10b). Daughter cells from a single mutant central zone cell might possibly displace adjacent central zone cells in the same layer of the meristem, but this is a less likely scenario because of the multiple steps involved.

When all central zone cells are replaced with mutant tissue, the result is an entire layer (LI, LII, or LIII) of mutant tissue growing adjacent to other layers of non-

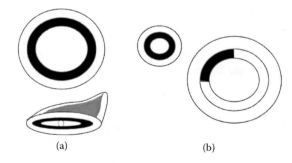

FIGURE 16.10 (a) Diagram of longitudinal section of shoot tip meristem showing mutant central zone cells in LII and their derivatives, extending into the mature leaf. (b) Diagram of transverse section across a shoot tip meristem (large circle) showing mutant central zone cells in LII (mericlinal chimera) with axillary shoot (small circle) originating in the center of the mutant streak, forming a periclinal chimera.

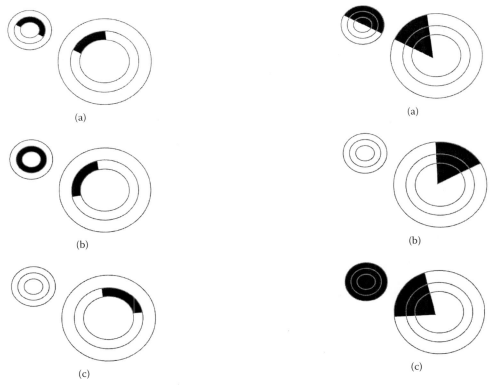

FIGURE 16.11 (a) Diagram of transverse section across a shoot tip meristem (large circle) showing mutant central zone cells in LII (mericlinal chimera) with axillary shoot (small circle) originating at the junction of mutant and nonmutant tissue. The resulting shoot will maintain its mericlinal nature. (b) Diagram of transverse section across a shoot tip meristem (large circle) showing mutant central zone cells in LII (mericlinal chimera) with axillary shoot (small circle) originating in the center of the mutant streak, forming a periclinal chimera. (c) Diagram of transverse section across a shoot tip meristem (large circle) showing mutant central zone cells in LII (mericlinal chimera) with axillary shoot (small circle) originating in a region with no mutant tissue. The resulting shoot will lose its chimeral nature.

FIGURE 16.12 (a) Diagram of transverse section across a shoot tip meristem (large circle) showing mutant central zone cells in LII (mericlinal chimera) with axillary shoot (small circle) originating at the junction of mutant and nonmutant tissue. The resulting shoot will maintain its sectorial nature. (b) Diagram of transverse section across a shoot tip meristem (large circle) showing mutant central zone cells in LII (mericlinal chimera) with axillary shoot (small circle) originating in nonmutant tissue. The resulting shoot will become nonchimeral. (c) Diagram of transverse section across a shoot tip meristem (large circle) showing mutant central zone cell in LII (mericlinal chimera) with axillary shoot (small circle) originating in the center of the mutant tissue. The resulting shoot will be entirely mutant.

mutant tissue. The periclinal chimera is called a hand-in-glove chimera because a complete layer of one tissue (the "glove") can surround a core of genetically different tissue (the "hand"). In some cases, the phenotype of the glove can mask the phenotype of the hand.

BRANCHING AND STABILITY IN CHIMERAS

Sectorial and mericlinal chimeras are less stable than periclinal chimeras during growth. Depending on where an axillary bud forms relative to a sector of mutant tissue, the meristem of the resulting shoot will maintain, lose, or change its chimeral nature. A mericlinal chimera, for example, might remain a mericlinal chimera if an axillary bud forms at the junction of mutant and nonmutant tissue (Figure 16.11a); it might become a periclinal chimera if an axillary bud forms in the center of the mutant streak (Figure 16.11b); or it might lose its chimeral status if it forms in a region that contains no mutated tissue (Figure 16.11c). Sectorial chimeras will either remain sectorial or become nonchimeral (Figure 16.12). Periclinal chimeras, however, will reproduce themselves reliably because the layer of mutant tissue is continuous (Figure 16.13).

EXAMPLES OF CHIMERAS

There are numerous examples of chimeras in horticulture. Many variegated plants are chimeral. In this case, the chimera is easily detected because the mutated tissue is a different color than the original tissue. Hostas are a good example of variegated plants that are chimeras (Figure 16.14). Hostas are monocots and have two-layered meristems. The LI forms the epidermis of the leaf as well as a border around the edge of the leaf, while the LII forms the central part of the leaf. A leaf that has a green center (Figure 16.14a) surrounded by a white margin can

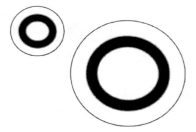

FIGURE 16.13 Diagram of transverse section across a shoot tip meristem (large circle) showing mutant central zone cells in LII (periclinal chimera) with axillary shoot (small circle). No matter where the axillary bud forms, the resulting shoot will always be a periclinal chimera.

FIGURE 16.14 (a) Hosta leaf, green center with white margin (a WG chimera). (b) Hosta leaf, light center with green margin (a GW chimera).

be termed a WG chimera (W or white comes from the LI, while G or green comes from the LII). A leaf having a green margin around a white center (Figure 16.14b) is termed a GW chimera.

Not all chimeras are readily apparent. 'Thornless Evergreen' blackberry (*Rubus laciniatus* Willd.) is a periclinal chimera in which a thornless "glove" (LI) masks the phenotype of a thorny "hand" (LII and LIII) (Figure 16.15a). A mutation in the LI of the thorny parent resulted in a thornless form with a Ttt arrangement in the three-layered meristem (T = thornless, while t = thorny). Because thorns are epidermal appendages (i.e., derived from epidermal cells) and the gene is expressed only in the epidermis of the plant (LI), a Ttt periclinal chimera is a thornless plant. However, the underlying tissues (LII and LIII) still contain the gene for the thorny character. Because gam-

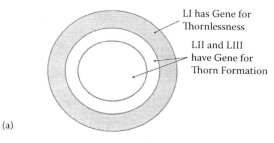

FIGURE 16.15 (a) 'Thornless Evergreen' chimeral structure. The LI contains the mutant gene for thornlessness. The LII and LIII retain the wild-type gene for thorn formation. (b) 'Thornless Evergreen' chimeral cane (bottom) and 'Thornless Evergreen' root sucker (top), which has reverted to a nonchimeral state.

etes form from the LII, which is t, 'Thornless Evergreen' breeds as though it were still a thorny plant. In addition, adventitious shoots from root cuttings are thorny because they arise from roots, which are formed from thorny genotype interior tissue (LIII) (Figure 16.15b). Blackberries are commercially propagated asexually, so their roots, which are adventitious, originate from the interior tissues of the cuttings.

PROPAGATING CHIMERAS

MAINTAINING A CHIMERA

Because of their unique arrangement of genetically different tissues, chimeras can only be propagated in a limited number of ways. Seed propagation will not perpetuate a chimera because gametes form only from the LII. Asexual propagation techniques that avoid adventitious bud formation are the only way to successfully propagate a chimera.

The various ways of propagating plants asexually can be divided into the following two general categories: techniques that produce new shoots from preformed meristems and techniques that produce new shoots adventitiously. Leaf-bud cuttings, division and layering, grafting, shoot cuttings, and tissue culture methods, which multiply plants from axillary buds, are techniques by which shoots are produced from preformed (axillary) buds. Leaf cuttings, root cuttings, and tissue culture processes involving callus, single-cell organogenesis, or somatic embryogenesis result in adventitious shoot formation. Adventitious buds tend to arise from one to a few adjacent cells and, as a result, they generally do not maintain the chimeral arrangement of tissues. Techniques that use preformed buds are the only methods that will reliably reproduce a chimera.

Axillary buds are fully formed meristems that include all the layers found in the apical meristem and develop very early in the differentiation of a shoot tip. Recall that the first signs of leaf formation occur at the base of the apical dome (see Figure 16.3). The first leaf primordium appears as a slight bulge near the apical dome. Its accompanying axillary bud begins to form shortly thereafter. Axillary buds originate from many cells and incorporate all layers of the meristem and, as a result, they are capable of maintaining the chimeral arrangement of the shoot apex.

Stabilizing a Chimera

Unstable chimeras can be stabilized by judicious propagation. For example, a leaf-bud cutting or a division taken from a mericlinal chimera can be stabilized as a periclinal chimera if the axillary bud arises from the center of the streak of mutant tissue (see Figure 16.11b). Nonchimeral plants can be recovered from sectorial or mericlinal chimeras (see Figure 16.11c, Figure 16.12b, Figure 16.12c). This is a technique frequently utilized in hosta breeding and propagation. Many variegated hostas begin life as "streaky" seedlings. Some of their cells contain plastids incapable of forming chlorophyll, and these patches of albino tissue form sectors or "streaks" that can be seen on the leaves. As the seedling matures, offsets form from axillary buds. Some of these offsets will arise from the center of a streak forming a stable periclinal chimera.

Destabilizing a Chimera

Destabilization of a chimera can be desirable. 'Thornless Evergreen' blackberry, which was discussed previously, is a stable periclinal chimera. However, the plant produces suckers (LIII tissue) during the growing season. Since these suckers arise adventitiously from roots, the new growth is thorny (see Figure 16.15b). Growers find it labor-intensive to remove these thorny suckers. In the case of a harsh winter during which canes freeze back to the ground, all new growth from the previously thornless plant will be thorny. A grower may find his entire planting of thornless blackberries has become thorny over a single winter.

Using tissue culture techniques, the 'Thornless Evergreen' periclinal chimera was induced to form adventitious shoots. Thornless shoots were selected and tested for their chimeral status. Eventually a nonchimeral thornless plant was selected and named 'Everthornless.' Because it is no longer a chimera, all its root suckers are also thornless.

Tissue Culture

Tissue culture has become a standard tool for the nursery trade. Plants produced in tissue culture arise in the following two different ways: growth and enlargement of preformed shoots (axillary buds and apical meristems) and by adventitious shoot formation. Chimeral shoots propagated at moderate to slow rates in tissue culture on medium that minimizes callus formation tend to be relatively stable because they arise from intact preformed buds that include all the layers of the apical meristem (LI, LII, and LIII). When shoots are propagated rapidly or on medium with high levels of growth regulators, a portion of the shoots that develop *in vitro* arise adventitiously. Adventitious shoots develop from either single cells or small groups of cells. Chimeras propagated on high growth regulator medium tend to segregate into pure types (e.g., green and albino plants) and off types (rearranged chimeras).

Rearranged chimeras may have value as a new cultivar, but for a nursery grower interested in clonal integrity, such off types can lead to loss of revenue, distrust among customers, and even lawsuits. Because the danger of chimeral segregation is high, micropropagation nurseries maintain clonal stability by propagating their plants at moderate to slow rates on medium that minimizes callus formation and by selecting only shoots originating from axillary buds.

Breakdown of formerly stable chimeras may be one source of somaclonal variation, or tissue culture-induced variability. For example, some hostas have a variegation pattern that is difficult to see in tissue culture, and the variegation can be lost during prolonged culture. This is an example of undesirable destabilization of a chimera.

EXERCISES

Experiment 1: Propagation of Color Variegated Chimeral Plants

Color variegated chimeras are the most readily identifiable chimeral plants because the source of the different tissues is identified by the tissue color. In this exercise, we will propagate the green and white variegated oval-leaf peperomia (*Peperomia obtusifolia* L. 'variegata,' Figure 16.16) by the following two different methods: leaf cuttings, which produce new shoots and new roots adventitiously, and two-node stem cuttings, which produces shoots from preformed meristems (in this case, an axillary bud). *Peperomia obtusifolia* 'variegata' is a GWG periclinal chimera; that is, the LI and LIII layers of the meristem have normal, green tissue, but the LII tissue has mutated to albino. The leaves of this plant have a white margin with a green center. The LI tissue is very thin and contributes little tissue to the mature leaf. The contribu-

FIGURE 16.16 Segregating adventitious shoots from a leaf cutting of the periclinal chimera, *Peperomia obtusifolia* L. 'variegata.'

tions of the LII and LIII layers of the meristem are apparent by the pattern of variegation.

The objective of this exercise is to demonstrate the effect of two different propagation techniques, leaf cuttings and stem cuttings, on the stability of the periclinal chimera *P. obtusifolia* 'variegata.'

Variegated *Sanseveria trifasciata* Prain. may be used instead of variegated peperomia, if desired. *S. trifasciata* can be propagated from leaf cuttings (adventitious shoots) or by division of offsets (preformed shoots). Nurseries sell-

ing *P. obtusifolia* 'variegata' and several variegated forms of *S. trifasciata* can be easily located on the Internet.

Materials

The following materials will be required to complete this experiment:

- Peperomia plants (the number of plants depends on class size and number of cuttings used)
- Knife
- Commercial rooting powder (0.1% to 0.3% auxin)
- Propagation mix and trays
- Humidity dome and greenhouse

Follow the experiment protocol outline in Procedure 16.1 to complete this experiment.

Anticipated Results

Two-node cuttings should root readily. Because the shoot tips have been removed, apical dominance has been broken and axillary buds will break. Shoots from the axillary buds would be expected to maintain their chimeral arrangement. Leaf cuttings will take longer to root and form shoots because the shoots must form adventitiously. Shoots would be expected to be either

---

Procedure 16.1
Propagation of the Color-Variegated Periclinal Chimera *Peperomia obtusifolia* L. 'variegata' by Leaf and Stem Cuttings

Step	Instructions and Comments
1	Obtain a single, multistemmed plant of variegated oval-leaf peperomia. Make 10 two-node cuttings, trimming the lower part of the stem to just below the first node and the upper part of the stem just above the second node.
2	Remove the lower leaf on each cutting by snapping off the petiole close to the stem. These leaves will be used for leaf cuttings. The upper leaf should remain on the stem cutting.
3	Plant the stem cuttings by dipping the lower part of the stem in a commercial rooting powder containing a low concentration (0.1% to 0.3%) of auxin. Make sure the lower node is coated with powder. Tap off the excess and stick the cutting into premoistened propagation medium. Cover with a humidity dome and move to a greenhouse. The cuttings should be kept moist.
4	Plant the leaf cuttings by sticking the petioles into premoistened propagation medium. The entire petiole up to the base of the leaf should be in the soil. Do not use rooting powder for the leaf cuttings. Cover with a humidity dome and move to a greenhouse. The cuttings should be kept moist.
5	Observe the cuttings for signs of shoot formation. This will take several weeks, especially for the leaf cuttings. Note the location where new shoots form and whether or not the shoots are variegated, albino, or entirely green.
6	Graph your results using a bar graph, comparing the number of variegated, albino, and green shoots for each cutting type.

completely albino or completely green because of their adventitious origin.

Questions

- Where did new shoots form on leaf cuttings? On stem cuttings? Were these axillary or adventitious shoots?
- Were any nonvariegated shoots formed on the stem cuttings? How might this happen?
- Were any variegated shoots formed on the leaf cuttings? How might this happen? Is the pattern of variegation the same as that of the mother plant?
- If a unique chimeral plant appeared in a propagation bench, how would you stabilize it?

EXPERIMENT 2: SEPARATION OF PHENOTYPES IN *PELARGONIUM GRAVEOLENS* L'HERTIER EX. AITON 'ROBER'S LEMON ROSE'

The scented geranium 'Rober's Lemon Rose' (Figure 16.17) is a periclinal chimera in which the morphological phenotype of the internal tissue is masked by that of the outer tissue. Normally, the cultivar has lobed leaves and small, asymmetrical violet flowers with a two-parted stigma and crinkled petals. The flowers are sterile. However, plants forming from internal tissues have highly dentate leaves and flowers with smooth petals and a five-parted stigma.

Separation of the inner phenotype is possible by propagation from regions of the plant which are derived from internal tissue. Adventitious shoots from root or leaf cuttings will express the internal phenotype.

The objective of this exercise is to separate the inner phenotype in 'Rober's Lemon Rose' by forcing adventitious shoots from internal tissue by means of root and leaf cuttings.

FIGURE 16.17 Leaf forms of segregating adventitious shoots from the periclinal chimera, *Pelargonium* X *domesticum* Bailey "Rober's Lemon Rose." Top to bottom, left to right: parental lobed, segregant dentate, segregant hairy, segregant dwarf, segregant flat leaf, segregant hairy dwarf, segregant lobed.

'Rober's Lemon Rose' is readily available by mail order. An Internet search will reveal numerous nurseries that carry this cultivar.

Materials

The following materials will be required for this experiment:

- Scented geranium 'Rober's Lemon Rose' plants (the number of plants depends on class size and number of cuttings used)
- Knife

Procedure 16.2	
Separation of Phenotypes in the Scented Geranium 'Rober's Lemon Rose'	
Step	Instructions and Comments
1	Remove 25 leaves with attached petioles from a stock plant of 'Rober's Lemon Rose.' Make sure that all traces of the axillary bud are removed. The stock plant should be retained to serve as a control.
2	Stick the leaf cuttings into a flat of premoistened propagation mix and cover with a humidity dome. Place in a greenhouse. The cuttings should be kept moist at all times.
3	Make 10 root cuttings from the stock plant. Cuttings should be at least 3 mm in diameter and about 5 cm long. Maintain root polarity by cutting the proximal end of the root straight across and the distal end at a slant.
4	Plant root cuttings horizontally in a flat of premoistened propagation medium. Cover the flat with a humidity dome. Place in a greenhouse. The cuttings should be kept moist at all times.
5	Observe cuttings for the appearance of adventitious shoots. Note the leaf morphology of each new shoot and compare it to the stock plant.

- Propagation mix and trays
- Humidity dome and greenhouse

Follow the experimental outline in Procedure 16.2 to complete the experiment.

Anticipated Results

All shoots arising from root and leaf cuttings are adventitious. Adventitious shoots from the leaves can arise from either epidermal (lobed phenotype) or internal (dentate phenotype) and may exhibit either morphology. 'Rober's Lemon Rose' stock plants are vegetatively propagated from cuttings, and their roots (which are adventitious in origin) come from LIII tissue. Therefore, all shoots arising from root cuttings would be expected to show the dentate phenotype only.

Questions

- All shoots arising from leaf and root cuttings are adventitious. Were any shoots produced that resemble the mother plant? How do you account for this?
- If a plant is not variegated and shows no obvious differences in morphology, how could you determine if it is a periclinal chimera?

LITERATURE CITED

Dermen, H. 1960. Nature of Plant Sports. *Am. Hort. Mag.* 39: 123–173.

Fitting, H., Jost, L., Karsten, G., and Schenk, H. 1921. *Strasburger's Text-Book of Botany*, 5th English ed. Macmillan and Co., New York. p. 302.

Mauseth, J.D. 1988. *Plant Anatomy*. The Benjamin/Cummings Publishing Co., Inc., Menlo Park, CA. 560 pp.

McPheeters, K. and R.M. Skirvin. 1983. Histogenic layer manipulation in chimeral 'Thornless Evergreen' trailing blackberry. *Euyphytica*. 32: 351–360.

Neilson-Jones, W. 1934. *Plant Chimeras*. Methuen & Co. Ltd., London. 123 pp.

Skirvin, R.M. and J. Janick. 1977. Separation of phenotypes in a periclinal chimera. *J. Coll. Sci. Teaching* 7: 33–35.

Steeves, T.A. and I.M. Sussex. 1989. *Patterns in Plant Development*. Cambridge University Press, Cambridge, UK. 388 pp.

Tilney-Bassett, R.A. 1986. *Plant Chimeras*. Edward Arnold, Ltd., London. 199 pp.

Part VII

Propagation by Stem Cuttings

17 Cloning Plants by Rooting Stem Cuttings

John M. Ruter

CHAPTER 17 CONCEPTS

- Vegetative propagation of plants by stem cuttings is the most commonly used method for producing herbaceous and woody landscape plants in many parts of the world.

- Stem cuttings can be classified as follows: deciduous hardwood cuttings, narrow-leaf evergreen hardwood cuttings, semihardwood cuttings, softwood cuttings, and herbaceous cuttings.

- When taking cuttings, plants should be true-to-type, stock plants should be pest free, and cuttings must be taken at the correct time of year.

- Mist systems are critical for propagation of most stem cuttings taken during the growing season.

- Care should be given to the selection of proper propagation substrates, fertility, water quality, and containers.

- Most woody cuttings benefit from the application of rooting hormones.

- Proper sanitation in propagation areas is essential for commercial success.

Vegetative propagation of plants by cuttings is the most commonly used method of propagation for herbaceous and woody landscape plants in many parts of the United States and around the world. A cutting can be described as a detached plant part that, when placed in a suitable environment, will regenerate its missing vegetative parts. These parts may be roots, shoots, or both. The three major advantages of propagation by cuttings are the following: (1) cuttings provide a more successful vegetative method of propagation compared to budding or grafting, (2) the cuttings will have the same genetics as the parent plant, and (3) a number of cuttings may be taken from one stock plant. Since many of our important fruit and landscape species are clonal (same genetic makeup as the parent), cuttings are an important method for propagating such taxa.

PROPAGATION OF PLANTS USING STEM CUTTINGS

Stem cuttings are a section of stem tissue that consists of lateral or terminal buds. Cuttings can be placed into the following categories:

- Deciduous hardwood cuttings
- Narrow-leaf evergreen cuttings
- Semihardwood cuttings
- Softwood cuttings
- Herbaceous cuttings

This type of classification is based on the maturity of the wood used as compared to basing cutting type on the part of the plant from which the cutting came (e.g., stem cuttings, leaf cuttings, leaf-bud cuttings, or root cuttings).

Deciduous hardwood stem cuttings are taken from the previous season's growth in late winter or early spring while the plant is dormant. In general, the best hardwood cuttings are taken from the midpoint of a stem and may be 15 cm to 30 cm in length. Cuttings of this type usually have a stem diameter equal to or greater than a pencil and contain at least three vegetative buds. Direct field sticking of deciduous cuttings in raised beds is a common practice. Examples of some plants that are propagated by deciduous hardwood cuttings are figs (*Ficus carica*), grapes (*Vitis sp.*), mulberries (*Morus sp.*), pomegranates (*Punica granatum*), and roses (*Rosa sp.*).

FIGURE 17.1 Variation in rooting of the narrow-leaf evergreen plant x*Cuprocyparis leylandii*. Note callused cutting with no roots on the left and a well-rooted cutting on the right.

Narrow-leaf evergreen stem cuttings are generally taken in the winter, although certain species, such as prostrate junipers and certain cypresses, may be propagated during periods of summer growth dormancy. Cuttings are usually 10 cm to 20 cm in length, are less than pencil thickness in diameter, and the bottom branches or leaves are removed. Narrow-leaf evergreen cuttings are responsive to bottom heat, wounding, and rooting hormones, yet may still take several months to root. The three different types of cuttings used for propagation of most conifers are the following: (1) stem cuttings (Figure 17.1), (2) heel cuttings, and (3) mallet cuttings.

Most narrow-leaf evergreen cuttings are taken in the late fall or early winter because they generally root best after some exposure to chilling temperatures. Many growers will stick conifer cuttings in late winter so the plants do not sit in the propagation house for several months before callusing and root formation occurs. Certain types of conifers are considered easy to root. These are *Chamaecyparis*, *Platycladus*, *Podocarpus*, *Taxus*, *Thuja*, and the low-growing forms of *Juniperus*. Taxa considered more difficult to propagate from cuttings include *Cedrus*, *Cupressus*, and *Pinus*.

Semihardwood stem cuttings of broadleaf evergreen are taken in late spring to midsummer when the wood has matured, but is not yet woody. The cuttings are often 8 cm to 16 cm in length, and the basal thickness of the stem will vary. The lower one-third of the leaves may be removed, and cuttings with large leaves can be cut back to reduce transpirational water loss. Plants propagated by semihardwood cuttings can include boxwood (*Buxus sp.*), camellias (*Camellia sp.*), citrus (*Citrus sp.*), hollies (*Ilex sp.*), and evergreen rhododendrons (*Rhododendron sp.*). (Figure 17.2)

Softwood stem cuttings are usually tip cuttings taken from new growth. The same procedures are followed as for

FIGURE 17.2 Camellias propagated from semihardwood cuttings under intermittent mist.

semihardwood cuttings. These cuttings are often quick to root, usually in eight weeks or less. Many plants flower from this new growth, so all flowers and flower buds should be removed. Terminal growth is often removed from the cuttings to prevent wilting or decay under mist. Plants rooted from softwood cuttings include deciduous magnolias (*Magnolia* spp.) and deciduous azaleas (*Rhododendron* sp.).

Herbaceous cuttings are handled the same as softwood cuttings. These cuttings can often be taken throughout the year. Some plants propagated by this method are chrysanthemums (*Dendranthema* sp.) and geraniums (*Pelargonium* sp.).

Some points that need to be considered before taking cuttings are the conditions of the stock plant (chapter 22), the time of year at which the cuttings are taken, and care of the cuttings once they have been taken. It is important that the stock plant is insect and disease free. Contaminated plants will not perform well and may infect other plants that are being propagated. Cuttings should be taken from plants in good nutritional health. In general, cuttings taken from nonvigorous or rapidly growing plants do not root well.

The time of year in which cuttings are taken may vary from year to year because of different environmental conditions. Propagators should keep records from year to year of when certain plants were rooted and their performance (e.g., rooting percentage, vigor). Although a recommendation may call for sticking cuttings after the first frost of the year has occurred, this does not mean that the cuttings are stuck on the same calendar date each autumn.

Growth phase (juvenile versus adult; chapter 15) and location of the cutting on the plant (topophysis) have a great influence on cutting success and subsequent growth. In general, the closer the cutting to the root:shoot interface of the plant, the more juvenile the cutting will be. Many plants, including some conifers, root best from juvenile cuttings versus cuttings taken from flowering or adult parts of the plant. If juvenility is a problem, once a few cuttings are rooted from a difficult cultivar, those plants can be used as stock plants. After several generations of selection, cuttings should be juvenile and root with better consistency. Conifers are also good examples of plants that show orthotropic (vertical) growth and plagiotropic (lateral) growth. When propagating plants, such as *Cunninghamia lanceolata* or *Sequoia sempervirens*, if lateral shoots are used for cuttings, then the resulting plants will grow horizontally for several years. Many conifers prone to plagiotropic growth (growth at an angle in response to gravity) benefit from having cuttings taken only from vertically growing shoots. Growing stock blocks of hedged plants that produce large quantities of vertical shoots are often used.

PROPAGATION STRUCTURES, BENCHES, AND MIST SYSTEMS

Environmental factors to consider once the cuttings have been stuck are air temperature, humidity, light intensity, and substrates. Not all plants can be propagated outdoors without protection; therefore, it is necessary to provide structures where environmental conditions can be maintained. Structures can be as elaborate as modern greenhouse ranges or as simple as small poly huts in the field. Items that should be taken into consideration when deciding which type of structure to use include (1) the type of structure, (2) cost, (3) covering to use, (4) structure and bench/bed orientation, and (5) heating and cooling needs (chapter 2).

A greenhouse can be defined as an enclosed structure designed to provide ideal conditions of light, temperature, and humidity for the growing of plants. In general, propagation greenhouses are no more than 11 m wide by 30 m in length, in order to provide optimum environmental conditions (Figure 17.3).

A variety of coverings are available for propagation houses. Many types of plastic are used instead of glass. Many types of rigid plastics are preferred because they will withstand many of the environmental abuses that

FIGURE 17.3 A typical propagation greenhouse covered with shade cloth.

FIGURE 17.4 A simple overwintering structure for rooted cuttings.

glass will not. Plastic is preferred in areas that are prone to hurricane damage because it can be easily removed.

Poly houses are popular structures for propagating cuttings because they have a lower initial construction cost. The frame is considered permanent, but the covering may be replaced yearly or once every three to five years, depending on the longevity of the plastic used.

The purpose of a shade house is to provide protection of nursery plants from environmental factors, such as wind, temperature, hail, heavy rain, and solar radiation. Shading may be provided using lath or shade cloth. These structures are usually permanent and can be used for propagation or hardening-off of rooted cuttings or newly potted plants. Saran or shade houses have replaced lath houses in most areas in part due to initial lower costs. Shade cloth can have a useful life of several years. Red shade cloth has become more popular in recent years and has been shown to increase growth once cuttings are rooted.

Full sunlight works well for many species once rooted because it allows for greater photosynthesis and thus greater food reserves and hormone production during the rooting-out process. High temperatures may be a problem under conditions of high light intensity. If this is the case, shading compound can be applied to greenhouse structures, or shade cloth may be used to reduce the amount of sunlight by up to 70%.

A coldframe is among the simplest and most economical of all propagation structures. In years past, coldframes were built low to the ground, were covered with glass sashes and were used for rooting cuttings, germinating seeds, and hardening-off tender plants. Simple coldframes covered with white or opaque plastic are used for propagation beds directly in the field. Many growers now use hoop houses covered with plastic and/or shade cloth as propagation structures (Figure 17.4).

Mist tents are commonly used inside greenhouses that have existing mist systems. The purpose of such tents is to increase and/or keep the relative humidity constant while rooting cuttings. Polyethylene sheeting is often draped over a structure constructed above the mist bed. For cuttings that are prone to rot under mist, the same structure can be used as a humidity tent, where mist is applied infrequently, just enough to keep sufficient relative humidity within the tent. Care must be taken to prevent excessive build up of heat in such tents.

Raised propagation benches are often preferred over ground beds for propagation of cuttings. Benches allow for greater control of the environment around the cuttings (improved drainage and isolation from soilborne diseases and pests) and also allow work to proceed in a more efficient manner. Bench size is often determined by the type of propagation (sticking in flats or individual containers), size of the propagation house, location of benches within the structure, and number of cuttings to be propagated. When designing a bench system, items to be taken into consideration include the following: (1) longevity of the bench material, (2) ease of assembly, (3) versatility, (4) weight, (5) air circulation, and (6) cost. While propagation benches can be made from many materials, many growers now prefer benches made from galvanized metal.

In general, the ideal air temperature for rooting cuttings is 18°C to 24°C (~65°F to 75°F). Ideal substrate temperatures for rooting are between 21°C and 24°C (70°F to 75°F) for most plants. These temperatures may be difficult to maintain, especially air temperature. Remember the adage when it comes to rooting cuttings: Cools Hands—Warm Feet!

Bottom heat provides a number of advantages in propagation, especially for the rooting of cuttings. These advantages include the following: (1) bottom heat is economical, (2) increased rooting percentages, (3) quicker rooting, (4) less hormone may be necessary, (5) fewer disease problems, and (6) higher-quality rooted cuttings. Bottom heat is especially beneficial for rooting conifers and evergreens

FIGURE 17.5 An overhead mist system in action.

during the winter months. Several methods are available for providing bottom heat. Heating cables and heat mats are useful for small areas. On a greenhouse scale, heat pipes under beds and benches, hot air under benches, and hot water pumped through pipe or plastic tubing are methods commonly used. Circulating hot water also helps eliminate electrical hazards, as heating cables and mats can break down over time.

A functioning, well-maintained mist or fog system (chapter 4) is a necessity because it prevents excessive transpiration and respiration while allowing the cuttings to photosynthesize. The purpose of a mist system is to maintain cutting turgidity by increasing atmospheric relative humidity at the surface of the leaf and to decrease the temperature of the leaves on the cuttings through evaporation. The two most popular mist systems in use are intermittent mist and fog systems (chapter 4). Fog systems have become popular because they attempt to provide 100% relative humidity while resolving the problems of poor aeration and excessive moisture associated with conventional mist systems. In hot, humid climates, lack of ventilation with fog systems often leads to excessive heat build-up in propagation structures.

A typical mist system (Figure 17.5) consists of the following components: (1) mist line, (2) water filter, (3) solenoid valves, (4) gate valves, (5) time clocks, and (6) mist nozzles. Polyvinyl chloride (PVC) pipe is the most common material used for mist lines because of its ease of installation, cost, and longevity. Flexible black plastic pipe available in large rolls is also cost effective and is often used for overhead mist systems. The size of pipe necessary for a mist system will depend on the type of nozzle used, water pressure (pounds per square inch: psi), liters per minute of water needed, and the size of the system. For fog systems, copper tubing is often preferred. Mist lines can be placed in the propagation beds or can be placed directly overhead. Mist heads are generally placed a minimum of 30 cm above the cuttings. If mist beds are greater than 90 cm across, it is advisable to install more than one mist line per bench. Many modern mist systems consist of two mist lines in propagation houses that are 6 m to 11 m in width.

A filter is a good idea when installing a mist system so that mist heads will not get clogged with particulate matter. Solenoid valves are installed as a means to electronically control the mist system. Plastic solenoid valves work well for low-pressure systems, but brass valves should be used if high-pressure water lines are utilized. Solenoid valves are available as "normally open" and "normally closed" valves. The "normally open" type is considered best for rooting cuttings because it remains open if the electricity goes off, thus the mist line can continue operating as long as there is water pressure. Always put a gate valve in the line so the system can be manually operated should power not be available. And design mist systems with low voltage to increase worker safety.

Three general types of mist nozzles that can be utilized for rooting cuttings are the following: (1) whirl or pressure jet types, (2) deflection or baffle type, and (3) fog nozzles. Whirl type nozzles (Figure 17.6) spin when water under pressure exits the openings on the ends of the nozzles. As water is forced through the grooves in the nozzle, it is reduced to smaller water particles. Many whirl type systems can use a low amount of water per hour compared to other systems. Deflection or baffle-type nozzles are the most commonly used nozzles for rooting cuttings. Because deflection nozzles require a larger volume of water, they are less prone to clogging. The orifice size on a fog nozzle is very small and is prone to clogging, especially with hard water. Water particle sizes of 7 to 10 µm are achieved by running the fog system at pressures greater than 450 psi.

Control devices for mist systems can be either preset (time clocks) or variable (environmentally controlled). The preset system usually consists of two time clocks, a 24-h clock and an interval clock. The 24-h clock con-

FIGURE 17.6 A whirl-type mist nozzle used for rooting cuttings. A cut-off valve is often placed on each riser.

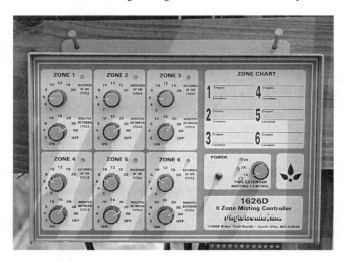

FIGURE 17.7 A six-zone mist controller. Duration of time between cycles and the number of seconds each cycle comes on can be set for each of the six stations.

trols when the system comes on in the morning and when the system shuts down in the evening. The interval time clock controls the frequency and duration of the mist cycle while the system is on. The interval clock is generally set to keep the foliage moist throughout the day, but this system often requires manual adjustment during the day for optimum effectiveness. Many modern electronic mist systems control both the 24-h clock and the interval cycles (Figure 17.7).

Several types of variable or environmentally controlled mist devices are available. The electronic leaf works on the principle of electrical conductivity between two electrodes. When a simulated leaf made from nonconducting material is misted, an electrical current is carried between two electrodes. As the simulated leaf dries out, the current is broken and the solenoid on the mist system is activated. Once enough moisture collects on the simulated leaf, current flow is reestablished and the solenoid shuts off.

The electronic leaf (Figure 17.8) is based on the principle of evaporative loss from a real leaf. A simulated leaf is actually a screen that collects water when the mist system is on. Once the screen leaf is weighted down, it shuts off the solenoid. As water evaporates, the screen leaf rises until it again triggers the solenoid to be activated. Calcium and other hard water mineral deposits on the screen leaf can be a problem with this type of system.

A third system is based on a photocell that activates the solenoid once a predetermined quantity of light (photons) has been absorbed. New computer controlled systems are also available. Such systems are useful for automating mist systems, but the importance of manually checking the propagation systems several times during the day should be stressed.

PROPAGATION SUBSTRATES, FERTILITY, AND CONTAINERS

A variety of propagation substrates are presently used, and they vary from nursery to nursery (chapter 5 and chapter 6). The most important factors to consider are that the

FIGURE 17.8 Example of an electronic leaf used for controlling mist frequency.

substrate should firmly hold the cuttings in place; be free of diseases, insects, and weeds; have a high water-holding capacity; and yet provide good aeration and drainage. Some of the most commonly used propagation substrate components are coarse sand, perlite, vermiculite, peat moss, and pine bark.

The functions of propagation substrates are to (1) provide anchorage and support for the cutting, (2) provide for the storage of substrate water, (3) act in a storage capacity for nutrients, and (4) supply adequate aeration for good root growth. A good propagation substrate for rooting cuttings should possess the following characteristics:

- Large, noncapillary pores for aeration and drainage
- Small capillary pores for water-holding capacity
- Good stable structure that will not break down
- Adequate porosity to allow for easy root penetration
- Buffered against rapid changes in chemical properties
- Ability to hold nutrients for plant use
- Light weight, but able to provide stability for plants

Three main factors that should be taken into consideration when choosing a propagation substrate are the following: (1) economics, (2) physical properties, and (3) chemical properties. Economic considerations to evaluate are the cost of the substrate, availability, and reproducibility. Plant propagators should always use the best substrate available that works for their operation. Commercially prepared propagation substrates are available and very reliable. Some substrate components can be difficult to find. Hardwood bark may be available in some areas, whereas only Douglas fir bark may be available in another part of the country. Once an acceptable propagation substrate has been found, reproducibility becomes an important factor. Having to change components, grades, or particle sizes can have deleterious effects on physical and chemical properties of a substrate.

Physical factors (chapter 5 and chapter 6) that influence the properties of a substrate are the following: (1) water holding capacity, (2) aeration, (3) bulk density, (4) particle size distribution, (5) uniformity, and (6) shrinkage. Nurserymen can perform their own tests to determine water-holding capacity and air space at container capacity, or commercial laboratories can determine these physical properties as well. An ideal propagation substrate will have 25% to 35% air space at container capacity. Altering particle size, uniformity, or using substrate components that will shrink will alter the amount of air space in the substrate. In general, air space decreases and water-holding capacity increases as particle size decreases.

The bulk density of a substrate is its weight compared to the volume of space that it occupies. Bulk density is usually expressed in terms of grams per cubic centimeter (g/cc). The bulk density of a substrate should be considered because the weight of a propagation substrate will influence shipping and handling costs. Uniformity of particle size and resistance to shrinkage are also important considerations. Hardwood bark and sawdust are examples of substrate components that will shrink if they are not composted first. During the course of a crop cycle, such shrinkage will lead to a decrease in the depth of the substrate in the container and a decrease in air space, both of which will adversely influence plant growth.

Chemical factors that contribute to the success or failure of a propagation substrate include the following: (1) cation exchange capacity (CEC), (2) pH, (3) soluble salts, (4) nutrient levels, and (5) sterility. The CEC of a substrate refers to its ability to retain positively charged cations against leaching from rainfall and/or irrigation. Organic matter will generally have a greater CEC than inorganic matter such as sand. Calcined clays or mineral soils have greater CECs than sand and other inorganic components such as perlite.

The pH is a measure of the acidity or alkalinity of the rooting substrate solution. Although different plants may have different preferences in regard to pH, the general accepted range of pH for most plants in organic-based substrates is 5.2 to 6.5. Adjustment of pH in the rooting substrate has been shown to influence the availability of nutrients, rooting of different species, and the incidence of certain disease organisms. Nutrient concentrations in propagation substrates are influenced by CEC and pH. Most micronutrients are available at a pH of 6.5 or less. Some debate exists over the application of fertilizers to the propagation substrate, but most research clearly shows that, while controlled release fertilizers do not help rooting percentages, growth after the cuttings root is improved. Considerable research has been conducted with different methods of nutrient application, such as liquid feed, top dressing fertilizer, incorporation of fertilizer, and nutrient mist. So far, no one method has proven superior for all plants. A variety of slow or controlled-release fertilizers have been used successfully on a number of crops.

Soluble salt levels and sterility of substrate components can be influenced by the component itself or the irrigation water used. When considering water quality, city and/or deep well water is often preferred over pond or stream water because of fewer disease organisms, algae, and suspended solids. Collection basins may have high soluble salt levels and organisms which can cause damage to plants and clogging of irrigation and mist lines. Table 17.1 provides an interpretation of water quality for nursery production and plant propagation.

Some of the problem elements and ions in irrigation water are the following: calcium and magnesium, sodium, boron, chloride, bicarbonate, fluoride, iron, and sulfur. Calcium and magnesium are found in water considered to be hard and, therefore, can cause buildups on mist equipment that can lead to mist system components not working. These two elements along with bicarbonate are also often responsible for the high pH and/or alkalinity found in irrigation water. Sodium and chloride add to the soluble salt problem and can occur in toxic concentrations. Chloride levels of less than 140 mg/L are not a problem, but levels greater than 350 mg/L are unsuitable. Boron and fluorine can also cause toxicity problems. Boron is not usually a problem if found at levels less than 0.5 mg/L, but at higher levels (>2.0 mg/L), irrigation water is not suitable for use. Fluoride is found in fluorinated city water, superphosphate, and perlite. Many foliage plants and monocots are sensitive to fluorides. If fluorides are a problem (>0.2 mg/L), this can be avoided by increasing shading, adjusting the pH of the substrate and/or irrigation water, or by not using fertilizers or substrate components that contain fluorides. Iron and sulfur can leave residues on leaves and also cause the formation of bacteria that can plug mist nozzles.

The functions of a container for rooting cuttings are the following: (1) provide protection for the root system, (2) facilitate transplanting and maximal survival, (3) provide ease of handling and/or shipping, and (4) provide consumer appeal for the end user. The criteria for the selection of a good container are the following attributes:

- Adequate drainage
- Ability to hold sufficient volume of substrate
- Lightweight
- Easy to handle
- Durable
- Free of toxic substances
- Prevent root circling

Containers are often used for rooting cuttings instead of growing plants in ground beds because transplanting losses are often minimized along with other advantages, such as fewer shipping and handling costs. One of the main problems associated with rooting plants in small containers is root circling and girdling. Different methods have been used to in attempt to solve problems associated with root circling. Some of the successful methods for many species have included air pruning of the roots (Figure 17.9), use of bottomless containers, containers with mesh bottoms that prune the roots as they enlarge, stair-stepped container, or treatments with root pruning chemicals such as copper hydroxide (Spin Out®).

Growers use several types of containers to root plants. Some examples include the following: flats (wooden, plastic, or galvanized metal), clay pots, plastic pots known as liners, compressed fiber pots, peat pots, plastic cell packs, paper pots, poly bags, and others. Growers should select containers that work best for their system of propagation and marketing and produce the highest quality plants. Plastic liner pots or cell packs are preferred by many large commercial operations. A 32-count or 40-count cell pack may work well for a short-term crop (softwood cuttings, propagation time of two to three months), whereas larger

TABLE 17.1
Estimating Greenhouse Water Quality Based on Conductivity

Conductivity Reading (dS/m or mmhos/cm)	Water Quality
0.0–0.25	Very Good
0.25–0.75	Good
0.75–1.5	Fair
1.50–2.0	Questionable
Above 2.00	Unsuitable

Cloning Plants by Rooting Stem Cuttings

FIGURE 17.9 An Accelerator® propagation container. Root circling is prevented by air-root pruning holes in the container.

individual liner pots are preferred for rooting of certain tree species that are slower to root (Figure 17.10).

USE OF ROOTING HORMONES AND STICKING CUTTINGS

Cuttings are commonly treated with growth-regulating compounds (chapter 14) to promote the formation of adventitious roots. The reasons for the application of such root-promoting chemicals are the following: (1) to hasten root formation, (2) to increase the percentage of rooting, (3) to increase uniformity of rooting, and (4) to increase the quality and number of roots per cutting.

The first root-promoting chemical to be discovered in the 1930s was the naturally occurring auxin indole-3-acetic acid (IAA). IAA is not used for the propagation of plants via cuttings because it is sensitive to environmental factors and is easily broken down by light. The two commonly used synthetic auxins are IBA (indole-3-butyric acid) and NAA (naphthalene acetic acid). These two chemicals can be applied to cuttings in two forms, as talc (powder) or as liquid (dip or spray).

IBA and NAA can be purchased as reagent-grade powders or in their salt formulations. The potassium salt of IBA and the sodium salt of NAA are water soluble, whereas the reagent-grade forms must be dissolved in alcohol or other organic solvents. NAA often dissolves best when a few drops of ammonium hydroxide (NH_4OH) or potassium hydroxide (KOH) are added. Concentrations greater than 20,000 mg/L (2%) are generally difficult to keep in solution. Formulation of auxins are sensitive to light and break down, thus they should be kept in amber or dark containers that limit exposure to light. In general, auxin solutions can be kept for up to six months if refrigerated. Precipitation problems can be avoided by using distilled or deionized water instead of tap water. Commercially formulated rooting hormones can also be dissolved in carboxymethylcellulose to form a gel. Lower concentrations of hormone can be used since more gel is retained on the cutting compared to a liquid dip.

When making calculations for the preparation of rooting hormones, the most useful values to remember are

$$1 \text{ mg/L} = 1 \text{ ppm}$$

and

$$1\% = 10,000 \text{ ppm or } 10,000 \text{ mg/L}$$

To prepare a liquid formulation that contains 3000 ppm or 0.3% IBA, one would dissolve 3000 mg of IBA in 1 L of solvent. Always remember to check the amount of active ingredient in your chemical and compensate when it is less than 100%. For example, if your chemical is only

FIGURE 17.10 Rooted cutting propagated in a compressed fiber container.

75% active ingredient, then it will require 1.33 (1/0.75) times as much of that chemical compared to a formulation that is 100% active ingredient. Several commercially available formulations of liquids and powders can be purchased.

Propagation of cuttings can be broken down into several stages, such as preparation and taking cuttings, sticking cuttings, callusing of cuttings, development of roots, and hardening-off of cuttings. Cuttings should be true-to-type; that is, they should have the same look and genetic make-up. Variegated plants may not be stable and, therefore, uniform cutting selection is important. Also, pay attention to rules regarding propagation of patented plants. Such plants should only be propagated after a license has been granted from the patent owner or suitable assignee (chapter 43). All cuttings should be disease and pest free, should come from healthy stock plants, and should be similar in size and maturity. Remember that many plants have optimum windows of opportunity for rooting. Good recordkeeping is essential.

Before the cuttings arrive in the propagation area, make sure the area where the cuttings will be rooted has been disinfected and is free of pests and weeds. Make sure all environmental and mist controllers are functioning properly. Propagation containers or flats should be selected and sanitized if necessary. Be sure there are sufficient numbers of containers on hand. Fill all containers in advance with appropriate substrate, and check to make sure sufficient labor is available to stick the cuttings when they come in.

Cuttings are often harvested the morning they are to be propagated. If this is not possible, many cuttings can be packed moist and stored overnight in a cooler. Certain cuttings, such as *Ilex xattenuata* "Foster No. 2," often defoliate if stored overnight during the growing season. Make sure stock plants are pest free. Cuttings being harvested should be uniform in stem diameter and size. Be sure to harvest cuttings at the correct stage of maturity (softwood, semihardwood, or hardwood). Once cuttings are taken, they are often keep in plastic bags and placed in portable ice coolers. Do not leave plastic bags out in the sun; they heat up very quickly and damage can occur to the cuttings. Keep cuttings cool until they are stuck. This may be several hours from the time the cuttings were taken.

Often the lower leaves are removed before rooting hormones are applied. Some cuttings respond best to wounding before they are treated with hormone and placed in the propagation substrate. Light wounding can be a slice in the low part of the stem, or stripping leaves can also cause wounding to occur. Heavy wounding often refers to cutting a 2.0-cm to 4.0-cm slice of the base of cutting on one side. Many conifers, hollies, magnolias, and evergreen rhododendrons benefit from wounding. Once the cutting is wounded, cells begin to divide and callus formation begins. As callus formation takes place, certain cells divide and root development is initiated. Length of dipping cuttings in rooting hormones and depth of sticking cuttings in the hormone solution varies by species. Once cuttings are stuck, misting should be initiated and shade provided, if necessary. Some cuttings, such as *Magnolia grandiflora,* benefit from a combination treatment of a quick dip followed by a talc application before the cutting is stuck.

If substrate temperature is appropriate (21°C to 24°C), callusing often occurs within a few weeks. High relative humidity must be maintained during this period so cuttings do not wilt. Mist frequency and longevity should be adjusted according to environmental conditions during the course of each day. Mist may be required to softwood or herbaceous cuttings for the first few nights, but should then be adjusted so the cuttings do not have wet foliage through the night. Once cuttings have developed callus, mist frequency and duration can be reduced.

As roots are initiated and begin to elongate, mist cycles can again be reduced and light intensity increased. Cuttings often benefit from the application of liquid or controlled-release fertilizers at this point. Check cuttings for disease problems and treat as necessary (chapter 8 and chapter 10). For some herbaceous species, plant growth regulators can be applied at this stage to reduce stretching. Fertilizers low in phosphorus should be used as well to prevent stretching of cuttings under conditions of low light intensity. If cuttings have been in the propagation areas for several weeks, begin to check pH and electrical conductivity of the substrate, and adjust fertility regimes as required.

Finally, cuttings need to be hardened-off before they are shipped or potted up and placed in production areas. Cuttings should be well rooted such that the root system remains intact when the plant is removed from the propagation container. Rooted liners are either moved to a different area or mist frequency is reduced to lower the relative humidity in the propagation area. Light intensity is often increased at this point in time. Increasing light intensity decreases internode elongation, increases leaf thickness, and often improves branching of rooted cuttings.

Some cuttings do not have good overwinter survival unless a new flush of growth is forced during the summer the cuttings were rooted. Some taxa with this problem include *Acer, Betula, Cornus, Hamamelis, Magnolia, Quercus, Rhododendron, Stewartia,* and *Viburnum*. Placing rooted cuttings under long-day conditions often helps plants break bud. Growers often use incandescent light bulbs spaced 1.0 m^2 and 1.0 m above the crop. Lights can be turned on for 5 min. every hour during the night or for a continuous period from 10:00 p.m. to 2:00 a.m. High concentrations of rooting hormones applied to enhance the rooting process often suppress bud-break, thus reducing winter survival. Research has shown that using lower concentrations of rooting hormones in conjunction with the antioxidant ascorbic acid (1.5% to 2.5%) can improve

Cloning Plants by Rooting Stem Cuttings

rooting and overwinter survival. Ascorbic acid also helps improve rooting of cuttings, such as camellias, which have high levels of phenolic compounds.

SANITATION DURING PROPAGATION

An effective propagation nursery requires a good disease and pest management program. The following advantages can be realized by the grower when a good sanitation program is followed during the rooting of cuttings:

- Improved plant quality
- Decreased plant mortality
- Reduced production costs
- Fewer production delays
- Increased profits
- Return customers

The pests that nurserymen often have to combat in the nursery are weeds, fungi, insects, bacteria, nematodes, algae, viruses, and mycoplasmas. These pests reach the propagation area in a variety of ways that may include (1) transmission by insects, (2) entering through openings in propagation structures, (3) splashing of contaminants by water, (4) hoses dropped on the ground, (5) contaminated water or hormone solutions, (6) infested soil or substrates brought into the propagation area by foot or nonsterile equipment, and (7) use of previously infected plant material.

The two major nursery areas that must be considered are the stock plant area and the propagation area. The stock plant area may consist of actual blocks of plants being grown in the field that are used exclusively as parent material for cuttings, or the stock area may be a production area in the nursery from which cuttings are taken. Since field-grown stock plant areas can be costly to maintain, many nurseries use cuttings taken from container production areas. Some production nurseries allow propagators to take cuttings from container or field-grown stock, and then later purchase the rooted cuttings back from the propagator.

PREVENTATIVE PRACTICES FOR STOCK PLANTS

The following lists some best management practices for the maintenance of stock plants—also consult chapter 22 for more details.

- Stock blocks should be physically separated from the propagation area
- Field-grown stock should be established in well-drained areas to prevent soilborne root diseases
- Container stock should be grown on raised sloping beds for ideal drainage
- When preparing a field, as much native plant material and debris as possible should be removed in order to control root diseases and hosts for foliar diseases and insects
- If planting in the ground, controls for noxious weeds, nematodes, and plant diseases should be applied before planting
- Allow adequate spacing between rows of plants. This will aid in the application of pesticides, allow for good air circulation and rapid drying of foliage, and slow the spread of insects and diseases among the plants
- Irrigation should be scheduled as early in the day as possible so the foliage is not wet during the evening hours. Irrigation scheduling should also take into account natural rainfall so the soil does not become waterlogged, a situation that is ideal for root rot organisms
- Weed growth should be controlled to prevent the spread and multiplication of new and existing weeds, either by seed or by vegetative means. Cover crops and turf around the nursery and in roadways should be kept mowed to prevent movement of pests into stock areas
- Container-grown plants should be grown in pest-free substrates
- Used containers should be disinfected before reuse if susceptible species are being grown
- Preventative applications of fungicides, insecticides, and preemergent herbicides can be used to keep pest populations low
- Proper cultural practices (pruning, fertilization, etc.) will help ensure that stock plants remain healthy

PREVENTATIVE PRACTICES IN THE PROPAGATION AREA

The following lists some of the best management practices for the propagation area:

- Plan propagation cycles for the most rapid rooting period for a given species or selection, thereby limiting exposure time to potential pests
- Propagation substrates can be treated with steam or a fumigant to eliminate disease and pest organisms
- Beware of potential problems areas, such as canals, wells, and collection basins, as potential sources of disease pathogens
- Do not use water in the propagation area that is high in soluble salts or has high levels of certain

elements, such as boron or chlorine. High soluble salt levels may cause necrosis of plant tissue or predispose the cuttings to other plant diseases
- If possible, take cuttings from the upper portions of the stock plant away from the soil where disease organisms cannot be splashed onto the foliage
- Always collect cuttings that are free of insects and diseases
- Do not mix batches of plant material that have been collected from different locations or were purchased from another nursery
- Treat cuttings that are prone to diseases with appropriate chemicals before propagation
- Check to make sure a proper environment is provided for plant material and not the pest organism
- Select a proper mist cycle that does not saturate the propagation substrate
- Check propagation areas daily for signs of diseases or other pests
- Select a propagation substrate that has appropriate chemical and physical properties
- Use raised beds or benches whenever possible. Flats or containers placed on the ground should be placed on supports so they do not remain in standing water
- Group plants according their needs

Care should be taken to clean all equipment and facilities before they are utilized. Household bleach is a good general-purpose chemical that can be used to clean equipment and benches, although it can be corrosive to certain metals and paper products, such as greenhouse cooling pads. Several copper-based products, as well as others, are available for use in propagation areas. Denatured alcohol can also be used as a good general disinfectant for pruners and other metal tools.

SUGGESTED READINGS

Blythe, E.K., J.L. Sibley, K.M. Tilt, and J.M. Ruter. 2007. Methods of auxin application in cutting propagation: A review of 70 years of scientific discovery and commercial practice. *J. Environ. Hort.* 25: 166–185.

Crawford, M. 2006. Improving root initiation of cuttings. *Amer. Nurseryman.* 201: 36–38, 40.

Davis, T.D., B.E. Haissig, and N. Sankhla. (Eds.). 1988. *Adventitious Root Formation in Cuttings.* Dioscorides Press. Portland, OR.

Dirr, M.A. and C.W. Heuser. 2006. *The Reference Manual of Woody Plant Propagation: From Seed to Tissue Culture,* 2nd ed. Varsity Press, Inc. Cary, NC.

Dole, J.M. and J.L. Gibson. 2006. *Cutting Propagation: A Guide to Propagating and Producing Floriculture Crops.* Ball Publishing. Batavia, IL.

Hartmann, H.T., D.E. Kester, F.T. Davies, Jr., and R.L. Geneve. 2002. *Plant Propagation: Principles and Practices,* 7th ed. Prentice-Hall, Inc. Upper Saddle River, NJ. 880 pp.

Jones, R.K. and D.M. Benson (Eds.). 2001. *Diseases of Woody Ornamentals and Trees in Nurseries.* The American Phytopathological Society. St. Paul, MN.

McDonald, B. 1986. *Practical Woody Plant Propagation for Nursery Growers,* vol. 1. Timber Press. Portland, OR.

Nelson, P.V. 1998. *Greenhouse Operation and Management,* 5th ed. Prentice Hall, Inc. Upper Saddle River, NJ.

Wilson, P.J. and D.K. Struve. 2004. Overwinter mortality in stem cuttings. *J. Hort. Sci. Biotechnol.* 79: 842–849.

18 Anatomical and Physiological Changes That Occur during Rooting of Cuttings

Ellen T. Paparozzi

CHAPTER 18 CONCEPTS

- Rooting of stem cuttings provides a commercial method of producing genetically identical plants (clones).

- Anatomically, adventitious roots on stem cuttings arise either from preformed root primordia or from dedifferentiated cells located near the vascular cambium.

- The ability of a plant to form adventitious roots is limited by internal and external factors. Internal factors that affect adventitious root formation include genetics, cell competence, phytohormones (plant growth regulators), nutrients, water, and carbohydrate status.

- Physiological processes that are affected when stem cuttings are taken and adventitious roots are induced include wounding, transpiration, photosynthesis, respiration, protein, carbohydrate, and phytohormone synthesis.

- External factors that affect adventitious root formation include light, air and rooting medium temperature, humidity, and rooting medium moisture.

- Various techniques are employed by the commercial plant propagator to manage to external (environmental) factors.

Plants recognize damage via signals and make changes in key cellular and metabolic processes. When a cutting of a plant is taken—say a chrysanthemum—the stock plant and the cutting both respond to that damage. In this section, we will focus on the signals the cutting will produce as well as how it responds anatomically and physiologically to being severed from the parent plant. The study of signal transduction and plant response to damages is in its infancy. This chapter represents the best sense of how stem cuttings form adventitious roots, given the current evidence.

Probably the two most important phenomena that occur when you clip off a piece of a plant stem are that the plant senses the wound and its transpiration stream is severed. When the leaves and the cut stem no longer have access to water, xylem will cavitate, and live cells will lose turgor. Water is very important to the plant physiologically because it is involved in the first step in the light reactions of photosynthesis. It is the substance that controls the temperature of leaves and carries nutrients. Water keeps cells turgid so that they may perform other metabolic functions, such as producing and storing carbohydrates and respiration, including the secondary metabolic pathways that spring from respiration. Guard cells in the leaves and stems must be turgid so that stomata remain open and gas exchange can occur (basically CO_2 in and O_2/H_2O vapor out).

Wounding causes a large difference in turgor between wounded cells and intact ones that are proximal to the cut edge. Additionally, an open cut or wound makes the plant more vulnerable to insects and disease-causing organisms. When a plant is stressed like this, it responds at the most basic level, the gene, as well as the more readily visible level via leaf wilt.

When stresses, such as wounding and dehydration occur, signals are produced. Some of these signals travel throughout the entire plant, whereas others trigger reactions in the cells close to the wound. When the signals are received, genes are expressed, resulting in new proteins/enzymes. Some of these signals are thought to directly affect certain compounds and their syntheses, for example, some phytohormones (plant growth regulators [see chapter 14]). Ethylene activity is an example of this because its synthesis is increased when a plant is wounded. For wounding specifically, current research indicates that there is a signaling cascade that produces another hormone, jasmonic acid, which activates genes. Ethylene and other compounds are also involved in this activation.

Thus, a role of phytohormones may be as "receivers and/or transmitters" of signals. Another phytohormone, abscisic acid, increases in concentration under drought stress conditions. Concurrently with this, other metabolic changes occur. These are just two examples of the molecular and biochemical events that are triggered by wounding.

Once genes are expressed, the synthesis machinery of the plant, particularly protein synthesis, is turned on. This will produce compounds that may send more signals and result in the synthesis of additional proteins/enzymes (e.g., peroxidases), phytohoromones (e.g., auxin, ethylene), and other compounds or cofactors (both promoters and inhibitors), as well as stimulate other cellular processes. There is a multiplication and magnification effect such that many things are happening at the same time. Wounding, for example, causes secondary compounds, such as alkaloids, to be synthesized, as well as catabolic enzymes that degrade cell walls and membranes. These are thought to be part of the plant's chemical defense system, and certain enzymes appear to be synthesized just for this pathway.

The ultimate purpose of the cascade of signals and the multistep processes that follow is to mobilize and direct to the wound the resources necessary to protect it from further water loss and pest invasion and promote new root formation. Many plants seal the wounded cut end of the stem with compounds, such as suberin and phenolics, whereas others will produce callus, which is a proliferation of relatively undifferentiated cells. Sometimes, both will occur. However, for all these steps to happen, plant cells must be at the right stage to receive and respond to a signal(s). When a plant cell is able to receive and respond to signals, it is said to be "competent."

Once the cut end is sealed, root induction and initiation will become a priority, and a sink for phytohormones, water, nutrients, carbohydrates and other components necessary to form roots is established. In order for roots to be initiated, cambial and parenchyma cells in, near, and around the vascular bundles and cortex must dedifferentiate and be induced to divide (Figure 18.1 and Figure 18.2). This is the logical place for a new root to be initiated, since it must have a vascular connection with the main stem in order for the cutting to receive water and nutrients for growth (Figure 18.3). Auxins, the phytohormone group that stimulates cells to divide, are keys in the dedifferentiation, induction, and initiation phases of this process. Internally, the auxin, indoleacetic acid (IAA), produced in shoot tips and young expanding leaves, moves downward (basipetally) in the stem toward the cut end. Often natural and synthetic auxins placed on the cut end of the stem are used to stimulate mitotic activity that initiates root primordia. Recent research suggests that another group of compounds, polyamines, specifically putrescine, may also be involved in the induction/initiation phase of adventitious root formation. Cytokinins are thought to be

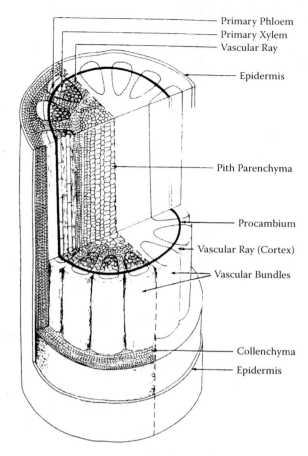

FIGURE 18.1 Three-dimensional cutaway view of the stem of an herbaceous dicot plant showing the vascular bundles. (From Hartmann, H.T., W.J. Flocker, and A.M. Kofranek. 1981. *Plant Science. Growth, Differentiation and Utilization of Cultivated Plants.* Prentice-Hall, Upper Saddle River, NJ. 676 pp. With permission).

synthesized in roots and also stimulate cell division, by affecting cytokinesis. However, a high-auxin, low-cytokinin ratio promotes cell division and elongation during adventitious root initiation. Recently, brassinosteroids—the steroid phytohormone group—in addition to auxins, have been found to stimulate cell division and elongation. While gibberellins have been found to increase cell division particularly in the vascular cambium, most research indicates that they inhibit adventitious root initiation. Ethylene is probably also indirectly involved in the formation of adventitious roots, either by interacting with auxins or in facilitating plant response to environmental factors.

As more cells divide, plant genotype, as well as compounds, such as phytohormones, will determine what cell types (such as xylem and phloem) will be formed and where they will be located in order to organize the new root primordia. This process is called "determination." Under stress, plants will change carbohydrate assimilation and partitioning by solubilizing and mobilizing stored carbohydrates, such as starch. This, combined with

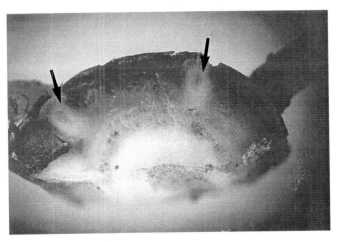

FIGURE 18.2 Cross-section through the stem of *Aphelandra*, zebra plant, showing two root primordia (arrows) emerging from the area of the vascular cylinder.

FIGURE 18.3 Longitudinal section through an emerged root of *Aphelandra*, zebra plant, showing the vascular connection between the vascular bundles of the stem and the stellar arrangement in the root (arrow).

production of new photosynthate, enables processes, such as cell wall synthesis, to occur. Other compounds, such as flavonoids (e.g., anthocyanin), may also contribute to adventitious root formation. As long as environmental and internal factors are favorable, cells will continue to divide, elongate, and differentiate until the new root protrudes to access water, nutrients, and oxygen. Often, once roots have emerged, new shoot growth will occur.

Roots that form in atypical places (i.e., form from any place other than the root pericycle) such as on stems are termed "adventitious." Roots on stem cuttings can originate at various points along the stem. They can arise around a node, at the base of the cutting, or by penetrating the stem epidermis. All of these adventitious roots originated from cells located in or near the vascular cambium. For example, in certain species of *Liriodendrum*, the vascular bundles themselves and the cortical area around the vascular bundles are the origin of the adventitious roots (Figure 18.1 and Figure 18.2). These cells either dedifferentiated and were induced to divide to become root primordia, or were preformed, but arrested in development. This latter type is found in some species, such as forsythia and willow (*Salix fragilis*), and can arise from cells that are laid down near the node very early in the plant's growth right after the shoot internode has ceased elongation (Figure 18.4). Adventitious roots can also originate from the aforementioned tissues higher up in the stem, but only protrude through the base of the stem, such as with mature, woody cuttings of forsythia (Figure 18.5). In this case, there is a ring of fibers located between the cork cambium (phellogen) and the vascular cylinder that prevents the roots from breaking radially through the stem (Figure 18.6a,b). Wounding the sides of the stem is a way to circumvent this problem and can speed up or increase propagation success (Figure 18.7a–c). Plants such as poinsettia produce extensive callus on the end of the stem and the roots will protrude through the callus (Figure 18.8).

Initiation, development, and growth of adventitious roots are controlled by a balance of compounds produced by the plant. These include water (turgor pressure

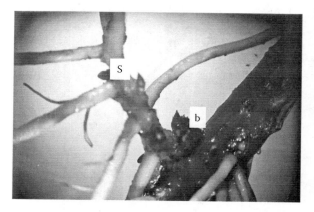

FIGURE 18.4 Roots can form at nodes on stem cuttings. An axillary bud (s) has also started to sprout as well as form roots. Forsythia has multiple axillary buds (b) at each node.

FIGURE 18.5 Root primordia (r) formed at base of a woody stem cutting of forsythia.

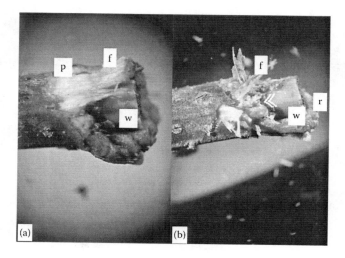

FIGURE 18.6 (a) A ring of fibers (f) forms a sheath between the phellogen, the periderm (p), and the wood (w) inside the stem. (b) Root primordia (<<) are present just below that sheath and at the cut edge of the stem (r).

is necessary for cells to expand), phytohormones, carbohydrates, and nitrogenous compounds. The new roots that are produced on the stem cutting have the same anatomy as a typical root—the xylem and the phloem are in the center of the root in a stellar arrangement surrounded by a pericycle (from which lateral roots will arise), an endodermis, cortex, epidermis with root hairs, and a root cap.

Vegetative propagation via cuttings or other plant parts is a way to commercially produce plants genetically identical to the parent, but success relies on the ability of the cutting or plant part to regenerate roots. Even if cells are competent to form roots, sometimes rooting may not occur. Thus, understanding the basic anatomical and physiological response of cuttings to environmental factors is essential for successful rooting. Factors that influence root formation are humidity around the cutting and moisture of the rooting medium, light, air, and root temperature. All of these factors can be manipulated by the plant propagator.

ENVIRONMENTAL FACTORS

The usual venue for plant water uptake is roots and for water loss is leaves, mainly through pores of the stomata via transpiration. Whenever the rate of transpiration is greater than the rate of water uptake, the plant dehydrates and is under stress. For example, because a softwood cutting has no roots, but its leaves are actively losing water through transpiration, it will tend to dehydrate unless either the environment around the leaves, or the leaves themselves, are modified. A common way to reduce water loss from the leaves is by misting them (chapter 4). This raises the relative humidity, allowing the stomata to stay open, and permitting photosynthesis and respiration to occur. The relative humidity, or the amount of water vapor in the air, around the plant is influenced by light (as heat), as well as air temperature. Remember, a set of genes is turned on that responds to these signals. In response to dehydration, the plant accumulates various proteins and smaller molecules

Anatomical and Physiological Changes That Occur during Rooting of Cuttings

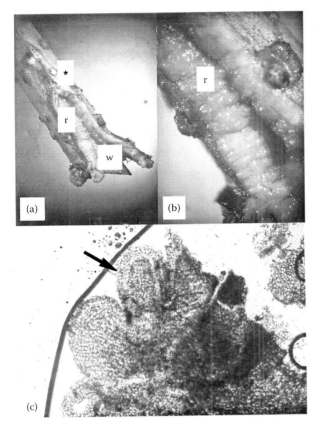

FIGURE 18.7 Root primordia (r, arrow) on a wounded stem at low (a) and high (b) magnification and in cross-section (c). w = wood; * = new root emerging from stem.

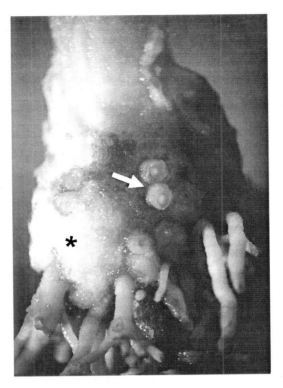

FIGURE 18.8 Callus formation on a stem cutting of poinsettia showing callus (*) and root primordia (arrow).

including amino acids (e.g., proline, glycine, and betaine) abscisic acid, and sugars.

Another method to decrease transpiration in leaves is by decreasing air temperature. This can be done by decreasing sunlight through the use of shade cloth, lowering the thermostat in the greenhouse, limiting the leaf surface from which the water is escaping by cutting larger leaves in half, or using antitranspirants.

The amount of moisture in the rooting medium is critical for growth of new roots after initiation. Thus, some time will elapse in which there must be enough moisture to hold the mix together and support the cutting, but the rooting medium must not be too wet or the end of the cutting will rot instead of forming roots. The temperature of the rooting medium is also important because if the temperature is too low there will be an increase in resistance of water flow to the roots. Combine cool rooting medium temperature with high moisture and you will create the perfect environment for disease (chapter 8). Keep in mind that before the plant can seal the cut end of the stem, sugars, amino acids, and other substances have probably leaked out. This can also attract bacteria and fungi and provide a medium for their growth. Soil temperature will influence initiation as well as growth of roots and will interact with moisture, which affects oxygen content. While the minimum medium temperature for root growth is around 5°C, generally, a higher temperature is required for root initiation. In fact, temperatures that favor root initiation inhibit growth. Commercial propagators will often provide bottom heat (30°C) during the root initiation period. As soon as roots are visible, medium temperatures should be lowered to 20°C to promote new root development. As a plant's genetics are also involved in this interaction, different genera of plants will benefit from different media temperatures. For some plants, higher temperatures may reduce rooting time.

Light may not be essential for root induction and initiation, but it promotes good rooting. This is due to its effect on photosynthesis and thus carbon fixation as well as its role in increasing air temperature. However, if light is too high, leaf temperature will increase, promoting transpiration, which will put a demand on the roots for more water. If a cutting has no or few roots, water stress occurs and leaves may dehydrate, fall off, or turn yellow or brown. This is due either to high air temperature or light intensity. A way to modify this is to increase the frequency of the mist cycle so that the humidity around the leaves stays high, but the leaves do not stay wet or by reducing incident light with shade cloth.

Many times after following all the correct directions, cuttings of certain plants just do not root. Part of this may be due to the environmental factors, but part may be due to the plant species itself. Here again, genetics plays a significant role in adventitious root formation, as the ease of rooting (rooting ability) varies across genera. Also, despite the

```
Take stem cutting.
        ⬇
Wound (possibly water) stress occurs.
        ⬇
Signals are produced.
        ⬇
Plant responds by:
    –altering gene expression.
    –altering phytohormone synthesis.
        ⬇
Changes occur in molecular and biochemical events
in living cells in order to mobilize resources
to seal the wound and induce adventitious roots.
        ⬇
If environmental and internal factors are supportive, plant
initiates new or preformed root primordia via cell division,
elongation and expansion.
        ⬇
Root primordia start to grow; cells differentiate and organize into
typical root anatomical structures.
        ⬇
Root primordia emerge; root hairs initiated to access water and
nutrients.
```

FIGURE 18.9 General scheme for how adventitious roots form on a stem cutting.

best environment, the cells within the plant may no longer be competent to receive or send all the signals as well as synthesize new compounds needed to complete the complex process of creating adventitious roots. Sometimes this loss or decline in cell competence is linked with aging. As a plant undergoes the phase change from juvenile to mature, it loses the capacity to form adventitious roots easily. For example, research has shown that for *Hedera helix*, English ivy, as lignin content increases, adventitious root formation decreases. That is why, in general, softwood tip cuttings are often most successfully rooted.

In summary, there are basically three overlapping phases for adventitious root formation on stem cuttings (Figure 18.9). First, taking a cutting induces stress to which the plant responds by mobilizing physiological processes to seal the wound. Second, competent stem cells, whether from preformed root primordia or cells are induced to dedifferentiate and divide. These new cells are then induced to divide. Both of these processes of cell division are affected by a number of internal factors (e.g., hormones [particularly auxin], water, carbohydrates, nutrients) as well as external factors (light, media and air temperature, humidity, and type of rooting medium). Third, as more and more cells are formed, expand, and are organized, their position and function in the root primordium is determined, and the morphology of a new root takes shape. Signals to initiate adventitious rooting and the genetic control over these events may be the key to why stem cuttings of some plants root more easily than others.

SUGGESTED REFERENCES

Buchanan, B., W. Gruissem, and R.L. Jones. 2000. *Biochemistry and Molecular Biology of Plants.* American Society of Plant Physiologists, Rockville, MD. 1367 pp.

Cockshull, K.E., D. Gray, G.B. Seymour, and B. Thomas (Eds.). 1998. *Genetic and Environmental Manipulation of Horticultural Crops.* CABI Publishing. New York. 225 pp.

Davis, T.D. and B.E. Haissig (Eds.). 1994. *Biology of Adventitious Root Formation.* Plenum Press, New York. 343 pp.

Davis, T.D., B.E. Haissig, and N. Sankhla (Eds.). 1988. *Adventitious Root Formation in Cuttings.* Dioscorides Press. Portland, OR. 315 pp.

Hawkesford, M.J. and P. Buchner (Eds.). 2001. *Molecular Analysis of Plant Adaptation to the Environment.* Kluwer Academic Publishers, Dordrecht, The Netherlands. 269 pp.

Khripach, V.A., V.N. Zhabinskii, and A.E. de Groot. 1999. *Brassinosteroids. A New Class of Plant Hormones.* Academic Press, New York. 456 pp.

Rennenberg, H., W. Eschrich, and H. Ziegler (Eds.). 1997. *Trees—Contributions to Modern Tree Physiology.* Backhuys Publishers, Leiden, The Netherlands. 565 pp.

Taiz, L. and E. Zeiger. 2006. *Plant Physiology,* 4th ed. Sinauer Associates, Inc., Sunderland, MA. 764 pp.

Wilkinson, R.E. (Ed.). 2000. *Plant-Environment Interactions,* 2nd ed. Marcel Dekker, Inc. New York. 456 pp.

19 Use of Auxins for Rooting Cuttings

Ellen G. Sutter and David W. Burger

Although cuttings can be rooted without auxins, rooting of cuttings often benefits from treatment with auxins and, in some cases, absolutely requires auxins to induce root formation. Auxins are a group of plant growth regulators that, in their natural form, are involved in many plant processes (chapter 14). These processes include apical dominance, fruit development, root initiation, phototropism, leaf formation, and abscission. Auxins occur naturally as indoleacetic acid (IAA) primarily and in less frequency as indolebutyric acid (IBA). Synthetic auxins often used in horticulture are naphthaleneacetic acid (NAA) and 2,4-dichlorophenoxyacetic acid (2,4-D) (Figure 19.1). The most commonly used auxins for rooting cuttings are IBA and NAA. IAA degrades easily in the plant, whereas 2,4-D does not induce root formation.

AVAILABILITY OF AUXINS

Auxins are available commercially as either a dry powder or as a liquid. Formulations use IBA or NAA, talc, or solvents for liquid formulations. Several brand-name products are Rhizopon™, Dip 'N Grow™, Vita Grow™ (formerly known as Wood's™ Ready to Use Rooting Hormone), Hormex™, Hormodin™, and Rootone™ (Figure 19.2). Commercial rooting products are comprised of a variety of different kinds and concentrations of auxins, as well as other compounds (Table 19.1).

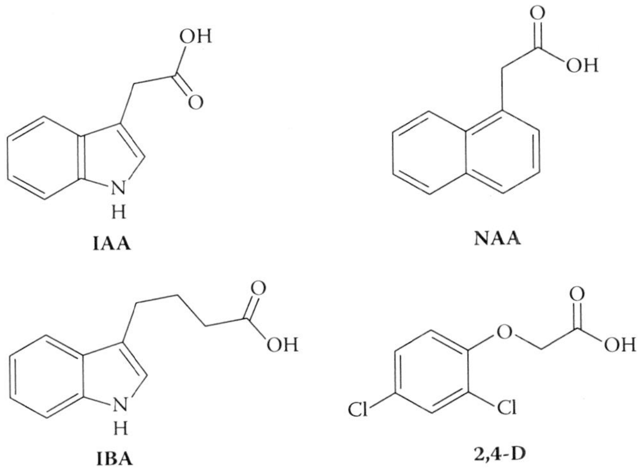

FIGURE 19.1 Chemical structures of auxins commonly used in rooting of cuttings.

In addition to using a preformulated commercial product, a propagator can prepare a formulation of auxin from the pure compound of choice. IBA and NAA can be purchased from chemical supply houses, such as Sigma Chemical Corp., Carolina Biological Supply, Phytotechnology Labs, and Caisson Labs. The powders are available in the acid form as indolebutyric acid or 1-naphthalene acetic acid or as the potassium salt of the acid.

There are advantages and disadvantages to both forms (free acid versus potassium salt) of the auxins. The free acid form of auxin is only soluble in water in extremely low concentrations. Thus, ethanol or dilute potassium hydroxide (KOH) must be used for dissolving the IBA or NAA. Ethanol, which generally is 50% of the final solution (see below for instructions on how to prepare the compound), may be toxic to plants and cause the base of the cuttings to die. On the other hand, in some cases, alcohol is thought to enhance the uptake of auxin into the base of the cutting and is used preferentially in such situations. The potassium salt of the auxin is freely soluble in water and mixes easily. Since there is no alcohol present, the solution is less damaging to some plants.

SELECTION OF CUTTINGS

For our discussion of rooting cuttings, we will focus on stem cuttings. One of the most important aspects of rooting with auxins is matching the cutting with the appropriate auxin formulation and concentration. Some kinds of cuttings do not need auxin to form adventitious roots. These generally are herbaceous and known as "easy to root." Cuttings in this group include *Salvia officinalis* (sage), *Perlargonium* cultivars (especially some scented geranium varieties), *Coleus blumei* (coleus), and *Penstemon digitalis* 'Husker Red' and other garden cultivars. Many houseplants, such as Pothos and succulents *Kalanchoe blossfeldiana* and *Echeveria,* fall in this category.

Most stem cuttings do need auxin; however, auxin may not be sufficient by itself to stimulate rooting. Selecting the proper part of the stem also may have a significant effect on rooting. Stems taken from a shrub or perennial plant are composed of different kinds of wood (Figure 19.3).

Cuttings from the first 3 to 5 cm of the tip, known as terminal cuttings, are considered to be softwood. They bend easily and are quite pliable. The section of the stem below the terminal cutting is known as subterminal. It is stiffer than

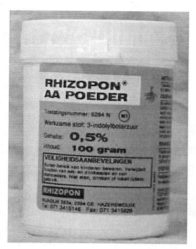 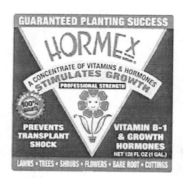  

FIGURE 19.2 Several commercial auxin products used for rooting cuttings. Rhizopon—https://www.leafman.nl/; Hormex—http://louisvillehydro.com/; Dip 'N Grow—http://www.dipngrow.com/; Rootone—http://www.gardentech.com/.

TABLE 19.1
Commercial Rooting Products Containing Auxins

Commercial Product	Manufacturer	Formulation	Active Ingredients
Rhizopon	Rhizopon	Powder and water soluble tablets	0.1 % IBA 0.3 % IBA 0.8% IBA
Dip 'N Grow	Dip 'N Grow, Inc.	Liquid	IBA and NAA
Vita Grow (formerly known as Wood's Ready to Use Rooting Compound)	Vita Grow, Inc.	Liquid	IBA and NAA
Hormex	Brooker Chemical Corp.	Powder	0.1 % IBA 0.3 % IBA 0.8% IBA
Hormodin	OHP, Inc.	Powder	0.1 % IBA 0.3 % IBA 0.8% IBA
Rootone	GardenTech, Inc.	Powder	Naphthaleneacetamide plus Thiram

Use of Auxins for Rooting Cuttings

FIGURE 19.3 Stem cuttings used in propagation.

the terminal section, but still can be bent. The lower part of a stem is the basal section and is often woody and difficult or impossible to bend. A cutting taken from this section is known as a basal or hardwood cutting. Stems of deciduous trees and shrubs taken during winter, when the leaves have abscised, are also known as hardwood cuttings. Depending on the plant, the specific kind of stem cutting may be critical for success. In addition, the time of year during which a cutting is taken may also be a crucial factor influencing success in obtaining rooted cuttings. Unfortunately, there are no rules at present that will predict the kind of cutting that will best form roots. A suitable method for determining the most appropriate kind of stem to use for rooting as a cutting is to research requirements of closely related species. The text, *Plant Propagation* by H.T. Hartmann et al. (2002), contains a great deal more information on this subject.

PREPARING AUXINS

COMMERCIAL POWDERS

Auxin formulations that are manufactured as powders are very easy to use. To use, transfer a small amount of the powder to a container, such as a weighing tin, weighing boat, or small jar about 2 in. in diameter. The powder should always be transferred from the original container in order to avoid contamination of the remaining compound. Do not return any powder that has been used back into the original container. Take cuttings, 2.5 to 4 in. long, and dip the cut end in tap water. Then dip in the powder. Cuttings should be tapped to remove excess powder. The cuttings are then stuck into a flat or cell pack containing a mixture of soilless components (see below). It is very difficult to change the concentration of powders. If they are mixed with talc, one must be sure that the original and additional talc are well mixed and homogeneous so that the concentration of IBA throughout the powder is uniform. Powders are usually used in their original concentrations

A different approach is taken by Hortus Inc., which sells potassium IBA under the name Hortus IBA Water Soluble Salts™. Instructions are included on how to measure the proper amount of the salt. Potassium IBA (K-IBA) is freely soluble in water and ethanol is not needed. They also manufacture potassium IBA in tablets (Rhizopon Rooting Hormone™). This allows the propagator to dissolve a specified number of tablets in water to obtain the desired concentration of IBA.

LIQUID PREPARED FORMULATIONS (SUCH AS HORMEX, DIP 'N GRO, VITA GROW)

Directions on the container of liquid formulations indicate the proper dilution and liquid to use for plants that are easy, moderate, or difficult to root.

PREPARING PURE IBA

IBA, as purchased, is insoluble in water at the concentrations required for rooting cuttings. Thus, IBA is first dissolved in 100% or 95% ethanol (propanol may be substituted, but do NOT use methanol) and then diluted by half with water to obtain a 50% alcohol solution, which is then used for dipping the cuttings. Calculations for several different concentrations are provided in Table 19.2.

The preparation of K-IBA is much more straightforward. The proper amount of K-IBA is weighed and added

TABLE 19.2
Amounts of Material to Use for Preparing Different Concentrations of IBA in 50% alcohol

Plant Ease of Rooting	IBA Concentration	1. Measure Out — Amount of Ethanol	2. Add IBA and Dissolve Completely — Amount of IBA	3. Add Water — Amount of Water	Total Volume
Easy to root	1000 ppm (= 0.1%)	500 mL	1 g	500 mL	1000 mL
Moderate to root	3000 ppm (= 0.3%)	500 mL	3 g	500 mL	1000 mL
Difficult to root	8000 ppm (= 0.8%)	500 mL	8 g	500 mL	1000 mL

directly to water. It will dissolve in extremely high concentrations, up to 40,000 ppm. All liquid auxin solutions should be kept in a dark container at 4°C to avoid degradation of the compound.

ROOTING MEDIUM

An excellent medium for rooting is 3 to 4 parts perlite to 1 part peat moss or vermiculite. This medium might appear to be very "light," but it retains the proper amount of moisture without becoming too wet, which would result in the base of the cuttings rotting.

SELECTING AN EFFECTIVE AUXIN CONCENTRATION

Auxin concentrations of 1000, 3000, or 8000 ppm (mg/L) will satisfy most rooting needs. Commercial powders are sold in these concentrations, and dilutions of liquid commercial formulations and IBA can easily be prepared in the same concentrations.

With this background, several laboratory exercises can be developed to study the effects of the following on rooting:

1. Different formulations of auxin
2. Different concentrations of auxin
3. The same auxin concentration on different parts of a stem
4. Different auxin concentrations on different plant species

EXERCISES

EXPERIMENT 1: EFFECTS OF DIFFERENT FORMULATIONS OF AUXIN ON ROOTING OF FAMILIAR HOUSEPLANTS

Materials

The following materials will be required to complete this study:

- Plants that can be used include coleus, scented geranium, pothos, and begonia (Angel Wing type)
- Healthy stem cuttings from the terminal and subterminal sections of stems
- Market pak, 3 × 5 × 8 in. or larger flats, depending on the size of the class
- Rooting medium—4 parts perlite, 1 part peat moss; be sure the peat moss is screened and moistened prior to use; perlite dust is dangerous to inhale; one way to avoid dust is to pour water into the plastic bag of perlite first and mix it, prior to scooping it out; another method is to have one person, wearing a mask, mix water into the dry perlite, which has been poured out onto a bench or mixing table
- Powder formulation of IBA
- Liquid formulation of IBA in the same concentration
- Clippers
- Water for dipping cuttings

Follow the instructions in Procedure 19.1 to complete this experiment.

Procedure 19.1
Effect of Different Concentrations of Auxin on Rooting of *Dianthus caryophyllus* (Carnation) and *Dendranthema morifolium* (Chrysanthemum) Shoots

Step	Instructions or Comments
1	Fill the flats with the rooting mix to within 0.5 in. of the top. Water the mix thoroughly and let it drain.
2	For coleus, use a 1000 ppm concentration of IBA in both liquid and powder. Take half of the terminal, subterminal, and basal cuttings and dip in water. Then dip in the powder and stick in the flat filled with the rooting mix. Take the other half of the cuttings and dip in the IBA solution for 5 sec. (quick dip). Stick the cuttings in the flat as well. Be sure that all treatments are well labeled and the date that the experiment was done is also on the label.
3	Take coleus plants and clip off the terminal 2.5 to 3 in. Cut again 2.5 to 3 in. lower on the stem, and then make a cutting from a woody part of the stem. There are now three types of cuttings: terminal, subterminal, and basal. Prepare 10 to 20 sets of cuttings per student or group, depending on the number of cuttings available.
4	Take half of the terminal, subterminal, and basal cuttings and dip in water. Then dip in the powder and stick in the flat filled with the rooting mix. Take the other half of the cuttings and dip in the IBA solution for 5 sec. (quick dip). Stick the cuttings in the flat as well. Be sure that all treatments are well labeled and the date that the experiment was done is also on the label. Data should be recorded in two to four weeks.

Anticipated Results

There may or may not be a difference in the treatments. The terminal cuttings treated with the 50% ethanol are more likely to rot if they are very succulent. The woody basal cuttings should root less well with the powder treatment. This experiment may also be performed using *Buddleia davidii* (butterfly bush). This plant is not available in the northern part of the country. Other possible plants to use include succulents such as those in the family Crassulaceae.

Questions

(Questions will depend on which of the factors were examined.)

- Was there any difference between the powder form and liquid form of auxin used?
- Was there a difference in rooting response with different concentrations of auxin? If so, what was the difference? In rooting percentage? In the morphology of roots formed?
- Did different types of cuttings (e.g., terminal, subterminal, basal) respond differently?
- Were there differences between different species tested?

EXPERIMENT 2: EFFECT OF DIFFERENT CONCENTRATIONS OF AUXIN ON ROOTING OF *DIANTHUS CARYOPHYLLUS* (CARNATION) AND *DENDRANTHEMA MORIFOLIUM* (CHRYSANTHEMUM) SHOOTS

Different plant species have different responses to auxin applied for rooting. A concentration that may be ideal for one species may be too high or too low for another. In this experiment, the effect of two concentrations of IBA on the rooting of terminal cuttings of carnation and chrysanthemum will be determined.

Materials

The following items are required to complete this experiment.

- Chrysanthemum cuttings from a commercial grower
- Carnation cuttings from a commercial grower
- Sixty cuttings each of carnation and chrysanthemum are required for each group of students. It is best to have several extra in case some cuttings are damaged or destroyed
- Flat large enough to hold 120 cuttings
- Rooting medium of 4 parts perlite:1 part peat moss or vermiculite; be sure the peat moss is screened and moistened prior to use; the dust of the perlite is dangerous to inhale; one way to avoid dust is to pour water into the plastic bag of perlite first and mix it, prior to scooping it out; another method is to have one person, wearing a mask, mix water into the perlite, which has been poured out dry
- IBA in powder formulation at 1000 and 3000 ppm
- Water for dipping cuttings
- Clippers
- Talc

Follow the instructions presented in Procedure 19.2 to complete this exercise.

Anticipated Results

Roots will likely begin to appear within 7 to 10 days. Soon after that rooting percentages and individual root lengths can be measured. With these species, it is likely you will see increasing rooting percentages with increasing auxin concentrations. At high auxin concentrations, phytotoxic symptoms (browning stems and/or roots, stunted root elongation) may be apparent.

Questions

- Did the two species (carnation and chrysanthemum) respond differently to the auxin treatments?
- Were any phototoxicity symptoms observed? If so, what might be some of the causes of those symptoms?

REFERENCES

Hartmann, H.T., D.E. Kester, F.T. Davies, Jr., and R.L. Geneve. (2002) *Plant Propagation: Principles and Practices*, 7th ed. Prentice Hall, Inc., Upper Saddle River, NJ. 880 pp.

WEB SITES

Dip 'N Grow. http://www.dipngrow.com/
GardenTech. http://www.gardentech.com/
Leafman Trading. https://www.leafman.nl/
Louisville Hydroponics Garden Supply. http://louisvillehydro.com/

Procedure 19.2
Effect of Different Concentrations of Auxin on Rooting of *Dianthus caryophyllus* (Carnation) and *Dendrathema morifolium* (Chrysanthemum) Shoots

Step	Instructions or Comments
1	The cuttings are used just as they come. Do not trim the bases. It is preferable to use the cuttings within one week of their delivery. Fill the flats with the rooting mix to within 0.5 in. of the top. Water the mix thoroughly and let it drain. Place wooden or plastic stakes in the flat with the proper treatment date and species.
2	The treatments are the same for carnations and chrysanthemums and are as follows: (1) Control (no IBA), (2) 1000 ppm IBA, and (3) 3000 ppm IBA. Take 20 cuttings of carnation, dip in water, and then in the 1000 ppm hormone powder. Stick in the flat in the proper row. Repeat with 20 cuttings for the 3000 ppm treatment. For the control, the cuttings are dipped in water, then talc, and lastly stuck in the flat. Repeat with the chrysanthemum cuttings. Place the flat in the greenhouse.
3	After two weeks, 10 cuttings of each treatment are removed and the amount of rooting recorded on a data sheet (see example Table 19.3). One method of quantifying root formation is to set up a scale ranging from 0 to 5, with 0 being no roots and 5 being a large number of roots.
4	After an additional two weeks, the remaining cuttings are removed and the amount of rooting is recorded. At this time, it is possible that students have already rated some root masses as 5 and now see even larger root masses. They can use higher ratings without affecting the results of the experiment. Note, however, that the same student or students must do the ratings for them to be reliable over time. The root ratings are combined with the percentage of shoots rooted to produce a rooting index that will enable the students to compare results across treatments and species (Table 19.4).

TABLE 19.3
An Example of Data Collection

Cutting Number	Week 2 Rating	Week 4 Rating
1	2	4
2	1	6
3	0	4
4	1	4
5	3	6
Mean	1.4	4.8
% rooted	80%	100%

TABLE 19.4
Calculating the Weighted Root Rating That Takes into Account the Number of Cuttings at Each Root Rating

A. Rating	B. Week 2 # Cuttings	C.	D. Rating	E. Week 4 # of Cuttings	F.
0	1	0	0	0	0
1	2	2	1	0	0
2	1	2	2	0	0
3	1	3	3	0	0
4			4	3	12
5			5	0	0
6			6	2	12
Rooting index		1.4			4.8

20 Adventitious Rooting of Woody and Herbaceous Plants

Lori A. Osburn, Zong-Ming Cheng, and Robert N. Trigiano

Generally speaking, any method of multiplying plants other than by seed is considered asexual reproduction. Where sexual reproduction generally results in unique, individual plants from two parents, asexual reproduction results in replicated plants (or clones) from one parent. Some common methods of asexual reproduction used to propagate plants include layering (chapter 26), grafting (chapter 27), budding (chapter 27), division and separation, micropropagation (chapter 31) and cuttings (chapter 17). Propagation of plants by rooted cuttings is used for a very large number of herbaceous and woody ornamentals and some fruit and nut crops. Whether or not a plant is propagated by cuttings is determined by the ease of which they root or if other propagation methods are not as successful. Cuttings can be taken from roots, leaves (chapter 23), or stems (chapter 17). Because cutting propagation is such a broad topic, this chapter will focus on the use of stem cuttings to propagate woody and herbaceous plants.

When using stem cuttings for propagation, the resulting progeny are identical to the parent plant from which they were taken and are called clones. Some of the reasons to use stem cuttings for propagation are the following:

1. To propagate plants that cannot be easily reproduced via seed. This is because some seed are difficult to germinate or have too strict or exacting germination requirements. Some seed are recalcitrant, meaning they have a short storage life.
2. To bypass the juvenile stage of growth and more quickly attain the desired mature phase of growth that allows for flowering and fruit bearing. This is the case with some annuals, many perennials, and some fruit, nut, and ornamental trees and shrubs.
3. To retain the genetic characteristics that allow one to create or maintain clones of a particular plant. Many clones are desired for special attributes that might be lost through sexual reproduction. Asexual reproduction retains unique characteristics by maintaining the genotype, whereas sexual reproduction consists of combining DNA from two parents and results in progeny that have new genotypes (chapter 37). Some unique attributes might include special foliage or flower color, leaf variegation, plant size, fruit type, etc.

Leaf and root cuttings are not sufficient to obtain a true clone—stem cuttings that contain at least one bud are required. Unlike stem cuttings that possess one or more buds, leaf and root cuttings do not contain a layered meristem. Adventitious buds are formed that may or may not retain all of the desired characteristics of the parent plant. Thus, some desired mutations (variegation of flowers, leaves, etc.), such as with chimeras (chapter 16), will not be expressed in the progeny. Plant chimeras are organisms which, because of a genetic mutation (natural or induced), contain two different genotypes within one individual and are phenotypically expressed as variegation or some pattern of distinction, such as thornless blackberries (McPheeters and Skirvin, 1983). A good example of a chimeric plant is *Sansevieria trifasciata*, which typically has a variegated leaf. This plant can be easily propagated via leaf cuttings, but the variegation is lost because the leaves do not have organized buds that have the different tissue layers (chapter 16). In this instance, rhizome (modified stem) cuttings would be more appropriate starting material. Buds are also important because they contain an auxin signal that causes rooting to initiate. In many cases, roots will not form on a stem cutting that has had all of the buds removed before rooting is initiated (Haissig and Davis, 1994; Went, 1934).

Rooting requirements can be very species specific, especially for woody plants. Based on the species being propagated, a specific type of cutting is usually required. For example, some species root more easily with hardwood cuttings, whereas others require softwood cuttings. Still others may root easier when semihardwood cuttings are used. Softwood cuttings are taken after the first flush of new spring growth while the wood is green and tender. Semihardwood cuttings are taken in summer after a flush of growth when the wood begins to mature, but still retains its leaves. Hardwood cuttings are taken during dormancy, after growth has ceased and the leaves have abscissed. Of course, there are advantages and disadvantages related to each of these types of cuttings. Softwood and semihardwood cuttings are similar to herbaceous cuttings—they require protected structures and a misting or humidity

TABLE 20.1
Facility Requirements for All Cutting Experiments

Greenhouse—maintain temperatures of about 25°C day/20°C night.

Shadecloth (47%–52%)—place over house or individual benches at least until the cuttings acclimatize or longer, if needed.

Overhead bench mist system—programmed to supply mist or fog at a set duration and interval as needed, depending upon the crop and the climate. Approximate settings for leafy cuttings are 10 sec. of mist every 15 min., beginning 1 h postdawn and ending 1 h predusk. When rooting leafless hardwood cuttings, the mist interval may be longer and the duration may be shorter than that for herbaceous cuttings. As the cuttings acclimate and develop a root system, the mist duration should be gradually decreased and the mist interval gradually increased.

Bottom heat system, if available: heating mat, warm water circulation tubing, etc.

TABLE 20.2
Materials Needed for All Cutting Experiments

Propagation trays (e.g., Dyna-Flat™F1020, Hummert Int.)	Black sharpies
Petri dishes (15 × 100 mm, Fisher Scientific, Atlanta, GA) or small cups	Rulers
Peat-based soilless potting medium (Pro-Mix BX)	Pruners or strong scissors
4-in. plastic labels	Latex gloves
Paper towels	Dust masks

Surface disinfectant for pruners or scissors (10% Clorox®, 70% isopropyl alcohol, Physan 20®, Greenshield®, Triathalon®, etc.)

Plants: Those listed in each experiment are merely suggestions and may not be appropriate for your class, depending on the climate, season, and availability.

---

control system to reduce transpiration until the cuttings produce their own root system (chapter 2 and chapter 3). Hardwood cuttings are much more versatile and economical because transpiration is not an issue, but not all species will produce roots on hardwood cuttings.

As with other propagation methods, there are many factors to consider when selecting a plant to propagate by stem cuttings, including the health, age, growth stage and maturity of the stock plant, time of year that the cuttings are taken, type of cutting used (in woody plants), plant growth regulators (PGRs) to be used, and the availability of an appropriate rooting environment (media, light, temperature, and water requirements). In many cases, after a cutting is taken from a woody stock plant, it is usually treated with an auxin and "stuck" into a suitable rooting medium to produce adventitious roots. Auxins are plant growth regulators, both natural and synthetic, that induce rooting (chapter 17). Auxins such as indole-3-butyric acid (IBA) and naphthalene acetic acid (NAA) are used because they are chemically more stable than auxins such as indoleacetic acid (IAA), an endogenous hormone. Overall, PGRs are widely used to increase rooting speed and uniformity. The longer a cutting remains under mist without roots, the weaker it can become. During this time, nutrients are leached from the leaves, and its susceptibility to pathogens increases.

In many species, polarity (top versus bottom) must be observed when the cutting is placed into the rooting medium. This is because roots will develop naturally on the end of the cutting (proximal end) that was closest to the crown of the parent plant from which it was taken. If polarity is reversed, several outcomes are possible. Rooting may not occur, or roots will occur on the distal (stuck) end, or occur on the proximal end (unstuck) end. Regardless of outcome, the cuttings are not likely to develop normally and would not be economically advantageous for a propagator. Even with leaf and root cuttings, polarity must be observed, or new roots and shoots will not develop.

The following laboratory/greenhouse experiments are designed to provide some experiences with rooting cuttings of various common plants. Each of the experiments requires some common facilities (Table 20.1), materials and supplies (Table 20.2), and procedures (Procedure 20.1). A listing of some commercially available rooting compounds and their compositions (auxins and supplements) is shown in Table 20.3.

EXERCISES

Experiment 1: The Effect of Various Exogenously Applied Auxins on the Adventitious Rooting of Selected Woody and Herbaceous Cuttings

Does Exogenously Applied Auxin Improve Adventitious Rooting of Selected Woody and Herbaceous Stem Cuttings?

Once a cutting is detached from the parent plant, several physiological processes occur, ultimately resulting

Procedure 20.1
Common Preparation for All Cutting Experiments

Step	Instructions and Comments
1	For details specific to each experiment, refer to the appropriate section number.
2	Work in small groups, each consisting of four to five members.
3	Prepare one propagation tray per experiment per group: Wet the potting media, mix, and test the moisture content by squeezing a handful. It should slightly clump. If not, add more water or dry media. Retest: Fill the tray to within 1/2 in. of the rim.
4	Disinfect pruners or scissors by dipping them into one of the recommended disinfectants listed in Table 20.2. Initially soak them for 10 min, then use a quick dip in the solutions between cuts.
5	Choose two plant species for each experiment. Select nonflowering stems.
6	Prepare plant labels with the following information for each species and treatment: group no., species, treatment, and date, then insert into the prepared propagation tray.
7	Label both the tops and bottoms of the petri dishes for water and the various auxins.
8	Fill one petri dish 3/4 deep with water to moisten cut stems prior to auxin talc application.
9	If using Hormex™ liquid, fill the petri dishes about 3/4 deep. A water dip is not needed.
10	While wearing gloves and a dust mask, measure three to four tablespoons of each auxin-talc from the stock container and place into its appropriate petri dish. Pour gently to avoid dust.
11	Referring to the appropriate experiment for the cutting length and number of nodes needed, obtain uniform cuttings of each plant by making your cuts as described below: make a straight cut at the distal end of each cutting, and make an angled cut at the proximal end of each cutting (Figure 20.5).
12	Place your prepared cuttings between sheets of moist paper towels until you are ready to treat and stick them. Write the plant name on the paper towel in pencil to avoid mix-ups.
13	When ready to treat and stick cuttings, remove the leaves from the lower half of the stem and trim any large leaves on the upper portion of the stem (Figure 20.1). Note: In Experiment 2, some treatments will require the upper leaves to be removed instead.
14	Moisten the proximal end of each stem in water and dip that end into the auxin talc. Lightly tap the cutting on the side of the petri dish to remove any excess clumping talc (Figure 20.2 and Figure 20.3). Note: In Experiment 2, some treatments will require a distal end dip.
15	Treat all control plants with a water dip.
16	Stick 1/3 of the cutting length into the medium within its prelabeled row (Figure 20.4).
17	After sticking all of the cuttings for the experiment, place the tray on a mist bench and lightly water. Avoid excess moisture, as it will encourage some disease organisms to grow.
18	Clean up: place lab supplies in a designated location, dispose of any trash, and wipe off your now-empty table.
19	Data collection: Each group elects one of its members to check the cuttings for roots at three or four weeks after sticking, depending upon time constraints. Gently tug on the cutting. If it resists, gently push away the surrounding medium and carefully remove the cutting. If it has any roots, then it is counted as having rooted. Return the cutting to the tray. Record the data in the appropriate table and calculate the rooting percentage.
20	Once they are rooted, the cuttings may be transplanted into individual pots.

FIGURE 20.1 Removing the lower leaves from a herbaceous stem cutting.

FIGURE 20.2 Dipping the trimmed cutting into the auxin talc after a water dip.

FIGURE 20.3 The trimmed cutting is now ready to stick.

Adventitious Rooting of Woody and Herbaceous Plants

FIGURE 20.4 Sticking the prepared cutting into the propagation tray.

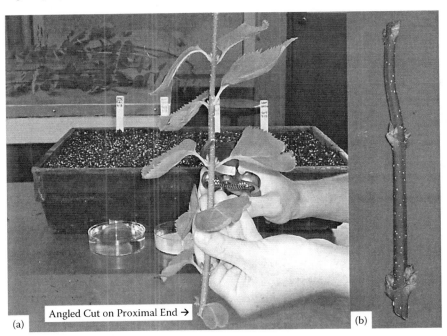

FIGURE 20.5 (a) Cutting the main stem into segments for treating and sticking. Notice the angled cut on the proximal end and that the student is making a straight cut on the distal end. (b) A close-up of the final cuts. Leaves are removed to show details.

in the development of adventitious roots (chapter 18). However, the capacity of different plant species to form adventitious roots varies greatly, from very easy to nearly impossible. In order to optimize rooting during cutting propagation, certain stock plant management practices (chapter 22) are utilized (Hartmann et al., 2002). Above and beyond those practices, auxins are used to enhance adventitious rooting in several ways. Auxins can increase the rooting uniformity within a crop, the number of roots formed, and the speed at which rooting occurs. Although auxin plays a role in many plant processes, this experiment will focus only on its function in rooting.

The primary auxin produced by plants is indole-3-acetic acid (IAA). However, this natural auxin is not commercially utilized because it is less chemically stable than other available auxins. The two most common auxins found in rooting preparations are indole-3-butyric acid (IBA) and 1-naphthalene acetic acid (NAA) (chapter 14). While many species readily respond to IBA, some may respond better to NAA. In addition, the specific formulation, combination, and concentration of auxin can also influence adventitious rooting.

The goals of this experiment are to learn the procedures used in cutting propagation and to compare the effect of various exogenously applied auxin formula-

TABLE 20.3
Some Commercially Available Rooting Compounds

Trade Name	% IBA	% NAA	Type
Hormo-Root[a]	0.1–2.0	—	Talc
Hormodin 1	0.1	—	Talc
Hormodin 2	0.3	—	Talc
Rhizopon AA #2	0.3	—	Talc
Hormodin 3	0.8	—	Talc
C—Mone K	1.0	—	Liq.
C—Mone	1.0–2.0	—	Liq.
Rootonea	—	0.20	Talc
Strike	—	0.5	
Rootone F[a,b]	0.057	0.067	Talc
Hormex	0.013	0.240	Liq.
Dip'N Grow	1.0	0.5	Liq.
C-Mone Plus	1.0	0.5	Liq.
Dip'N Gro	1.0	0.5	Liq.
Woods Rooting Compound	1.0	0.54	Liq.

[a] These products also contain the fungicide Thiram.
[b] This product contains two additional forms of NAA.

tions on the rooting of selected woody and herbaceous stem cuttings.

Materials

Students working in teams of four or five will require the following materials to complete the experiment:

- Materials listed in Table 20.1 and Table 20.2
- Water
- Rhizopon AA #2 talc (0.3% IBA)
- Hormex liquid (0.013% IBA + 0.240% NAA), full strength
- Rootone talc (0.20% NAA)
- Woody plant suggestions: *Hydrangea macrophylla* (florists' hydrangea), *Ficus benjamina* (weeping fig), and *Ficus elastica* (rubber tree)
- Herbaceous plant suggestions: *Solenostemon scutellarioides* (coleus), *Pelargonium* spp. (scented geranium), and *Plectranthus australis* (Swedish ivy)

Follow the instructions outline in Procedure 20.2 to complete this experiment.

Anticipated Results

Most cuttings will root to some degree in response to an auxin application. Therefore, one could expect that the rooting percentage will be increased with the treatment of a higher auxin concentration. However, the higher concentrations required for woody plants may be too strong for the herbaceous plants and could inhibit rooting or cause stem damage. Some of the tender herbaceous cuttings may produce roots without an auxin application. It will be interesting to see which species respond better to NAA than IBA.

Questions/Discussion

- Plant propagation by rooted cuttings is an example of what type of reproduction?
- What is the purpose of using intermittent mist while rooting cuttings?
- What advantages and disadvantages did you observe between the auxin talc and liquid form?
- If you were a large grower and needed to propagate 1000 *Buddleia davidii* liners for Bob's Nurs-

Procedure 20.2
Exogenous Auxins and Adventitious Root Formation on Selected Woody and Herbaceous Plants

Step	Instructions and Comments
1	See Procedure 20.1 for preparation of trays, labels, auxins, and cuttings.
2	Choose one herbaceous and one woody species.
3	For each species, use five cuttings per treatment. Example: 5 cuttings × 4 treatments = 20 cuttings per species.
4	Prepare cuttings 3 to 4 in. long with 3+ nodes. See Figure 20.1 through Figure 20.4 and Procedure 20.1 for complete instructions.
5	Apply treatments: Dip the proximal end of each stem into one of the following: (1) Water only, (2) Hormex liquid, (3) Rhizopon AA #2 talc, and (4) Rootone talc.
6	Stick cuttings into the prepared trays as directed in Procedure 20.1.
7	After all treatments are stuck, place tray on the mist bench as directed in Procedure 20.1.
8	Collect data in three to four weeks, and record in Table 20.4. See Procedure 20.1 for complete instructions for collecting data.

TABLE 20.4
Data Collection for Experiment 1: Rooting Woody and Herbaceous Species with Auxins

Plant	Trt #	Experimental Treatments	# Cuttings	# Rooted	% Rooting[a]
Species 1	1	No auxin—water only	5		
Species 1	2	Hormex liquid	5		
Species 1	3	Rhizopon AA#2 talc	5		
Species 1	4	Rootone talc	5		
Species 2	1	No auxin—water only	5		
Species 2	2	Hormex liquid	5		
Species 2	3	Rhizopon AA#2 talc	5		
Species 2	4	Rootone talc	5		

[a] # of rooted cuttings / # of cuttings used.

ery, which auxin formulation would you use—a liquid or a talc? Justify your answer.
- Why do some cuttings root so easily without an auxin application? Would the use of an auxin on that species be a justifiable expense?
- Under what situation might auxin inhibit rooting in a plant that it at one time promoted rooting?

EXPERIMENT 2: THE EFFECT OF POLARITY AND EXOGENOUSLY APPLIED AUXIN ON THE ADVENTITIOUS ROOTING OF SELECTED WOODY AND HERBACEOUS STEM CUTTINGS

If Polarity Affects the Production of Adventitious Roots on Stem Cuttings, Can an Exogenously Applied Auxin Overcome That Effect?

During commercial rooted cutting propagation, stems are typically treated with an auxin that is applied to the proximal (base) end prior to being inserted into a rooting medium. Propagators are careful to ensure that they treat and stick the proper end of the cutting because many plant species exhibit a phenomenon called polarity. Polarity dictates that roots will form on the proximal end of the cutting, i.e., the end closest to the crown of the plant from which it was taken (Hartmann et al., 2002). To distinguish between proximal and distal, one end of the stem is cut straight and the other is cut at an angle. If the cutting is inserted into the medium with the distal end down (inverted), several not-so-desirable outcomes are possible (Nanda et al., 1969). With some exceptions, the cutting will not produce roots, then die; produce roots on the inverted proximal end, then die; or root on the inserted distal end with impaired growth. In some cases, an exogenously applied auxin may be able to overcome reversed polarity, and the cutting will root on the inserted (distal) end (Cristoferi et al., 1988; Went, 1941), and in some cases polarity cannot be overcome (Sheldrake, 1974).

The goal of this experiment is to observe the effect of polarity on the rooting percentage of selected woody and herbaceous stem cuttings and to determine if an exogenously applied auxin can overcome the influence that polarity exerts on adventitious rooting capability.

Materials

Students will need the following materials to complete this experiment.

- Materials listed in Table 20.1 and Table 20.2
- Hormodin 3 talc (0.8% IBA)
- Water
- Woody plant suggestions: *Rosa* spp. (rose), *Salix erythroflexus* (contorted weeping willow), and *Buddleia davidii* (butterfly bush)
- Herbaceous plant suggestions: *Solenostemon scutellarioides* (coleus), *Schefflera arboricola* (schefflera), and *Dendranthema x grandiflorum* (chrysanthemum)

Follow the instructions outlined in Procedure 20.3 to complete this experiment

Anticipated Results

Based on polarity alone, it would be expected that those cuttings inserted into the medium distal end down will not root. Some species have a very strong polar influence and will root at the proximal end even if they are inverted. Nevertheless, it may be possible for auxin to overcome polarity such that if the auxin-treated distal end of an herbaceous species or an easy-to-root woody species is inserted, roots might be produced. However, the addition

Procedure 20.3
Effects of Auxins and Polarity on Adventitious Rooting

Step	Instructions and Comments
1	See Procedure 20.1 for the preparation of trays, labels, auxins, and cuttings.
2	Choose two woody species.
3	For each species, use five cuttings per treatment. Example: 5 cuttings × 4 treatments = 20 cuttings per species
4	Prepare cuttings 3 to 4 in. long with 4+ nodes. See Procedure 20.1 for complete instructions with one exception: For each species: If plants have leaves, instead of removing all of the lower leaves, remove the lower leaves from half of the cuttings and remove the upper leaves from the other half of the cuttings. Example: 20 cuttings of rose, remove the lower leaves from 10 cuttings and remove the upper leaves from the other 10 cuttings. Repeat for second species.
5	Prepare to apply all treatments (1 through 4) to both of the species. Refer to Figure 20.6 to observe proximal versus distal sticking.
6	Treatment 1: Dip proximal end first into water, then into Hormodin 3. Stick cuttings into the prepared trays as directed in Procedure 20.1.
7	Treatment 2: Dip proximal end into water only (no auxin). Stick cuttings into the prepared trays as directed in Procedure 20.1.
8	Treatment 3: Dip distal end first into water, then into Hormodin 3. Stick cuttings into the prepared trays as directed in Procedure 20.1.
9	Treatment 4: Dip distal end into water only (no auxin). Stick cuttings into the prepared trays as directed in Procedure 20.1.
10	After all treatments are stuck, place tray on the mist bench as directed in Procedure 20.1.
11	Collect data in three to four weeks and record in Table 20.5. See Procedure 20.1 for complete instructions.

FIGURE 20.6 Orientation of cuttings (leaves have been removed for illustration). (a) Proximal end in medium. (b) Distal end in medium.

TABLE 20.5
Data Collection for Experiment 2: Effects of Polarity and Auxin on Adventitious Rooting

Species	Trt #	Experimental Treatments	Cuttings	# Rooted	% Rooting
Species 1	1	Proximal end down—auxin	5		
Species 1	2	Proximal end down—no auxin	5		
Species 1	3	Distal end down—auxin	5		
Species 1	4	Distal end down—no auxin	5		
Species 2	1	Proximal end down—auxin	5		
Species 2	2	Proximal end down—no auxin	5		
Species 2	3	Distal end down—auxin	5		
Species 2	4	Distal end down—no auxin	5		

of auxin would not likely be able to overcome polarity in recalcitrant plants.

Questions/Discussion

- In what tissues is endogenous auxin produced, and which direction is it transported throughout the plant?
- Why might one plant species to be responsive to exogenously applied auxin, whereas another species is unresponsive?
- What physiological process(es) occur when roots form on an inverted cutting without auxin?
- How can auxin overcome the effect of polarity?
- Based on the results of this experiment, which of the conditions tested would you recommend a propagator use to maximize rooting percentage?

EXPERIMENT 3: THE EFFECT OF BUD REMOVAL AND EXOGENOUSLY APPLIED AUXIN ON THE ADVENTITIOUS ROOTING OF SELECTED WOODY STEM CUTTINGS.

If the Presence of Buds on a Stem Cutting Is Required for the Production of Adventitious Roots, Can an Exogenously Applied Auxin Fulfill That Requirement?

Endogenously produced auxin, primarily in the form of indole-3-acetic acid (IAA), plays a role in many plant functions and might be best known for its role in rhizogenesis (root initiation and formation). Auxin is produced in the leaves and meristematic tissues (shoot and root apices, lateral buds), then transported basipetally (downward) via the phloem. Should those auxin-producing tissues be removed during the first few days after cutting, auxin will not be synthesized and rooting will not be initiated, even with the application of auxin (Haissig and Davis, 1994; Went, 1934).

The objective of this experiment is to observe the influence of an exogenously applied synthetic auxin on the rooting of both disbudded and intact hardwood deciduous and evergreen stem cuttings.

Materials

Students will require the following materials to complete this study:

- Materials listed in Table 20.1 and Table 20.2
- Hormodin 3 talc (0.8% IBA)
- Water
- Sharp knife (disinfected)
- Deciduous plant suggestions: *Salix erythroflexus* (contorted weeping willow), *Lagerstroemia indica* (crape myrtle), and *Hydrangea quercifolia* (oak-leaf hydrangea)
- Evergreen plant suggestions: *Euonymus* spp., *Nandina domestica*, and *Ligustrum* spp. (privet)

Follow the instructions outline in Procedure 20.4 to complete this experiment.

Anticipated Results

Because buds are required to at least initiate rooting, the disbudded deciduous species without the exogenously auxin treatment are least likely to root. It is always possible that application of auxin to those disbudded deciduous species will provide sufficient hormone to overcome the loss of the endogenous auxin source. The evergreen species might be more likely to root in both treatments due to the presence of leaves just prior to their removal for this experiment.

Procedure 20.4
Effects of Bud Removal and Exogenously Applied Auxin on the Adventitious Rooting of Selected Woody Stem Cuttings

Step	Instructions and Comments
1	See Procedure 20.1 for preparation of trays, labels, auxins, and cuttings.
2	Choose one evergreen and one deciduous species.
3	For each species, use five cuttings per treatment. Example: 5 cuttings × 4 treatments = 20 cuttings per species.
4	Prepare cuttings 3 to 4 in. long with 4+ nodes. See Procedure 20.1 for complete instructions. Remember to make angled and straight cuts to designate the proper polarity.
5	For each species: In addition to the instructions in Procedure 20.1, carefully remove all buds and leaves from half of the cuttings using a sharp, disinfected knife (Figure 20.7). Some species' buds will pull off easily by hand. Examples: 20 cuttings of willow, remove all buds from 10 cuttings. Repeat for second species.
6	Treatment 1: Buds removed, dip proximal end into (1) water, then (2) Hormodin 3. Stick cuttings into the prepared trays as directed in Procedure 20.1.
7	Treatment 2: Buds removed, dip the proximal end into water only (no auxin). Stick cuttings into the prepared trays as directed in Procedure 20.1.
8	Treatment 3: Buds intact, dip proximal end into (1) water, then (2) Hormodin 3. Stick cuttings into the prepared trays as directed in Procedure 20.1.
9	Treatment 4: Buds intact, dip the proximal end into water only (no auxin). Stick cuttings into the prepared trays as directed in Procedure 20.1.
10	After all treatments are stuck, place tray on the mist bench as directed in Procedure 20.1.
11	Collect data in three to four weeks and enter in Table 20.6. See Procedure 20.1 for complete instructions.

FIGURE 20.7 Removing the buds from a hardwood stem cutting.

TABLE 20.6
Data Collection for Experiment 3: Disbudding

Species	Trt #	Experimental Treatments	# Cuttings	# Rooted	% Rooting
Species 1	1	Buds removed—auxin	5		
Species 1	2	Buds removed—no auxin	5		
Species 1	3	Buds remain intact—auxin	5		
Species 1	4	Buds remain intact—no auxin	5		
Species 2	1	Buds removed—auxin	5		
Species 2	2	Buds removed—no auxin	5		
Species 2	3	Buds remain intact—auxin	5		
Species 2	4	Buds remain intact—no auxin	5		

Questions/Discussion

- Describe the difference(s) between an endogenous auxin and an exogenous auxin. Where does each of these come from?
- Why must a leaf or bud be present to form adventitious roots on a stem cutting?
- If endogenous auxin is produced in the buds and the leaves, how does it promote roots at the proximal end of a stem cutting?
- Name another endogenous compound that affects the rooting ability of stem cuttings.
- Hardwood stem cuttings were used in this lab. What other categories of stem cuttings are there?
- Based on the results of this experiment, what conditions tested would you recommend a propagator use to maximize rooting percentage?

LITERATURE CITED

Cristoferi, G., N. Filiti, and F. Rossi. 1988. The effects of reversed polarity and acropetal centrifugation on the rooting of hardwood cuttings of grapevine rootstock Kober 5bb. *Acta Hort.* (ISHS) 227: 150–154.

Haissig, B.E., and T.D. Davis. 1994. A historical evaluation of adventitious rooting research to 1993, in *Biology of Adventitious Root Formation*. Davis, T.D. and B.E. Haissig (Eds.), Plenum Press, New York. pp. 275–331.

Hartmann, H.T., D.E. Kester, F.T. Davies, Jr., and R.L. Geneve. 2002. *Plant Propagation: Principles and Practices*, 7th ed. Prentice Hall, Upper Saddle River, NJ. 880 pp. Web site: http://wps.prenhall.com/chet_hartmann_plantpropa_7

McPheeters, K. and R.M. Skirvin. 1983. Histogenic layer manipulation in chimeral 'Thornless Evergreen' trailing blackberry. *Euphytica*. 32(2): 351–360.

Nanda, K.K., A.N. Purohit, and V.K. Kochhar. 1969. Effect of auxins and light on rooting stem cuttings of *Populus nigra*, *Salix tetrasperma*, *Ipomea fistulosa*, and *Hibiscus notodus* in relation to polarity. *Physiol. Plant.* 22: 1113–1120.

Sheldrake, A.R. 1974. The polarity of auxin transport in inverted cuttings. *New Phytol.* 73: 637–642.

Went, F.W. 1934. A test method for rhizocaline, the root-forming substance. *Proc. Kon. Ned. Akad. Wet.* 37: 445–55.

Went, F.W. 1941. Polarity of auxin transport in inverted *Tagetes* cuttings. *Bot. Gaz.* 103: 386–390.

21 Rooting Cuttings of Tropical Plants

Richard A. Criley

While many people think of tropical plants as something grown only in warm climates, many of the plants grown as houseplants and for interiorscapes originated in tropical climates. The techniques used to propagate herbaceous and woody tropical plants are much the same as for similar plants of the temperate zones. For a plant propagation class, however, the quantities of stock material may be limited in northern areas, as a small greenhouse or conservatory may be the sole environment in which to grow the plants year-round. Examples of tropical plants that may be used in the laboratory exercises are listed in Table 21.1.

Cutting types used in the propagation of tropical plants include terminal (or tip) cuttings of varying degrees of maturity (herbaceous, softwood, greenwood, semi-hardwood, and hardwood), stem cuttings with no terminal bud (often called cane cuttings), and stem cuttings with one or two nodes or with opposite buds and the stem split between them. For the most part, cuttings should retain at least some foliage, but some foliage plant propagators will use single- or double-node cuttings without an attached leaf to save space in the propagation flat. Removal of some foliage from cuttings is done for the following reasons: (1) to save space in a flat, (2) to reduce the "umbrella" effect of large leaves over neighboring cuttings, (3) to reduce transpiring surface areas and minimize desiccation, (4) to ease insertion of the cuttings into the medium, and (5) to eliminate foliage that might rot in the medium or reduce air circulation among densely planted cuttings. Leaves provide the photosynthate and hormones (endogenous plant growth regulators) needed for initiating and developing roots.

CALCULATING THE ROOTING INDEX (RI)

While rooting can be evaluated by counting the number of roots, cutting them off and weighing them wet or dry, or measuring root lengths, another method is to calculate the rooting index (RI). This is an easy method and basically sorts the rooted cuttings into categories. Arrange a replicate of 10 cuttings from the largest (visually) to smallest rooted cuttings, including those that are alive but have no roots and those that are dead. When you have done this for several replicates, it should become obvious what constitutes heavy, medium, or light rooting. The cuttings in these categories are each assigned weights: 5 for heavy rooting, 4 for medium rooting, and 3 for light rooting. Alive, but not rooted, cuttings receive a score of 2 and dead cuttings a 1. The weighted values for the 10 cuttings are added together and divided by 10. This value is the RI for that replication. Since the number falls anywhere between 1 and 5, it can be related to the original category; a 4.3 RI, for example, suggests that most of the cuttings were medium to heavily rooted. If there are several replications of a treatment, the RI values for each are averaged together for a treatment RI. Treatment RI values can be compared to distinguish between treatment effects using statistics or represented graphically. Table 21.2 shows an example of RI calculations. If using a spreadsheet, such as Excel, a formula can be entered into the RI column that will automatically calculate the RI. A column for percent rooted is also included.

Several exercises are offered that use different tropical plants, mostly from the "foliage plant" grouping of tropical plants because tropical fruits and many woody ornamentals are less available in temperate climates. For each exercise, an objective is suggested that illustrates an important aspect of plant propagation. At a minimum, record the percent rooting, root quality, and time from planting the cuttings to rooting.

EXERCISES

EXPERIMENT 1: DETERMINATION OF THE POSITION ON THE STEM WHERE ROOTS DEVELOP MOST READILY

For this exercise, choose plants that produce relatively long, unbranched stems, so that wood can be selected of differing maturities along the length of the stems (Figure 21.1). If the terminals are still rather soft and succulent, and the lower portion of the stem has matured and thickened, differences in both the extent and quickness to root may occur. It is, of course, problematic to compare a terminal cutting against a stem cutting, but in this case, the age of the wood is the principal feature. Rooting compounds can be used, but may mask some of the differences in "rootability" among the different portions of the stem.

Materials

(The instructor may reduce the quantity of material to be used if plant material is in short supply.)

TABLE 21.1
Tropical Plant Materials for Various Cutting Propagation Exercises

Botanical Name	Common Name	1 or 2 Nodes	Position of Basal Cut	Response to Leaf Area	Maturity of Wood	Cane, Stem Pieces
Abutilon X *hybridum*	Flowering maple				X	
Acalypha wilkesiana, A. hispida,	Beefsteak plant, Chenille plant		X	X		
Aglaonema spp. and hybrids	Aglaonema					X
Allamanda cathartica	Allamanda		X		X	
Anthurium andraeanum	Anthurium	X				X
Aphelandra squarrosa	Zebra plant	X				
Bougainvillea spectabilis, B. glabra	Bougainvillea	X			X	
Brunfelsia pauciflora	Yesterday, today & tomorrow				X	
Cissus rhombifolia, C. discolor, C. antarctica	Grape ivy	X				
Clerodendrum thomsonae, C. X *speciosum*	Bleeding heart vine, Glorybower		X		X	
Codiaeum variegatum	Croton	X		X		
Cordyline fruticosa	Ti					X
Costus speciosus	Crape ginger					X
Dieffenbachia maculata, D. amoena	Dumbcane					X
Dischidia major		X				
Dracaena marginata, D. fragrans, D. fragrans 'Warneckii'	Madagascar dragon tree, Corn plant, Striped dracaena					X
Dracaena reflexa (Pleomele reflexa)	Song of India	X				X
Epipremnum pinnatum	Pothos	X				
Euphorbia pulcherrima	Poinsettia		X			
Ficus elastica	India rubber tree	X			X	
Ficus pumila	Creeping fig				X	
Gardenia augusta (G. jasminoides)	Gardenia			X		
Graptophyllum pictum	Caricature plant	X	X	X		
Hedychium spp., *Zingiber* spp.	Ginger	X				
Hibiscus rosa-sinensis, H. schizopetalus, H. tiliaceus	Hibiscus			X	X	
Hoya carnosa	Wax plant	X				
Ixora coccinea, I. chinensis	Ixora			X		
Jasminum sambac	Arabian jasmine		X			
Justicia brandegeana	Shrimp plant		X			
Justicia carnea	Jacobinia		X			
Ligustrum japonicum	Japanese privet			X		
Mandevilla X *amabilis, M. splendens*	Mandevilla	X				
Maranta leuconeura	Prayer plant	X				
Marsdenia floribunda	Stephanotis	X			X	
Nerium oleander	Oleander		X			
Odontonema cuspidatum	Fire spike		X			
Passiflora spp.	Passionflower	X				
Peperomia obtusifolia, P. crassifolia	Peperomia	X				
Philodendron scandens var *oxycardium*	Philodendron	X				

TABLE 21.1 (Continued)
Tropical Plant Materials for Various Cutting Propagation Exercises

Botanical Name	Common Name	1 or 2 Nodes	Position of Basal Cut	Response to Leaf Area	Maturity of Wood	Cane, Stem Pieces
Pilea cadierei, P. nummariifolia	Aluminum plant, Creeping Charlie	X				
Polyscias guilfoylei, P. fruticosa, P. scutellaria	Panax, Ming aralia, Balfour aralia					X
Pseuderanthemum carruthersii	False eranthemum	X	X	X		
Rhaphidophora celatocaulis	Shingle plant	X				
Rhododendron spp.	Vireya rhododendrons		X			
Rhoicissus capensis	Cape grape, evergreen grape	X				
Saritaea magnifica	Purple bignonia		X			
Schefflera actinophylla	Octopus tree, Umbrella tree					X
Schefflera arboricola	Dwarf brassaia	X	X		X	
Schefflera elegantissima	False aralia (Dizygotheca)				X	
Solenostemon scutellarioides	Coleus	X	X	X		
Strobilanthes dyerianus	Persian shield	X	X			
Syngonium podophyllum	Arrowhead vine	X				
Tecoma capensis (Tecomaria capensis)	Cape honeysuckle		X			
Yucca elephantipes	Yucca					X

TABLE 21.2
Calculating the Rooting Index

Treatment	Replication	# Cuttings Set	Number of Cuttings Heavy (×5)	Medium (×4)	Light (×3)	Alive (×2)	Dead (×1)	Sum of Weights	Rooting Index	Percent Rooted
Control	A	10				9	1	19	1.9	0
	B	10				10		20	2.0	0
	C	10				10		20	2.0	0
	Mean								1.97	0
1000 ppm	A	10	2	2	4	2		34	3.4	80
	B	10	2	1	3	2	2	29	2.9	60
	C	10	1	3	2	2	2	29	2.9	60
	Mean								3.07	66.7
3000 ppm	A	10	4	1	1	2	1	38	3.8	60
	B	10	3	2	5			38	3.8	100
	C	10	2	2	3	3		33	3.3	70
	Mean								3.63	76.7

FIGURE 21.1 From tip to base, different carbohydrate and auxin gradients occur. Single-node cuttings of the *Ficus longifolia* (or similar plant materials) can be rooted to show readiness to root and extent of rooting along such gradients.

For each student or team of students, provide the following items:

- Thirty stems, two feet or longer, of species, such as allamanda, bougainvillea, glorybower, croton, Indian rubber tree, and hibiscus
- Three flats of medium (coarse vermiculite or a equal mixture of vermiculite and perlite)
- Pruning shears or knives to cut the stems into pieces
- Labels and markers
- Rooting hormone (if used, try 0.1% or 1000 ppm IBA or NAA)

Each flat will constitute one replication using similar parts of 10 stems. For example, a row of 10 terminals, a row of 10 greenwood subterminals, one or more rows of 10 intermediate-wood stem pieces, and a row of 10 basal stem pieces.

Follow Procedure 21.1 to complete this exercise.

Anticipated Results

Speed of rooting, percent rooting, and quantity of roots should show a gradient from tip to base, but they may also be optimum for midsections of the stem, tapering off toward the tip and base.

Questions

- Did the different portions of stem begin to root at different times?
- Is there a portion of the original stem that gave the best results for root initiation and percent rooting?
- What internal factors are responsible for the good rooting that occurs? The poor rooting?
- How might the stock plants be managed to improve the amount of stem from which you can get good rooting?

EXPERIMENT 2: THE EFFECT OF BASAL CUT POSITION ON ROOTING FROM STEMS

Some plant materials root readily in the middle of an internode, whereas others produce more roots when the basal cut is made just below or just above the node (Figure 21.2). For this exercise, it is best to choose plant materials with relatively long internodes so that a clear distinction can be made in the position of the basal cut (Figure 21.3a). Cuts will be made in the middle of an internode, just above a node, and just below a node (Figure 21.3b). The cuts should be squared off rather than slanted. The stem cuttings can be either terminals (tips) or stem pieces of several nodes. A variant of this experiment for plants with long internodes that do root in the internode is to examine the differences among different internode lengths.

Materials

(The instructor may reduce the quantity of material to be used if plant material is insufficient.)

The following materials will be needed to complete this experiment:

- Ninety cuttings, 30 of each type (location) of basal cut, from plant materials, such as *Acalypha*, allamanda, glorybower, poinsettia, caricature plant, shrimp plant, oleander, coleus, *Pseuderanthemum*, *Schefflera arboricola*, cape honeysuckle
- Three flats of medium (coarse vermiculite or an equal mixture of vermiculite and perlite)
- Pruning shears or sharp knife to make the basal cuts
- Labels and markers

Procedure 21.1
Position on Stem Where Roots Develop

Step	Instructions and Comments
1	Place moistened medium in each flat to a depth of 7 to 8 cm, and level the surface. Flats will be designated A, B, and C.
2	Divide the 30 stems into three groups of 10 stems so that each group is comparable to the others.
3	Determine how to divide a stem into pieces of differing maturity. For example: green terminal, next 10 cm, next 10 cm, and so on. Remove basal foliage that might be under the medium after insertion.
4	Gather the first 10 terminals and insert the basal 3 to 4 cm into the medium in one row in flat A. (If an auxin treatment is used, apply it to all cuttings.) Do the same for the 10 stem pieces that were just below the terminals; do the same for each set of 10. Repeat for the other two groups of 10 stems, using flats B and C.
5	Label the rows in each flat: A-1, A-2, A-3, etc.; B-1, B-2, B-3, etc.; C-1, C-2, C-3, etc. Add plant name, propagator's name, and date. Record the characteristics (e.g., diameter, degree of woodiness, stem or bark color) that distinguish the cuttings from terminal to base by their numbers into a notebook.
6	Place flats of prepared cuttings into your propagation area under intermittent mist or fog set to come on for 6 sec. every 6 min.
7	Evaluate cuttings for evidence of root initiation about two to three weeks after planting them. When the earliest signs of rooting are observed, record the date and wait another two to four weeks before carefully removing them to evaluate the extent of rooting and percent rooting. For each 10 cuttings, calculate the rooting index (Table 21.2) and percent rooted.
8	Average the data for the three replications of each cutting type, and represent these graphically.

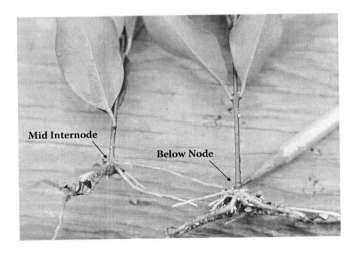

FIGURE 21.2 Influence of position of basal cut on rooting.

FIGURE 21.3 Select material with long internodes, such as Allamanda, to demonstrate effect of basal cut location. (a) Single-node cuttings are made from a long stem. (b) Cut just above node. Middle: cut at mid-internode, Right: cut just below node.

Each flat will have rows of 10 cuttings, each row representing a different position for the basal cut. Where more than one plant material is used, two kinds of plants can share a flat.

Follow Procedure 21.2 to complete this exercise.

Anticipated Results

Many of the suggested species will root anywhere along an internode, but some, such as poinsettia with its hollow stems, will root better with the basal cut made just below a node.

	Procedure 21.2
	Effect of Basal Cut Position on Rooting from Stems
Step	Instructions and Comments
1	Place moistened medium in each flat to a depth of 7 to 8 cm, and level the surface. Flats will be designated A, B, and C.
2	Select cutting materials such that similar stem lengths and diameters and leaf areas can be compared. Cuttings can be terminals or stem pieces, but the two types should not be mixed.
3	Make 30 cuttings in which the base of the cutting is made about 8 mm above a node. Make another set of cuttings with the base just below a node. A final set of cuttings is made with the base in the middle of the internode. (If an auxin treatment is used, apply it to all cuttings). Except in the case of single- or double-node cuttings, your cuttings will be about 7 to 12 cm long. Remove any foliage that might be under the medium after insertion.
4	Insert 10 cuttings of each type in a row in each of the flats (A, B, C).
5	Label the rows in each flat: A-1, A-2, A-3; B-1, B-2, B-3; C-1, C-2, C-3. Add plant name, propagator's name, and date. Record the characteristics (above or below the node cuts, middle of internode) that distinguish the cuttings by their numbers into a notebook.
6	Place flats of prepared cuttings into your propagation area under intermittent mist or fog set to come on for 6 sec. every 6 min.
7	Evaluate cuttings for evidence of root initiation about two to three weeks after planting them. When the earliest signs of rooting are observed, record the date and wait another two to four weeks before carefully removing them to evaluate the location, extent of rooting in terms of root number and lengths, and percent rooting. For each 10 cuttings, calculate the rooting index (Table 21.2) and percent rooted.
8	Average the data for the three replications of each cutting type, and represent these graphically.

Rooting Cuttings of Tropical Plants

FIGURE 21.4 Influence of leaf area on rooting of cuttings. Single-node croton cuttings with leaf blade intact or trimmed to 50% or 25% of the original area. Greater root production develops on intact cutting.

Cape honeysuckle is a plant that roots mainly at the node.

Questions

- Do the different sites for the basal cuts alter the quickness or extent of rooting?
- Where differences in rooting among positions of the basal cut were evident? What factors were responsible?
- What practical impact does making the basal cut in a certain position have on the propagation operation?

EXPERIMENT 3: THE RELATIONSHIP BETWEEN THE AMOUNT OF LEAF SURFACE REMAINING ON THE CUTTING AND THE ABILITY TO ROOT

There are two ways to approach the reduction in leaf surface—cutting leaves to ½, ⅓, or ¼ of their leaf areas or removing whole leaves. Leaf removal can be one leaf of each pair on opposite-leaved plants, removal of upper leaves, or removal of lower leaves. Cutting leaves may lead to leaf abscission from ethylene production caused by injury or from an opportunistic microorganism infection. This approach should be reserved for large-leaved plant materials. The number or area of the remaining leaves will often influence the speed of rooting or extent of rooting (Figure 21.4). Once plant material is in hand, the student should decide which approach is most appropriate and design an experiment that represents a gradient of leaf areas or counts. There should be 30 cuttings for each choice, with at least 3 choices representing, for example, all leaves remaining, ⅓ leaves (area) remaining, ½ leaves (area) remaining, ⅔ leaves (area) remaining, no leaves. Other options include selective removal of top leaves, one each of a pair of opposite leaves, or basal leaf. Note that cuttings with zero leaves may fail to produce roots, or rooting will be much delayed and a poor root system may result.

Materials

(The instructor may reduce the quantity of material to be used if plant material is insufficient.)

The following materials will be needed to complete this experiment:

- Sufficient plant material (such as croton, gardenia, caricature plant, hibiscus, privet, and *Acalypha*) that a minimum of 90 cuttings (30 for each treatment) can be produced
- Three flats of medium (coarse vermiculite or an equal mixture vermiculite and perlite)
- Pruning shears or knives to cut leaves
- Labels and markers

Each flat will have rows of 10 cuttings, each row representing a different leaf area treatment. Where more than one plant material is used, two kinds of plants can share a flat.

Follow Procedure 21.3 to complete this experiment.

Anticipated Results

In general, better rooting is obtained with more foliage left on the cutting. However, some cuttings have sufficient reserves in the stem that they will develop roots even if totally defoliated, but this usually is accompanied by lateral bud break and new shoot and leaf development. The shock of cutting off a portion of the leaf can lead to rapid senescence of that leaf and the loss of its contributions to the rooting process. Large leaves can provide an "umbrella" over neighboring cuttings, leading to their dehydration. Too much foliage can also reduce air circulation in the cutting flat, leading to *Botrytis* (chapter 8) or other diseases that attack the cuttings.

Questions

- Was there a difference in speed or extent of rooting that could be related to leaf area?

	Procedure 21.3 Amount of Leaf Surface Remaining on the Cutting and the Ability to Root
Step	Instructions and Comments
1	Place moistened medium in each flat to a depth of 7 to 8 cm, and level the surface. Flats will be designated A, B, and C.
2	In this example, using Japanese privet, all cuttings must be very uniform to start. For example, make cuttings with three pairs of leaves and a leafless basal node. Leaf size and area should be equivalent across all the cuttings.
3	Divide the cuttings into groups of 30. Leave all foliage on one set, and insert 10 cuttings into each flat. Remove the basal pair of leaves on another 30 cuttings, and insert 10 cuttings into each flat. Remove all except the top pair of leaves on another set of 30 cuttings, and insert 10 cuttings into each flat. Another group of cuttings with no leaves or with one leaf per cutting could also be prepared and distributed similarly.
4	Label the rows in each flat: A-1, A-2, A-3, etc.; B-1, B-2, B-3, etc.; C-1, C-2, C-3, etc. Add plant name, propagator's name, and date. Record the characteristics (full leaf, 2/3 leaves, 1/3 leaves, etc.) that distinguish the cuttings by their row numbers into a notebook.
5	Place flats of prepared cuttings into your propagation area under intermittent mist or fog set to come on for 6 sec. every 6 min.
6	Evaluate cuttings for evidence of root initiation about two to three weeks after sticking them. When the earliest signs of rooting are observed, record the date and wait another two to four weeks before carefully removing them to evaluate the location and extent of rooting and percent rooting. For each 10 cuttings, calculate the rooting index (Table 21.2) and percent rooted.
7	Average the data for the three replications of each cutting type and represent these graphically.

- Do the results of this experiment support the hypothesis that more rooting occurs with greater foliage retention? Why or why not?
- Would the results be the same if the upper rather than lower leaves were removed? Why or why not?
- What impact does the practice of reducing foliage (number of leaves or leaf area) have on the propagation operation?
- If similar leaf areas from a leaf removal treatment and leaf cutting treatment were compared, were the results similar? Why or why not?
- If your humidity control environment was different, would you expect different results? Why?

Experiment 4: Comparison of the Effectiveness of Single- and Double-Node Cuttings on Rooting

A number of foliage plants with a vining habit of growth are propagated by single- or double-node cuttings (Figure 21.5). The technique is not limited to vines, however, as many woody tropicals are propagated by single nodes, leaf-bud cuttings, or split-node (stem) cuttings (Figure 21.6 and Figure 21.7). The retention of a leaf also improves rooting, but on double-node cuttings, the basal leaf is usually removed. Recently matured stems are preferred to younger or older cutting materials. The time to produce a salable plant is often longer with small starting units. The stem length of the cutting unit is usually 2.5 to 5 cm. Cut-

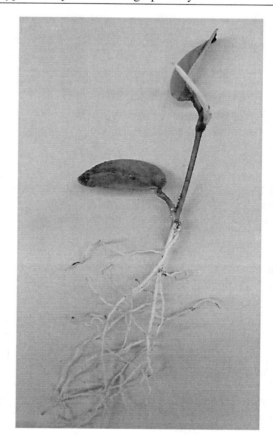

FIGURE 21.5 Some tropical vines have preformed roots at their nodes and can be propagated by single- or double-node cuttings. Rooted double-node cutting of *Scindapsus exotica*.

Rooting Cuttings of Tropical Plants

FIGURE 21.6 Two cuttings can be made by splitting the stem at the node of opposite-leaved plant materials. *Sanchezia speciosa* (Acanthaceae) is representative of a tropical plant easily propagated by split-node cuttings.

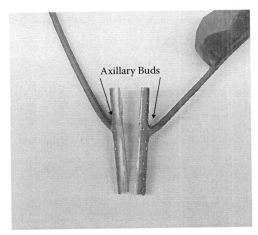

FIGURE 21.7 Opposite-leaved plants can often be propagated by splitting the stem at the node to include the axillary bud and leaf blade.

tings are inserted into the medium so that the bud is just below or at the medium surface, as rooting often occurs at the base of the new shoot.

Materials

(The instructor may reduce the quantity of material to be used if plant material is insufficient.)

Students will need the following items to complete this project:

- Sufficient plant material (such as philodendron, pothos, *Syngonium*, grape ivy, Indian rubber tree, caricature plant, *Hoya*, mandevilla, stephanotis, passionfruit, coleus, *Pseuderanthemum*, *Schefflera arboricola*) is needed to provide 30 uniform cuttings of each type
- Three flats of medium (coarse vermiculite or an equal mixture of vermiculite and perlite)
- Pruning shears or knives to prepare the cuttings
- Labels and markers
- Rooting compounds (0.1% to 0.3% or 1000 to 3000 ppm IBA or NAA)

Each flat will have rows of 10 cuttings, each row representing a different type of cutting. Three flats are used so that three replications of 10 of each type of cutting can be produced. Where more than one plant material is used, two or more kinds of plants can be placed in the same flat.

Follow Procedure 21.4 to complete this exercise.

Anticipated Results

In general, nodes with a leaf left on will root faster and with more roots than leafless cuttings. Cuttings with two nodes, the bottom one in the medium, will often root faster than single-noded cuttings. Cuttings with a complete piece of stem will retain more reserves to support rooting than leaf-bud cuttings with only a sliver of stem. Split-node (split-stem) cuttings (Figure 21.6 and Figure 21.7) may rot because of the exposed tissue, but where successful, they double the number of plantlets that can be obtained from a stem.

Questions

- Does rooting occur from the original stem or from the base of the elongating shoot?
- Does the number of nodes on the cutting make a difference in your results? Why or why not?
- Does the presence of a leaf on the cutting make a difference in the results? Why?
- If split-node cuttings (stem split between a pair of opposite leaves) were used, where did roots develop?
- What is your opinion of the value of the practice of single-, double-, or split-node cuttings in a nursery operation?

EXPERIMENT 5: THE EFFECTS OF LENGTH AND ORIENTATION ON ROOTING OF CANE CUTTINGS

A number of foliage plants are propagated from long, leafless stem pieces known as canes (Figure 21.8). For the most part, these are straight stems, but some plant materials are marketed with stems in spirals or other contortions. Cane pieces can be as short as a single node or as long as up to 1.5 m in length. Since the axillary buds have not begun to push, rooting is often delayed until hormones produced by these buds are available and translocated to the rooting zone. Succulent materials,

	Procedure 21.4 Comparison of the Effectiveness of Single- and Double-Node Cuttings on Rooting
Step	Instructions and Comments
1	Place moistened medium in each flat to a depth of 7 ot 8 cm, and level the surface. Flats will be designated A, B, and C.
2	**Vine example:** Cut a nephthytis, pothos, or philodendron vine into 30 pieces about 4 cm long, containing a leaf with one node. Prepare a similar set of 30 cuttings, but remove the leaf. Prepare a third set of two-node cuttings and remove the lower leaf. Prepare a fourth set of two-node cuttings from which you remove both leaves. For all cuttings, leave at least 18 mm of stem below the bottom node. This end will be inserted into the medium. **Stems with opposite leaves:** Cut a caricature or coleus plant stem into 30 single-node and 30 two-noded units. The length of stem should be about 1.2 cm above the top node and about 2.5 cm below the bottom one. A third batch of single-node cuttings will be made and split down the middle between the two leaves. It may be necessary to cut the leaves in half transversely for efficient use of space. Treat the basal ends with auxin. **Stems with alternating leaves:** Cut croton or India rubber tree stems above and below the node such that the axillary bud and leaf have 8 to 12 mm of stem above and below them. Treat the basal end with auxin. (The large leaves of the India rubber tree can be rolled under and held with a rubber band. A wooden skewer or small bamboo stake in the middle can be used to hold the leaf upright when inserted into the medium.)
3	**Vine:** Insert the stem into the medium so that the bud (or a lower bud) is just at or just below the surface of the medium. **Stems:** Insert the stem into the medium so that the leaf–stem juncture is just above the medium surface.
4	Label the rows in each flat: A-1, A-2, A-3, etc.; B-1, B-2, B-3, etc.; C-1, C-2, C-3, etc. Add plant name, propagator's name, and date. Record the characteristics that distinguish the cuttings by their row numbers into a notebook.
5	Place flats of prepared cuttings into your propagation area under intermittent mist or fog set to come on for 6 sec. every 6 min.
6	Evaluate cuttings for evidence of root initiation about two to three weeks after sticking them. When the earliest signs of rooting are observed, record the date and wait another two to four weeks before carefully removing them to evaluate the extent of rooting and percent rooting. For each 10 cuttings, calculate the rooting index (Table 21.2) and percent rooted.
7	Average the data for the three replications of each cutting type, and represent these graphically.

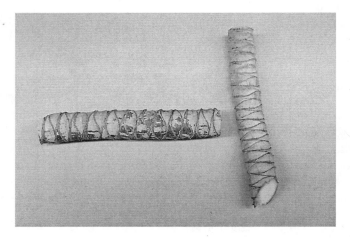

FIGURE 21.8 *Dracaena marginata* is a classical example of a woody tropical plant propagated by stem pieces called "canes." A slant cut is made at the base to indicate the "down" end of the cutting. Cuttings can be inserted vertically in or laid horizontally on the medium. Roots develop at the basal end and buds and shoots develop toward the original apical end, no matter which placement orientation is used.

such as *Dieffenbachia*, require a period of time to suberize and form a protective periderm layer over the cut surface before being placed in a moist medium. The ends of cane pieces of *Yucca* and *Dracaena* from some producers are often waxed to prevent desiccation, but this also prevents the uptake of moisture needed for rooting, so recutting the base may be necessary. Making vertical cuts above the base permits root development farther up the stem to provide better anchorage after potting. Length and diameter of the stem influence bud break and rooting by virtue of the moisture and carbohydrate contents of the cane. Short cane pieces may be laid on their sides, nearly covered in medium, with a prominent bud on the upper side at the apical end. Even if the propagator does not recognize which end of the stem is basal, the cane will always root at the basal end because of polarity. Thus, the orientation of planting the cutting is important for vertical insertion, but it is of less concern for horizontal placement of the cane piece. If terminals are available, compare results of cane piece versus similar length terminals.

Materials

(The instructor may reduce the quantity of material to be used if plant material is insufficient.)

The following materials will be needed by students to complete this exercise:

- Cane material (Aglaonema, Anthurium, Ti, Dieffenbachia, Dracaena species, Polyscias species, octopus tree, Yucca)
- Flats of medium (coarse vermiculite or an equal mixture of vermiculite and perlite)
- Pruning shears or knives to prepare the cuttings
- Labels and markers
- Rooting compounds (0.1% to 0.3% or 1000 to 3000 ppm IBA or NAA)

Follow the protocol outline in Procedure 21.5 to complete the experiment.

Anticipated Results

While long, thick cane pieces might be expected to produce the most roots because of the amount of reserve nutrients in the cane, the base can also be hard to root because of little parenchyma. Cane pieces from the middle portion of the stem may root more quickly than from the base. Whether a cane piece is laid horizontally or placed vertically in the medium, the roots will always emerge from the basal end, while bud break and new shoots will develop at the distal end (the original top) of the cane. Terminals of canes usually retain foliage that aids in rooting, but differences can be observed when the terminal tissue is too soft (it usually rots); stem tips of such material need some recently matured woody tissue at the base.

Procedure 21.5	
Effects of Length and Orientation on Rooting of Cane Cuttings	
Step	Instructions and Comments
1	Place moistened medium in each flat to a depth of 7 to 8 cm, and level the surface. Flats will be designated A, B, and C. (Depending on availability of material, three flats may not be needed).
2	Cut pieces of dracaena or ti cane into 4-, 8-, or 12-in. lengths, keeping in mind which end is basal and which is distal. Longer cane pieces can be used if available. One useful technique uses a sloping cut at the base to indicate the end to insert into the medium.
3	One-half of the cane pieces of each length will be inserted vertically into the medium, and one-half will be pressed horizontally into the medium so the top edge is barely out of the medium. On long cane pieces, make 5- to 7.5-cm longitudinal cuts above the base. If diameters differ among canes, distribute similar diameters equally between the two orientations.
4	Label the rows in each flat: A-1, A-2, A-3, etc.; B-1, B-2, B-3, etc.; C-1, C-2, C-3, etc. Add plant name, propagator's name, and date. Record the characteristics that distinguish the cuttings by their row numbers into a notebook.
5	Place flats of prepared cuttings into your propagation area under intermittent mist or fog set to come on for 6 sec. every 6 min.
6	Evaluate cuttings for evidence of root initiation about two to three weeks after planting them. When the earliest signs of rooting are observed, record the date and wait another two to four weeks before carefully removing them to evaluate the extent of rooting and percent rooting. For each 10 cuttings, calculate the rooting index (Table 21.2) and percent rooted. If time permits, evaluate shoot emergence, which is often later than rooting.
7	Average the data for the three replications of each cutting type, and represent these graphically.

Questions

- Does cane length or diameter make a difference in speed or extent of rooting? Why?
- If horizontal and vertical insertion of the cane pieces is compared, what differences in root and shoot development were observed?
- Does cane size affect the number of buds that develop into shoots?
- What is the practical value of using canes for propagation when terminals often root more readily?
- If cane pieces versus terminals are compared, what were the differences in speed and extent of rooting?

SUGGESTED READINGS

Auld, R.E. 1988. The basics of propagating bougainvillea. *Comb. Proc. Intl. Plant Prop. Soc.* 36: 211–213.

Avery, J.D. and C.A. Beyl. 1986. Caliper of semihardwood cutting influences footing of kiwifruit. *Plant Propagator* 32: 5–7.

Calma, V.C. and H.W. Richey. 1930. Influence of amount of foliage on rooting of coleus cuttings. *Proc. Amer. Soc. Hort. Sci.* 27: 457–462.

Chadwick, L.C. 1949. The influence of the position of the basal cut on the rooting and arrangement of roots on deciduous softwood cuttings. *Proc. Amer. Soc. Hort. Sci.* 53: 567–572.

Conover, C.A. 1994. Storage temperature and duration affect propagation of *Dracaena fragrans* 'Massangeana.' *Foliage Digest.* 17: 1–4. (See also *Proc. Fla. State Hort. Soc.* 104: 331–333).

Conover, C.A. and R.T. Poole. 1993. Propagation of *Dracaena fragrans* 'Massangeana' affected by cane position on stock plants. *Foliage Digest.* 16: 1–3.

Davies, F.T., Jr. and J.N. Joiner. 1978. Adventitious root formation in three cutting types of *Ficus pumila* L. *Comb. Proc. Intl. Plant Prop. Soc.* 28: 306–313.

Griffith, L.P., Jr. 1998. *Tropical Foliage Plants—A Grower's Guide.* Ball Publishing, Batavia, IL. 318 p.

Henley, R.W. 1979. Tropical foliage plants for propagation. *Comb. Proc. Intl. Plant Prop. Soc.* 29: 454–467.

Higaki, T. 1981. Single node propagation of *Dracaena goldieana*. *Plant Propagator* 27: 8–10.

Jackson, H.C. 1981. Propagation of *Dieffenbachia* by node cuttings. *Comb. Proc. Intl. Plant Prop. Soc.* 31: 248–249.

Lane, B.C. 1987. The effect of IBA and/or NAA and cutting wood selection on the rooting of *Rhaphiolepis indica* 'Jack Evans.' *Comb. Proc. Intl. Plant Prop. Soc.* 37: 77–82.

Larkman, B. 1988. Single node vs. double node cuttings for the propagation of *Pyrostegia venusta*, *Hardenbergia violacea* 'Happy Wanderer' and *Clytostoma callistegiodes*. *Comb. Proc. Intl. Plant Prop. Soc.* 38: 106–109.

Lee, C.W. and D.A. Palzkill. 1984. Propagation of jojoba by single node cuttings. *HortScience.* 19: 841–842.

Mahlstede, J.P. and E.P. Lana. 1958. Evaluation of the rooting response of cuttings by the method of ranks. *Proc. Amer. Soc. Hort. Sci.* 71: 585–590.

Marlatt, R.B. 1969. Propagation of *Dieffenbachia*. *Econ. Bot.* 23: 385–388.

McConnell, D.B. 1983. Students learn by evaluating commercial foliage propagation techniques. *Foliage Digest.* 6: 4–5.

Morgan, J.V. and H.W. Lawlor. 1976. Influence of external factors on the rooting of leaf and bud cuttings of *Ficus*. *Acta Hort.* 64: 39–46.

Neel, P.L. 1979. Macropropagation of tropical plants as practiced in Florida. *Comb. Proc. Intl. Plant Prop. Soc.* 29: 468–480.

Oka, S. 1979. The effect of IBA and NAA on rooting cuttings of selected *Dracaena* species and cultivars. *Plant Propagator* 25: 11–12.

O'Rourke, F.L. 1944. Wood type and original position on shoot with reference to rooting in hardwood cuttings of blueberry. *Proc. Amer. Soc. Hort. Sci.* 45: 195–197.

O'Rourke, F.L. and M.A. Maxon. 1948. Effect of particle size of vermiculite media on the rooting of cuttings. *Proc. Amer. Soc. Hort. Sci.* 51: 654–656.

Petersen, T.H. 1981. Propagation of philodendrons from node cuttings. *Comb. Proc. Intl. Plant Prop. Soc.* 31: 219–220.

Poole, R.T., A.R. Chase, and L.S. Osborne. 1986. Yucca. *Foliage Digest.* 9: 4–6.

Poole, R.T., A.R. Chase, and L.S. Osborne. 1991. Schefflera. *Foliage Digest.* 16: 4–8.

Poole, R.T., A.R. Chase, and L.S. Osborne. 1993. Dracaena. *Foliage Digest.* 16: 1–6.

Poole, R.T. and C.A. Conover. 1982. Propagation of spineless yucca. *Foliage Digest.* 5: 14–15.

Poole, R.T. and C.A. Conover. 1984. Propagation of ornamental *Ficus* by cuttings. *HortScience* 19: 120–121.

Poole, R.T. and C.A. Conover. 1987. Vegetative propagation of foliage plants. *Comb. Proc. Intl. Plant Prop. Soc.* 37: 503–507.

Poole, R.T. and C.A. Conover. 1990. Propagation of *Ficus elastica* and *Ficus lyrata* by cuttings. *Foliage Digest* 8: 3–5.

Wang, Y.-T. 1987. Effect of warm medium, light intensity, BA, and parent leaf on propagation of golden pothos. *HortScience* 22: 597–599.

WEB SITE

Commercial Foliage Crop Production Notes: http://www.mrec.ifas.ufl.edu/Foliage/folnotes/folnotes.htm

22 Care and Management of Stock Plants

Gary R. Bachman

CHAPTER 22 CONCEPTS

- Selection of plants having the characteristics desired that are relatively easy to root from cuttings is important for propagators.

- Maintenance of selected stock plants through manipulation of cultural and environmental inputs to optimize propagation success is essential.

- Maintaining juvenility will also optimize the potential for success of rooted cuttings.

Selecting and maintaining plants having desirable characteristics from which to collect cuttings have been important parts of asexual plant propagation since the beginnings of agriculture. Plants originally selected for propagation were those that provided the greatest benefit from food and fiber production and were easy to root. The ability of our early ancestors to establish food and fiber resources in one place contributed greatly toward the transformation from predominately hunter/gatherers to our modern society. Today, besides the importance of food and fiber considerations, plants are selected for aesthetic reasons, such as flowering, leaf color, growth habit, medical and chemical compounds, to mention a few.

The reproduction of plants through asexual propagation relies on a few key management practices. First, the parent plant or plants (this will depend on the individual production practices) should be true-to-type and free of pests. Starting with good, clean stock for cutting collection will help ensure production of high-quality cuttings. Propagators use the saying "start clean, stay clean" to highlight the need of good sanitation to eliminate insects and diseases on cutting stock. The second management practice does not have so much to do with the health of stock plants, but with having adequate stock plants to ensure the production of cutting material in sufficient quantities to meet production goals. Sufficient source material must be available to obtain the necessary quantities of cuttings. Collection of cuttings can be from different sources, typically dedicated stock plants or from production plants during normal culture and management operations.

Traditionally, cuttings for propagation have been collected from plants grown strictly for this purpose. These stock plants are usually grown in the field, but can be maintained in the greenhouse. Another important aspect of having dedicated stock plants is recordkeeping. The propagator is able to maintain, through the use of record keeping, an accurate history of each stock plant, from optimum rooting percentages to timing of cutting collection, and preserving true plant identity. Since these plants are dedicated stock plants, it is much easier to employ cultural techniques to increase the number of cutting per stock plant. However, maintaining stock plants specifically for cutting collection may occupy valuable land or greenhouse space, and this must be considered by the grower or propagator in relation to other operations occurring at the nursery or greenhouse.

Today, many ornamental crops are grown in containers, and there is a good chance that these growers do not have dedicated stock plants. In this situation, the grower usually collects cuttings from production stock during normal pruning operations. These cuttings, when rooted, become the source of the next generation of cuttings. This practice of collecting cuttings from previously rooted, container-grown plants is called serial propagation. While this practice may not be ideal when considering the optimum time of cutting collection, these cuttings are usually similar in quality and acceptable rooting percentage. If the timing for making cuttings is absolutely inappropriate, usually due to environmental considerations, they may be stored at low temperature for propagation at a more acceptable time. Never collect cuttings from production culls. Growers are sometimes tempted to sell all of their best stock and use leftover plants as the source of their cutting stock. There was a reason that these plants were not purchased, and collecting cuttings from obviously inferior plants is not a desirable practice and will lead to diminished overall quality of nursery stock.

Stock plants are maintained in good condition by manipulating the growth management practices and environmental conditions, within acceptable parameters, with the intent of producing cuttings of high quality. The grower has control over the following environmental and cultural practices:

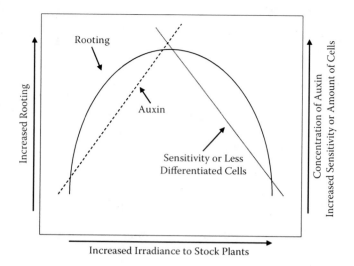

FIGURE 22.1 Schematic illustration of the generalized effect of photon flux on stock plants. (From Bertram, L. 1992. *Acta Hort.* 314: 291–300.; http://www.actahort.org. With permission.)

- Water status
- Light
  - Level
    - Etiolation
    - Banding
  - Photoperiod
- Nutrient status
  - Nitrogen nutrition
  - Carbon/nitrogen ratios
- Timing of cutting collection

WATER STATUS

The water status of stock plants is critical to successfully rooting cuttings. Cuttings should be collected when the stock plants are in turgid condition. This is most easily accomplished early in the morning or during cloudy periods when water stress is most likely to be reduced.

LIGHT

PHOTON FLUX

Photon flux measures the number of photons that strike a plant. This is used to record the total amount of light a plant receives during the course of the day. The amount of light a stock plant receives has a direct effect on how successfully a cutting can produce new root tissue. Generally, the ability of cuttings to produce roots increases with greater light levels that the stock plant is exposed to; however, not all plant species have the same reaction to increasing light levels. Plants that normally are grown in the shade generally will produce cuttings that root better at lower levels of light than do cuttings from plants grown under higher light conditions. Most plants do have an optimum light level that will result in the greatest rooting success. Low light levels generally cause a reduction in rooting percentage because of the reduced carbohydrate production by the stock plant. However, if a stock plant is exposed to a photon flux intensity greater than its light saturation point, then rooting percentages of the cuttings taken from it decrease. With some plant species, this rooting reduction is related to a change in plant hormones, specifically a reduction in endogenous auxin levels because of photodestruction of indole acetic acid (IAA) (Figure 22.1).

An example of reduction in rooting percentage caused by supraoptimum light is illustrated by *Hibiscus* (Figure 22.2). Stock plants exposed to full sunlight produced cuttings that had a rooting success of 67.5%. When stock plants were grown in 65% shade, the cuttings from them rooted at 97.5%. When cuttings from the stock plants grown in the full sun were treated with 0.8% IBA powder, the rooting success increased to 92.5%. IBA did not increase the rooting success of cuttings taken from the shade-grown plants. This is illustrative of the endogenous auxin being photodegraded in the full sun plants and the exogenous IBA treatment restoring the auxin levels required for optimum rooting success.

Etiolation and Banding

Reduction of the light levels that stock plants receive can be beneficial to rooting of some plant species that are difficult to root. Two techniques are most commonly used, etiolation and banding. Etiolation in the purest sense consists of growing the stock plant in the total exclusion of light and, in some cases, propagation production schemes of maintaining stock plants under a heavy shade can also have similar effects on rooting success. After the stock plants have been grown under these light-exclusion or heavy-shade treatments, the plants are allowed to grow under normal light conditions. The propagator then col-

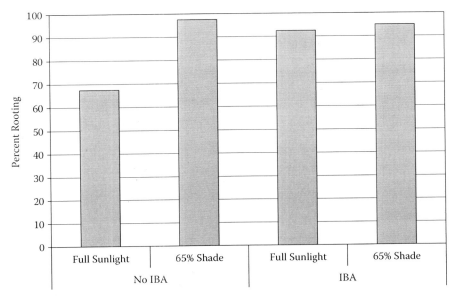

FIGURE 22.2 Effects of *Hibiscus rosa-sinensis* stock plant irradiance and IBA treatment of cuttings on rooting success. (From Johnson, C.R. and D.F. Hamilton. 1977. *HortScience.* 12: 39–40. With permission.)

lects propagation material from the shoots that have elongated under the total-light-exclusion or heavy-shade-grown stock treatments.

Banding consists of applying materials to exclude light from localized portions of the stems of stock plants. When the cutting is removed from the plant, the banded area becomes the base of the cutting. Banding materials may consist of black tape or 2.5 cm (1 in.)-wide black Velcro, which is easily removed at the end of the treatment period. Banding is effective when applied to etiolated stems or stems of light-grown stock plants. Many times, etiolation and banding are used in sequence to enhance rooting. Typically the stock plants are grown in the exclusion of light, allowing the initial growth to be etiolated, followed by removal of the shading materials and banding for an additional eight weeks prior to collection of cuttings. There many examples of the effectiveness of both etiolation and banding promoting greater rooting percentages, and the practice is considered essential for successful rooting of some species. An extensive listing of plant species showing improved rooting percentages after banding, etiolation, etiolation plus banding, and shading was compiled by Maynard and Bassuk (1988).

Both light exclusion techniques promote an increased responsiveness to either endogenous auxins or to applied auxin treatments. There also are physiological changes occurring in the stems themselves in response to the reduced light levels. Among these changes are reduced or no chlorophyll content, increases in stem internode length, increased succulence, and decreased mechanical strength of the treated softwood stems.

Normal stems have a high number of sclerenchyma cells. These cells, with their thick cell walls, give a plant stem strength and rigidity to help support the plant. This stem strength can also be a hindrance to the formation of adventitious roots by creating a physical barrier impeding latent root emergence. Etiolated or band-treated stems have a much lower preponderance of sclerenchyma cells and an increase in relatively simple parenchyma cells, which present less of a physical barrier to adventitious root formation.

Photoperiod

Photoperiod has an indirect influence on the rooting of cuttings by affecting the photosynthetic and morphogenic responses in plants. Daylength is important for stock plants for maintaining photosynthesis and accumulation of carbohydrates at adequate levels. Carbohydrates must be in sufficient quantity for adventitious root formation during the rooting process.

Photoperiod can have a major effect on morphogenic responses of plants, such as the vegetative growth versus floral initiation. Flowering is antagonistic to successful rooting because there is a large requirement for the same carbohydrates needed for root formation. A photoperiod that encourages vegetative growth and, thus, inhibits floral initiation, would act indirectly to promote rooting. An example that is useful for this discussion is the short-day flowering species *Euphorbia pulcherrima* (poinsettia—Table 22.1). Stock plants of 'Freedom Red' and 'Monet' exposed to a four-hour night break (a common technique used by growers to keep plants vegetative during natural short daylength conditions) produced cuttings that rooted at greater percentages than those exposed to shorter daylengths.

TABLE 22.1
Percentage of Total Cuttings with Visible Roots from *Euphorbia pulcherrima* 'Freedom Red,' 'Monet,' and 'V-17 Angelika Marble' Stock Plants Exposed to a Four-Hour Night Break and Natural Photoperiod[a]

Cultivar	Photoperiod	Percentage of Total Cuttings Rooted (days after stick)				
		14	18	21	26	31
Freedom Red	Natural	30 b[b]	62 b	78 b	90 a	100 a
	Night Break	44 a	88 a	100 a	100 a	100 a
Monet	Natural	6 a	20 b	40 b	46 b	78 a
	Night Break	10 a	36 a	64 a	86 a	96 a
V-17 Angelika Marble	Natural	38 a	72 a	86 a	96 a	100 a
	Night Break	28 a	70 a	92 a	94 a	100 a

[a] *Source:* Data from Roll, M.J. and S.E. Newman. 1997. *HortTechnology* 7: 41–43. With permission.

[b] Means within columns followed by different letters are different at $P<0.05$ as determined by pairwise t-test. Means based on 10 cuttings.

NUTRIENT STATUS

Nutrition and Nitrogen/Carbohydrate

Mineral nutrition is an important part of stock plant management. Nutrients must be available for plant uptake and overall plant growth and development. Fertilization of stock plants should be managed according to the type of cutting collected from these plants: softwood, semi-softwood, or hardwood cuttings. Of particular interest is whether the cutting is capable of photosynthesizing at the time it is collected or not.

Softwood cuttings having high carbohydrate/moderately high nitrogen content tend to root readily. These cuttings would normally be leafy and producing new shoot tissue. Excessively high nitrogen content promotes vegetative growth at the expense of new roots. The ability to photosynthesize (i.e., able to produce carbohydrates) during the rooting of softwood cuttings is critical, as it alleviates the competition for carbohydrates between new shoot and root tissue formation. Additionally, fertilization at high rates of nitrogen can also inhibit rooting of hardwood cuttings. High nitrogen rates stimulate shoot growth, and this would be at the expense of the cutting producing adventitious roots, with both shoots and roots competing for carbohydrate reserves in the cutting.

There is evidence that woody species that are fertilized at minimal nitrogen levels produce optimum rooting. Reductions in applied nitrogen have a twofold effect. First, the rate of growth of shoot tissue is slowed. The reduction in nitrogen slows growth by limiting protein and nucleic acid biosynthesis. Second, there is an increase in the amount of carbohydrate being stored in the cutting material if the plants are grown at their optimum light level. The production of new tissue, in this case the production of new roots, requires a great deal of energy, and these stored carbohydrates serve as the energy source. Conversely, plants grown for production receive high levels of nitrogen to encourage growth. Cuttings from production plants sometimes root more poorly than those from dedicated stock plants because stock plants receive lower nitrogen fertilization. For this reason, nurseries grow dedicated stock plants.

We can examine this relationship between nitrogen and carbohydrates with the rooting of *Ilex crenata* 'Rotundifolia' (Figure 22.3). As nitrogen fertilization increased, the total nonstructural carbohydrate (TNC) in the leaf tissue increased, and rooting percentage decreased. As more nitrogen was available for metabolism, more of the photosynthate stream was diverted from storage for stem growth at the expense of adventitious root formation.

TIMING OF CUTTING COLLECTION

When cuttings are collected can have a significant effect on propagation. Cuttings collected from stock plants at the proper time of the year have optimum rooting success, unlike cuttings collected from the container nursery during the normal course of pruning operations. Cuttings from deciduous trees and shrubs, broad-leaved evergreens, and narrow-leaved evergreen plants are collected at different times of the year to ensure propagation success.

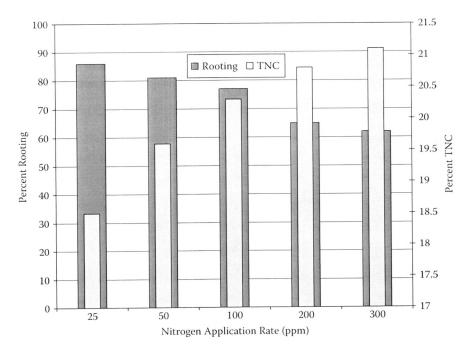

FIGURE 22.3 Influence of increasing nitrogen nutrition on decreasing rooting success. (From Rein, W.H., R.D. Wright, and D.D. Wolf. 1991. *J. Environ. Hort.* 9: 83–85. With permission.)

Deciduous Plants

Cuttings are typically collected from deciduous plants while the plant is dormant during late fall, after the leaves have fallen, to bud swelling in the spring. These are called hardwood cuttings and are mature stem sections produced during the previous growing season. The plant has been storing energy in the stems and branches for next year's growth and will provide the energy required for adventitious root formation.

Broad-Leaved Evergreen Plants

Cuttings from these plants root with greater success when collected at the semihardwood stage. At this point, the stems and leaves have stopped elongating, and no new vegetation is actively being produced by the plant. This stage typically occurs in mid to late summer for most species of broad-leaved evergreens.

Narrow-Leaved Evergreen Plants

Optimum rooting success occurs with narrow-leaved evergreen plants when cuttings are collected in mid-fall to mid-winter. These cuttings need to include about 1.25 cm (½ in.) of mature stem and are considered hardwood cuttings.

It should be pointed out that the timings of cutting selections described above are general statements about these plant groups and are not intended to imply these are the only times successful propagation of these plants can be accomplished.

The effects of plant physiological age, not chronological age, have considerable consequences on the success of rooting cuttings. As a stock plant ages, it undergoes a change from the juvenile to the adult phase, and the ability to produce adventitious roots decreases. Cultural management to maintain at least a portion of the stock plant in the juvenile phase to collect cuttings is useful. Typically, this is done by removing the outer, more mature portions of the plant and releasing the lateral buds from the growth inhibition imposed by the apical bud through apical dominance. These lateral buds, located in the more juvenile area of the tree or shrub, begin to grow, providing juvenile shoot tissue that has a greater potential for rooting. This area is commonly termed the "cone of juvenility" and is the portion of the stem that marks the division between mature and juvenile tissue (Figure 22.4). Plants produce flowers from mature tissue, typically found on the outside or ends of branches, which is the youngest tissue in terms of chronological age, but growing on a much older, more mature plant than when the juvenile tissues were formed. It is advantageous to remove the mature tissues from stock plants to expose the juvenile tissue for collection of cuttings. This is done by such nursery practices as heading back stock plants, hedging, and mound layering.

Care and management of stock plants for propagation is really about using sound horticultural principles to optimize the potential to obtain propagules that root at high percentages. Adequate irrigation, proper nutrition, regulating light levels, and good cultural techniques all have a profound impact on the ultimate success a propagator has rooting plants.

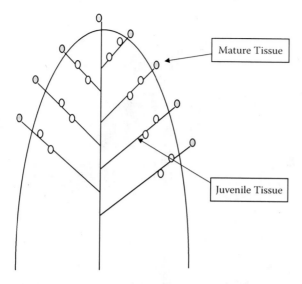

FIGURE 22.4 Representation of areas of a stock plant indicating locations of juvenile tissues.

LITERATURE CITED AND SUGGESTED READINGS

Bassuk, N. and B. Maynard. 1987. Stock plant etiolation. *HortScience*. 22: 749–750.

Bertram, L. 1992. The physiological background for regulation of adventitious root formation by the irradiance to stock plants: A hypothesis. *Acta Hort*. 314: 291–300.

Hansen, J. 1987. Stock plant lighting and adventitious root formation. *HortScience*. 22: 746–748.

Hartmann, H.T., D.E. Kester, F.T Davies, Jr., and R.L. Geneve. 2002. *Plant Propagation: Principles and Practices*. Prentice Hall, Inc., Upper Saddle River, NJ. 880 pp.

Heins, R.D., W.E. Healy, and H.F. Wilkins. 1980. Influence of night lighting with red, far red, and incandescent light on rooting of chrysanthemum cuttings. *HortScience*. 15: 84–85.

Johnson, C.R. and D.F. Hamilton. 1977. Rooting of *Hibiscus rosa-sinensis* L. cuttings as influenced by light intensity and ethephon. *HortScience*. 12: 39–40.

Maynard, B.K. and N. Bassuk. 1988. Etiolation and banding effects on adventitious root formation, in *Advances in Plant Sciences Series Vol. 2: Adventitious Root Formation in Cuttings*. Davis, T.D., B.E. Haissis and N. Sankhla (Eds), Dioscoriodes Press, Portland, OR. pp. 29–46.

Rein, W.H., R.D. Wright, and D.D. Wolf. 1991. Stock plant nutrition influences the adventitious rooting of 'Rotundifolia' holly stem cuttings. *J. Environ. Hort.* 9: 83–85.

Roll, M.J. and S.E. Newman. 1997. Photoperiod of poinsettia stock plants influences rooting of cuttings. *HortTechnol*. 7: 41–43.

Sun, W.-Q. and N.L. Bassuk. 1991. Stem banding enhances rootings and subsequent growth of M.9 and MM.106 apple rootstock cuttings. *HortSci.* 26: 1368–1370.

Veierskov, B. 1988. Relations between carbohydrates and adventitious root formation, in *Advances in Plant Sciences Series Vol. 2: Adventitious Root Formation in Cuttings*. Davis, T.D., B.E. Haissis, and N. Sankhla (Eds), Dioscoriodes Press, Portland, OR. pp. 70–78.

Part VIII

Propagation by Leaf and Root Cuttings

23 Adventitious Shoot and Root Formation on Leaf and Root Cuttings

Caula A. Beyl

CHAPTER 23 CONCEPTS

Root cuttings

- Root cuttings are a useful means of propagating species that do not come from true seed or root poorly when stem cuttings are used. They are also useful for obtaining clones of trees which have been selected for propagation based on mature characteristics, such as flowering, fruiting, or sex, but have lost the ability to root easily from stem cuttings.

- Many hardwood species that exhibit natural suckering are capable of being propagated by root cuttings. Suckers are shoots that arise from roots either adventitiously or from preformed buds.

- In general, thicker cuttings of 5 to 15 cm in length that have adequate carbohydrate reserves are more successful for root cutting propagation with a greater percentage of rooting and more vigorous shoots.

- Most root cuttings typically have a definite "window of opportunity" for success and should be taken from October to March during the dormant season. Cuttings taken during the summer usually will not develop shoots and root well.

- Polarity of a root cutting is important, and which end of the cutting was nearest the crown could easily be confused. This is alleviated by using a straight cut across the cutting at the proximal end (nearest the tree trunk) and a slanted cut across the cutting at the distal end (farthest from the trunk).

- New shoots arise from the proximal end and new roots from the distal end of the root cutting or from the base of the new shoot.

Leaf cuttings

- Some plants can be propagated by cuttings made from detached leaves without any part of the original stem or the axillary bud being used.

- Plants like African violet (*Saintpaulia ionantha* Wendl.) are propagated using the leaf with the petiole still attached. Other species, like jade plant (*Crassula ovata* (Miller) Druce) or burro tail (*Sedum morganianum* E. Walther), have leaves that can be separated from the parent plant and merely inserted as they are into the rooting medium.

- With some species, leaves are cut into pieces, sections, or strips to prepare them as cuttings. Examples are *Sansevieria* (Snake Plant) and *Streptocarpus* (Cape Primrose).

- Most plants propagated by leaf cuttings regenerate roots from secondary meristems induced as part of the wounding response. Other species, such as pick-a-back plant (*Tolmeia menziesii*), mother of thousands (*Bryophyllum daigremontianum* (Raym.-Hamet & E. P. Perrier) Berger), and walking fern (*Asplenium rhizophyllum* L.), develop plantlets from primary meristems that are preformed.

- Cuttings that are only pieces of the original leaf will work for some plants, such as Rex begonia (*Begonia rex* Putz.), snake plant (*Sansevieria trifasciata* hort ex. Prain), and streptocarpus (*Streptocarpus formosus* Hilliard & Burt). Begonia leaves can be cut into squares about 3 to 5 cm on the side and containing a vein. *Sansevieria* and *Streptocarpus* leaves can be cut transversely into strips.

- Roots are formed more easily on leaf cuttings than shoots. With some species, roots will form, but shoots may never develop; examples are *Ficus elastica* Roxb. (rubber tree) and some of the *Lilium* species.

In conventional vegetative plant propagation, much attention is focused on the use of various types of stem cuttings, with much less attention paid to use of structures, such as roots and leaves. For many species, particularly hardwood trees, some fruit species, and some herbaceous perennials, root cuttings present a viable alternative to seed or stem cutting propagation. Leaf cuttings also receive less attention than stem cuttings for propagation of tropical foliage plants, but for those plants that root easily from leaves, either with or without petioles, leaf cuttings remain a useful technique for obtaining additional plants. There are many surprising ways in which plants can be successfully propagated.

ROOT CUTTINGS

Species that naturally sucker and form clumps of trees that are clonally related generally make good candidates for propagation from root cuttings. Examples of genera that tend to form large clonal stands are *Acacia, Ailanthus, Diospyrus, Populus, Quercus, Robinia, Salix, Sassafras, Ulmus,* and *Rhus*. Clonal stands typically arise when an original seedling tree dies or is destroyed by fire or another event. The term used to describe the original seedling tree is the *genet*. The stems that arise asexually in a cluster from the surviving root system are the ramets. One male clonal stand of *Populus tremuloides* Michx. (quaking aspen) consisted of over 47,000 ramets spreading across more than 43 hectares (107 acres) (Figure 23.1). Clearly, these clonal stands arising from suckering at the root crown can become very large. Plant propagators have taken advantage of this natural ability to regenerate new plants from roots or root pieces in these species and others, including many woody and herbaceous perennials.

ADVANTAGES AND LIMITATIONS OF USING ROOT CUTTINGS

Root cuttings are most advantageous for selections of plants that cannot be produced from seed or are difficult to root using other types of cuttings. Seedlings of many species display considerable variability, making it desirable to select and clonally propagate individual plants with desirable characteristics. A variety of techniques have been discussed in earlier chapters to clonally propagate elite selections, but many times plants are recalcitrant (uncooperative) to these methods and fail to root successfully. Root cuttings often succeed where other methods have low or nonexistent success rates.

Many species are difficult to root from stem cuttings once they have gone beyond their juvenile period (chapter 15). Native American pawpaw (*Asimina triloba* (L.) Dunal) is a good example. Stem cuttings taken from seedlings up to two months old will root, but beyond that time, this species becomes progressively more difficult

FIGURE 23.1 Male aspen clone in Utah composed of more than 47,000 stems of genetically identical *Populus tremuloides* that developed as a consequence of prolific suckering. The clone was named "Pando," which translated means "I spread" in Latin. (Photo courtesy of Jeff Mitton, University of Colorado, Boulder.)

to almost impossible to root from cuttings from mature trees. Root cuttings, however, generally work well. For other species, shoot cuttings derived from the aerial portion of the tree will not root, but if they are forced from root cuttings, they root more easily. For example, mature silktree or mimosa (*Albizia julibrissin* Durz.) is extremely difficult to root from shoot cuttings, but if shoots are harvested from root cuttings, these can be rooted easily. Although these new shoots have the bipinnate leaf form of the mature tree, they will root even without application of IBA. They cannot be said to be "juvenile," but they do exhibit "juvenile characteristics." With other species, such as tree of heaven (*Ailanthus altissima* (P. Mill.) Swingle), only the female tree is a desirable ornamental because the flowers on male trees smell foul. Unfortunately, once the tree has flowered and the sex of the tree can be identified from its flowers, the tree is mature and hard to propagate from stem cuttings. Seeds of *Ailanthus* germinate readily, but because the sex of the seedlings is unknown, it is far better to propagate this species by taking root cuttings from female trees. Other species, such as shining sumac (*Rhus copallina* L.), smooth sumac (*R. glabra* L.), and staghorn sumac (*R. typhina* L.), have seeds that germinate slowly or inconsistently, but will propagate successfully from root cuttings. The cutleaf cultivars of *R. typhina* like 'Laciniata' will not root from stem cuttings.

Use of root cuttings offers other advantages, in that only simplistic technology is needed for success. Very little expertise is involved in collecting root cuttings compared to performing other propagation techniques such as grafting, so skilled labor is not a requirement. Root cuttings are also easily transported and stored. The requirements of the facility for producing root cuttings successfully is also much less elaborate than that needed for rooting softwood stem cuttings, which require a mist or high-humidity system to prevent desiccation.

One distinct disadvantage is the difficulty of obtaining large numbers of cuttings. Typically cuttings are taken when plants are lifted from the field or by digging and harvesting roots while they are still in place; however, this is often the time when nursery operations slow down. For unknown reasons, the success rate of root cuttings varies from year to year, particularly with some species, such as *Paulownia*. Predicting the numbers of finished liners that will be produced by taking root cuttings from a certain number of stock plants is often difficult, and experiences may vary from location to location and year to year.

Root cuttings are not a good technique to propagate plants that are chimeras (chapter 19) because the origin of shoots is from tissue that may have a different genotype than the one desired. This could be a problem with some plants. Thornless blackberries are an example of a periclinal chimera chapter 16), where the mutation occurred in the outermost layer of the developing apical meristem. This layer produces the epidermis, which in the thornless blackberry, does not have "thorns," but shoots produced from root cuttings have the thorny genotype. Root cuttings are also not a good choice for propagating a plant that has been grafted, unless you wish to propagate the rootstock.

OPTIMIZING SUCCESS WITH ROOT PROPAGATION

STOCK PLANTS

Healthy, robust stock plants with adequate carbohydrate reserves stored in root tissues are good sources of root cuttings. In *Paulownia*, secondary roots branching off primary laterals were as good a source for new tissues to develop as primary lateral roots. Roots may be harvested from bare root plants that have been collected and placed in cold storage prior to being sold if care is taken not to take too many cuttings from each plant. Another easily harvested source of root cuttings is containerized plants, particularly if the container is large enough for good growth of the root system without the contortions associated with being root bound.

In nurseries where some species are routinely propagated by root cuttings, stock plants are lined out specifically for this purpose. They are lifted or dug annually and their roots harvested. A piece of equipment that does this job nicely is a spring-trip cultivator with shanks having long narrow bull-tongue type shovels. They are then replanted in the same beds or in nearby sites. If the amount of root development is not enough to allow annual collection of roots, a biennial cycle of collection and replanting may be followed. Problems occur when extensive root systems develop, making it difficult to distinguish which roots belong to which plant in the nursery, and sometimes the root cuttings left behind become a new "weed problem" when they develop shoots and grow.

Although juvenile plants or trees are better sources for obtaining root cuttings, sometimes the plant to be propagated is a rare or established mature tree. In that case, taking root cuttings from the juvenile zone closest to the crown (chapter 18) optimizes the chances for success and allows the propagator to ensure that the roots being taken actually belong to the desired tree.

SIZE OF THE ROOT CUTTING

In general, the thicker the cutting, the more likely it is to produce shoots and subsequent roots. The largest cuttings produce the most vigorous root and shoot development. No shoots form on pawpaw root cuttings with diameters of less than 5 mm. For royal paulownia (*Paulownia tomentosa* (Thunb.) Sieb. & Zucc. Ex Steud.), thicker cuttings with diameters between 10.1 and 15.0 mm produced the most shoots, and those with diameters between 5.1 and 10.0 mm producing the tallest shoots. *P. taiwaniana* Hu & Cheng cuttings with diameters less than 9 mm regenerated poorly no matter what the cutting length. The effect of the diameter of the root cutting on its likelihood for success may be related to its having adequate carbohydrate reserves to support meristem growth and shoot development. The size of the root cutting may also have an impact on how long it takes before shoots appear. In black locust (*Robinia pseudoacacia* L.), shoots appeared on root cuttings with diameters 5 mm and above in as little as 15 days. With fall-flowering *Anemone* x *hybrida*, cutting size had no effect on initiation of new shoots and roots nor on how much time was required for that regeneration, but it did have an impact on the size of the plant ultimately resulting from the root cutting. Larger cuttings resulted in larger plants.

POLARITY

Tissues of root cuttings exhibit an important trait—polarity. Each end of the cutting behaves differently, with new shoots nearly always arising at the proximal end of the cutting (the end nearest the main stem or trunk) and roots

FIGURE 23.2 Root cuttings of oakleaf hydrangea (*Hydrangea quercifolia* Bartr.) illustrating the horizontal cut made at the end nearest the crown and the slanting cut made at the end farthest from the crown of the plant.

at the distal end (the end farthest from the main stem or trunk) no matter how the cutting is oriented in the rooting medium. This polarity occurs no matter what the original orientation (horizontal, diagonal, or vertical) of the root was on the original root system. Root polarity is very strong in the lateral direction (along the length of the cutting), but there is no corresponding transverse (across the root section) polarity, since shoots will arise from all positions around the circumference of the cutting at the proximal end. To keep track of polarity, when root cuttings are taken, the proximal end is cut perpendicular to the root and the distal end is separated with a slanting cut so that the two ends can be differentiated (Figure 23.2). Auxins are transported from the shoot apices in a polar manner and regulate development, with high concentrations promoting roots and suppressing shoots.

INFLUENCE OF JUVENILITY

It has long been recognized that the closer to the trunk that the root cuttings are taken, the more likely they are to regenerate successfully. This is due to a phenomenon called juvenility, discussed more thoroughly in chapter 15. The early life of some plants, particularly woody plants is the juvenile phase, characterized by rapid growth, the ability to root easily, and sometimes even morphological differences, such as plant form and leaf shape. Juvenile tissues, although they are the oldest chronologically, are the youngest physiologically. The "cone of juvenility" encompasses the tissues of the tree closest to the crown or the transition zone between shoot and root (Figure 15.3). For the root system, some authors depict the cone of juvenility to be maximal at soil level and decreasing to a certain soil depth. The shape and depth of the juvenile zone may vary from species to species. When the plant becomes capable of reproduction, it is deemed mature. Mature plants have slower growth rates, root more reluctantly or not at all, and may have different morphologies and leaf shapes. Juvenility can be taken into account either by choosing to take cuttings from younger plants or by taking cuttings as close to the crown as possible.

The profound effect of juvenility as a function of the distance from the crown that the root cutting is taken is usually seen in the percentage of the cuttings exhibiting shoot formation. In other species that still produce shoots even if cuttings are taken some distance away from the trunk, the effect of juvenility may be seen in the number of shoots produced per cutting, the growth or vigor of those shoots, or with how extensive a root system is developed. For example, *P. tomentosa* cuttings taken from horizontal roots 101 to 200 mm away from the trunk regenerated more extensive root systems than those from more than 200 mm. On the other hand, in the case of crab apple (*Malus* spp.) trees, it does not seem to matter if cuttings are taken from roots that spread horizontally in the soil or those that are oriented more vertically or diagonally.

SEASONALITY

Nurserymen who commercially propagate plants by root cuttings typically collect the cuttings from December through March, although some species have smaller windows of success with root cuttings. If cuttings are to be directly stuck in the field, root cuttings can often be taken as early as October and November. This gives them a longer time to develop when soil and temperature conditions are suitable for shoots to form and for roots to become established. The seasonal variation in the suckering capacity of root cuttings of some species may be related to endogenous auxins. For trembling aspen (*P. tremuloides*), the number of suckers produced from root cuttings was inversely related to the quantity of endogenous auxin in the roots, highest in June, a time outside of the best window for success with root cuttings. Treatment

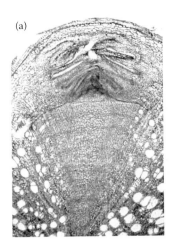 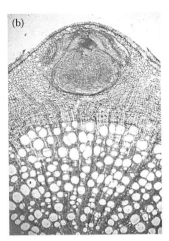

FIGURE 23.3 (a) Cross section of a sassafras root showing a root bud that is located at the level of the vascular cambium. This "additional bud" formed when the root was young, so as the root grows in diameter, the bud grows outward with the vascular cambium, leaving a clearly visible vascular trace connecting the bud to the center of the root. Also pictured are a "reparative bud" and its associated sphaeroblast (spherical structure with cambium and xylem). (b) Sassafras roots also produce buds that are located at more external positions in the bark. These "reparative" buds are found primarily on roots of older trees, do not appear to require direct root injury for their formation, and show no connection with the vascular tissues of the root, but they do form spherical vascular cambia associated with these buds, resulting in spherical nodules of wood in the bark. In sassafras, these buds appear incapable of sprouting in field conditions. (Images courtesy of Michael J. Bosela, Indiana University–Purdue University at Fort Wayne, IN.)

with an antiauxin caused an increased number of suckers to be produced on each root. A similar relationship was found between auxin levels in roots and sucker production for European aspen (*P. tremula* L.). This is not surprising, since endogenous auxins are known to suppress growth of lateral buds in shoots.

RESPONSES TO GROWTH REGULATORS

Root cuttings lack root tips that produce cytokinins, hormones that promote shoot formation, but suppress root formation. In tissue culture (chapter 31 through chapter 35), cytokinins in the medium stimulate the formation of buds on root segments, and it may be that exogenous application of cytokinins could stimulate bud break of preformed buds as well in root cutting segments. Cytokinins inhibit root formation when applied at the distal end of the cutting or that farthest from the crown. In general, another class of hormones, auxins, suppresses bud development, but stimulates root development. In some cases, as with *Anemone*, application of low concentrations of the auxin, IBA, was helpful in increasing the size of the plant resulting from root cuttings, but in other cases, application of IBA has not resulted in significant improvement or has been detrimental in terms of shoot initiation or growth.

ORIGIN OF THE SHOOTS AND ROOTS THAT EMERGE FROM THE CUTTING

New roots can arise from the distal end of the cutting so that the cutting remains a part of the newly developing plant (*Rhus* spp.) or they can arise from the base of the new shoots emerging from the proximal end of the cutting. In this case, typically the original root cutting withers.

In roots, two types of buds occur, each of which is capable of giving rise to shoots. They are different in several characteristics, such as when they form, the morphological origin of the bud, and often the likelihood of the bud to sprout and grow. "Reparative buds" form in response to damage, injury, or senescence, and their tissue of origin may be phellogen, secondary phloem, or callus. The other type, "additional buds," form very early in the growth of the root, when the plant is young. The bud is visible at the level of the vascular cambium, and it remains connected with the center of the root by a bud trace through the secondary xylem. The trace can be seen in the cross section for sassafras (*Sassafras albidum* (Nutt.) Nees.) (Figure 23.3a), which has both types of bud, but only the additional buds give rise to root sprouts from root cuttings. The reparative buds have no connection with the interior tissues of the root (Figure 23.3b). In other species, buds that form *de novo* in response to root injury appear to be the primary, if not only source of root sprouts. For some species, such as black locust, additional buds form adjacent to the poles of the protoxylem of the root; this is the reason that shoots sprout from these root cuttings in longitudinal rows.

PREPARATION OF CUTTINGS AND THE ROOTING MEDIUM

Cuttings may be handled in a number of ways once they have been collected; they may be stored, allowed to callus on the cut ends, stuck in containers in a greenhouse or cold frame, or outplanted directly. Aftercare ensures that adequate moisture is available, temperature extremes are avoided, and aeration is sufficient. When cuttings are taken in late fall, winter, or early spring, they are sometimes allowed to dry for several days to remove excess moisture before being placed in the rooting medium. *Paulownia* is a good example of application of this practice. The rooting medium must be well drained and aerated. In other cases, the root cuttings can be dug, cleaned, trimmed, and after being treated with an appropriate fungicide, stored in moist (not wet) medium until time to plant in the field. The fungicide can be easily applied by placing the root cuttings in a Ziploc bag containing the fungicide and shaking the bag until the cuttings are well coated.

Root cuttings can be sown either vertically or horizontally. When sown horizontally in flats or in the field in shallow trenches and lightly covered with about 1 to 2 cm of medium or soil, the polarity of the cutting is not as important as when sowing root cuttings vertically. In that case, the end of the cutting that was nearest to the crown or the proximal end with the cross-cut should be just below the soil line or just above it. The distal end with the slanted cut should be planted more deeply in the well-aerated medium. The medium that the cuttings are placed in must remain moist to prevent the root cuttings from drying out, but not so wet as to encourage rotting of the cutting. Some propagators recommend waxing the cut ends of the cuttings when they are first taken to minimize loss of moisture.

Root cuttings are sometimes bundled in groups of 10 to 25 for ease of counting and handling and placed in flats containing a well-drained medium, such as peat and either sand or perlite for callusing. Callusing is the development of wound tissue at the cut surfaces, and this protects the cuttings when they are later lined out in the field in early spring. Callusing proceeds best at cool temperatures above freezing (approximately 5°C).

Cuttings may also be propagated in a cold frame by direct sticking, the more traditional approach to root cuttings, in a prepared soil bed or by placing root cuttings vertically in deep flats or containers with their proximal ends just below the surface. The cuttings are then covered with a thin layer of sand or perlite. Root cuttings that are placed in the cold frame during the winter or early spring are grown out throughout the spring and summer. Those stuck in containers may require fertigation and lifting sooner than those grown directly in the prepared bed. Using the traditional approach, root cuttings would be left in the cold frame to develop until fall of the first growing season, when they would be lifted and potted in containers or lined out in the field.

If cuttings are stuck in flats or containers in the greenhouse, temperature control is extremely important. Initially the temperature range is held at 2°C to 7°C, but as the season progresses, temperatures in the greenhouse will also begin to rise. If this happens too quickly, cuttings will not have developed root systems sufficient to survive water loss and will die. From March to April, the temperatures can be allowed to range between 7°C and 10°C. Plants are typically ready to go to the field by July.

Species Most Suitable for Use of Root Cuttings

A common observation is that any plant that suckers readily in the wild is likely to be successfully propagated with root cuttings. Root cuttings can be used for commercially important hardwood species, such as *Paulownia* spp., *Populus* spp., and *Robinia* spp. One of the fastest-growing tree species in the world, *Paulownia* is valued for its strong, light-weight wood. Its ability to produce commercial crops of timber in six years makes it advantageous to be able to clone selected individuals with desirable traits. In the U.S., most of the *Paulownia* trees are produced from containerized seedlings, but in other countries, root cuttings are used extensively. *Paulownia* is cross-pollinated, so even though it is easily propagated by seeds, there is no guarantee that seedlings from an outstanding tree will share all of its traits. Only clonal propagation can guarantee a tree with an identical genotype, but *Paulownia* does not propagate easily from stem cuttings. Root cuttings, on the other hand, give rise to shoots readily and root relatively easily.

Several *Populus* species are important commercially including quaking aspen and European aspen. Quaking aspen is valued for its light wood, which is resistant to shrinkage and used for lumber and matches. It is also a fast-growing pioneer species commonly found in pure clonal stands. It does not propagate easily from softwood cuttings, but suckers from root cuttings can be induced to form roots relatively easily. The same is true for most of the *Populus* species. Root cuttings also provide an inexpensive method to propagate a large number of plants for reforestation and bank stabilization purposes. The beauty of this is that they can be planted in the soil directly at the site where establishment is desired.

Root cuttings can be used for ornamental woody species, such as crape myrtle (*Lagerstroemia indica* L.), panicled goldenraintree (*Koelreuteria paniculata* Laxm.), lilac (*Syringa vulgaris* L.), flowering quince (*Chaenomeles* spp.), hypericum (*Hypericum calycinum* L.), European mountain ash (*Sorbus aucuparia* L.), roses (*Rosa* spp.), wisteria (*Wisteria floribunda* (Willd.) DC.), and lacebark elm (*Ulmus parvifolia* Jacq.). Only a very few of the many species that can be propagated with root cuttings are listed here. Oakleaf hydrangea (*Hydrangea quercifolia* Bartr.) is another of these, and in the landscape setting, it suckers readily (Figure 23.4). Root cuttings of forsythia (*Forsythia* x *intermedia* Zab.) can be taken in fall, overwintered in nursery flats containing perlite, set out in April, and will grow to 3- to 4-ft plants by October. Some ornamentals,

Adventitious Shoot and Root Formation on Leaf and Root Cuttings

FIGURE 23.4 Root used for the root cuttings in Figure 23.2a showing the root sucker origin of the ramet on the right and its root connection to the original oakleaf hydrangea plant (ortet) on the left.

such as *Euonymus* spp., *Spiraea* spp., and *Viburnum* spp., can be grown from summer softwood cuttings taken from the shoots that emerge from the original root cuttings. If an ornamental plant is grafted, cuttings taken from the root system (rootstock) will not be of the same genotype as the ornamentally desirable scion (portion of the plant above the graft). Root cuttings are only useful when the root system is of the same genotype as the shoot (as in own-rooted shrubs and trees). For some ornamentals that are notoriously difficult to propagate asexually, such as Georgia plume (*Elliottia racemosa* Elliott), root cuttings are still the only practical method of propagation.

Root cuttings can also be used to propagate fruit species, such as kiwifruit and its relatives (*Actinidia* spp.), apples (*Malus* spp.), and stone fruits (*Prunus* spp.), but is most commonly used for *Rubus* spp., such as blackberry and raspberry. Blackberry and raspberry plantings have been established easily using root cuttings placed directly in the field, but the ability to regenerate successfully at a high percentage may vary with cultivar. Remember though, with thornless blackberry, the new shoots that form from root cuttings lack the thornless characteristic of the original shoots because the shoots arise from tissue of the internal layers. For other fruit species, a specialized application for root cuttings is for obtaining dwarf fruit trees by propagating genetically dwarf selections on their own roots rather than using the more common approach of grafting standard cultivars onto dwarfing rootstocks.

Probably the best known group of plants for which root cuttings are particularly well suited are herbaceous perennials, such as windflower (*Anemone* spp.), bellflower (*Campanula* spp.), coneflower (*Echinacea spp.*), blanket flower (*Gaillardia aristata* Pursh), oriental poppy (*Papaver orientale* L.), sages (*Salvia* spp.), thrift (*Phlox subulata* L.), and some *Geranium* species. If the roots are thin as with yarrow (*Achillea millefolium* L.), flowering spurge (*Euphorbia corollata* L.), gaillardia, thrift, and Stokes aster (*Stokesia laevis* (J. Hill) Greene), they can be cut into pieces about two inches long, scattered on the rooting medium, and lightly covered with a thin layer of the medium. If the roots are fleshier like those of peony (*Paeonia* spp.) and oriental poppy, the cuttings can be inserted into the medium with the distal end pointing downward. Some propagators advocate allowing the top ¼ in. to stick out of the medium, whereas others suggest burying the entire cutting about an inch under the surface of the rooting medium.

LEAF CUTTINGS

Many different parts of the plant can be used as cuttings for propagation, even roots as seen in the first part of this chapter. Use of leafy stems has been discussed in chapter 24, but for some plants, only the leaf or the leaf with petiole is needed to propagate a new plant that is genetically identical to the original plant. These plants can regenerate new ones without any portion of the original stem still attached.

ADVANTAGES AND LIMITATIONS OF USING LEAF CUTTINGS

Leaf cuttings are a relatively simple way to propagate some plants. Very little expertise is needed, and leaf cuttings will root without elaborate specialized facilities. The medium should be one that will retain moisture, but still drain freely. A typical rooting medium is one consisting of ½ peat to retain moisture and ½ sand, perlite, or vermiculite to ensure good drainage. A variety of containers can be used, as long as some means of restricting moisture loss and ensuring high humidity around the cuttings is established. This can be as simple as the base of a plastic liter soft drink container cut in half and inverted over the pot—a "cloche" donated by the soft drink industry.

In some cases, the way the cutting is taken determines the numbers of plants that can be obtained. For example, with *Begonia rex*, the leaf can be separated from the parent plant at the petiole and inserted into rooting medium or even water. New plantlets will form at the base of the petiole (Figure 23.5a). Sometimes the petiole is trimmed off the leaf, and the leaf blade is rolled and then inserted into the medium (Figure 23.5b). However, if a larger number of plants are desired, the Rex Begonia leaf can be subdivided into squares about 3 to 5 cm on a side, each containing a portion of a vein, and this will result in a much larger number of plants (Figure 23.5c). Another good example is *Streptocarpus*, which can be propagated using an intact leaf cutting resulting in one plant being formed. Cutting the leaf into chevrons (v-shaped pieces) will allow more

FIGURE 23.5 Rex begonia showing different methods for conducting leaf propagation including (a) leaf blade and petiole rooted in water, (b) leaf with petiole and part of the blade, rolled and inserted into the medium, and (c) leaf cut into squares secured by florists' pins to hold them in firm contact with the rooting medium.

plantlets to develop from the cut surfaces, but cutting the leaf into two halves with the midrib removed and embedding the cut surface firmly into the rooting medium will result in the largest numbers of plantlets formed. *Sansieveria* can be propagated by division, which results in fewer plants being obtained compared to being propagated by 2- to 3-in. pieces of the leaf (Figure 23.6), but this technique cannot be used with the variegated forms, otherwise the variegation will be lost.

Leaves of some plants will easily form roots when used as leaf cuttings, but they do not form shoots. *Hoya, Clematis,* and *Mahonia* leaves can be rooted, but either fail to form adventitious shoot buds or only rarely do so. Obviously, successful propagation by leaf cuttings is dependent not only upon adventitious root initiation, but also the ability to form adventitious shoots. Although plants may be related as closely as within the same genera, one may have the ability to form adventitious shoots and roots and can be propagated with leaf cuttings, whereas others may not. Sometimes the ability differs among cultivars of the same species, as has been demonstrated with patchouli (*Pogostemon patchouli* Pelleto), a plant valued for its fragrant oil. Young leaves of sugar beet (*Beta vulgaris* L.) can be rooted with the use of IBA, but adventitious bud formation occurs only at very low frequency.

This peculiarity of rooting without shoot formation has also been used in plant research. Leaves of common bean (*Phaseolus vulgaris* L.) will root, but no shoots form. After adventitious roots form at the end of the petiole, they continue to develop (increasing root to shoot ratio), and the original leaf blade photosynthesizes, allowing researchers to investigate the effects of immersion in various growth regulators and changes in endogenous cytokinins without the complications of shoot buds. Leaves

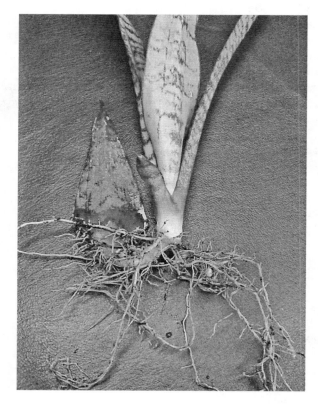

FIGURE 23.6 Original leaf section cutting of *Sansevieria trifasciata* showing the new plant originating from the base of the leaf cutting.

of English ivy (*Hedera helix* L.) root from cuttings consisting of only a leaf blade and petiole. This feature has been used to investigate root initiation in reciprocal grafting experiments using mature leaf lamina (blades) grafted onto juvenile petioles and vice versa. Root initiation in response to IBA was more rapid for combinations with

Adventitious Shoot and Root Formation on Leaf and Root Cuttings

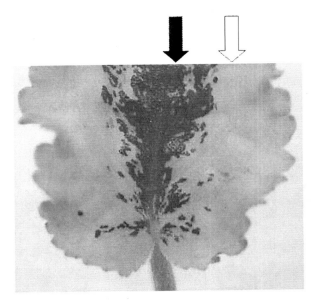

FIGURE 23.7 Cross section of a variegated African violet leaf showing pigmented (solid arrow) and albino (white arrow) tissues. Shoots arising from nonpigmented tissue will be albino and unable to survive on their own.

juvenile petioles. The capacity to form adventitious shoots has also been explored as a means to obtain whole (nonchimeral) mutants as a result of ionizing radiation when the adventitious bud is the result of only one cell—in the case of Saintpaulia, one epidermal cell.

Plants that are chimeral may not propagate with the desired appearance, depending on the origin of the newly formed shoots and roots. In African violets (*Saintpaulia ionantha* Wendl.), new shoots arise from newly dedifferentiated cells lying just beneath the epidermis. If those shoots arise from the albino portion of the variegated *Saintpaulia* shown in Figure 23.7, albino plantlets could be formed, which cannot survive on their own.

THE ORIGINS OF NEW SHOOTS AND ROOTS ON LEAF CUTTINGS

Plants propagated by leaf cuttings can be divided into two groups based on the histology of the shoots and roots that form, whether they form from new adventitious meristems or from preformed primary meristems. In the first group, plants such as *Saintpaulia* or *Begonia*, meristematic cells must form from mature differentiated tissue. In the case of new adventitious roots, this means the parenchyma located near the xylem–phloem boundary of the vascular bundle, and in the case of new adventitious shoots, epidermal cells sometimes near a large vein. Leaves of some species, such as *Sedum*, detach from the parent plant very easily, and this stimulus causes them to form secondary meristems that develop from vacuolated, differentiated parenchyma. These secondary meristems then give rise to roots and shoots. The location where the new meristems form on the leaf may also vary. For *Saintpaulia*, *Peperomia*, and *Streptocarpus*, and the monocots *Sansevieria* and *Lilium*, new meristems form at the base of the petiole or where the leaf was severed. For the carnivorous plant sundew (*Drosera*), new meristems develop randomly on the leaf blade.

In the second group, embryos are preformed early in the development of the leaf and they retain their meristematic activity. These primary meristems can be located on the periphery of the leaf in notches (*Bryophyllum*), at the ends of the leaf (*Asplenium*) or at the base of the leaf blade where it joins the petiole (*Tolmeia*). This ability to form buds and often roots on leaves that are still attached is not as prevalent as the ability to form roots and shoots on leaves detached from the parent plant. Still more species can form roots on detached leaves, but cannot form shoots.

The family Crassulaceae contains the genus *Bryophyllum*, many of which are well known for the ability to produce plantlets from preformed meristems (sometimes called foliar embryos) located in the indentations at the leaf margins. These foliar embryos usually develop a couple of leaf primordia first and then a couple of roots form. This ability is the reason Bryophyllums are sometimes referred to as "maternity plants." Not all the *Bryophyllum* species have this ability, and some produce plantlets on the floral stem instead. Plantlets (Figure 23.8a) on this *Bryophyllum daigremontianum* develop 1 to 2 roots (Figure 23.8b), then detach from the leaf margins, falling to the surface

FIGURE 23.8 Leaf of mother-of-thousands showing (a) the plantlet forming from preformed meristems at the periphery of the leaf and (b) the roots that form on the plantlets after the leaves have formed (abaxial surface).

FIGURE 23.9 Pick-a-back plant with new plantlets forming at the juncture of the leaf blade and the petiole. (Image © 2005 Barry Glick—www.sunfarm.com. Used with permission from Barry Glick, Sunshine Farm & Gardens, Renick, WV.)

of the growing medium and becoming established. Often older leaves, whose meristems are quiescent, drop from the plant, fall to the growing medium, and as the leaf senesces, new shoots and then roots are formed from the indentations at the margins of the leaf.

For pick-a-back plant (*Tolmiea menziesii* (Pursh) Torr. & A. Gray), new plantlets form at the juncture where the petiole meets the leaf blade on mature leaves (Figure 23.9), which explains another of its common names, "youth on age." If the leaf is separated from the parent plant and placed on moist medium, the new plantlet can further develop. The original leaf may stay green while the new plantlet begins to grow, but eventually shrivels and turns brown.

Walking fern (*Asplenium rhizophyllum* L.) is a hardy fern found in moist shady limestone or calcareous outcroppings (Figure 23.10a) where there is enough soil to retain moisture and a nearly neutral pH. It has the ability to propagate by forming plantlets at the apex of the leaf (Figure 23.10b). The long fronds arch, and when new plantlets form, the weight of the plant allows the frond to touch the ground and root. This results in a network of new plants spaced around the original plant and still connected to it by the long tapered fronds, giving the impression of the fern 'walking' across the ground.

RESPONSES TO GROWTH REGULATORS

Benzyladenine, a cytokinin, increased the number of shoots obtained from leaf cuttings of *Sedum*, but reduced the number of roots formed. Treatment with auxins, such as IBA or naphthaleneacetic acid (NAA), promotes root formation. Rieger elatior begonias (*Begonia* x *hiemalis* Fotsch) of the Aphrodite group do not consistently produce adventitious buds at the base of the petiole, but an application of 6-(benzylamino)-9-(2-tetrahydropyranyl)-9H-purine, a cytokinin, as either a foliar spray or a petiole dip stimulated bud and shoot development.

FIGURE 23.10 (a) Walking fern in its normal habitat on limestone outcroppings on Monte Sano Mountain, Huntsville, Alabama, showing the long slender fronds. (b) New plantlet forming at the extreme tip of the slender attenuated frond of walking fern. This plantlet will root easily once it has made contact with the ground. (Photos courtesy of Christopher Sutton, Huntsville, AL.)

SPECIES SUITED FOR PROPAGATION BY LEAF CUTTINGS

Some of the most common examples of species that propagate easily from leaf cuttings are members of the Crassulaceae family, the succulents. Examples include *B. daigremontianum* (mother of thousands), *Crassula ovata*

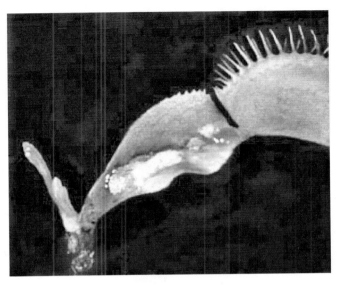

FIGURE 23.11 Specialized leaves of Venus Flytrap can be separated from the parent plant and inserted into shredded sphagnum for rooting. Best results are obtained if the lightly pigmented base of the leaf is retained when the cutting is made. (Photo courtesy of David Webb, University of Hawaii, Honolulu.)

(jade plant), some members of the genera *Echevaria* and *Pachyphytum,* and *Sedum morganianum* (burro tail). *C. ovata* and *S. morganianum* are both propagated easily from leaves broken off to separate them from the parent plant and placed cut surface downward, in firm contact with a well-drained rooting medium (½ peat, ½ perlite works well).

A surprising example of a species that can be propagated by leaf petiole cuttings is Koreanspice viburnum (*Viburnum carlesii* Hemsl.), a woody ornamental shrub valued for its fragrant flowers. The entire process takes several months, but after the petiole has rooted, a new shoot will form. Many of the carnivorous plants can be propagated using leaf cuttings, including *Cephalotus, Dionaea, Drosera,* and *Sarracenia.* Although each of these species has highly specialized leaves (*Cephalotus* and *Sarracenia* have leaves like a pitcher containing fluid to trap insects, *Drosera* has leaves with sticky trichomes, and *Dionaea* has leaves that can envelope then digest the insect), they can be separated carefully from the plant and induced to form roots. Venus Flytrap (*Dionaea muscipulata* Ellis) has highly specialized leaves that trap and digest insects that land on them, stimulate the trigger hairs, and cause leaf closure. If the leaf is pulled off the plant so that it retains some of the original white tissue at the base of the leaf (Figure 23.11) and its base is inserted into rooting medium, new plants will form several months later.

Some members of the Liliaceae family root from leaf cuttings, but rather than form shoots directly, go through a bulblet formation stage first. Cape cowslip or *Lachenalia* will root from leaf cuttings, but new plants arise only after it has formed a bulblet at the base of the rooted leaf (Figure 23.12). *Lilium longiflorum* Thunb. can also be propagated from leaf cuttings that form bulblets followed by adventitious roots. There are selected examples from other plant families that follow a similar pattern. *Zamioculcas zamiifolia* (Lodd.) Engl, sometimes known as ZZ plant, is a member of the Araceae family, although it resembles a cycad. When leaflets of this plant fall to the ground, they easily root and form bulblets. This ability is not shared by the other members of the Araceae family, like *Spathiphyllum* and *Aglaonema.*

CONCLUSION

In summary, propagation by either leaf or root cuttings may seem simplistic and somewhat old fashioned. Advances in tissue culture have presented viable alternatives for the propagation of many difficult-to-root species; however, even with the considerable amount of research devoted to *in vitro* regeneration, not all species respond favorably. This fact argues that there is still a role for root-cutting propagation for some of these species. The ease of producing new plants from leaf cuttings and the lack of need for high technology virtually ensures that leaf cutting propagation will continue to be in use for selected species in the future.

FIGURE 23.12 *Lachenalia,* a member of the Liliaceae family, roots easily from leaf cuttings, but only forms new plants after it has formed a bulblet at the base of the rooted leaf. (Photo courtesy of Mark Mazer, Intarsia, Ltd., Gaylordsville, CT.)

REFERENCES AND SUGGESTED READINGS

Bosela, M.J. and F.W. Ewers. 1997. The mode of origin of root buds and root sprouts in the clonal tree *Sassafras albidum* (Lauraceae). *Amer. J. Bot.* 84: 1466–1481.

Davies, Jr., F.T. and B.C. Moser. 1980. Stimulation of bud and shoot development of Rieger begonia leaf cuttings with cytokinins. *J. Amer. Soc. Hort. Sci.* 105: 27–30.

Del Tredici, P. 1995. Shoots from roots: A horticultural review. *Arnoldia.* 54: 11–19.

Dirr, M.A. and C.W. Heuser. 1987. *The Reference Manual of Woody Plant Propagation—From Seed to Tissue Culture.* Varsity Press Inc., Athens, GA.

Ede, F.J., M. Auger, and T.G.A. Green. 1997. Optimizing root cutting success in *Paulownia* spp. *J. Hort. Sci.* 72: 179–185.

Geneve, L., M. Mokhtari, and W.P. Hackett. 1991. Adventitious root initiation in reciprocally grafted leaf cuttings from the juvenile and mature phase of *Hedera helix* L. *J. Exp. Bot.* 42: 65–69.

Hartmann, H.T., D.E. Kester, F.T. Davis, Jr., and R.L. Geneve. 2002. *Plant Propagation: Principles and Practices,* 7th ed. Prentice-Hall, Upper Saddle River, NJ. 880 pp.

Miedema. P, P.J. Groot, and J.H.M. Zuidgeest. 1980. Vegetative propagation of *Beta vulgaris* by leaf cuttings. *Euphytica.* 29: 425–432.

Stoutemeyer. V.T. 1968. Root cuttings. *Plant Propagator* 14: 4–6.

Togood, Alan. 2003. *Plants from Cuttings.* American Horticultural Society, New York.

24 Propagation by Leaf Cuttings

John L Griffis, Jr., and Malcolm M. Manners

Successful vegetative propagation from virtually all types of cuttings depends on the ability of the plant to make some organ system adventitiously—stem cuttings must produce adventitious roots; root cuttings must form adventitious shoots. However with leaf cuttings, there is no root or stem tissue present, so both primary organs must be made adventitiously. Most plant species are not capable of doing this, but in those species that have the ability, the method allows the production of large numbers of new plants from relatively little starting material—at the least, one leaf will produce one new plant, and in some cases, several new plants can be made from a single leaf. Plants commonly propagated from leaf cuttings are listed in Table 24.1.

Leaf cuttings should be made from relatively young, healthy leaves that are fully expanded. Leaves that are damaged or diseased are more likely to rot than they are to form plantlets. Leaves that are not yet fully developed often do not have enough stored nutrients to complete the maturity process and produce new plantlets. In preparing leaf cuttings, an auxin-based growth regulator treatment is generally not used since, while it would promote rooting, it would at the same time inhibit formation of adventitious shoots. Therefore, no pretreatment is generally used.

EXERCISES

EXPERIMENT 1: PLANTLETS FORM FROM LEAF CUTTINGS WITH PETIOLES

The most popular method for producing new plantlets from leaf cuttings is to use the entire leaf with the petiole still attached. Far more species have the ability to form both roots and shoots adventitiously from petiole tissue than from leaf blade or vein tissues. If there is a disadvantage to this strategy, it is that relatively few new plants will develop from each leaf. If the stock plant has plenty of leaves that might be removed and used for propagation, then low plantlet yield per leaf hardly matters.

Materials

The following materials are needed for each student or team of students:

- Trays or pots of moistened perlite
- Clippers, knife, or some implement to cut leaves cleanly from the stock plants
- Stock plants of almost any *Peperomia* spp. Ruiz & Pav., African violets (*Saintpaulia ionantha* Wendl.), florist's gloxinias (*Sinningia speciosa*

TABLE 24.1
Plants Commonly Propagated from Leaf Cuttings

Family	Scientific Name	Common Name
Agavaceae	*Sansevieria* spp.	Snake plant, mother-in-law's tongue
Araceae	*Zamioculcus zamiifolia*	ZZ
Begoniaceae	*Begonia* spp.	Begonia (Rex, Angel Wing)
Crassulaceae	*Bryophyllum* spp.	Mother-of-thousands
	Kalanchoe spp.	Kalanchoe
	Sedum morganianum	Burro's tail, donkey's tail
Gesneriaceae	*Saintpaulia ionantha*	African violet
	Sinningia speciosa	Florist's gloxinia
	Streptocarpus spp.	Streptocarpus
Piperaceae	*Peperomia* spp.	Peperomia
Saxifragaceae	*Tolmeia menziesii*	Piggyback plant

	Procedure 24.1
	Plantlets Form from Leaf Cuttings with Petioles
Step	Instructions and Comments
1	Each student or team should cut three to five leaves from the stock plants, taking some petiole with each cutting. (Figures 24.1a,b).
2	Carefully insert the petiole end of the first cutting into the perlite, so that the blade of the leaf is just above the surface of the perlite, but not quite touching it.
3	Insert the other cuttings into the perlite in a similar manner, being sure to space the multiple cuttings about 2 to 3 cm apart in the container to allow light to shine on all the cuttings (Figure 24.1c).
4	Set the container of cuttings in a warm, shaded spot (e.g., under a greenhouse bench), and keep the perlite damp at all times.
5	Make observations on the number of new plants obtained from each cutting and the appearance of the new plants. Note when the new plants appeared and the origin of the new plant on the original cutting. Once data have been collected, potting up the new plants into containers of potting medium (with or without the original leaf cutting still attached) will complete this exercise.

(Lodd.) Hiern), almost any *Bryophyllum* spp. Salisb., and/or piggyback plant *(Tolmeia menzeisii* (Pursh) T. & G.) with mature leaves
- Leaves of *Hoya carnosa* (L. f.) R. Br. (for comparison); other choices for the comparison are leaves of *Coleus* spp. Lour., *Impatiens* spp. L., rubber tree *(Ficus elastica* Roxb.)
- Moistened vermiculite or moistened soilless media can be substituted for perlite in any of these experiments, but the water-holding capacity of either of these is significantly greater than that of perlite and may lead to a greater loss of leaves to disease unless the watering schedule is closely monitored; as conditions from greenhouse to greenhouse vary, it may also be necessary to apply a fungicidal drench to the medium to discourage disease problems (chapter 8)

Follow the protocol outlined in Procedure 24.1 to complete this experiment.

Anticipated Results

Root formation should occur within three to four weeks, and plantlets should become visible above the perlite in four to seven weeks. (Figure 24.1d,e). After the new plants have a few leaves, they may be transplanted to individual pots of regular potting medium. Note that the florists' gloxinia may take longer to produce visible plantlets, as the leaf will first produce a small tuber at the cut end of the petiole, and the new plantlets will sprout mostly from the new tuber rather than arise directly from the petiolar tissue. Note also, if the piggyback plant is used as the leaf source, it is important to select mature leaves that do not already have plantlets forming on them. The same would be true if *Bryophyllum pinnata* (Lam.) Kurz, *B. diagre-*

montianum Hamet & Perr., or *B. fedtschenkoi* Hort. were used as a leaf source. All of these plants actually have certain leaf cells that are predetermined to form asexual embryos within the leaf tissue. A variety of stimuli, including removing a leaf from the plant for propagation purposes, may cause these cells to grow and divide and produce embryonic plantlets. For comparison, it may be noted that the embryonic plantlets will arise from various areas of the leaf blade tissues rather than from the petiolar tissues that will yield the adventitious plantlets formed by African violets or the various peperomias. *Hoya, Coleus, Impatiens,* or *Ficus* leaves are included in this experiment as examples of species that easily form roots from leaf cuttings, but that usually do not form adventitious shoots. So, the result is a rooted leaf. It may survive for many months, but it won't usually produce a plant (Figure 24.1f).

Question

- Most plants that are propagated from leaves tend to be rather succulent. What advantage(s) would a succulent leaf have over other leaves in this method?
- Exogenous PGRs are not usually used with leaf cuttings. Since root formation requires auxin, what would be the source of auxin for root formation on leaf cuttings?
- Why do shoots usually form only after roots have already been formed?

EXPERIMENT 2: PLANTLETS FORM DIRECTLY FROM LEAF BLADE TISSUES

This method for producing new plantlets from leaf cuttings works well for certain plants that have little or no petiole tissues. If the plants are chimeric, as are some variegated cultivars of snake plant, such as *Sansevieria trifasciata*

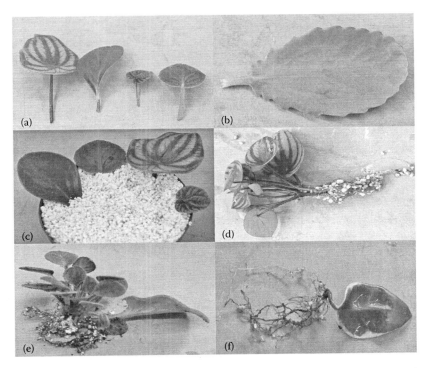

FIGURE 24.1 (a) Leaf cuttings from three different peperomias and from an African violet. (b) Leaf cutting of *Bryophyllum fedtschenkoi*. (c) Leaf cuttings with petioles inserted in perlite. (d) Peperomia leaf with adventitious roots and plantlets. (e) African violet leaf with adventitious roots and plantlets. (f) Hoya leaf with only adventitious roots.

'Laurentii', Thunb., then the resulting plantlets may not display the desired, striped pattern like the parent. Therefore, this experiment may also be used to determine whether or not a particular plant is a chimera (chapter 16).

Materials

The following materials will be required to complete this experiment:

- Trays or pots of moistened perlite
- Clippers, knife, or some implement to cut leaves cleanly from the stock plants (and with certain plants, to cut the leaves into pieces)
- Stock plants of snake plant (*Sansevieria trifasciata*), burro's tail (*Sedum morganianum* Walth.), panda plant (*Kalanchoe tomentosa* Bak.), or Cape primrose (*Streptocarpus* spp. Lindl.) that can provide mature leaves. If either *S. morganianum* or *K. tomentosa* is used as the stock plant, the entire leaves will easily break off of the parent plant, so no cutting is necessary

Follow the instructions listed in Procedure 24.2 to complete this exercise.

Anticipated Results

Roots should form in about three to four weeks, and shoots from the leaves by about the seventh week (Figure 24.2d,e). At the end of the experiment, remove the leaf cuttings from the perlite, and observe where the plantlets have formed. Note whether the upside-down cuttings made any plantlets, and if so, from what areas the plantlets arose. If a variegated cultivar of burro's tail is selected as the parent plant, whether or not the parent plant is chimeric can also be determined using this method (chapter 16).

Questions

- If a variegated variety of snake plant was used, notice the pattern of variegation on the plantlets (or lack thereof); does the pattern match that of the parent leaf?
- Is the variegation of the parent plant chimeric?

EXPERIMENT 3: PLANTLETS THAT FORM FROM LEAF-BLADE VEINS

Some types of begonias, as well as some *Streptocarpus* spp., have the ability to make adventitious plantlets from leaf pieces that contain major or large leaf veins. Since the leaves may be quite large, one can often make several cuttings from a single leaf. Each leaf cutting may produce more than one plantlet, so the main advantage to this type of propagation is that numerous plantlets can be made from a single, mature leaf. Since this method of propagation utilizes small leaf pieces that are prone to rotting, clean tools and equipment, and sterile media should be

	Procedure 24.2
	Plantlets Form Directly from Leaf-Blade Tissues
Step	Instructions and Comments
1	When using the dwarf cultivars *S. trifasciata* 'Hahnii', 'Golden Hahnii', *S. morganianum*, or *K. tomentosa*, remove entire leaves from the stock plant (Figure 24.2a).
2	When using one of the longer-leafed cultivars of *S. trifasciata*, cut horizontally across the leaf, creating leaf pieces about 7 to 8 cm long (Figure 24.2b). When using strap-leafed *Streptocarpus* spp., cut horizontally across the leaf, creating leaf pieces about 5 cm long.
3	Keep track of the original orientation of the cut leaf pieces by writing the word "top" or "up" or making a small arrow on the distal (upper) part of the cutting.
4	Insert the bases of the cuttings into the perlite, leaving about 2 to 3 cm of space between cuttings. (Figure 24.2c). Insert most of them upright, but also insert one or two of the leaf cuttings upside-down, for comparison. Entire leaves from the burro's tail or the kalanchoe may also be planted upside down or even left to rest sideways on the medium.
5	Set the container in a warm, shaded spot (e.g., under a greenhouse bench), and keep the perlite damp at all times.
6	Make observations on the number of new plants obtained from each cutting and the appearance of the new plants. Note when the new plants appeared and the origin of the new plant on the original cutting. Once data have been collected, pot up the new plants, with the original leaf cutting still attached, in a container of potting medium.

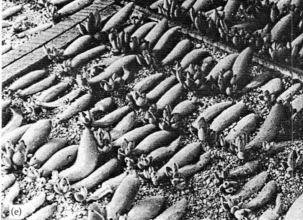

FIGURE 24.2 (a) Stock plant of *Sansevieria trifasciata* 'Hahnii' with entire leaves removed and inserted in perlite. (b) Leaves of *Sansevieria trifasciata* 'Moonshine' whole and cut into sections for propagation. (c) Leaf cutting of *Sansevieria trifasciata* 'Moonshine' inserted in perlite. (d) Snake plant leaf with adventitious plantlets. (e) *Kalanchoe tomentosa* leaves with adventitious plantlets.

Propagation by Leaf Cuttings

FIGURE 24.3 (a) Leaf cuttings from a rex begonia and an angel wing begonia. (b) Leaf cutting from a rex begonia with the petiole removed. (c) Leaf cutting from a rex begonia with the outer leaf edge tissue removed. (d) Leaf cutting from a rex begonia with the central leaf tissue cut up into small, triangular pieces with a major vein running down the center of each. Pieces without large veins usually do not form plantlets. (e) Small, triangular leaf pieces from a rex begonia and an angel wing begonia inserted into perlite. (f) On the left, a small plantlet develops from leaf piece of a rex begonia. Slightly older plant on the right developed quickly when provided proper nutrition and growing conditions.

	Procedure 24.3 Plantlets That Form from Leaf-Blade Veins
Step	Instructions and Comments
1	Remove a mature leaf from the begonia stock plant (Figure 24.3a).
2	Carefully cut around the petiole to remove it, leaving the palmately arranged major veins of the leaf not quite touching each other (Figure 24.3b).
3	Trim the upper part of the leaf so that the distance from the base of the veins to the top of the leaf is about 3 to 5 cm all around (Figure 24.3c).
4	Cut the remaining leaf piece into multiple cuttings so that each sectional leaf cutting contains a major vein, creating what looks like small, triangular leaves, with a major vein as a midrib (Figure 24.3d).
5	Bury the lower 1/3 to 1/2 of the cuttings in 10-cm (4-in.) pots containing medium. If desired, several cuttings may be placed closely together in the same pot; doing so will result in a much fuller plant in less time, making the final product marketable at an earlier date—a commercial advantage (Figure 24.3e).
6	Place the pots in an intermittent mist system for about a week, and then transfer them to a shaded area in a greenhouse. Alternately, they can be placed under a greenhouse bench and hand-misted several times a day for the first week.
7	Make observations on the number of new plants obtained from each cutting and the appearance of the new plants. Note when the new plants appeared and their points of origin on the original cutting. Once data have been collected, repot the new plants with the original leaf cutting still attached.

	Procedure 24.4 Plantlets That Form from Leaflets of a Compound Leaf
Step	Instructions and Comments
1	Remove a mature but healthy, deep green leaf from the ZZ stock plant (Figure 24.4).
2	Using a clipper, knife, or just your fingernails, cut off an individual leaflet from the rachis and allow the cut surface to air-dry (Figure 24.5a,b).
3	Bury the lower 1/4 to 1/3 of the leaflet (the end that was attached to the petiole) in the perlite.
4	Repeat the procedure with other leaflets, lining them up in the tray, leaving 3 cm or more space between them to allow light to shine between the cuttings. (Figure 24.5c).
5	Place the tray in an intermittent mist system or, alternately, in a humid, shady place (e.g., under a greenhouse bench).
6	Make observations on the number of new plants obtained from each cutting and the appearance of the new plants. Note when the new plants appeared and the origin of the new plant on the original cutting. Once data have been collected, potting up the new plants into containers of potting medium will complete the exercise (Figure 24.6).

used at all times. It may or may not be necessary to drench the potting mix with a general purpose fungicide, depending on greenhouse conditions.

Materials

The following items will be required for the exercise:

- A number of 10-cm (4-in.) pots of 1 peat:1 perlite (by volume) potting mix, moistened
- Stock plants of *Begonia* spp. L. or *Streptocarpus* spp. that can provide mature, fully expanded leaves (*B.* x *rex-cultorum* and other rhizomatous and angelwing cultivars of begonias work very well; tuberous types of begonias and *B. semperflorens* Hort. are not recommended)
- Razor blade, knife, box cutter, scissors, or other cutting device

Follow the instructions outlined in Procedure 24.3 to complete this exercise.

Anticipated Results

Roots should form in about two to three weeks, and small plantlets should appear after six to seven weeks. Given proper growing conditions and nutrition, plants can reach saleable size rather quickly (Figure 24.3f).

Question

- Do plantlets always form from veins, or do some also arise from the leaf blade between the veins?

Experiment 4: Plantlets Form from Leaflets of a Compound Leaf

The "ZZ" plant (*Zamioculcas zamiifolia* (Lodd.) Engl.) has recently become quite popular as a shade-tolerant indoor foliage plant. It has long, pinnately compound leaves that resemble those of a palm or cycad and is unusual in that it is propagated from individual leaflets cut or broken from the leaf.

Materials

The following materials are needed for each student or team of students:

- Trays or pots of moistened perlite
- Clippers, knife, or some implement to cut leaflets from the stock plants
- Stock plants of *Z. zamiifolia*

Follow the protocol outlined in Procedure 24.4 to complete this experiment.

Anticipated Results

Root formation should occur within three to four weeks, and plantlets should appear visible above the medium in six to nine weeks (Figure 24.5d). Like gloxinia, the cuttings will also produce a large corm by the time leaves are produced. After the new plants have a few leaves, they may be transplanted to individual pots of regular potting medium.

Propagation by Leaf Cuttings

FIGURE 24.4 A plant of *Zamioculcas zamiifolia* to be propagated.

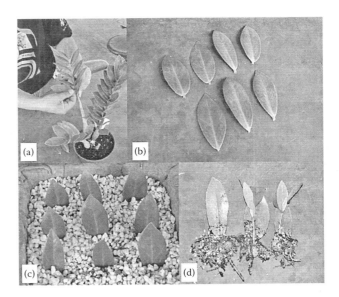

FIGURE 24.5 (a) Leaflets of *Zamioculcas zamiifolia* may be snapped or pinched off of the midrib. (b) Leaflets, ready to be stuck as cuttings. (c) A tray of leaflets planted with the lower 25% of the leaflet in the medium. (d) Several weeks later, plantlets have formed with roots and corms.

FIGURE 24.6 Plantlets of *Zamioculcas zamiifolia* formed from leaflets are potted.

Questions

- Do plantlets always form after the corms or do some also arise directly from the leaflet bases?
- Very few plants can be propagated from leaflets; why do you think *Zamioculcas* can be propagated this way?

SUGGESTED READINGS

Browse, P.M. 1979. *Plant Propagation*. Simon and Schuster, New York. 96 pp.

Davis, T.D. and B. E. Haissig (Eds.). 1994. *Biology of Adventitious Root Formation*. Plenum Publishing. New York. 358 pp.

Dirr, M.A. and C.W. Heuser. 1987 *Reference Manual of Woody Plant Propagation: From Seed to Tissue Culture*. Varsity Press, Inc., Athens, GA. 223 pp.

Hartmann, H.T., D.E. Kester, F.T. Davies, Jr., and R.L. Geneve (Eds.). 2002. *Plant Propagation: Principles and Practices,* 7th ed. Prentice Hall, Inc., Upper Saddle River, NJ. 880 pp.

MacDonald, B. 1992. *Practical Woody Plant Propagation for Nursery Growers,* Vol. 1. Timber Press, Portland, OR. 669 pp.

25 Propagation of Sumac by Root Cuttings

Paul E. Read

Many species and cultivars are extremely difficult to propagate from stem cuttings, but can be propagated quite successfully from pieces of roots or root cuttings. Two examples of plants propagated commercially by root cuttings are *Albizia julibrissin* Durazz and *Elliottia racemosa* Muhlenb. ex Elliott. Others are presented in Table 25.1. Plants that sucker readily in nature and produce shoots from roots are generally excellent candidates for propagation by root cuttings. This technique is not commonly used for propagation of plants because of the time and expense of obtaining the root pieces to be used as cuttings, particularly if the donor plant is in a field setting. Additionally, excising and harvesting of root cuttings may seriously injure or kill the stock or donor plant. You may have already inadvertently experienced the technique, and a familiar example illustrates this point—that of dande-

TABLE 25.1
Examples of Plants That May Be Propagated by Root Cuttings

Species	Common Name
Actinidia deliciosa [Chev.] Liang et Ferg.	Kiwifruit
Albizia julibrissin Durazz	Silk tree
Anemone hupehensis Hort. Lemoine	Japanese anemone
Aralia spinosa L.	Devil's walking stick
Armoracia rusticana P. Gaertn.	Horseradish
Campsis radicans (L.) Seem. ex Bur.[a]	Trumpet vine
Elliottia racemosa Muhlenb. ex Elliott	Georgia plume
Ficus carica L.	Fig
Gymnocladus dioica (L.) C. Koch	Kentucky coffee tree
Gypsophila paniculata L.	Baby's breath
Hydrangea quercifolia Bartr.	Oakleaf hydrangea
Ipomoea batatas (L.) Lam.	Sweet potato
Papaver orientale L.	Oriental poppy
Phlox paniculata L.	Garden phlox
Populus spp.	Aspens, poplars
Pyrus calleryana Decne.	Callery pear
Rhus glabra L.[a]	Smooth sumac
Robinia pseudoacacia L.	Black locust
Rosa spp.	Several rose species
Rubus spp.[a]	Red raspberries, blackberries
Sambucus spp.	Elderberries
Sassafras albidum (Nutt.) Nees[a]	Sassafras
Syringa vulgaris L.	Lilac
Ulmus parvifolia Jacq.	Lacebark elm
Wisteria floribunda (Willd.) DC	Japanese wisteria

[a] These species may be easily substituted for staghorn sumac.

FIGURE 25.1 *Rhus typhina* (sumac). (Photo courtesy of Dr. Tom Ranney, North Carolina State University.)

lion (*Taraxacum officinalis* Wiggers)—when even a small piece of root is left in a lawn after pulling the dandelion, a new plant will soon grow in its place from the remaining root piece.

STOCK PLANT CHARACTERISTICS

As is true for other cutting types, physiological stage, season of the year, and stored food in the propagules are important factors when propagating by root cuttings (chapter 22). Juvenility is an important consideration, with more successes resulting if cuttings are taken from young plants or from within the "cone of juvenility" of more mature, older plants, such as trees (chapter 15). The cone of juvenility for the root region is almost a mirror image of the above-ground cone of juvenility for the aerial portion of the plant. The best season of the year for rooting cuttings varies with species. Excellent success with red raspberry (*Rubus idaeus* L.) was achieved when cuttings were taken from fall through spring, but little success when root cuttings were taken during the summer months (Hudson, 1955). In contrast, root cuttings of Oriental poppy (*Papaver orientale* L.) excised during the summer were easy to propagate, whereas horseradish (*Armoracia rusticana* P.Gaertn.) can be propagated throughout the year by root cuttings. With many species, the size of the cutting is important, in part because of the stored food and the amount of surface available for adventitious bud formation. Note that the goal of propagating by root cuttings is to obtain new shoots that arise adventitiously or directly from root tissues—by definition, roots cannot have preformed buds. Because these adventitious buds arise from internal root tissue, there may be instances where this form of asexual propagation does not result in a faithful reproduction of the parent plant. For example, root cuttings of some thornless cultivars of blackberry result in production of new plants that are thorny. This phenomenon occurs when the characteristic, in this case thornlessness, is a result of the stock plant being a chimera (chapter 16). Similarly, variegated geraniums, aralias, and bouvardias propagated by root cuttings will produce plants with solid green leaves rather than exhibiting the original variegated leaf pattern. Also, root cuttings are an inappropriate way to faithfully reproduce a plant that has been grafted onto a rootstock. Taking a root cutting would result in a new plant resembling the rootstock and not the scion (chapter 27). Root cutting is only a viable alternative for propagation if the plant is own-rooted (the roots are the same genetically as the aerial portion or scion).

EXERCISE

EXPERIMENT 1: ROOT CUTTING EXPERIMENT

Propagators make use of several aids and techniques, such as the use of bottom heat, high humidity, and chemical treatments, such as cytokinins and auxins, to enhance the success of the process. They also often employ root cuttings to generate a crop of softwood cuttings (adventitious shoots) that can then be propagated by standard softwood cutting protocols (chapter 17). Forcing shoots from root cuttings is an excellent way to reintroduce juvenility, and shoots harvested from root cuttings will often display juvenile characteristics, including the ability to root more easily. Propagators can take advantage of this when they have an older mature specimen that they would like to propagate, which is impossible or very difficult to root from traditional stem cuttings.

The following experiment will focus on cutting size and orientation of root cuttings of staghorn sumac (*Rhus typhina* L.; Figure 25.1). *Rhus typhina* readily develops shoots from roots (Figure 25.2) and is a good candidate for use of root cuttings. Alternative species amenable to root cutting are noted in Table 25.1

Propagation of Sumac by Root Cuttings

FIGURE 25.2 Adventitious shoots formed from roots of *Rhus typhina* 'Tiger eyes'. (Photo courtesy of Dr. Tom Ranney, North Carolina State University.)

Harvest Storage

Unless the experiment is to be done immediately following excision of the cuttings, it may be convenient to collect the cuttings in the fall and store them until they are to be used as propagules. With some species, such as *Robinia* and *Rhus*, fall collection of roots for root cuttings and then cold storage has resulted in greatly reduced success. Care must be taken to store cuttings collected in the fall in a cool place and in a medium that is not too wet; otherwise rotting can occur.

Materials

Students or teams of students will require the following materials to complete this experiment:

- Cuttings from sumac or one or more of the substitute plants listed in Table 25.1. If storing the roots for prolonged periods, dipping the cuttings in captan (see label for rate) is advisable. Always wear protective clothing and follow all safety protocols
- Moist (but not wet) sphagnum peat moss for storing cuttings
- Refrigerator or cold room (5°C to 10°C or 41°F to 50°F) for storing cuttings
- Sharp knife or other utensil used for cutting the stock roots
- Potting medium (vermiculite, heat-treated sand, or commercial soilless medium), flats, labels, and marking pens
- Greenhouse or growth room at 21°C (70°F)

Follow the instructions outlined in Procedure 25.1 to complete this experiment.

Anticipated Results

Within two to three weeks, the ends of the cuttings should exhibit callusing, and shoots will appear as early as three to four weeks. More shoots will emerge from the proximal ends of the cuttings, since the cuttings exhibit polarity. Shoots will continue to emerge from the root cuttings (Figure 25.2) for many months, so a time should be specified for data collection by the group.

Questions/Discussion

- Did horizontal or vertical cutting orientation result in more shoots? Roots?
- Were there differences between horizontal and vertical orientation in shoot or root length?
- What influence did cutting diameter have on propagation? Did it affect shoot or root numbers or size?
- Explain any differences observed and suggest how information derived from this exercise could benefit a commercial propagator.

Procedure 25.1
Adventitious Shoot Formation from Root Cuttings

Step	Instructions and Comments
1	Dig up the parent (stock) plants in sufficient numbers to provide an adequate number of cuttings based on student numbers and how many experiments will be performed.
2	Cut roots from the stock plant. It is preferable to obtain roots of 1/4 to 1/2 in. (6 to 12 mm) in diameter and one to two feet in length (30 to 60 cm).
3	Remove soil adhering to the roots by shaking or washing with a strong spray of water, then allow the roots to dry.
4	If experiments are to be performed immediately, skip to Step 7. If roots are to be stored, follow the instructions in Steps 5 and 6.
5	Treat the root pieces with an appropriate fungicide, such as captan, taking care to follow label directions and to use protective clothing and equipment as required.
6	Place the root pieces in moist (not wet) sphagnum moss, and store in a cool place at about 5°C to 10°C (41°F to 50°F) until needed for the laboratory exercise.
7	When ready to begin experiments, remove root pieces from storage and cut into pieces of equal length, about 3 in. (~7.5 cm). Cut the proximal end perpendicular to the length of the cutting and make a slanting cut on the distal end. This will ensure that the proximal end can be placed up and the distal end down, as indicated in Steps 12 and 13.
8	Sort cuttings by diameter to obtain 40 cuttings of 1/4 in. (6mm) and 40 cuttings of 1/2 in. (12 mm) diameter. The sap of sumac may be irritating to persons with sensitive skin, so wash hands thoroughly after handling the cuttings.
9	Fill four flats with clean, moist propagating medium. A good propagating medium is 1/2 peat to 1/2 perlite, giving a medium with good water-holding capacity, but with excellent aeration.
10	Place 20 root cuttings of 1/2 in. diameter horizontally on the medium in flat #1, about 1 in. apart, and cover with 1 in. of medium. Firm the medium gently to ensure good contact with the root cutting and to eliminate air pockets.
11	Place 20 root cuttings of 1/4 in. diameter in the second flat, and cover as in Step 10.
12	Place 20 root cuttings of 1/2 in. diameter vertically in the medium, about 1 in. apart and with about 1 in. above the soil line. Be sure to push the distal (slanted cut) end into the medium.
13	Place 20 root cuttings of 1/4 in. diameter vertically in the medium in the same fashion as in Step 12.
14	Label each flat with your name, date, and treatment (horizontal 1/2 in., horizontal 1/4 in., vertical 1/2 in., and vertical 1/4 in.).
15	Water flats to thoroughly moisten the medium, and place in a greenhouse or room at about 21°C (70°F). Monitor flats carefully to ensure that the medium does not dry out.
16	Record the date when the first shoots are seen. Three weeks later, remove cuttings from flats, and record number and length of adventitious shoots per cutting and number and length of adventitious roots per cutting. Note where along the cutting that the shoots emerged. Present mean shoot and root lengths numbers in a table.
17	If sufficient root cuttings are available, alternative experiments could include a treatment with the root cutting placed with the proximal end down into the medium and the distal end sticking up out of the medium against the direction of polarity, or treatment of the root cuttings with various growth regulators.

LITERATURE CITED AND SUGGESTED READINGS

Creech, J.L. 1954. Propagating plants by root cuttings. *Proc. Intl. Plant Prop. Soc.* 4: 164–167.

Cross, R.E. 1981. Propagation and production of *Rhus typhina* 'Laciniata,' cutleaf staghorn sumac. *Proc. Intl. Plant Prop. Soc.* 31: 524–527.

Flemer, W. III. 1961. Propagating woody plants by root cuttings. *Proc. Intl. Plant Prop. Soc.* 11: 42–47.

Fordham, A.J. 1969. Production of juvenile shoots from root pieces. *Proc. Intl. Plant Prop. Soc.* 19: 284–287.

Hudson, J.P. 1955. The regeneration of plants from roots. *Proc. 14th Intl. Hort. Congr.* 2: 1165–1172.

MacMillan-Browse, P.D.A. 1980. The propagation of plants from root cuttings. *The Plantsman.* 2: 54–62.

Neilson-Jones, W. 1969. *Plant Chimeras,* 2nd ed., The Camelot Press, London. 123 pp.

Part IX

Layering

26 Layering

Brian Maynard

CHAPTER 26 CONCEPTS

- Simple layering: Covering a bent stem with soil, also known as tip layering.

- Compound layering: Bending a flexible stem in several places and covering with soil.

- Continuous (trench) layering: Laying a shoot in a shallow trench before bud break and allowing new lateral shoots to grow up through the soil.

- Mound layering: Forcing numerous new shoots from a cut-back plant and covering new shoots with successive layers of sawdust, leaf mould, or soil.

- Air layering: Wrapping a girdled shoot with sphagnum moss and plastic.

- Drop layering: Replanting a shrub deep in soil to promote rooting of shoot tips.

- Basic goals in successful layering include the following:

  - Stimulating root formation by restricting the flow of sugar and auxins in the phloem.

  - Excluding light while providing moisture, warmth, and proper aeration in a suitable root growth medium.

  - Allowing time for suitable root development.

Layering is a method of propagation that encourages adventitious roots to form on stems while they remain attached to the parent or "stock" plant. This ancient propagation method takes advantage of the natural propensity of plant stems to initiate roots when they are broken or bend and touch the ground, as is often observed with rhododendron stems and blackberry canes. Layering can range from simple layering, in which a stem is bent over and partially buried in the ground, to air layering, in which the stem is wounded and wrapped with a moist medium for rooting. By whatever method, plant propagation by layering is a relatively slow process, requiring 1 to 18 months, and does not often come to the mind of the plant propagator as a commercially viable technique. However, for difficult-to-root plants, technologically simple propagation operations, or when only a small number of quality plants are needed, layering can offer a high rate of success. Familiarity with the propagation technique of layering is a valuable horticultural skill.

Propagation by layering has a rich history in European nurseries, particularly in the production of hazels and magnolias, which can still be seen in England and Holland, and of fruit tree rootstocks in the states of Oregon, Washington, and the province of British Columbia. In the past, layering was the preferred method of propagating woody plants that needed to be propagated vegetatively (cloned) and on their own root systems, but that were difficult to root from stem cuttings. For certain plant species or cultivars, grafting was either not convenient or not desirable if the incidence of suckering from the rootstock was high. Seed (sexual) propagation (see chapter 36 and chapter 37) of important plant selections typically yields unacceptable variation among the progeny. Many of the shrubs that are now rooted under mist and grown in containers were layered in the past. Most growers now consider layer propagation too labor intensive. Yet for apple and pear rootstocks, in particular, it is common to hear of tens of thousands of well-rooted stems being produced in stool beds thousands of feet long. The mounding and harvesting processes that are considered too laborious for many crops, in this instance, are accomplished with tractors equipped with simple implements.

For noncommercial propagators and the hobbyist, layering offers a low-tech method of cloning a favorite plant without risking the loss of the stock plant. While it may take months longer than other propagation methods, success is almost guaranteed and a larger plant with a better-developed root system can be had for very little cost. Furthermore, there is no risk of delayed graft incompatibility or rootstock suckering.

Because layered stems remain attached to the stock plant throughout the rooting period, they are not as sensitive to variations in humidity, temperature, and moisture as are stem cuttings set to root in a humidity tent or under mist. Through its attachment to the stock plant, the layered stem receives a continuous supply of water, nutrients, and carbohydrates. Water stress, the bane of many propagation methods, is kept to an absolute minimum.

Layering is particularly useful for plants that produce trailing branches capable of developing roots naturally where they touch the ground or that produce many suckers or stool shoots when cut back hard (i.e., stooled or coppiced). Indeed, the most successful ground cover plants, such as *Hedera helix* L. (English ivy), layer naturally where their stems contact the soil. However, even specimen trees, such as *Fagus* (beech) and *Carpinus* (hornbeam), which are notoriously difficult to root from stem cuttings, will root where the branches touch the ground. This can result in a handsome grove. Many of the layering techniques, except for air- and drop-layering, take advantage of flexible stems that can be bent down to the ground. This may be accomplished easily or might require some pruning in advance to stimulate vigorous basal growth, which is also more likely to form adventitious roots. Bear in mind, however, that the number of propagules obtained from each layered stock plant will range from one to only a dozen or so, not the hundreds that are available through cutting propagation or tissue culture. On the other hand, layered plants will often be larger and have better-developed root systems than plants started by other vegetative propagation methods, reducing the time needed to produce a *salable* plant.

While layering is wonderful in its simplicity, it is important to understand how layering works and what the basic requirements are for successful layering. The first stage of layering is the initiation of adventitious roots. This is accomplished by restricting the flow of sugars and hormones in the phloem, not by removing the stem entirely as with stem cuttings, but by constricting or girdling the stem. Full girdling is usually done only in air layering, where a strip of bark (i.e., phloem) is removed just below the point where roots are expected to form. Constricting the stem may be accomplished in various ways, ranging from strongly bending the stem, as in simple layering, to "wrenching" (twisting) the stem to damage the phloem, or placing a metal ring or wire around the stem so that it girdles the stem as it increases in girth over the growing season. In response to restricting the flow of materials from the buds and leaves, sugars and auxin build up distal (on the side of the apical meristem) to the constricted or girdled point and stimulate root initiation. It is common for adventitious roots to emerge first at a stem node where leaf and bud traces exit the stem, though exceptions to this do occur. Take advantage of this by wounding or constricting the stem just below a node.

Another function of layering is to exclude light from the tissues where roots are desired (chapter 15 and chapter 22). As you learn about the various layering methods, consider how each includes some aspect of light exclusion from the portion of the stem where roots are expected to form. If light is excluded by shading a previously formed green stem, it is called "blanching." If a new shoot develops in darkness, the process is known as "etiolation." Generally etiolation promotes better rooting or successful rooting of a wider variety of plants, but is harder to maintain because etiolated shoots are very delicate. Blanching green stems works for many plants, is easier to accomplish, and is more commonly used.

Once roots have initiated, they must elongate (grow out) into a healthy root system. At this stage good horticultural practice comes to bear. Key goals include maintaining light exclusion from the roots, providing adequate (but not excessive) moisture, suitable warmth, and, of course, providing a growing medium that is well drained, yet high in organic matter. If the soil were prepared properly beforehand, all of these goals will be met. Finally, perhaps because the layered stem is attached to the parent plant, layering takes time, anywhere from 1 to 18 months, depending on the plant species, time of year, and size of plant desired at the end of the process.

METHODS OF LAYERING

SIMPLE LAYERING

In simple (common) layering (Figure 26.1), the stem to be layered is bent over into a small hollow in the soil, and the tip of the branch is bent back up again at a point 15 to 30 cm (6 to 12 in.) behind the tip. The sharp bend in the branch, usually just below a stem node, restricts the movement of sugar and auxin, and redirects the stem to a nearly vertical orientation. It is also possible to wound the stem lightly by making a diagonal cut up and forward from the bend, no more than one-third of the way through the stem. A matchstick or other sliver of wood may be used to hold the slice open, and rooting powder may be added if deemed necessary. Alternatively, a simple wire twist or hog ring staple may be placed just above the bend and below a node, to girdle the stem as it expands. The stem usually must be pegged into the soil at the base of the depression, just behind the bend, to prevent the shoot pulling back out of the soil. The bent stem and peg are then covered with 8 to 15 cm (3 to 6 in.) of soil firmed into place. If necessary, the emerging shoot can be supported with a short stake. The layers should be watered well and irrigated as needed during drought. Once roots have

Layering

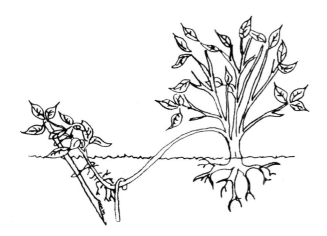

FIGURE 26.1 Simple or common layering.

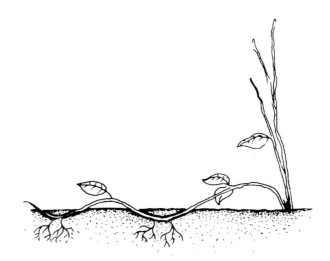

FIGURE 26.2 Compound or serpentine layering.

formed, the layer may be dug up, cut free from the parent stem, and grown on, or simply cut free behind the buried portion and left in place to develop a larger root system. Examples of plants (among many) that may be layered by common or tip layering include the following: *Abelia, Buddleia, Cotoneaster, Deutzia, Ilex, Magnolia, Rhododendron, Rosa, Rubus* (tip layering), and *Viburnum*.

COMPOUND LAYERING

Compound (serpentine) layering (Figure 26.2) is virtually identical to simple layering, except that more flexible stems (such as grape, wisteria, or clematis) are run in and out of the ground several times with stem nodes buried in subsequent depressions in the soil. Buried stems are bent or wounded as described under simple layering. Rooted layers can then be separated by cutting between each point in the system, closest to the next buried stem section. Examples of plants that are well suited to compound layering include the following: *Campsis* (trumpet creeper), *Clematis,* English ivy, *Hydrangea petiolaris* Sieb.& Zucc. (climbing hydrangea), *Vitus* (grape), and *Wisteria*.

TRENCH OR CONTINUOUS LAYERING

Trench or continuous layering (Figure 26.3) is slightly more involved and takes some advanced preparation, but is the method used to start most commercial apple root stock layering operations. One to two years before propagation, a stock plant is established and cut back or pinched repeatedly to develop many vigorous basal stems. In one variation, the stock plant is initially planted at an angle of 30 degrees or so above the horizontal and allowed to establish for a season. The layering process is started just before bud break in the spring by digging a 15 cm (6 in.) deep trench in which the stems and/or entire plants are laid down and covered with 5 to 13 cm (2 to 5 in.) of soil. Shoots grow up through the soil and are mounded once again as they develop. The method is thought to etiolate

FIGURE 26.3 Continuous or trench layering.

the base of the layered shoots as they develop, thereby increasing the rooting potential of hard-to-root plants. Layered stems may be harvested by running a rotating saw blade just above the buried stem; the soil is then swept free from the horizontal stem, and it is ready to be layered again as buds burst the following spring. Trench layering is used commercially in Europe for the propagation of hazels, magnolias, lindens, and many other plants.

MOUND LAYERING

Mound layering of stooled plants (Figure 26.4) takes advantage of the juvenile tissues associated with sucker shoots arising from roots of plants that have been hedged or cut back hard (stooled or coppiced). A plant established especially for stooling is cut back hard to 2.5 to 5 cm (1 to 2 in.) above the soil in early spring. Shortly after new shoots begin to emerge from basal buds, soil, sawdust, wood shavings or leaf mould is mounded to an 8 to 15 cm (3 to 6 in.)

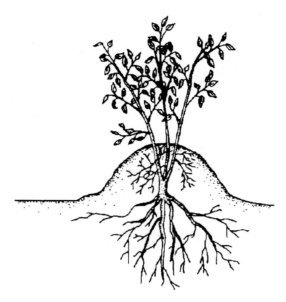

FIGURE 26.4 Mound layering.

height around the growing stems. If the stem is ringed with wire or a hog ring, this stricture would be applied just before the first mounding. Three to four weeks later the shoots are mounded again, to a final depth of 15 to 23 cm (6 to 9 in.). In commercial application, the stool beds may be enclosed in a 30 cm (12 in.) deep wood frame. Some nurseries have experimented with foliar applications of auxin to stimulate rooting, such as 200 ppm indolebutyric acid (IBA) in an aqueous solution sprayed once or twice to drip or run-off, in the hope that the additional auxin taken up by the foliage will move to the rooting zone of the layered stems. Commercial mound layering is used for production of *Pyrus* (pear) and *Malus* (apple) dwarfing rootstocks, and is used in the Europe and the Pacific Northwest for propagation of fruiting and ornamental cultivars of *Corylus avellana* L. Mounding as also successful with *Cotinus* spp., *Syringa vulgaris* L. cultivars, and many *Prunus* species (almond, cherry, and plum).

AIR LAYERING

Air layering, also known as pot layering, Chinese layering, or even by the Chinese name "Gootee" (Figure 26.5), might be the first type of layering that comes to the mind of the average person. This method is usually reserved for plants that either have large or mature stems that cannot easily be bent to the ground or that do not sucker sufficiently to permit stool layering. Because of the intensive nature of this method, only a few branches can be air layered at a time. Air layering is accomplished by ringing or slice wounding the stem, sometimes treating the wound with root promoting auxin powder, wrapping with moist sphagnum peat or milled sphagnum, and sealing with saran or other plastic wrap. Ringing is usually reserved for dicotyledonous plants that possess a ring of vascular cambium. Monocots

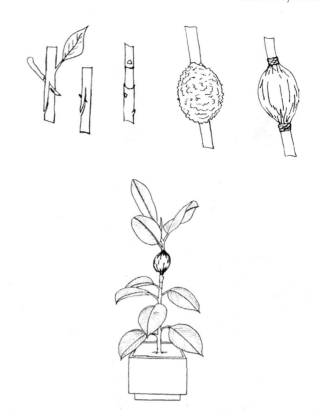

FIGURE 26.5 Air layering.

are best wounded by slicing the stem about 1/3 to 1/2 the way through in an upward slanting motion. Key factors for success include keeping the sphagnum moss moist, but not wet, maintaining a good seal around the stem, and allowing sufficient time for new roots to form. Sometimes an additional wrap of aluminum foil is used to exclude light and reduce solar heating of the rooting area. The term "pot" layering derives from the use of open-bottom containers to contain the rooting medium around the stem. The wound must remove a large enough ring to prevent regrowth of the bark over the girdle, or must be held open with a toothpick or wooden match stick to prevent healing if a slice wound is used. A bark girdle 1.3 to 2.5 cm (½ to 1 in.) wide is usually sufficient. When roots have formed, the wrappings are removed, and the shoot cut free below the root zone and potted up to grow on. Newly potted layers should be placed in a humidity tent or covered by a plastic bag and held in the shade for several weeks before being weaned to ambient light and humidity. Air layering is often cited in the propagation of indoor landscape plants, such as ornamental figs, *Ficus elastica* Roxb. ex Hornem. (rubber plant), *Croton*, *Dieffenbachia,* and large slow-growing landscape plants, such as *Magnolia,* holly, *Camellia,* and azalea.

DROP LAYERING

Drop layering (Figure 26.6) is a relatively uncommon method of propagation in which an entire stock plant is

Layering

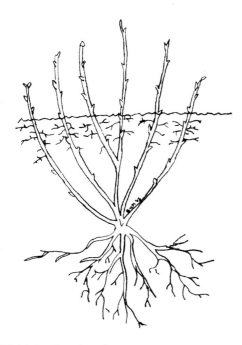

FIGURE 26.6 Drop layering.

planted 15 to 23 cm (6 to 9 in.) deep, so that only the shoot tips are exposed. Because of the depth of the receiving hole that is required, this method is limited to small, well-branched shrubs, such as heath, heather, and woody herbs. The setup of this method lends itself well to girdling the stem with wire or a hog ring, so that the stem is gradually constricted as the stem increases in girth. Drop layering requires excellent soil drainage to avoid low oxygen stress and prevent rotting of the plant. The stock plant is usually discarded after the procedure. Drop layering is reserved for small, fine-textured shrubs, such as *Buxus* (dwarf boxwood), *Calluna* (heather), *Daphne cneorum* L. (garland flower), *Erica* (heath), woody herbs (*Lavandula, Thymus*), and dwarf rhododendrons.

STEPS IN SUCCESSFUL LAYERING

STOCK PLANT HEALTH

As with all other methods of vegetative plant propagation, the health of the stock plant is crucial for success. Plants that are under water or severe nutrient stress will not form adventitious roots as readily. On the other hand, high levels of nitrogen in the stem tissue can also inhibit rooting. Depending on the method of layering, stems should either be actively growing (for stool mounding) or formed the previous growing season (simple, compound, trench, and drop layering). Air layering is done with older wood and, accordingly, takes longer to produce roots.

The act of cutting the stock plant back hard and repeatedly also leads to the development of more juvenile growth, which usually roots more easily. While the various layering techniques may be attempted on virtually any plant (even certain monocot species), the best results will be obtained when well-established stock plants are cultivated for several years, by pinching and cutting back, to produce many vigorous basal shoots (chapter 22). A well-maintained layering bed can yield thousands of propagules each year.

SOIL

Prior to establishing the stock plants to be layered, the soil should be screened to remove stones that might interfere with digging receiving holes and trenches, or with harvesting rooted layers. Drainage should be improved if necessary and organic matter added to improve characteristics of moisture retention, soil tilth, and soil warming. A sandy loam soil is a good starting point, but given time, any soil may be suitably prepared as a layering bed so long as drainage and moisture retention are optimized.

In some instances, layering operations have been compromised by the buildup of soilborne pathogens, such as *Agrobacterium tumefaciens* (crown gall), *Thielaviopsis* (black rot), *Verticillium,* and *Phytophthora* (chapter 8). This highlights the importance of selecting clean stock plants, pasteurizing or fumigating the soil, and occasionally rotating or reestablishing the layering bed.

TIMING

There is general agreement that the best time to start layering stems is late winter to early spring, before growth begins. Air layering of tropical plants may be initiated almost any time. The rooting capacity of many woody plant species begins to decline about six to eight weeks after bud break, as the stems lignify. Indeed, the rooting capacity of stems of *Syringa vulgaris* (common lilac) and *Chionanthus* spp. (fringetree) drops off within two to four weeks of bud break. Taking advantage of this "window of opportunity" might be the key to successful layering of some plant species. Similarly, if blanching or etiolation of the stem is an important part of the layering program, consider that the sooner light is excluded from the stem, the greater the rooting response will be. Blanching fully expanded shoots will have little or no effect on layering success. Light can be excluded from breaking shoots, using a shade enclosure, to yield truly etiolated stems and leaves. Mounding or air layering could then be applied and the remaining nonlayered part of the shoot gradually acclimatized to higher light levels. Be forewarned that etiolated shoots can be quite brittle and burn readily if exposed to full sunlight, so extreme care must be taken if this method is attempted. Recent research has shown that heavy shading (60% to 80% reduction of light) can be nearly as effective as full etiolation, and the resulting shoots are much less delicate.

STEM TREATMENTS

As stated previously, restricting the flow of auxins and sugars away from the shoot tip is fundamental for successful layering. Bending the stem, girdling by cutting or by wrapping wire around the stem, applying hog rings, slicing the stem, or simply twisting (wrenching) the stem are all acceptable means of restricting transport of materials in the phloem while allowing free movement of water from the roots in the xylem. Which method works the best with each layering method is mainly a matter of convention, e.g., removing a strip of bark during air layering, even though a wire girdle might work just as well. Experimentation is always encouraged. With any method of stem constriction, additional auxin rooting compound may be added, either as a powder applied to the girdle or slice wound, as a foliar spray, or even as auxin-soaked toothpicks inserted into the stem or slice wound. One modification of air layering uses 2.5 cm (1 in.) wide bands of Velcro™ hook and loop fabric dipped in rooting powder to exclude light, wound the stem, and apply auxin, all at the same time.

MAINTENANCE OF THE LAYERING PLANT

Part of the attraction of layering relative to other propagation methods is the lack of attention that is needed during the rooting process. Unlike cutting propagation, the layer is supported by the parent plant and is much less likely to experience water or nutrient stress. However, drought, poor stock plant nutrition, or poor drainage can all reduce root formation on layered stems. Light mulching, irrigation, and moderate fertilization of the stock plant might all promote better layering. Regular inspection of layered plants is absolutely necessary to evaluate soil moisture (particularly when air layering), to ensure the rooting area remains covered and secure, and to identify and respond to rodent damage and insect or disease problems.

Monitoring Root Development

Roots should not be expected to form on woody plant layers sooner than 8 to 12 weeks after the layer is initiated. Herbaceous layers can be expected to start forming roots within two to four weeks. Root development appears to accelerate during the late summer and early fall as carbohydrate storage increases. One of the attractions of air layering is the ability to see roots as they reach the outside of the sphagnum wrap, but resist the temptation to unwrap the rooting stem, as the newly formed roots are weakly attached at first and also are susceptible to drying and injury from strong light. Simple layers may be inspected by digging gently in the soil, but take care not to dislodge the stem or fully expose the roots for more than a few minutes, as they can dry out very quickly.

HARVESTING THE LAYER

As in evaluating rooting development, it is best to err on the side of caution when harvesting newly layered stems. Heavy root development, particularly if the stems were ringed or girdled and snap off easily below the root zone, is a sign that the plant is ready to be harvested and potted or replanted in the field or garden. Newly harvested layers should be potted into a moist soil and shaded for several weeks as new roots develop, depending on the time of year. Well-rooted layers should be less susceptible to winter damage than newly rooted cuttings, but stem splitting may still be a risk in some climates. It is quite acceptable to recover a layering stem or sever the connection to the parent plant and leave the rooted layer in place for an additional three to nine months. The rooted layer will be well established by the end of the following growing season, usually much sooner. Normal precautions for transplanting should be taken when moving rooted layers.

PREPARING THE STOCK PLANT FOR THE NEXT YEAR

Ideally, the layering process will not severely damage the stock plant, and new shoots will emerge the following spring. Soil or sawdust used in the mounding process should be removed so that the plant crown can dry out to prevent bark rot or suffocation of the root system. In mound and trench layering, a number of stems are left unringed or just allowed to grow on to produce the next crop of stems for the following year. These should be pruned or supported as necessary to prepare for the next layering event. A light mulch or application of pre-emergent herbicide will help prevent weed growth in the spring, though chemical herbicides should not be used if soil residues might interfere with root development in simple, compound, or trench layers.

EXERCISES

Layering is one of the simplest and lowest technology methods of vegetative propagation, but requires the most time for rooting to occur, for plants to be ready to harvest, and yields fewer propagules than most other methods. If time preparing soil and stock plants is included, the entire process can take up to three years. Of course, layering may be the only method that will reliably produce rooted shoots of particularly difficult-to-root or -graft plants. The following exercises use easily rooted plants to demonstrate the method of layering. They also allow the student to observe the effects of girdling and other means of wounding, auxin application, and where roots form on layered plants. In consort with the Concepts portion of this chapter, the student then should be able to devise a protocol for layering more difficult plants.

EXPERIMENT 1: DROP OR MOUND LAYERING WITH RINGING

Both the mound (stooling) and drop methods of layering involve covering growing stems with soil or growing medium (Figure 26.4, Figure 26.6). Mounding is applied to stock plants cut back to ground level to produce a number of vigorous sucker-like shoots, sometimes called stool shoots. These are covered by 15 to 23 cm (6 to 9 in.) of soil as they grow in early summer and remain mounded until the fall or the following spring. In the commercial mound layer propagation of *Corylus avellana* 'Contorta,' Harry Lauder's Walking Stick, hog ring staples are placed around the base of the young shoot to girdle the stem gradually as it expands during the growing season. In this exercise, a simple system will be constructed for observing the rooting process in mounded or drop-layered plants and the effects of ringing or wounding stems. The steps below are just a guideline—you should be able to identify new methods of wounding or otherwise stimulating root formation. (Can you design a method for etiolating the stock plant before setting up the layer?)

Materials

The following items are needed for each student or student team:

- A healthy, rapidly growing potted plant that has been pinched hard or cut back to within 2.5 to 3 cm (1 to 2 in.) of the soil surface and has produced 5 to 10 vigorous, upright-growing shoots (*Dendranthema* x *grandiflorum* Kitam (florists' chrysanthemum), *Coleus* spp., *Mentha* spp., *Lavandula* spp., or similar plant material)
- Similar-sized pot, can, or sleeve of stiff plastic to hold soil over the potted plant
- Suitable soilless growing medium that has excellent drainage yet holds some moisture, and which can easily be shaken off rooted stems (e.g., vermiculite, perlite, growing mix)
- 16- to 20-gauge florists' wire suitable for tying around a stem
- Grafting or similar knife
- Low-strength powdered rooting compound (e.g., Hormodin 1™, Hormex 1™, Rootone™) containing 1000 ppm IBA or NAA or a combination of both auxins
- Plastic or wooden labels and pencil for marking

To complete the following exercise, follow the protocols listed in Procedure 26.1.

Anticipated Results

After four to six weeks, roots should be visibly emerging above wounded areas and possibly at stem nodes. Students should be able to note the location and degree of root production and relate their observations back to the treatments applied. If other groups used different rooting media or stem treatments, more can be learned about the influence of these factors on root initiation and development.

Questions

- Was girdling necessary for root initiation in this method of layering? Was the rooting powder necessary or helpful?
- Which other plant species that you are familiar with might be able to be propagated in this manner? Would other propagation methods be more efficient for the species you used in this exercise?

EXPERIMENT 2: AIR LAYERING

Air layering is an excellent method for propagating large tropical plants that have lost their lower leaves or become leggy. While air layers are easy to set up, many fail because not enough attention is given to maintaining the proper moisture level in the sphagnum moss used for the layer. If the plastic wrap is tied well enough and the initial moisture was adequate, the layer should be in good shape; if the wrap is too loose, water may collect in the layer by running down the stem and drown any developing roots, or the moss may dry out. Rooting normally takes 8 to 16 weeks or longer—so be patient. The layer has succeeded when new white roots can be seen on the inside of the plastic wrap. Remember, when harvesting the rooted layer that there might not be enough roots formed to support the plant at first, so pot it into good growing mix and cover with a plastic bag tied loosely at the top and gradually wean the plant to ambient light and humidity.

Materials

The following items are needed for each student or student team:

- A healthy, rapidly growing green potted (foliage) plant (e.g., *Ficus, Cordyline, Dracaena,* etc.)
- Grafting or other similar knife
- Low-strength powdered rooting compound (e.g., Hormodin 1, Hormex 1, Rootone) containing 1000 ppm IBA or NAA or a combination of both auxins

	Procedure 26.1
	Drop or Mound Layering with Ringing
Step	Instructions and Comments
1	Carefully strip the leaves off of the lower ¾ of the stock plant.
2	Choose stems for control (no-treatment), ringing with wire, slice wounding, girdling with a knife or by twisting a single strand of wire about the stem. Draw a diagram showing the treatments or controls in your notebook—making sure to mark the pot to orient the plant to the top of your diagram.
3	Apply treatments to a point midway between the base of the shoot and the foliage. Either use the knife to make a slice wound upward into the stem about 1/3 of the way through, or wrap a strand of wire once around the stem and twist the ends together to cinch the wire firmly against the stem. Hold slice wounds open by inserting a small toothpick into the slice. Apply a small dusting of rooting powder to half of the slice wounds as an additional treatment; leave the rest untreated as a control.
4	Remove the bottom from the empty pot, and cut the pot or can down one side to allow it to be pulled open and slipped around the stock plant. Reduce the height of the container if necessary so that the foliage of the stock plant will be just above the rim. Push the container lightly into the soil surface of the potted plant. Wrap the container with tape or a heavy elastic band to keep it from opening up when soil is added. If necessary, stake the container to prevent it from tipping over.
5	Add growing medium to the container, working the medium down around the shoots with your fingers or a thin stick. Add enough medium to fill the container to a point just below the foliage. Water from the top until water drains from the bottom of the potted plant.
6	Monitor for insect or disease problems. Maintain adequate, but not excessive moisture in the upper container. If necessary, enclose the entire system in a large plastic bag tied loosely at the top.
7	Check for rooting every two weeks and harvest when ready by undoing the container and shaking the growing medium off the stock plant. Make observations of root location and number or extent of rooting for all control and treated stems, referring to your diagram as needed. Remove and pot up rooted stems as desired.

- Moistened, whole sphagnum peat, milled sphagnum moss, coarse coir fiber, or shredded newsprint
- Plastic wrap or similar 1 mil (0.0254 mm) thick clear plastic
- Waterproof tape, waxed string, or heavy elastic bands for tying the wrap
- Aluminum foil
- Plastic or wooden labels and pencil for marking

To complete the following exercise, follow the protocols listed in Procedure 26.2.

Anticipated Results

Roots will form at the upper edge of the girdled or sliced section of stem. Once roots are visible at the surface of the layer, the plant may be harvested and potted up in a suitable growing medium. It may take eight weeks or longer before roots become visible through the plastic and the air layer is ready to harvest.

Questions

- Did the type of wound or auxin treatment affect the type or numbers of roots formed? Did an additional wrap of aluminum foil affect rooting (if yes, why)?
- Was the air-layered stem ready to be harvested as soon as roots were visible or should you have waited longer for more roots to form?

	Procedure 26.2
	Air Layering
Step	Instructions and Comments
1	Remove leaves over a 15 to 30 cm (6 to 12 in.) section of stem to leave about 23 to 46 cm (9 to 18 in.) of leafy stem above the air layer.
2	Wound monocot or dicot stems by slicing 1/3 way through in an upward-directed cut. Hold the slice wound slightly open with a toothpick or a small amount of moss. Wound dicots, but not monocots, by removing a 1.3 to 2.5 cm (1/2 to 1 in.) wide strip of bark down to the outermost wood.
3	Dust a small amount of rooting powder onto the wounded section of stem. Alternatively, use toothpicks presoaked in a 1000 mg/L auxin solution to hold the slice wound open.
4	Squeeze a fist-sized ball of milled sphagnum moss to drain excess water and place around the wounded section of stem. Hold this in place while wrapping with plastic.
5	Wrap a sufficiently large piece of plastic wrap around the sphagnum moss, join the edges of the wrap, and roll together to tighten the wrap against the ball of moistened sphagnum. Tape or tie the lower and then upper edge of the plastic wrap firmly against the stem. If desired, place an additional wrap of aluminum foil around the finished plastic wrap. Twist in place, but do not tape or tie the foil.
6	Label the layered plant and record treatment data in your notebook.
7	Check the layer every three weeks by removing the foil wrap and observing for root development. Record data on time to first root visible and describe the roots that are seen.
8	Harvest the layer when ready by removing the wraps and tape. With the moss and roots intact, cut below the layered stem, and transplant into suitable growing mix. Cover the potted layer with a plastic bag loosely tied around the base of the plant. Gradually wean the rooted layer to ambient greenhouse conditions.

SUGGESTED REFERENCES

Browse, P.M. 1988. *Plant Propagation*. Simon & Schuster, New York. 96 pp.

Hartmann, H.T., D.E. Kester, F.T. Davies, Jr., and R.L. Geneve. 2002. *Plant Propagation: Principles and Practices*. 7th ed. Prentice Hall, Inc., Upper Saddle River, NJ. 880 pp.

Macdonald, B. 1986. *Practical Woody Plant Propagation for Nursery Growers*, vol. 1. Timber Press, Portland, OR. 669 pp.

Mahlstede, J.P. and E.S. Haber. 1957. *Plant Propagation*. John Wiley & Sons, New York. 413 pp.

Maynard, B.K. and N.L. Bassuk. 1988. Etiolation and banding effects on adventitious root formation, in *Adventitious Root Formation in Cuttings, Advances in Plant Science Series*, vol. 2. Davis, T.D., B.E. Haissig, and N. Sankhla, (Eds.), Dioscorides Press, Portland, OR. pp. 29–46.

Maynard, B.K. and N.L. Bassuk. 1992. Stockplant etiolation and stem banding effects on the cutting propagation of greening stems of *Carpinus betulus* L. *J. Amer. Soc. Hort. Sci.* 117: 740–744.

Thompson, P. 1992. *Creative Propagation: A Grower's Guide*, 3rd ed. Timber Press. Portland, OR. 220 pp.

Toogood, A.R. 1981. *Propagation*. Stein and Day, Briarcliff Manor, New York. 320 pp.

Toogood, A.R. 1999. *American Horticulture Society Plant Propagation*. DK Publishing, New York. 320 pp.

Wells, J.S. 1955. *Plant Propagation Practices*. MacMillan, New York. 344 pp.

Part X

Grafting and Budding

27 Grafting: Theory and Practice

Kenneth W. Mudge

CHAPTER 27 CONCEPTS

- Grafting refers to the natural or deliberate fusion of plant parts so that vascular continuity is established between them and the resulting composite unit functions as a single plant.

- Grafting involves uniting two (or more) distinct genotypes into a compound genetic system in which stock and scion are fused anatomically, but otherwise maintain their genetic identity throughout the life of the plant.

- Grafting has its broadest application with woody plants (tree fruit crops and ornamentals), but is also important in forestry and some herbaceous ornamentals and vegetables.

- Grafting is one of several methods that are commonly used today for asexual (clonal) propagation.

- When grafting is used for clonal multiplication, clonal scions are usually grafted onto seedling understock, but for some crops (notably fruit trees, e.g., apples) genetically improved rootstocks are used. The improved clonal rootstocks are selected for disease resistance, effects on the scion (e.g., dwarfing), and performance in different soil environments.

- If stock and scion are sufficiently closely related to form a functional and durable graft union, they are termed "compatible"; if not, they are termed "incompatible." In general, the more closely related the scion and rootstock are, the more likely the chances of creating a successful graft.

- For a graft to be successful, the vascular cambia of the scion and rootstock must be brought into intimate contact with each other.

PROPAGATION BY GRAFTING AND BUDDING

CONCEPTS AND DEFINITIONS

Innovations in propagation have profoundly influenced plant domestication since the beginning of agriculture. Ten to twelve thousand years ago, unintentional selection of self-pollinating species that reproduced true-to-type from seed gave rise to wheat, barley, and other staples. Four or five thousand years later the discovery of asexual propagation paved the way for the domestication of heterozygous species, such as grape, olive, and fig, which rooted easily when a broken branch was stuck in moist soil. The domestication of more difficult-to-clone pome and stone fruits awaited the relatively recent discovery of grafting. Despite considerable advances in the science and technology of plant propagation since then, particularly over the past 50 years, grafting still plays an important role in production of horticultural crops including fruits, ornamentals, and even some vegetables.

Grafting refers to the natural or deliberate fusion of plant parts so that vascular continuity is established between them and the resulting composite unit functions as a single plant. Although two adjacent intact plants can become naturally or intentionally grafted together (approach grafting), or different branches of the same plant (autografting), most "horticultural" grafting involves deliberately cutting a shoot from a donor plant and inserting it into an opening in another plant growing on its own root system (detached scion grafting). The shoot piece cut from a donor plant (Figure 27.1a) that will grow into the upper portion of the grafted plant is referred to as the scion, and the plant that receives and fuses with the scion and functions as the root system of the grafted plant is called the stock or understock. The term rootstock is similar, but not entirely synonymous with stock because rootstock implies that only the roots system is derived from the stock (Figure 27.1b), whereas the terms stock or understock are more appropriate when the lower portion of the grafted plant includes not only the root system, but also some portion of the shoot system on which

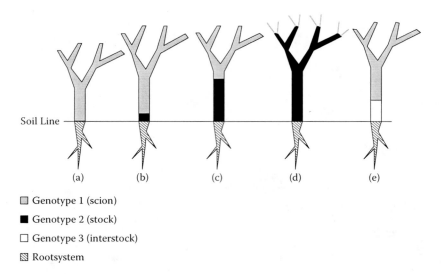

FIGURE 27.1 Grafted plant components and configurations. See text for explanation. (Graph courtesy of the author, © 2000.)

the scion is grafted (Figures 27.1c,d). In the simplest case, grafting consists of one stock and one scion genotype (the genetic identity or DNA sequence of an organism), but in some applications a third genotype (or more) is involved. Double working refers to a genetically compound plant (usually a tree) consisting of three genetically distinct parts, separated in linear sequence by two graft unions, as illustrated in Figure 27.1e.

Grafting involves uniting two (or more) distinct genotypes into a compound genetic system in which stock and scion are fused anatomically, but otherwise maintain their genetic identity throughout the life of the plant. In sexual reproduction (chapter 36), the genes from two parent plants recombine in a single cell (zygote) that grows into a plant made up of genetically uniform hybrid cells throughout. A grafted plant is not a hybrid in this sense, but rather more aptly is described as a constructed chimera. A natural chimera is a single plant composed of some configuration of genetically different tissues, either arranged in separate layers or sectors (chapter 16). Even though each part of the new plant resulting from the fusion and subsequent growth of stock and scion maintains its original genetic identity and much of its original phenotype, the stock and scion (and interstock in the case of double working) do have an influence on the growth and development of each other. For example, a dwarfing (size-controlling) rootstock results in a decrease in vegetative vigor and ultimate size of the scion growing on it, while at the same time, a more vigorous scion accelerates the growth of a less vigorous rootstock, causing it to grow larger than if it were not grafted. The term stion (stock + scion) was introduced in 1926, by H. J. Webber, to emphasize how this reciprocal influence results in a functionally unique composite organism that is fundamentally different from either of the original donors or any other combination of stock and scion genotypes.

More recently the term stion has been replaced by the wordier, but perhaps more descriptive term "compound genetic system."

Grafting qualifies as an asexual or clonal propagation method, in so far as the scion derived portion of the grafted plant is a genetic replica of the scion donor plant. Keep in mind, however, that the rootstock retains its unique genetic identity, different from the scion, and, as such, the rootstock, and the stion as a whole, is in no way a clone of the scion donor plant. Understocks themselves are either clonal if propagated asexually (by some method other than grafting) from a stock donor plant, or seedlings if propagated from normal zygotic (not apomictic) seed. This is an important distinction in terms of rootstock performance, as will be seen below, and from the standpoint of nursery production.

The term grafting or graftage can be used either in the broad sense to refer to the process resulting in fusion of any configuration of stock and scion that results in a genetically compound plant, or more narrowly, to refer to the use of a scion shoot piece that consists of multiple buds, while the term budding or bud grafting refers to the use of a scion that consists of only a single bud. Ultimately this is a minor distinction, since there are no fundamental differences in the postgrafting (or budding) growth or performance of a grafted or budded plant.

Grafting has its broadest application with woody plants, especially tree fruit crops and ornamental trees and shrubs, but it also plays an important role in modern clonal forestry and even, to a limited but increasing extent, with herbaceous plants including vegetables and a few ornamentals (e.g., cactus). In Japan and Korea, for example, much of the greenhouse production of cucurbit vegetables (melons, cucumbers) and tomatoes relies on grafting onto selected seedling rootstocks that confer resistance to root system–associated diseases,

Grafting: Theory and Practice

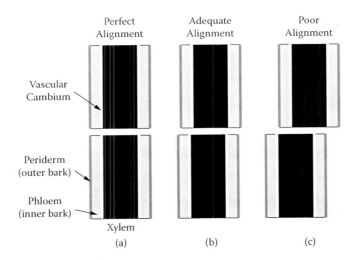

FIGURE 27.2 Alignment of vascular cambia of stock and scion. (Graph courtesy of the author, © 2001.)

such as *Fusarium* and nematode pests, thereby increasing the production period by as much as several months (Lee, 2003).

GRAFT UNION FORMATION

Although the anatomy and physiology of graft union formation varies among different species, methods of grafting, and seasons (timing), the process of graft union formation shares certain common features across all these variations. Within hours or days of the placement of a properly cut and aligned stock and scion, cell division occurs near the cut surface, in the vicinity of the vascular cambium, of the stock and scion (Figure 27.2), eventually filling any reasonable gap that might exist between the two with intermingled and, to some extent, interlocking callus cells. Differentiation of new meristematic cells proceeds from the cut edges of the existing cambia of stock and scion inwards, until a new vascular cambium cylinder forms a continuous bridge across the callus. Through the same process that produces new xylem and phloem from any established vascular cambium, cell division of this new bridging cambium generates cells to the inside that differentiate into new xylem and cells to the outside that differentiate into new phloem (inner bark). Along the outer perimeter of these bridging tissues, periderm (bark) formation occurs, and eventually the junction between stock and scion is (more or less) indistinguishable from normal (nongrafted) tissues. All these growth and developmental processes that differ in timing and spatial sequence among species and season are influenced by temperature, cellular water relations, phytohormone gradients, and poorly understood factors related to genetic compatibility. The resemblance of this process to wound healing in nongrafted plants is more than coincidental, as has been pointed out by several authors. It could be argued that the only fundamental difference between graft union formation and wound healing is the absence of periderm (bark) formation between stock and scion.

NATURAL AND HUMAN HISTORY OF GRAFTING

The question of who invented grafting is difficult to answer. It has been claimed by most modern authors that grafting was invented 3 or 4000 years ago in China, but definitive historical evidence is hard to find. The earliest written account of grafting in China may be no older that the book *Qimin Yaoshu* (*Essential Techniques for the Farming Populace*) written by Jia Sixie sometime between CE 500 and CE 600. Its only (weak) claim to documentation of anything approaching the origins of grafting is that it is written about agricultural practices from much earlier, though unspecified, times. He describes pear grafting and notes the superior characteristics of one rootstock species over another. Recently, it has been claimed by Harris et al. (2002) that grafting originated not in China, but rather in Mesopotamia (present-day Iraq) about 3800 years ago with grape vines, but the source cited is an unpublished personal communication (S. Dalley), and so awaits confirmation.

Whomever and wherever early farmers first began deliberate grafting, they did not so much invent the practice as discover it. No doubt they were acute observers of the natural world around them and must have noticed natural grafting, which is a fairly common occurrence between either the branches or the roots of many species of trees. Natural shoot grafting results in the fusion of two branches within the canopy of a single tree and even occasionally between the branches of adjacent trees. Natural root grafting, which is even more common than natural shoot grafting, occurs extensively within the root systems of many species including white pine (*Pinus strobus*), birch (*Betula alleghaniensis* Britt.), and white

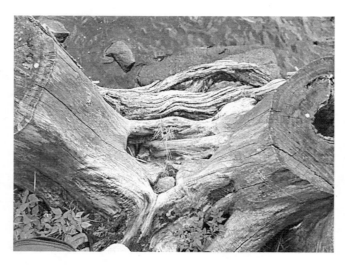

FIGURE 27.3 Looking down on two tree stumps of Atlantic white cedar, the roots of which are naturally grafted together. (Photo courtesy of the author, © 2001.)

cedar (*Chamaecyparis thyoides*). In some cases, natural root grafting results in the interconnection of adjacent trees (Figure 27.3), with potentially important ecological and agricultural consequences. In natural forest or managed landscapes root graft-interconnected trees may share water, nutrients, and even diseases. For example, Dutch elm disease can spread from tree to tree among American elms (*Ulmus americana*), and viral transmission across root grafts has been documented in citrus orchards. Having noticed natural shoot or root grafting, early agriculturists probably first mimicked it by some version of approach grafting in which the shoots of both trees remained attached to their own root systems while the graft union was forming. Eventually the more challenging practice of detached scion grafting enabled the cloning of pears, plums, and probably citrus by the Chinese.

In early Classical times (approx. 500 BCE), knowledge of grafting probably migrated along with the earliest domesticated apple trees, from Central Asia to Western civilization to Greece. In 323 BCE, the Greek philosopher Theophrastus (a student of Aristotle), who is sometimes called the "father" of horticulture, described cleft grafting accurately, but offered this odd description of the process of graft union formation, "The twigs [scion] use the stock as a cutting uses the earth … a kind of planting and not a mere juxtaposition." This would seem to imply that the scion roots (as a cutting) into the stock, which is certainly not the case. In about 29 BCE, the Roman poet Virgil wrote extensively about grafting in his epic poem, the *Georgics*, which functioned as a sort of extension bulletin for gentlemen farmers. The following description clearly shows his understanding of the essential elements of cleft grafting, and is indeed poetic in its expression, "… in knotless trunks is hewn a breach, and deep into the solid grain a path with wedges cloven; then fruitful slips are set herein, and—no long time—behold! To heaven upshot with teeming boughs, the tree strange leaves admires and fruitage not its own." On the matter of graft compatibility, however, Virgil was very much mistaken, suggesting that his information was secondhand, and he had little direct experience, "But the rough arbutus with walnut-fruit is grafted; so have barren planes [*Platanus* sp.] ere now stout apples borne, with chestnut-flower the beech, the mountain-ash with pear-bloom whitened o'er, and swine crunched acorns 'neath the boughs of elms."

REASONS FOR GRAFTING

There are almost as many reasons for grafting as there are ways to graft (see below), but most of them fall into either or both of two major categories. Grafting has been used for thousands of years to propagate plants asexually that cannot readily be cloned by other methods. As advances have been made in other vegetative propagation practices, grafting merely for the sake of propagation per se has declined in importance, while the use of grafting to take advantages of the unique combinations of characteristics associated with compound genetic systems (stions) has increased. These and several other reasons for grafting are discussed below.

GRAFTING FOR CLONAL MULTIPLICATION

Grafting is used as an alternative to cuttage or layerage for propagation of difficult-to-root species. Along with cuttage, layerage, division, and micropropagation, grafting is one of several methods that are commonly used today for asexual propagation. For some species, grafting is the only, or at least the most practical, method of cloning despite some dramatic improvements in cutting propagation technology (e.g., misting, synthetic rooting compounds, and polyethylene) and the introduction of micropropagation

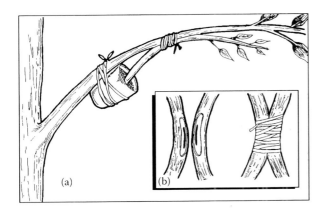

FIGURE 27.4 Approach grafting used for propagation of difficult-to-root species. (a) Potted understock plant tied to scion donor tree. (b) Splice graft method and wrapping used to attach scion and stock. (Drawings by J. Aluko and A. Gachucha.)

over the past several decades. Bud grafting onto seedling understocks is the most cost-effective way to propagate many clonal shade tree cultivars, such as Norway maple (*Acer platanoides* L.), red maple (*Acer rubrum*), sugar maple (*Acer saccharinum* L.), honey locust (*Gleditsia triacanthos* L.), ash (*Fraxinus* spp.), Ohio buckeye (*Aesculus glabra* Willd.), horse chestnut (*Aesculus hippocastanum*), and the few oak (*Quercus* spp.) cultivars in the nursery trade. Clonal selections of unusual growth forms of otherwise common woody ornamentals are often propagated by grafting to seedling understocks. Selections in this category include cultivars of spruce, fir, pine, and other conifers that have been selected for dwarfness, or other unusual growth habits, such as weeping, or variegation. Grafting is also used to propagate unusual growth forms of deciduous species including filberts (hazelnuts) (*Corylus* spp.), flowering cherries (*Prunus subhirtella*), Japanese maple (*Acer palmatum*), crabapple (*Malus* spp.), and others selected for weeping, "corkscrew" habit, fasciations, variegation, dissected leaf forms, etc.

Although grafting for propagation of otherwise difficult-to-clone species usually involves detached scion grafting or budding, approach grafting is occasionally used for especially difficult-to-graft species because graft union formation occurs while the scion is still attached to its own root system (Figure 27.4).

When grafting is used primarily for clonal multiplication, clonal scions are usually grafted onto seedling understocks rather than genetically improved clonal rootstocks. Investment in elaborate clonal rootstock breeding programs is not justified when clonal propagation is the only goal of grafting because the genotype of the understock is relatively unimportant and essentially any seedling will do as long as it is graft compatible with the scion.

In most cases discussed so far, and indeed for most grafted plants, the rootstock becomes a permanent part of the grafted plant for the remainder of its lifespan regard-

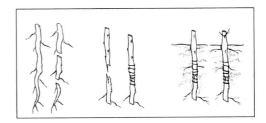

FIGURE 27.5 Nurse root grafting. Roots are cut into 4- to 8-cm sections (left), joined by whip and tongue graft (center), and planted deep, below the graft union, to encourage scion rooting (right). (Drawing by Vanessa Gray, 1999.)

less of the reason for grafting. In the special case of nurse grafting however, the scion's dependence on the stock is only temporary—long enough for the scion to become self-rooted. Hence, nurse grafting is used to facilitate asexual propagation of difficult-to-root species, such as lilac, clematis, and avocado. A difficult-to-root scion is grafted to a seedling understock (in many cases just a piece of root as illustrated in Figure 27.5) and planted unusually deep, so that the graft union is below ground level. Here the buried base of the scion becomes self-rooted in response to the dark, moist environment. This is really a layering process (chapter 26) that is facilitated by grafting and, hence, should be called nurse graft layering, but it is usually referred to simply as nurse grafting or nurse root grafting (when the stock is a root piece). Nurse graft layering is a rare case where graft incompatibility (described below) is an advantage rather than a disadvantage because the later stage of the process involves failure of the graft union and death of the stock while the scion lives on on its own root system. In the case of nurse root grafting of lilac (*Syringa vulgaris* L.), for example, delayed incompatibility is encouraged by using a stock of a different genus (California privet, *Ligustrum ovalifolium*) within the same family (Oleaceae). In this case, the stock and scion are related closely enough (same family or intrafamilial graft) to form a temporary graft union, but distantly enough related (different genera or intergeneric graft) that the graft is not likely to succeed in the long run because of genetic incompatibility. In fact, a problem associated with this propagation strategy is unintended survival of the stock, which, in the case of privet, may sucker; giving rise to vigorous privet shoots gradually replacing the original lilac. Because of this and related problems, lilacs are no longer commonly nurse grafted, but rather cloned by micropropagation.

GRAFTING USED TO AVOID PROBLEMS ASSOCIATED WITH CUTTING-DERIVED ROOT SYSTEMS

This reason for grafting has to do with the observation that some species that can be rooted relatively easily from cuttings develop adventitious root systems that are functionally inferior to seedling root systems. The president of Princeton Nursery, William Flemer, pointed out several examples of this including *Cornus florida* 'Rubra,' *Acer palmatum* 'Bloodgood,' and *Pyrus calleryana* 'Bradford' in his 1989 presentation to the International Plant Propagator's Society titled "Why we must still graft and bud."

GRAFTING FOR INDEPENDENT OPTIMIZATION OF STOCK AND SCION GENOTYPE

By combining the useful characteristics of scion and of stock genotypes, a compound genetic system (stion) can be created that may be more agriculturally useful than either genotype growing on its own root system (nongrafted). Compound genetic systems can be used to achieve multiple objectives that would be difficult if not impossible to find or breed into a single (nongrafted) genotype. When grafting is performed to take advantage of a compound genetic system, the specific genotypes of both the stock and the scion (and the interstock in the case of double working) are important to achieving the desired ends, in contrast to grafting for clonal multiplication in which the genotype of the stock is relatively unimportant and seedling rootstocks generally suffice. Hence, when grafting is used for independent optimization of stock and scion, the rootstock genotype is critically important. This can be accomplished in some cases by choosing rootstocks at the species level, i.e., rootstocks of a species different from the scion that confers one or more useful characteristics not present in the scion. Although of different species, obviously the stock and scion must be related closely enough (same genus or at least same family) that the two are compatible. Examples of interspecific grafting for independent optimization of stock and scion genotypes include the use of Northern California black walnut (*Juglans hindsii*) as a rootstock for English walnut (*J. regia*), peach (*Prunus persica*) rootstock for almond (*P. amygdalus*), and AH cucumber (*Cucumis metuliferus*) as a *Fusarium*-resistant rootstock for cucumber (*C. sativus*) (Lee, 2003).

In contrast to interspecific compound genetic systems, even greater optimization can be achieved through selection at the intraspecific level, i.e., breeding and/or selection for desirable rootstock clones within the same species. The most technologically advanced use of genetically optimized rootstocks has come about through long-term systematic clonal rootstock breeding programs for several of the world's most important temperate fruit crops including apples (*Malus domestica*), pears (*Pyrus communis*), several stone fruits (*Prunus* spp.), and grapes (*Vitus* spp.).

In fact, rootstock breeding would be a relatively simple task if genetic improvement of plant cultivars could be accomplished by selecting for a single desirable trait, such as rootstock control of scion vigor (size control) or rootstock disease resistance. More realistically, however, a successful rootstock (or scion) genotype is chosen based on a complement of several genetically controlled characteristics that give rise to a desirable set of phenotypic characteristics. Unlike scion cultivars that are selected for characteristics that affect fruit yield and quality and vegetative growth, a desirable rootstock genotype is selected for the following: (1) adaptation to below-ground environmental constraints, (2) effect on scion vigor and other scion phenotypic characteristics, and (3) resistance to root system diseases and pests, as well as to shoot (aboveground) diseases and pests during the pregrafting nursery production phase.

BIOTIC AND ABIOTIC STRESS RESISTANCE/TOLERANCE

In many crops, rootstocks have been chosen because of their natural resistance to or tolerance of various biotic stresses including bacterial, fungal, and viral diseases, insect and noninsect pests including nematodes and even small herbivorous mammals (voles, etc.), as well as tolerance of adverse environmental (stress) conditions. Examples for each of these categories are shown in Table 27.1. Grapes are a dramatic and historic example of the importance of pest-resistant rootstocks. Wine grape production (primarily *Vitus vinifera*) in Europe during the nineteenth century was nearly eliminated because of an aphid-like, root-feeding insect pest known as phylloxera (*Daktulosphaira vitifoliae*). The existence of the wine grape industry today in phylloxea-infested regions (most of the world where grapes are grown) is heavily dependent on the use of phylloxera-resistant rootstocks that have been bred from naturally phylloxera-resistant North American grape species (e.g., *Vitus labrusca*). Similarly, the apple industry in New Zealand and Australia and other parts of the world was based on the original East Malling (EM or M) series clonal rootstocks during the early part of the twentieth century, but it was seriously compromised because of infestations of wooly aphids (*Eriosoma lanigerum*), which feed on the shoots of apple rootstock plants during the pregrafting phase of nursery production. East Malling rootstocks, chosen primarily for their effect on size (vigor) of the scion cultivar, were highly susceptible to wooly aphid damage. The Malling Merton breeding program was undertaken largely to incorporate wooly aphid resistance by crossing susceptible EM cultivars with the highly resistant Northern Spy apple. The result was the Malling Merton (MM) series of size-controlling rootstock that are resistant to wooly aphids.

Rootstocks are also chosen for tolerance of adverse abiotic environmental stresses including low temperature and soils that are drought prone or excessively wet or poorly drained. Some examples can be seen in Table 27.1.

TABLE 27.1
Examples of Biotic Stress Resistance and Abiotic Stress Tolerance in Rootstocks

Stress Category	Stressor[a]	Crop	Susceptible Rootstocks	Resistant/Tolerant Rootstocks
Biotic (disease or pest)				
Microbial				
Bacterial	Fireblight (*Erwinia amylovora*)	Apple	Budagovsky (Bud.) 9, Ottawa (O) 3, East Malling (EM).9	Geneva (G)11, G.65 EM.7, Robusta.5
			Malling Merton (MM) 111	OHxF (Old Home x Farmingdale) series
		Pear	Quince	
	Bacterial canker (*Pseudomonas* spp.)	Cherry	Mazzard	Colt, Maheleb
	Crown gall (*Agrobacterium tumefaciens*)	Cherry	Colt	Mahaleb
		Walnut	Paradox	Northern California black walnut (*Jaglans hindsii*)
Fungal and Fungal-like organisms	Crown/root rot (*Phytophthora* sp.)	Apple	EM.26, MM.106	G.65
		Walnut	N. Cal. black walnut	Paradox
	Oak root fungus (*Armillaria* sp.)	Plum	Myrobalan 29-C	Marianna 26-24
Viral	Tomato ring spot virus	Apple	MM.106	
	Tristeza virus	Citrus	Sour orange (*Citrus aurantium*)	Rough lemon (*citrus jambhiri*), Citrange (*C. sinensis x Poncirus trifoliata*)
Pest				
Insects	Phylloxera (*Daktulosphaira vitifoliae*)	Grape	'Harmony' and 'Freedom' (*Vitus vinifera* hybrids)	*Vitus riparia x rupestris* hybrids
	Wood aphid (*Eriosoma lanigerum*)	Apple	EM.9, EM.26	MM.111
Nematodes	Root knot nematode (*Meloidogyne sp.*)	Peach	Lovell peach	Nemagard peach
		Walnut	N. Cal. black walnut	Pradox
		Coffee	*Coffea Arabica*	*C. canephora*
Mammal	Vole (*Microtus*)	Apple	Bud.9	Novole
Abiotic				
Low temperature		Apple	MM.111, EM.7	EM.26, Ottawa 3
		Peach	Nemagard	Lovell, Guardian
Wet soils		Peach	Nemagard, Viking	Citation
		Cherry	Mahaleb	Colt
		Apple	EM.26	EM7, MM.111
Dry soils		Apple	Mark	MM 104

[a] Name of disease (scientific name of disease organism), or common name of pest (scientific name of pest).

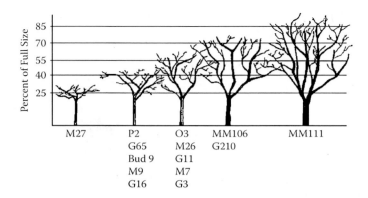

FIGURE 27.6 Effect of common apple rootstock on relative size of a fruiting scion. Abbreviations for rootstock series (breeding programs): M, East Malling; P, Polish; G, Geneva; Bud, Budagovsky; O, Ottawa; MM, Malling Merton. (Drawing by Vanessa Gray, 1999.)

ROOTSTOCK EFFECTS ON SCION VIGOR AND VICE VERSA

Rootstocks are selected not only for characteristics that directly affect the root system and/or the shoot system of the pregrafted rootstock plant in the nursery, but also for rootstock effects on scion growth and development. Rootstock genotype can affect scion phenotype in different ways, but most importantly scion vigor. Although less important, rootstock genotype can also affect fruit size and fruit quality.

Vegetative vigor and ultimate plant size, of course, can be influenced by many environmental and genetic factors. Selection for genetically controlled dwarfness is considered a desirable variant from the norm in many species. The centuries-old quest for dwarf fruit trees is motivated by the fact that small trees are easier to harvest, easier to apply pesticides to, and often have higher fruit yields per acre than full-sized trees. Dwarf selections of ornamental trees and shrubs are valued mainly as specimen plants, i.e., novelties. Genetic dwarfness is brought about through a decrease in vegetative vigor. With respect to grafting, it has long been understood that grafting a "normal" (nondwarf) scion onto a genetically dwarf rootstock will result in dwarfing of the scion. There, however, may be other causes of rootstock-associated scion dwarfing, such as mild incompatibility between stock and scion. As has been discussed above, rootstock effect on vegetative vigor (size control) of scion cultivars was the primary selection criterion for the EM and MM series apple rootstocks that launched the modern era of clonal tree fruit rootstock selection and breeding, and it continues to be an important selection criterion in essentially all apple rootstock and other tree fruit breeding programs today.

The terms dwarf, semidwarf, vigorous, and very vigorous have been used to describe the range of size-controlling rootstocks in apple, but they are misleading because the range of size control falls along a continuum rather than in discrete steps. Furthermore, the ultimate size of the grafted tree is the result of the interaction between the vigor of the stock and that of the scion. Each affects the other. For these two reasons, it is more useful to express the size-controlling effect of a given rootstock clone as a percentage of the size of the tree that would have resulted if the scion cultivar were not grafted (on its own roots) or grafted onto a seedling understock. Figure 27.6 categorizes the relative effect of a number of common apple rootstocks on the vegetative vigor of a scion cultivar. Keep in mind that size control is only one of the characteristics associated with common clonal apple rootstocks. Although rootstock effects on scion vigor as an orchard management tool is most developed with apples, size-controlling rootstocks are used for other tree fruit crops as well. For example, among traditional cherry (*Prunus avium* L., *P. cerasus* L.) rootstocks, sweet cherry cultivars grafted on Mahaleb seedling rootstocks grow considerably larger than the same cultivars on Mazzard seedling rootstocks. Newer clonal rootstocks for cherry and other stone fruits have been selected for size control along with other desirable characteristics (disease resistance, soil tolerances, etc.). Quince (*Cydonia oblonga*) seedlings are sometimes used as dwarfing rootstocks for pear, and a number of clonal quince rootstocks varying in size control are available. For oranges and other citrus, trifoliate orange (*Poncirus trifoliata*) is an especially cold-hardy stock that is used for dwarfing.

Not only does the vigor of the rootstock affect the vigor of the scion, but as noted above, the vigor of the scion affects the vigor of the root system. Nonetheless, size control of the rootstock by the scion is rarely considered a reason for grafting, except in the unusual case of Mukabit grafting of cassava. Although a rather obscure case, it is worth describing here in part because it is a modern confirmation of the point Liberty Hyde Bailey made in the *Standard Cyclopedia of American Horticulture* (1928) quoted at the beginning of the section below on the requirements for successful grafting. He maintained that, despite grafting's ancient history, anyone with a basic knowledge of the three basic requirements could literally invent a new grafting method to suit his needs. In 1952, a peasant farmer named Mukabit on the Indonesian island of Java discovered that the yield of the edible tuberous root crop, cassava (*Manihot esculenta*), an important dietary staple in Africa and Asian tropics, could be increased by at least twofold (100%) or more (five- to tenfold increases have been claimed) by grafting onto it a scion of the Ceara rubber tree (*Manihot glaziovii*) (de Foresta et al., 1994). This is rather extraordinary because the normal (nongrafted) root system of *M. glaziovii* produces no tubers at all. In fact, the percentage increase in growth of the cassava stock induced by the Ceara rubber scion far exceeds any claim regarding the effect of stock on scion yield for apple or any other commonly grafted crop. This profound growth stimulation of cassava root is apparently because of the fact that the Ceara rubber grows later than cassava into the dry season before going dormant, allowing more time for starch accumulation in the cassava tubers.

SPECIALIZED GROWTH FORMS

A special case of grafting to achieve independent optimization of stock and scion genotypes is the practice of top working for the construction of arborescent (tree-like) specimens from scions that would grown on their own roots into nonarborescent shrubs. Top working for this purpose is sometimes referred to as high working. This approach is used to construct such beauties or oddities (depending on your point of view) as tree roses (*Rosa* spp.), tree peonies (*Paeonia suffruticosa*), and weeping forms of flowering cherries.

GRAFTING TO INFLUENCE GROWTH PHASE AND FLOWERING

All plants grown from seed undergo a natural age-related maturation process before they are capable of flowering (chapter 15). Seedlings begin in the juvenile growth phase during which they are, by definition, incapable of flowering. In woody plants, this period of juvenility typically lasts several years in fruit trees to several decades in forest species, until the juvenile undergoes a transition to the mature (sometimes called adult) phase when they become capable of flowering under inductive environmental conditions (appropriate photoperiod, temperature, etc.). Even when propagated vegetatively by cuttage, graftage, layerage, etc., a mature-phase plant tends to retain its ability to flower. Grafting a mature scion onto any rootstock will produce a flowering tree several years sooner than growing the same species from seed. Unlike a seedling, the scion has already undergone its juvenile-to-mature phase transition. For example, grafted apple will flower in two to three years from the time it leaves the nursery, in contrast to a seedling apple, which will not flower until it is approximately seven years old. Hence, in many fruit crops, this is one, but rarely the only, reason for grafting. The duration of the nonflowering period before even a mature scion commences reliable fruit production is influenced by the growth phase and genotype of the stock. Some rootstock genotypes (cultivars) result in shorter delay in the onset of scion flowering than others, and in general it has been observed that the greater the dwarfing effect of the rootstock, the shorter the delay in fruiting of the scion. Rootstocks that accelerate the onset of flowering and fruit production are said to be precocious.

GRAFTING FOR MULTIPLE (MORE THAN TWO) GENOTYPES

The reasons for grafting described above pertain to two-part (stock/scion) grafted trees, but the same and/or additional advantages can also be conferred by grafting together more than two genotypes, either in "series" or in "parallel," by analogy with the terminology used to describe simple electrical circuitry. Double working is a multigenotype compound genetic system assembled "in series," with stock, interstock, and scion in linear sequence (Figure 27.1e). Two or more scion genotypes grafted at different places on a common stock or interstock could be considered a "parallel" compound genetic system (Figure 27.1d).

DOUBLE WORKING

Each of the three parts of a double-worked tree—scion, interstock, and rootstock—generally consists of separate genotypes. Typically, the interstock is several to many centimeters long, depending on its function. Double working is performed for any of several reasons. Some are similar to the reasons for grafting a two-part tree ("single working," although the term is not commonly used), i.e., to independently optimize the components of a compound genetic system. However, double working involves optimization of three component genotypes rather than just two. Whereas the components of a two-part tree correspond more or less to the root system (stock) and shoot system (scion) of the grafted tree, the components of a double-worked tree correspond to the root system and sometimes the lower portion of the trunk (stock), all or some portion of the trunk (interstock), and more or less of the upper trunk, branches, and leafy canopy (scion). Some desirable characteristics associated with an interstock-derived trunk include cold tolerance or resistance to specific diseases where the trunk of the nongrafted tree is especially vulnerable. Even the size-control function exerted on the scion by the rootstock of a two-part tree, as described above, can instead be assumed by an appropriately selected interstock. For example, an interstock rather than a rootstock is sometimes used to achieve size control in apples grown on wet soils that would be prone to lodging (falling over) under a heavy crop load if grafted onto a small poorly anchored dwarfing rootstock, such as M.9. Although trellising (mechanical support) is the most common solution to this problem, some growers, particularly in the past, resorted to a double-worked tree consisting of a relatively large, well anchored, vigorous rootstock, such as MM.111, with an M.9 dwarfing interstem underneath the fruiting cultivar. An advantage of this approach is that the degree of scion dwarfing imparted to the scion variety by the interstock is proportional not only to the genotype of the interstock (M.9, dwarfing versus MM.111, vigorous) but also to the length of the interstock, such that a longer M.9 interstock results in greater dwarfing of the scion.

Another reason for double working, related to independent optimization of genotypes, is what Robert Garner calls "stem building," in his classic work, *The Grafter's Handbook*. This is similar to the use of a two-part tree to create a special growth form, such as tree-like standard, by grafting a weeping cultivar on top of a tall straight stem (high working). An example of stem building is the case of a double-worked standard rose. A rootstock genotype, such as Fortuniana, might be selected (in part) for its nematode resistance, Grifferaie as an interstock, selected for its tall, straight, thornless trunk, and a scion of a bush-type rose high worked at 36 in.

Double working can also be used to overcome some cases of graft incompatibility by inserting a mutually compatible interstock between otherwise incompatible stock and scion cultivars. An example of this strategy to overcome the incompatibility between quince under stock and some pear cultivars is described in the section on compatibility below.

MULTIPLE-SCION CULTIVARS

Unlike double working, which involves a linear sequence of three genotypes separated by two graft unions (in "series"), it is possible to graft multiple scion cultivars onto different branches of the same understock (or interstock) in a "parallel" configuration (Figure 27.1d). Two or more scion cultivars grafted onto a single tree can be used to make interesting combinations of various fruit species, such as apple, e.g., Macintosh, Red Delicious, and Crispin, on the same tree, or multiple citrus species on the same tree, such as orange, lemon, and grapefruit. Using the methods described in the hibiscus grafting exercise below, it is possible to construct a single hibiscus tree with three or more different branches bearing red, yellow, or orange flowers simultaneously. Such combinations of multiple fruit or flowering cultivars are usually constructed as novelty specimen plants by or for amateur gardeners rather than for commercial crop production.

On a larger orchard scale, it is more practical to grow different cultivars as separate trees. Aside from its novelty value, placing more than one scion cultivar on certain fruit tree species has another advantage related to pollination and fruit set. Many tree fruit crops in the Rosaceae family, including apples, cherries, and pears, are sexually self-incompatible, i.e., the flowers of a single tree or even on multiple trees of the same clone cannot fertilize themselves and require pollen from another cultivar to set fruit. The requirement for a pollinizer cultivar is accomplished in most commercial-scale fruit orchards by planting pollinizer trees of a different genotype from the main crop (e.g., a crabapple), at strategic locations throughout the orchard. For smaller-scale "backyard" production, however, pollination and fruit set can be increased by top working a pollinizer cultivar onto one or more branches of the original tree. It is important to choose a pollinizer that flowers at the same time as the primary cultivar.

GRAFTING FOR DIAGNOSIS OF VIRAL INFECTIONS, AND THE PROBLEM OF GRAFT TRANSMISSION OF VIRUSES

Most plant viruses are not seed transmitted, but they are spread from the stock plant to the propagule via almost any form of asexual propagation including grafting. Viruses are readily translocated across a graft union from an infected plant to a noninfected one. For this reason, agricultural regulatory agencies in some countries and states attempt to protect regional commercial fruit or other crop industries by maintaining and enforcing virus indexed bud or scion wood certification programs. These involve rigorous indexing of new germplasm introductions to assure they are free from specific viral pathogens, maintaining virus indexed donor trees of stock and scion cultivars in foundation blocks under semiquarantined conditions to avoid reinfection, and careful regulation of bud/scion wood distribution to growers. Some definitive means of detecting virus in plants suspected of infection is necessary for the success of these clean stock programs because many crop species show no visually apparent viral disease symptoms even when infected, but nevertheless do respond adversely with reduced vigor and declining yield of harvestable crop. It may seem ironic, but fortunate, that graft transmission of plant viruses can be used to considerable advantage as a means of screening for and detecting virus-infected plants through a process called biological or graft indexing. Graft indexing involves grafting a scion from an asymptomatic crop species or cultivar suspected of being infected onto a stock of a more sensitive indicator species. If the virus is present and transmitted across the graft union, the indicator stock plant will display characteristic visual symptoms, such as streaking or mottling. This technique of viral detection is used not only with normally graft-propagated fruit tree crop species, such as citrus and cherry, but also with nonwoody crops, such as sweet potato and strawberry. Although graft indexing is still an important tool of the plant pathologist, it has declined in importance as more specific and in some cases more sensitive virus testing methods have been developed, most notably the immunological assay know as ELISA (enzyme-linked immunosorbent assay).

GRAFTING FOR REPAIR

Just as human organ transplantation (a type of grafting) can be used to replace failing organs and heal sick people, so too is plant grafting occasionally used to repair or prevent future structural problems in trees. Although in modern commercial orchard production it is usually more cost effective to replace a damaged tree, repair grafting is sometimes used as a last resort to rescue a high-value landscape specimen. Three kinds of damage that can be addressed by grafting are girdling of the trunk, damage to the root system, and splitting of narrow branch crotches. Girdling of young trees is often caused by voles or mice feeding on the inner bark (phloem) when other food sources are in short supply during the winter. Additionally, string weed trimmers can damage ornamental trees if the operator allows the string to remove the bark near the soil line. Girdling blocks downward translocation of photosynthate to the root system, via the phloem, eventually killing the tree. Figure 27.7 shows how bridge grafting can be used to restore a continuous living bark across a girdled area. As a general rule, one bridging "interstock" (although double working terminology is not usually applied to bridge grafting), about the diameter of a pencil, is inserted for each 2.5 cm (inch) of trunk diameter. Damage or decline of a root system from freezing injury, disease, mechanical damage, or even gradual (delayed) graft incompatibility is another prob-

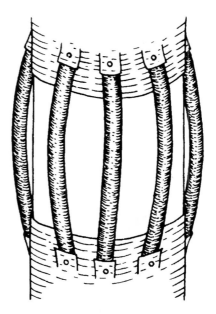

FIGURE 27.7 Bridge grafting using inlay bark graft to attach bridging pieces. (Courtesy of LH Bailey Hortorium, © 1978.)

lem that can be addressed by a type of approach grafting known as inarching (Figure 27.8). Similar to other types of approach grafting, formation of the inarched graft union occurs while both stock and scion are attached to and hence sustained by their own root systems. Unlike the use of approach grafting for propagation purposes (Figure 27.4) in which the potted rootstock along with its newly fused scion is removed for replanting, in the case of inarching to replace a damaged root system, the stock is planted beside the tree with the damaged root system and left in place permanently to serve as a new, surrogate root system for the remaining life of both trees. Brace grafting is occasionally used to mechanically reinforce a branch with a narrow crotch angle that is likely to split away from the main trunk. In this case, the scion is grafted between both the branch and the trunk to offer support and prevent breakage.

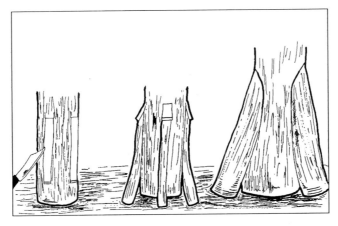

FIGURE 27.8 Inarching for replacement of damaged or incompatible root system, using an inlay bark graft to attach new rootstock to established trunk. Note: inarching is a type of approach grafting.

REQUIREMENTS FOR SUCCESSFUL GRAFTING

There are many different reasons for grafting, but there are even more ways (methods) to go about it. Liberty Hyde Bailey, remarked on the profusion of existing grafting methods and the development of new ones in his *Standard Cyclopedia of American Horticulture* (1928),

> The ways or fashions of grafting are legion. There are as many ways as there are ways of whittling. The operator may fashion the union of the stock and the cion to suit himself, if only he apply cambium to cambium, make a close joint, and properly protect the work.

Bailey's advice should be reassuring to the beginner who might otherwise be overwhelmed by all the dozens (at least) of different grafting and budding methods practiced at commercial nurseries, orchards, and backyard gardens and described in various books, articles, Web sites, etc. His advice is as true today as it was in 1928. If students of grafting pay careful attention to how the universal requirements for successful grafting are satisfied, while learning just a few basic methods, they will find it easier to learn additional methods as the need arises. Bailey's remark emphasizes three of the four universal requirements that can be stated with less oratorical flourish as cambial alignment, pressure, and avoidance of desiccation. The fourth requirement, not mentioned in the quotation from Bailey is genetic compatibility. In fact, the issue of genetic compatibility is so important that it will be discussed first because, if stock and scion are not genetically compatible, nothing else matters (except in the case of nurse grafting).

COMPATIBILITY

If stock and scion are sufficiently closely related genetically to form a functional graft union, they are said to be compatible. Conversely, a stock–scion combination that is not sufficiently closely related is described as incompatible, and a graft performed between them will display, either immediately or eventually, incompatibility symptoms, such as leaf drop, bulging and cracking of the initial graft union, and eventually death of the scion. Failure of grafting from poor technique or timing is not incompatibility. Delayed incompatibility refers to initial formation of an apparently normal graft union followed at some later time, typically one to several years, by gradual failure of the graft union accompanied by decline and eventually death of the scion. Incompatibility greatly limits the practice of grafting to closely related species. If the Roman poet Virgil (29 BCE) had been correct in his "observation" that, "… swine crunched acorns 'neath the boughs of elms" (interfamilial oak/elm compatibility), the agricultural possibilities would be endless. For example, most common apple cultivars can be grafted together, as can most common citrus species (oranges, lemons, grapefruit, etc.), but apples and oranges, which are in different families, cannot be successfully grafted.

The more closely the stock and scion candidates are related to each other, the more likely they will be compatible, but it is difficult to generalize beyond that point. Interfamilial (between different families) combinations are universally incompatible, except for occasional claims of a few rare, horticulturally unimportant exceptions. At the other end of the spectrum, intraclonal (within a clone) combinations (same species) are universally compatible, but when combinations between different clones, different species, or different genera are considered, the degree of "relatedness" necessary for compatibility, known as the limits of compatibility, varies with different possible combinations. Generalizations are of limited predictive value. As shown in Table 27.2, intraclonal combinations are universally compatible. Intraclonal grafting is uncommon except for repair grafting of a damaged tree by bridge or brace grafting with a piece of shoot of the same clonal cultivar. Interclonal combinations within the same species are nearly always compatible. Interclonal grafting is often the basis for nursery production of many difficult-to-root species, such as Norway maple and other shade trees. However, there are economically significant exceptions where interclonal grafting is not ultimately successful, such as the sporadic (unpredictable) delayed incompatibility between red maple (*A. rubrum*) cultivars and red maple seedling rootstocks. Interspecific combinations within the same genus are often compatible and frequently horticulturally important. For example, almond (*Prunus dulcis*) scions are successfully grafted on peach (*P. persica*) rootstock, but many interspecific combinations are not feasible, such as the incompatible combination almond and apricot (*P. armeniaca*). Note that, while some interspecific/intrageneric combinations of stock and scion are compatible, such as Myrobalin B plum (*P. cerasifera*) scion on Hale's early peach (*P. persica*) stock, the reciprocal combination of Hale's early peach scion on Myrobalin B stock is incompatible. Successful combinations among different genera (intergeneric) within the same family are relatively rare, but several of those that do succeed are horticulturally significant, such as some (but not all) pear (*Pyrus communis*) cultivars (e.g., 'Farmingdale') on quince (*Cydonia oblonga*) rootstock. Nonetheless, the closely related pear cultivar Old Home is incompatible with quince. As described above, double working a 'Farmingdale' interstock between the 'Old Home' and the quince stock can avoid the incompatibility entirely. Table 27.2 gives only a few examples in each category.

Since there are important exceptions to generalizations at the level of interclonal/intraspecific, interspecific/intrageneric, and intergeneric/intrafamilial combina-

TABLE 27.2
Compatibility

Relationship	Likelihood of Success[a]	Compatible Example	Incompatible Example[b]
Intraclonal	100% (a sure thing)	Bridge graft (scion from same tree, for repair of girdled trunk)	None
Intraspecific (interclonal)	High	'Macintosh' apple (*Malus domestica*)/EM 9 (clonal rootstock cultivar of *M. domestica*)	Red maple clonal cultivars (*Acer rubrum*)[c]/red maple seedling
Intrageneric (interspecific)	Moderate	Sweet orange (*Citrus sinensis*)/rough lemon (*C. jambhiri*) Sweet cherry (*Prunus avium*)/tart cherry (*P. cerasus*)	Almond (*Prunus amygdalus*)/peach (*P. persica*)
Intrafamilial (intergeneric)	Unlikely	'Old Home' pear (*Pyrus communis*)/quince (*Cydonia oblonga*) Blue spruce (*Picea pungens*)/Norway spruce (*Picea abies*)	'Bartlett' pear (*P. communis*, Rosaceae family)/Quince (*C. oblonga*, Rosaceae) Lilac (*Syringa vulgaris*, Oleaceae family)/California privet (*Ligustrum ovalifolium*, Oleaceae)[d]
Interfamilial	0% (considered impossible, despite a few claims to the contrary)		

[a] Probability of genetic compatibility, assuming that grafting technique and environmental factors are not the cause of failure.
[b] Scion/stock.
[c] Varies with scion cultivar, i.e., some clones exhibit essentially no incompatibility, while for others, as high as 25% of grafted individuals develop incompatibility.
[d] Delayed incompatibility is the desired outcome of nurse root grafting, used to obtain scion rooting.

tions, any potentially useful combination of unknown compatibility must be tested in order to have confidence in the outcome. Such empirical testing is often prohibitive in terms of time and money. It would be helpful to have access to a definitive listing of all or nearly all potentially horticulturally useful combinations. A number of worthy attempts have been made to publish the results of empirical tests (such as Andrews and Marquez, 1993), but there is currently no comprehensive guide available. Such a guide would be valuable, considering the substantial economic losses that some growers have incurred after investing several years of time, labor, and money in growing a crop of grafted trees, only to have some or all of it destroyed by a delayed incompatibility.

What is equally frustrating to propagation physiologists is that there is no consistent explanation for the causes (mechanisms), let alone a cure for incompatibility. Some incompatibilities appear to be related to the production of an organic compound by one member of the pair that is toxic to the other, whereas in other cases, the mechanism appears to be related to differences in the biochemistry of wood formation between stock and scion. One of the best understood incompatibility mechanisms is that between Bartlett pear (*Pyrus communis*) and quince (*Cydonia oblonga*) rootstock. In this case, a cyanogenic glucoside called prunasin, present naturally (harmlessly) in the quince understock is broken down chemically when it comes into contact with an enzyme produced by the pear tissue, releasing toxic cyanide that results in destruction of the pear phloem.

A final category of incompatibility worth mentioning is so-called induced incompatibility, which is related to a more or less asymptomatic virus in the stock or scion that triggers an incompatibility reaction at the graft union that would not otherwise occur. Black line of walnut (English walnut, *Juglans regia*, scion/Northern California black walnut, *J. hindsii*, stock) is induced by *cherry leafroll virus*, and apple union necrosis and decline (AUND) is induced by *tomato ringspot virus*.

CAMBIAL ALIGNMENT

The vascular cambium is a cylinder of living tissue a single cell layer, between the secondary xylem (wood) to the inside and secondary phloem (inner bark) to the outside (Figure 27.2). The "carpentry" or knife cuts made while performing any of the various grafting and budding methods, such as those illustrated in this chapter, result in exposure of a section of the cambial cylinder of both stock and scion. The grafter's carpentry must be sufficiently accurate so that the shapes of the cut surface of stock and scion complement each other to the point where the cambia of each come into contact, or nearly so, along as much of the perimeter of the cambial cylinders as pos-

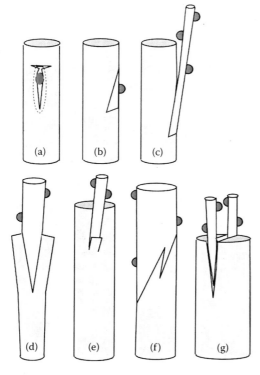

FIGURE 27.9 Examples of several rind and nonrind grafting methods. (a) T budding, (b) chip budding, (c) veneer grafting, (d) top wedge grafting, (e) bark grafting, (f) whip and tongue grafting, and (g) cleft grafting.

sible. In fact, some grafting methods, by the nature of their carpentry, expose more cambial surface than others, which allows for more cambial contact between stock and scion if the two are cut complementarily to each other. For example, the interlocking "tongues" of the whip and tongue graft (Figure 27.9f) increases the potential zone of cambial contact by almost 100%, compared to a splice graft, which involves the same diagonal cuts joining stock and scion without the interlocking tongues. Similarly the greater zone of cambial contact with a chip bud (Figure 27.9b) compared to a T-bud (Figure 27.9a) accounts at least in part for the fact that chip budding allows for a quicker, stronger, more winter hardy union and overall better performance than T budding.

As described above, the process of graft union formation involves formation of a callus bridge across any gap that exists between the stock and scion, followed by dedifferentiation of a new cambium across the callus bridge to connect the vascular cambia of the stock and the scion. This occurs in an inward direction from the stock and scion vascular cambium until they connect somewhere in between in the callus tissue. Without so-called "cambial contact" (reasonable cambial alignment), the two converging ends of the newly forming cambium do not "find" each other, and vascular continuity is never established. Some particularly difficult-to-graft species, such as temperate nut trees like hickories (*Carya* sp.) and walnuts (*Juglans* sp.), produce very little callus, and hence actual cambium-to-cambium contact (minimal gaps) is critical. In such cases, the carpentry must be especially well executed to assure a near perfect fit. Figure 27.2a represents perfect alignment that allows more or less direct cambial contact between stock and scion when the two are brought together, whereas in Figure 27.2b, the alignment is imperfect but still acceptable (sufficiently close); however, in Figure 27.2c, the gap between the cambia of the stock and scion is too great for the graft union to develop successfully.

PRESSURE

Once the stock and scion are fitted together so that their cambia are well aligned and preferably in direct contact, pressure is necessary to maintain this alignment; otherwise mechanical disruption from handling or wind in the nursery could easily cause the scion to move with respect to the stock. Movement of the scion could create or increase a gap, disrupting the cambial contact and/or alignment established when the stock and scion were initially joined by the grafter. When cut surfaces are exposed to air, as occurs when a gap is formed between stock and scion, the surfaces tend to dry out more readily, creating a problem related to the next universal requirement—avoidance of desiccation. Furthermore, as mentioned above, the greater the gap between stock and scion, the more callus must be generated to fill that gap, and once callus formation has occurred within weeks after placing the graft, further disruption will tear apart the extremely soft (nonlignified) callus cells. Firm pressure between stock and scion will minimize the chance of unintended movement between the two and thereby facilitate graft union formation. In addition to these mechanical considerations, pressure also influences alignment of the planes of cell division and dedifferentiation in the developing callus during the graft union formation, giving rise to spatially organized (properly lined up) tissues of cambium, xylem, and phloem (Barnett and Asante, 2000).

In the case of some grafting methods, pressure between stock and scion is generated by the carpentry itself related to elasticity of the wood of interlocking portions of the stock and scion. This can be seen for several nonrind grafts illustrated in Figure 27.9, including the top wedge graft (d), whip and tongue graft (f), and cleft graft (g). Little or no elastic pressure is generated in the case of other noninterlocking, nonrind grafting methods, such as the splice approach graft (Figure 27.4b), chip bud (Figure 27.9b), or the veneer graft (Figure 27.9c), nor in the cases of rind grafting methods including T budding (Figure 27.9a) or bark grafting used for top bark grafting (Figure 27.9e), bridge grafting (Figure 27.7), or inarching (Figure 27.8), in which the bark is peeled back and the scion is placed on the surface of the wood.

Regardless of whether or not elastic pressure is generated from the carpentry per se (nonrind grafting methods),

Grafting: Theory and Practice

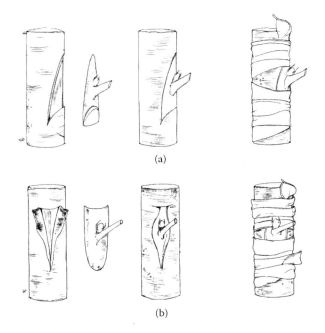

FIGURE 27.10 Two common budding methods. (a) Chip budding; (b) T budding. (Drawing by Eric Hsu.)

pressure is usually imposed by tying or wrapping the graft union (Figure 27.4b, Figure 27.5, Figure 27.10, and Figure 27.11b), or in some cases nailing (Figure 27.7). Many different materials are or have been used for this purpose including string, tape, natural plant fiber, various kinds of plastic strips, and latex rubber bands (Figure 27.11b), sometimes referred to as "budding rubbers."

AVOIDANCE OF DESICCATION AND SEASONAL CONSIDERATIONS

A freshly cut scion must survive without a direct, anatomically continuous, connection to the root system of the stock plant for at least several weeks, during which time it is extremely vulnerable to desiccation. Moisture can be lost from a scion by transpiration from exposed foliage and/or nonlignified stem tissue, as well as by evaporation from freshly cut (wet) surfaces. Without direct connection to a root system, there is limited opportunity for the scion to replace water once it is lost. If there is a deficit between water loss and uptake, drought stress results, which can retard graft union formation and eventually cause scion death if it is severe. The first step to avoiding desiccation, even before scion collection, is to begin with a well-hydrated scion donor plant. Once a leafy scion is severed from the scion donor plant it is critical to minimize transpiration by removal of leaves, use of moisture barriers after grafting to prevent water vapor loss (waterproofing), and/or by grafting at a time of year when the scion is dormant and natural transpiration is minimal.

SEASONAL CONSIDERATIONS (GRAFT TIMING) FOR MOISTURE MANAGEMENT

In addition to carpentry and other technique-related considerations, the timing (season) of grafting is an extremely important and complex part of the art and science of grafting. This is not only because of the influence of season and phenology (seasonal changes in plant growth and development) on potential evapotranspiration, but also because of the obvious relationship between season, temperature, and dormancy. The optimum time of year for performing a particular grafting method for each species is related to moisture management, stage of plant development, and temperature.

For any combination of grafting method and species, there is a critical balance between sufficient heat for growth and potential evaporation. Higher temperature (up to some species-specific optimum) promotes cell division and other temperature-dependent physiological processes

FIGURE 27.11 Top wedge grafting of tropical hibiscus. (a) Tapered wedge at base of scion inserted into vertical cleft (split) in understock. (b) Polyethylene bag humidity tent covering scion tied below stock–scion junction. Note overlapping turns of budding rubber wrapped tightly around stock–scion junction are visible inside the bag.

involved in graft union formation, while at the same time, transpiration from leafy, and especially nonlignified (succulent) shoots increases with increasing temperature. Hence, as a general rule, grafting is performed when the temperature is adequate, but no higher than necessary for maintaining the physiological processes associated with graft union formation while minimizing transpiration. For many deciduous woody plants, the optimum seasonal combination of these two factors occurs either in early spring when they are still leafless and just emerging from dormancy, or late summer/early fall when deciduous plants are still leafy, but shoots are lignified and have already set dormant buds.

DORMANT OR SEMIDORMANT SCION GRAFTING

Rind grafting refers to methods that involve peeling back the bark (rind) of the understock and placing the cut surface of the scion beneath the bark (rind), on the surface of, but not into the underlying wood of the stock. Examples include T budding (Figure 27.9a, Figure 27.10b), bark grafting (Figure 27.7, Figure 27.8, Figure 27.9e), four flap grafting, patch budding (not illustrated), etc. Nonrind grafting methods are those that involve inserting the scion through the bark and into a cleft in the underlying wood. Nonrind grafting methods include cleft grafting (Figure 27.9g), chip budding (Figure 27.9b, Figure 27.10a), veneer grafting (Figure 27.9c), top wedge grafting (Figure 27.11), etc. The distinction between rind and nonrind grafting is intimately related to seasonal grafting considerations, since peeling back the bark is only feasible when the bark is "slipping." This occurs only during periods of active shoot growth, from late spring through early fall/late summer, when the cells of the vascular cambium are actively dividing and their thin cell walls are easily pulled apart when the bark is separated from the underlying wood. Earlier in the spring or later in the fall, the bark tears rather than slips (peels) away from the wood, and nonrind grafting is the only alternative. Hence, rind grafting is feasible only during periods of active growth when deciduous plants are leafy and potential evapotranspiration is high.

Because potential evapotranspiration is less during periods of dormancy, nonrind, dormant scion grafting methods are more widely practiced in temperate climates than rind grafting methods. For example, dormant scion grafting in early spring is practiced for top working established trees for repair, reworking to a new cultivar, or inserting a pollinizer branch, etc. The timing is fairly critical, and the "window of opportunity" fairly narrow because the stock plant (established tree in most cases) must be coming out of dormancy and beginning to grow (bud swelling), but not be fully leafed out. At the same time, it is important that the scion still be as dormant as possible so that it will not leaf out and transpire excessively before callus formation and vascular (xylem and phloem) interconnection with the stock is well underway. Leafing out of the scion too early would quickly result in desiccation of the scion because transpiration from the newly expanded, succulent leaf surface would exceed water uptake. This seemingly contradictory requirement for more advanced growth activity in the stock than in the scion is managed by the timing of scion wood collection and handling. Scions are collected while fully dormant in mid-to-late winter and stored under refrigeration until growth activity of the stock resumes in the early spring. At that time, the still-dormant (tight bud) scion can be grafted onto a stock that is already beginning to resume growth (bud swelling). By the time the scion leafs out several weeks later, it is already well along in the process of forming a functional graft union with the understock.

Whereas moisture management during early spring grafting is accomplished largely by dormant grafting of a nonleafy scion, the strategy for "fall" budding (late summer through early fall), commonly practiced on tree fruit and ornamental nursery stock, is somewhat different. In this case, both the stock plants (field or container nursery stock) as well as the scion bud donor plants are still in leaf, but have set dormant buds in response to decreasing day length and cooler nights of later summer. Excessive transpiration from the bud scion is prevented simply by removing the leaf blade at the time of bud wood collection. A petiole stump is left in place as a "handle" for the grafter. These semidormant, debladed bud scions can be grafted onto the still leafy rootstock either by chip budding (Figure 27.10 a) or T budding (Figure 27.10 b) if the bark is still slipping, or somewhat later only by chip budding, for a week or two after the bark has ceased to slip. In fact, it is increasingly common for chip budding to be preferred over T budding even when the bark is still slipping because chip budding has been shown to result in better, quicker graft union formation, improved overwintering and overall better long-term performance. In either case, transpiration from the dormant scion bud is minimal, even as graft union formation is underway. The bud remains dormant until the following spring, when the shoot of the stock plant above the grafted bud is cut back, forcing the new bud to grow. For some nursery crops, especially stone fruits in regions with longer growing seasons (USDA Hardiness zone 6 or greater), "June" (early summer) budding is practiced. Moisture management is critical during June budding because the scion buds resume active growth shortly after insertion. The advantage of June budding is that the nursery tree is "finished" a year earlier than fall-budded plants.

Two additional dormant scion grafting strategies for temperate species include bench grafting of deciduous nursery stock and greenhouse container grafting of narrow leaved evergreens and some deciduous species. Bench grafting refers to any technique applied to bare root nursery stock, indoors, "at the bench," during the dormant period. For example, in the production of double-worked apple trees, a whip and tongue bench graft is performed in February or March to join stock and future interstock. Graft union formation occurs before the plant is lined out in the field in the spring. New growth from the intended interstock is then bud grafted, as described above, to a fruiting scion cultivar in the fall of the same year. In the case of temperate narrow leaved evergreens, dormant scion grafting is practiced in a cool greenhouse during the winter. Ornamental cultivars (dwarf and other unusual growth forms etc.) of conifers including pines (*Pinus spp.*), spruce (*Picea spp.*), firs (*Abies spp.*), Juniper, etc. are commonly grafted to seedling understocks because they are difficult to root from cuttings. Some deciduous species, such as selections of Japanese maple are propagated this way. Dormant containerized nursery stock is typically brought into a cool (ca. 50°F) greenhouse during midwinter and allowed several weeks to begin to emerge from dormancy. Just as growth of these stock plants is resuming, they are grafted with dormant scions using a side veneer graft and overwintered in a cool greenhouse.

NONDORMANT (GREENWOOD) SCION GRAFTING

In the case of tropical broad leaved evergreen plants (e.g., citrus, avocado, hibiscus, etc.) that do not undergo seasonal defoliation and winter dormancy like deciduous temperate species, a different moisture management strategy is required. Leaves are removed from nondormant scions and top wedge grafted as described in the laboratory exercise below. To minimize transpiration, the actively growing scions and the graft union itself are covered with a polyethylene tent to create a high-humidity environment.

COLLECTION OF SCION OR BUD WOOD

Depending on the seasonal strategy for grafting or budding, scion or bud (scion) wood is usually collected hours, days, or even weeks before it is used for grafting. Especially when grafting is performed during the growing season, water stress of the scion donor plant and the stock plant should be avoided by keeping them well watered and by collecting the scion wood during the cooler morning hours. Other practices used to minimize desiccation of scion wood include avoiding exposure to direct sunlight after cutting, removing leaves to minimize transpiration, and storing sticks in a cool, high-humidity environment (e.g., in wet burlap under shade and/or in an insulated cooler). Scions collected for bud grafting are stored as 10- to 20-cm long "bud sticks" and cut into individual scion buds only as they are needed. If scion wood is to be stored for more than a few hours before grafting, it should be wrapped in moist paper or burlap, placed inside a plastic bag and kept refrigerated until use. For grafting performed during early spring, before leaf-out of deciduous plants, dormant scions are usually collected weeks earlier, during midwinter when they are completely dormant, and stored under refrigeration until needed.

WATERPROOFING MATERIALS

The final moisture management strategy to minimize scion desiccation is coating and/or tying the stock–scion junction with some moisture-retardant material that minimizes evaporation from freshly cut surfaces. Wrapping with grafting tape, waxed string, or other semiporous materials makes a less effective moisture barrier than budding bands (budding rubbers), Parafilm™, and new specially formulated plastic grafting tapes. Wrapped or tied graft unions are sometimes coated with low-melting-point wax preparations. While Parafilm and grafting waxes serve primarily as moisture barriers, waxed string, various tapes, and elastic budding rubbers also serve to create pressure. Some relatively unyielding wrapping materials, such as string and regular polyethylene strips, have the potential for girdling the stem over time as it increases in diameter. Therefore, these must be deliberately removed several weeks after grafting. Traditional latex budding rubbers degrade naturally when exposed to sun-(ultraviolet) light. When used outdoors for field budding, they crack and fall off on their own, but when used inside a glass greenhouse where the ultraviolet component of sunlight is filtered out by the glass, budding rubbers must be removed by hand. The use of a polyethylene tent to enclose the entire scion is not commonly practiced except for a few situations, such as the top wedge grafting of tropical fruit tree nursery stock discussed above and in the laboratory exercise below.

METHODS OF GRAFTING

As Bailey said, there are as many different grafting methods as there are "ways to whittle a stick." In his 1672 treatise, *The History of the Propagation and Improvement of Vegetables ...*, the English cleric Robert Sharrock (1672) listed nine methods, while Robert Garner (1958) listed over 40 in *The Grafters Handbook,* and the seventh edition of Hartmann et al. (2002) comprehensive modern reference, *Plant Propagation, Principles and Practices* lists 16 grafting and 7 budding methods. Classifying these into several different functional and structural categories helps clarify the relationships among them. Structural classification based on the shape of the stock or scion cuts (carpentry) is the basis for naming many but not all grafting methods, such as cleft, top wedge, four flap, veneer, and saddle grafting, as well as T, chip, and patch budding. Names, such as approach grafting, inarching, brace, and bridge grafting, describe the function rather than the structure of the grafts and may not even imply a specific carpentry at all. For example, approach grafting might be performed using the (nonrind) splice approach method (Figure 27.4) or with an inlay bark graft in the case of the inarch approach graft illustrated in Figure 27.8. It is important to note that different authors and practitioners, historically or even currently, sometimes use different names to describe essentially the same method. For example, the most common contemporary name for the type of budding shown in Figure 27.9a and Figure 27.10b is T budding, but *Hortus III* (Bailey, 1976) refers to it as shield budding, while others call it T and shield budding. The nonrind graft referred to as top wedge grafting (Figure 27.9d) is quite similar to the cleft graft (Figure 27.9g), except for the relative size of scion and understock.

EXERCISE

EXPERIMENT 1: TOP WEDGE GRAFTING OF HIBISCUS (*HIBISCUS ROSA SINENSIS* L.)

This laboratory exercise will give you an opportunity to learn a nonrind grafting technique that incorporates many different strategies to satisfy three of the four universal requirements for successful grafting described above—cambial alignment, pressure, and avoidance of desiccation (Mudge et al., 2003). Keeping these in mind as you perform this exercise will help you apply what you learn about this method to other grafting and budding methods that you may wish to learn in the future. Tropical hibiscus is recommended for this exercise because it can be grown in a greenhouse or a bright window at any time of year without the interruption of seasonal dormancy. It is also relatively easy to graft, assuring you a positive learning experience. After learning top wedge grafting, you can use the same stock plant to practice other grafting and budding methods. Using different hibiscus cultivars, you can create your own unique compound genetic system that will bear differently colored flowers for years to come.

Background

Top wedge grafting is commonly used on tropical fruit tree nursery stock including avocado, mango, and sometimes citrus, although T budding is more commonly used with the latter. This and other grafting methods are used to propagate fancy, polyploid cultivars of this species, which tend to be difficult to root from cuttings.

Materials

The following materials will be needed to complete the laboratory exercise:

- Three hibiscus stock plants about 24–30 in. tall (four- to five-month-old rooted cuttings) in 4 to 6 in. diameter pots. Preferably each of the three plants should be a different cultivar
- Paper and string labels
- Sharp grafting knife
- Pruning shears
- Latex rubber budding strips, 5 in. (12.7 cm) × 0.25 in. (0.64 cm) × .016 gauge (FarmHardware.com, Sacramento, Calif.)
- Spray bottle containing water
- 3 × 6 in. polyethylene bags
- Twist ties

Follow the instructions in Procedure 27.1 to complete this exercise.

Anticipated Results

Within a week you should notice new scion growth, which is a good sign of anticipated success. After three weeks, a careful examination of the junction between the stock and scion should reveal a white fluffy callus along the cut surfaces, which is an early stage of graft union formation. By week four, the graft should be sufficiently strong to remove the wrapping and the plastic bag entirely, assuming that the scion was acclimatized during week three. One week later your grafted hibiscus will be able to grow in full sun. With sufficient sunlight and protection from cold winters (avoid temperatures below 60°F) by moving it indoors, you should get years of enjoyment from your multiflower-colored tropical hibiscus.

Questions

- How did your top wedge grafting of hibiscus satisfy each of the four requirements for successful grafting and budding?
- What other methods of grafting would be appropriate for this species if you wanted to end up with a plant with three or more different cultivars (flower colors) grafted to it?

Grafting: Theory and Practice

Procedure 27.1
Top Wedge Grafting of Hibiscus

Step	Instructions and Comments
1	The upper 5 to 7.6 cm (2 to 3 in.) of the shoot of each plant will be removed to serve as the scion for one of the other plants. Each of the three decapitated plants will serve as an understock for a scion from another plant.
2	Collect scion wood from each of the potted hibiscus plants by making a horizontal cut with sharp pruning shears to remove the distal (top) 7.6 cm (3 in.) or more of the shoot. Unless it is to be used immediately, the scion wood should be wrapped in wet paper towels to avoid desiccation. Note: A scion may be either a terminal section from another stock plant or a subterminal section from any point along the main shoot. Scions from lower sections of the shoot of the scion donor plant are progressively woodier toward the base and somewhat slower to resume growth when grafted onto another stock plant, but desiccation and wilting during the grafting process is less likely with woodier scions.
3	Use a sharp grafting knife or pruning shears to prepare the scion by trimming off any leaves that are anywhere from half to fully expanded. Regardless of whether it is a terminal section from the top of a scion donor plant or a subterminal section, it should be no longer than about 7 to 10 cm, and several nodes long. Note: Scions longer than recommended are more easily desiccated and more easily dislodged accidentally. Ideally the scion diameter should be the same as that of the horizontal cut at the top of the under stock. If necessary, it may be up to several mm narrower, but it should not be any wider than the understock.
4	With a sharp grafting knife, split the understock vertically, down the center, from the horizontal cut to a depth of approx. 2.5 to 4 cm (1 to 1.5 in.) as shown in Figure 27.11a. Do not cut a v-shaped wedge in the top of the understock; simply split it vertically. Note: The height on the stock plant where a scion from another plant is inserted is largely a matter of preference, as long as it is between 90 cm (near the original top) to about 20 cm from the soil line with at least several leaves remaining on the understock. A lower point of scion insertion will produce a more compact grafted plant, but a higher graft will leave room for insertion of other scions from the same or other cultivars by side wedge grafting or bud grafting (chip or T budding). The higher the scion is placed on the stock plant, the more likely it is to be dislodged accidentally.
5	At the basal end of the scion, cut a tapered, v-shaped wedge that comes to a sharp point by making two long, straight, flat cuts on opposite sides of the scion (Figure 27.9d). While being cut, the scion should be held firmly with the left hand (if right handed) above the end where the wedge is to be cut. The distance from the top to the bottom of the wedge should be no less than two but no more than 3 cm long. Note: Terminal nonlignified scions are easier to cut than more basal partially lignified scions, but it is easier to control the knife to make long straight cuts that meet in a sharp point on partially lignified scion wood than on those that are softer and nonlignified. Ideally, each side of the v-shaped wedge should be made by a single straight pass with the knife, rather than a concave, convex, and/or uneven "whittled" surface created by multiple passes of the knife.
6	Insert the tapered wedge of the scion into the vertical slit in the under stock until top of scion V is even with horizontal cut surface of the stock. If stock and scion are of equal diameter, the vascular cambia of the scion should be aligned with that of stock, but if the scion is slightly smaller in diameter than the stock, the scion should be placed to one side of the stock (rather than centered), so that the cambia align on that side only. Note: Leaving a crescent of the top portion of the scion wedge cut exposed above the horizontal cut of stock is a common mistake that permits evaporative water loss from the wet cut surface.

Continued

7. Wrap the entire stock/scion junction with a budding rubber (inside bag, Figure 27.11b). Begin at the top by holding one end of the band against the scion just above the vertical cut of the stock. Bring the remainder of the band around behind the scion to cross over itself, creating an "X" that prevents the top end of band from unraveling. Continue to wrap the band downward around the scion, while stretching the band to create pressure, and overlapping each turn slightly to prevent gaps that would permit evaporative water loss. Continue wrapping until the entire vertical cut in the stock is covered. Finish off the wrapping by pulling out a loop at the bottom turn of the band by pinching it between thumb and forefinger, and passing the free end of the band through this loop to trap the end and prevent unraveling. Note: Most wrapping of grafts (e.g., chip and T budding) is made from the bottom up, but in the case of top wedge grafting, the somewhat slippery (wet) cut surface of the scion wedge tends to be squeezed up and pushed out of the cleft in the understock if wrapping pressure is applied from the bottom up. Hence, it is often easier to wrap from the top down, making at least one turn around the scion before continuing to wrap around the top of the stock, downward.

8. Spray a minimal amount of water into a plastic bag and create a humidity tent by placing the inverted bag over the scion, being careful not to crush or dislodge the scion. Tie the bag firmly in place with a twist tie, about 2.5 cm (1 in.) below the stock–scion junction (Figure 27.11b).

9. Label each graft with the date, your initials, and the names of the stock and scion varieties. Place grafted plants in a designated greenhouse under approximately 70% shade cloth.

10. Observe your grafted hibiscus weekly. If water (from transpiration) accumulates at the bottom of the bag where it is tied against the stock, open the bag and drain off the water, then retie. Observe the timing and extent of new scion growth. After three weeks, cut a single 1-in.-long slit in the upper portion of the plastic bag to begin acclimatization of the scion to ambient relative humidity. One week later (week 4), make another 1-in. slit in the bag, and several days later, remove the plastic bag entirely. The budding rubber can be removed at this time to observe the stock–scion junction for callus formation. If the stock and scion are securely bound together by callus, the band need not be reapplied. Keep the grafted plant under shade for one additional week before moving to full sun. Note: Using the space between the terminal top wedge graft and the base of the stock plant, one or two additional scions may be placed on opposite sides of the stock plant using any of several different conventional grafting or budding methods (or invent your own). T budding and chip budding work well (Mudge et al., 2003), but are slightly more challenging than top wedge grafting. Morgan (2003) describes and illustrates several different side grafting methods for hibiscus. Regardless of method(s), keep in mind the universal requirements for successful grafting and you will succeed.

LITERATURE CITED

Andrews, P.K. and C.S. Marquez, 1993. Graft incompatibility. *Hort. Rev.* 15: 183–231.

Bailey, L.H. 1928. *Standard Cyclopedia of American Horticulture,* Macmillan, New York.

Bailey, L.H., 1976, *Hortus III,* LH Bailey Hortorium staff (Eds.), Macmillan, New York.

Barnett, J.R. and A.K. Asante. 2000. The formation of cambium from callus in grafts of woody species, in *Cell and Molecular Biology of Wood Formation,* R.A. Savidge, J.R. Barnett, and R. Napier (Eds.), BIOS Publ., Oxford, U.K. pp. 155–168.

Dalley, S. Personal communication.

de Foresta, H., A. Basri, and Wiyono, 1994. A very intimate agroforestry association. Cassava and improved home gardens: the Mukibat technique. *Agroforestry Today.* 6: 12–14.

Flemer, W., 1989. Why we must stillgraft and bud. International Plant Propagator's Society presentation.

Garner, R. J. 1958. *The Grafter's Handbook.* Oxford University Press, New York.

Harris, S.A., J.P. Robinson, and B.E. Juniper. 2002. Genetic clues to the origin of the apple. *Trends Genet.* 18: 426–430.

Hartmann, H.T., D.E. Kester, F.T. Davies, Jr., and R.L. Geneve. 2002. *Plant Propagation: Principles and Practices,* 7th ed. Prentice Hall, Inc., Upper Saddle River, NJ. 880 pp.

Lee, J.M., 2003. Advances in vegetable grafting. *Chronica Hort.* 43: 13–19.

Morgan, W. 2003. Grafting techniques, step by step pictures, 24 Aug. 2003, http://www.widebaytrader.com/graft/graft.htm.

Mudge, K.W., K. Hennigan, and P. Podaras, 2003. Use of tropical hibiscus for instruction in grafting. *HortTechnology* 13: 723–728.

Sharrock, R. 1672. *The History of the Propagation and Improvement of Vegetables by the Concurrence of Art and Nature.* R. Robinson, Oxford, U.K.

28 Grafting and Budding Exercises with Woody and Herbaceous Species

Garry V. McDonald

Grafting and budding are propagation techniques used when plants cannot be grown from seed or cuttings. This may be because the plants do not set seed or produce nonviable seed and cannot be normally propagated from cuttings because of difficulty in rooting or other factors. Grafting is the process of physically joining together two different plant parts (see chapter 27). The two parts are referred to as the scion and the stock or rootstock. The scion is a stem or branch of the plant that has traits or characteristics of interest. This could be fruit or nut quality, flower size, color, leaf pattern, or some other desirable trait. The stock is the root system that the scion will be grafted or budded onto. Stocks are usually chosen for vigor, disease resistance, a lack of thorns, and ease of rooting. The term grafting or graftage refers to using a scion that has two or more buds along a stem or branch. Budding, on the other hand, refers to a scion that is restricted to one bud. Other than the number of scion buds, the operation is essentially the same. Grafting is both an art and a science. It is important to understand the physiology and anatomy of the plant parts in order to make sure there is a proper alignment of cambial tissue so that a graft "takes" (chapter 27). It is also an art in that it takes much practice and skill to manipulate the plant parts to ensure a suitable number of grafts are successful. This chapter of laboratory experiments will detail two grafting and one budding technique commonly used to produce fruit trees, roses, and other ornamental plants.

EXERCISES

EXPERIMENT 1. ROSE T-BUDDING

Materials

The following materials will be needed for each student or team of students:

- Rooted cuttings of *Rosa multiflora* Thunb. ex J. Murr. thornless cultivars, such as 'Brook's 48,' 'Brook's 56,' 'K1,' 'Dan Whiteside', or 'Ginn 66'
- Bud sticks of any rose cultivar with well developed buds
- Paraffin film wrap (Parafilm™)
- Grafting or budding knife (Figure 42.3) or sharp pocket knife
- Rubber budding strips or rubber bands cut into strips about 4 in. long
- Bandages and first-aid supplies to treat attempted "finger grafts"

Follow the instructions found in Procedure 28.1 to complete this experiment.

Anticipate Results

After the T-bud is inserted and wrapped, the healing process should begin. The bud should remain green and turgid. The cambium layer around the "T"-shaped cut will begin to grow and cover the edges of the inserted bud. After about one week, the bud should still be green and turgid. The healing process should be completed after two weeks. At this time, the inserted bud should have formed a union with the surrounding tissue. If the inserted bud has turned brown or has shrunken, the bud probably did not take. After healing and the bud forcing steps have been done, the bud should initiate new growth and produce a new stem with the desired characteristics from the scion. The new shoot should be pruned to force further branching and to reduce the possibility of tearing of the bud union before sufficient secondary growth has occurred to make a strong graft union.

Questions

- Why is it advisable to disbud the rootstock before T-budding?
- Why should the bark be in the "slip" stage when budding?
- What is the purpose in "crippling" the stem above the T-bud?

	Procedure 28.1
	Rose T-Budding
Step	Instructions and Comments
1	**Rootstock preparation.** Rootstock cuttings should be rooted before needed in 4-in. pots and vigorously growing. Cutting length should be sufficient to allow manipulation of the required cut and insertion of the vegetative bud (10 to 12 in. long). The rootstock should be disbudded (removing lateral vegetative buds) except for the top two or three buds using a grafting or pocket knife. This will help to prevent suckering of the rootstock later after planting out. Rootstock suckers often are more vigorous than the selected scion cultivar. The vigorous growth will ensure that the bark will "slip" or that the cambium is actively growing. This may necessitate growing the plants in a greenhouse during the dormant season to force new growth.
2	**Bud stick preparation.** Bud sticks should be about the length and diameter of a #2 pencil. They can be harvested from any rose cultivar of choice. Bud sticks can be prepared beforehand, wrapped in moistened paper toweling, placed in a plastic resealable bag, and stored under normal refrigeration. If the budsticks have thorns, they may be removed prior to storage to facilitate handling.
3	**Preparing the rootstock to receive the bud.** Take a sharp knife and make a vertical cut about 1 in. long and deep enough to cut through the cambium layer (1/16 to 1/8 in.), but not into the wood. The second cut is a horizontal cut about 1/3 around the diameter of the stem at the same depth into the cambium as the first cut. Take the tip of the knife blade and insert it into the cut to slightly loosen the flaps of the cambium. The two cuts should resemble a "T" (Figure 28.1).
4	**Preparing the bud to insert into the stock.** Take a budstick and make an initial cut about 1/2 in. below the bud. The cut should be deep enough to get under the cambium layer, but not so deep as to cut into wood. The slanting cut should extend about 1 in. above the bud. The second cut is a horizontal cut about 3/4 inch above the bud through the cambium and into the wood to yield a shield-shaped bud with the curved part of the shield toward the bottom of the bud (Figure 28.2).
5	**Inserting the bud into the stock.** Take the bud and push it into the T-shaped cut under the two flaps of cambium. The horizontal top cuts of the bud and stock should line up (Figure 28.3).
6	**Wrapping the bud and stock.** After the bud has been inserted into the stock, it is necessary to wrap the bud to secure it into place and prevent the bud from drying out. Take a length of the rubber band strip, and leaving the bud exposed, wrap it securely around the two flaps of cambium and extend below and above the cuts. After securing with the rubber strip, take a small square (about 2 × 2 in.) piece of paraffin film and wrap over the entire T-bud section to prevent the bud and stock from drying out.
7	**Growing on.** After the T-budding operation has taken place, place the potted budded plant in a well-lit place protected from strong winds that might damage or dry out the protected buds. Rose buds will normally "take" in about two weeks. If the T-bud is successful, the bud will remain green and turgid during the healing process. After about two weeks, remove the paraffin wax film. Leave the rubber strip in place to prevent the bud from being damaged or accidentally dislodged before the cambium completely heals around the new bud.
8	**Forcing the bud.** When the T-bud has callused over and "healed," it is time to force growth in the new bud (Figure 28.4). This is done by "crippling" or breaking the stem of the stock plant growth above the inserted bud. The object is to bend the stem over and snap it without completely breaking the stem off. This will release the bud and allow it start to grow. Once the inserted bud has started to grow, the remaining stock growth may be removed right above the new bud union. Care should still be taken to prevent damage to the new shoot. If the new shoot is overly vigorous, it is sometimes desirable to prune it back to force lateral branching and prevent top-heavy growth, which could break off the new shoot before the union is completely healed.

Grafting and Budding Exercises with Woody and Herbaceous Species

FIGURE 28.1 T-shaped cut cambium loosened to receive bud.

FIGURE 28.2 Making a shield-shape cut under the bud.

FIGURE 28.3 The T-bud inserted into the rootstock.

FIGURE 28.4 Forcing the bud after healing.

EXPERIMENT 2: SPLICE GRAFTING HERBACEOUS PLANTS

Materials

The following materials will be needed to complete this experiment:

- Four-in. potted plants of coleus (*Solenostemon scutellariodes* (L.) L.E.W. Codd) to use as rootstocks
- Stock plants of desirable coleus cultivars to use as scion material
- Plastic straws (soda straws work well) similar to the diameter of the stems being used
- Paraffin film
- Bandages and first-aid supplies to treat cuts

Follow the instructions provided in Procedure 28.2 to complete this experiment.

Anticipated Results

If the scion and rootstock are properly aligned, the healing process should begin in a few days. The scion will probably wilt or lose turgor for the first few days, but should regain turgor once the healing process begins. This loss of turgor can be compensated by placing the plant under an intermittent mist system or by covering the whole plant with a plastic bag. When the scion or top piece begins to show new growth, it is safe to remove the straw piece used to join the two pieces. This is best done by using a sharp knife or razor blade to slit the straw. Care should be taken not to damage the new graft union.

Questions

- When making the cuts on the scion and rootstock, what is importance of making identical angled cuts?
- Why is paraffin film used to cover the straw and spliced stem pieces?
- Why is the whole plant either placed under a mist system or placed in a plastic bag during the healing process?

	Procedure 28.2
	Splice Grafting Using Coleus
Step	Instructions and Comments
1	**Stock plant preparation.** Take a 4-in. potted plant to use as a rootstock and make a diagonal 45° complete cut through the stem about 2 in. above the soil line. Take a section of plastic straw tubing about 1 in. long and slide it down the length of the stem of the rootstock (Figure 28.5).
2	**Scion preparation.** Take a scion stock plant and find a stem that is about the same diameter as the cut stem section of the rootstock. Cut the scion stem section into a length that has at least two buds (about 2 to 4 in. long). Make a diagonal 45° cut to match the one made on the stock plant.
3	**Splice graft.** While holding the scion carefully in place, pull the section of tubing up the stem of the rootstock over the inserted scion making sure the 45° angles match up. The tubing should be snug enough to ensure that the two stem sections are in intimate contact (Figure 28.6). The straw and exposed ends of the scion and rootstock should be wrapped with paraffin film to prevent drying out and give added support while the healing process occurs.
4	**Growing on.** If the stems are soft and succulent, it may be necessary to place the whole grafted plant under an intermittent mist system or enclose in a plastic bag for a few days until healing begins. After the graft has healed, the plastic tube may be removed by using a razor blade to slit the tubing lengthwise. Care should be taken not to damage the graft union.
5	**Alternative Exercise: Herbaceous Grafting of Vegetable Transplants.** Another example of herbaceous grafting is the splice grafting of vegetable transplants during the seedling stage. This is commonly used in commercially produced greenhouse cucumbers and increasingly with various types of field grown melons. Grafting is typically done to produce a plant that has a root system resistant to soilborne diseases or problematic soils with a scion that produces superior fruit. The procedure is similar to the splice graft described above, but uses seedlings at a young stage with the scion often having only the first set of true leaves present. The procedure is similar to that described above with various types of specialized tubing used to align the splice graft during the healing process.

FIGURE 28.5 Rootstock with angled cut and plastic straw to support graft.

FIGURE 28.6 The scion and rootstock joined by a length of plastic straw.

Experiment 3: Whip and Tongue Grafting

The whip and tongue graft is useful when the plant material is small in size, with stem diameters of ¼ to ½ in. in diameter. The success rate is high if correctly done, since the area of cambial contact is great. If aligned properly, the wound heals quickly and a strong union is formed. Ideally, the scion and rootstock should be of equal diameter, with the scion having two buds.

Materials

The following materials will be required to complete this experiment:

- Four-in. potted plants of Boxwood (*Buxus sempervirens* L.) or other woody plants with a stem ¼ to ½ inch in diameter and a length greater than 4 in. for use as rootstock
- Boxwood or other woody plant material using the same species as the rootstock for scion wood
- Grafting or budding knife or sharp pocket knife
- Rubber budding strips or rubber bands cut into a strip about 4 in. long
- Bandages and first-aid supplies to treat attempted "finger grafts"

Follow the instructions in Procedure 28.3 to complete this experiment.

Procedure 28.3	
Whip and Tongue Grafting Using Boxwood	
Step	Instructions and Comments
1	**Rootstock preparation.** A slanting cut from 1 to 2½ inches is made at the top of the stock, 2 to 4 in. above the soil line of the potted plant. With a sharp knife, a cut is made in one stroke to make an even straight cut. A straight cut will ensure better cambial contact between the scion and the rootstock. A second cut in the reverse direction is made about 1/3 down from the top of where the first cut ended and should extend about half the way down the original cut. The second cut should parallel the first cut to assure a close fit. When completed, the cut should resemble a sideways letter "Z."
2	**Scion preparation.** Choose scion wood the same diameter as the rootstock. Remove any attached leaves. The scion wood should be long enough to have two to three dormant buds. Two identical cuts as in the rootstock are made (Figure 28.7).
3	**Joining the scion and the rootstock.** The scion is inserted into the rootstock so that the two Z-shape cuts interlock. Care should be taken to ensure that the cambium of the two pieces line up and there are no gaps between the scion and rootstock. After the two pieces have been joined, they should be securely tied in place with a length of rubber stripping and then wrapped in paraffin film to prevent drying out of the scion wood while healing (Figure 28.8). If the graft takes, new growth will initiate from the scion buds. Any growth from buds below the graft union should be removed.

FIGURE 28.7 Making identical cuts on the rootstock and the scion.

FIGURE 28.8 Graft area wrapped with paraffin film ready for healing process.

Anticipated Results

If the scion is properly matched up with the rootstock, the healing process will begin. Wrapping the two sections with paraffin film will prevent excess drying out of the scion. Once the healing process is complete and the graft union has formed, the lateral buds along the scion should begin to develop and grow. Once new growth has started, the rubber budding strip and paraffin film may be removed. This healing process will depend on the vigor of the plant, temperature, and time of year. This period could range from three to six weeks, so patience is needed when attempting whip and tongue grafting.

Questions

- Why is a large amount of cambial surface area important in whip and tongue grafting?
- Why should the scion wood and rootstock be the same diameter?
- Why is it important to remove any shoot growth occurring below the graft union?

SUGGESTED READINGS

Bailey, L.H. 1911. Graftage, in *The Nursery Book,* 15th ed. Macmillan, New York, pp. 94–108.

Hartmann, H.T., D.E. Kester, F.T. Davies, Jr., and R.L. Geneve. 2002. *Plant Propagation: Principles and Practices,* 7th ed. Prentice-Hall, Inc., Upper Saddle River, NJ. 880 pp.

Part XI

Bulbs and Plants with Special Structures

29 Storage Organs

Jeffrey A. Adkins and William B. Miller

CHAPTER 29 CONCEPTS

- Geophytes are plants that form specialized storage organs that serve as overwinter survival structures.

- The term "bulb" is often used to represent all specialized plant storage organs.

- "True bulbs" have a compressed stem (basal plate) and modified fleshy leaves that serve as the primary storage tissue.

- Corms, tubers, rhizomes, and enlarged hypocotyls derive from stem tissue, but differ in their sites of origin, presence or absence of scales, and orientation.

- Tuberous roots develop from enlarging root radicles and secondary roots.

- Propagation of various storage organ may be accomplished by scaling, scooping, or scoring of bulbs, division, stem cuttings or adventitious shoot production.

Plant species evolve over time in response to changing environmental conditions to ensure survival and continued reproduction. These evolutionary changes are what provide us with the great diversity of plant forms and functions. Geophytes represent a unique group of plants that have developed adaptations allowing them to survive as underground storage organs until environmental conditions are favorable for growth. There are many types of underground storage organs; however, bulbs, corms, tubers, tuberous roots, tuberous stems (enlarged hypocotyls), and rhizomes are the most common in horticulture. Each type of storage organ is defined based on the origin of development on the plant.

THE INDUSTRY

Flowerbulbs are grown for the following two major purposes: (1) for use in commercial forcing of cut flowers and pot plants, and (2) as decoration in gardens, landscapes, and parks. The relative proportion of each of these uses varies by country, but worldwide, a significant majority of bulbs are used for forcing. Flowerbulb production is a specialized activity and requires unique equipment, industry infrastructure, farming practices, horticultural skill, and the proper climatic conditions for consistent success. While flowerbulbs are produced worldwide, the center of production is The Netherlands. Across a wide spectrum of the range of bulbous plants grown, the Dutch industry enjoys specific competitive advantages in each of the areas listed above. The major ornamental bulbous crops grown in The Netherlands include tulip (*Tulipa*), hyacinth (*Hyacinthus*), daffodil (*Narcissus*), hybrid lilies (*Lilium*), and a range of others including *Crocus, Muscari, Gladiolus, Iris, Allium,* and others too numerous to list.

The path of a bulb from a field in The Netherlands to market in North America is a long and complicated one. In most cases, bulbs are grown in fields (there are more than 25,000 ha [hectares] of bulbs produced in Holland annually), and harvested at or during leaf senescence, after current photosynthesis has "refilled" the bulb. Depending on the species, the basic procedure is that bulbs are cleaned, "peeled" (the process or removing side bulbs, often used for planting stock), sized (generally by circumference), counted, placed into sales packages (for the retail or commercial landscape market), or into "export crates" for transport to professional forcing markets around the world.

Bulbs are creatures of temperature, thus the industry is highly advanced in its understanding and use of controlled temperature environments for bulb storage and transportation. Bulbs exported from Europe are, with very few exceptions, transported via ocean vessel, in temperature-controlled shipping containers. In principle, from the moment the bulbs are loaded into the container until delivery to the final customer (this takes from two to five weeks for shipments to North America, depending on route and destination), temperature and ventilation can be controlled within the container.

Other than The Netherlands, there are significant production regions that emphasize specialty bulbs or bulbs with unique climactic requirements that give competitive advantages to production in those locations. To cite just a few examples, Israel is a major and specialized producer of paperwhite narcissus (*Narcissus tazetta*) and amaryllis (*Hippeastrum*). Significant numbers of amaryllis are also produced in South Africa and Brazil. The Pacific Northwest region of the United States is known for Easter lily bulb production (*Lilium longiflorum*) and tulip, daffodil, and iris. The development of the southern hemisphere as a production source has revolutionized the bulb industry, as it allows the production of bulbs that are six months "off cycle" from the northern hemisphere. This has important implications for scheduling and forcing of bulbs in seasons that are traditionally difficult or impossible (e.g., allowing tulip flowers to be available in summer and fall). The next 25 years will increasingly see bulb production activities move away from the physical borders of The Netherlands as pressures of a burgeoning population, stricter environmental and social regulations, and prohibitive costs are encountered.

BULBS

The term bulb is often used in general to describe several specialized storage organs. However, a "true bulb" has several unique characteristics that distinguish it from other storage organs. Bulbs are compressed stems with modified fleshy leaves, called scales, which serve as the primary storage site for carbohydrates, nutrients, organic compounds, and water (Figure 29.1). The compressed stem forms the basal plant and contains a growing point and adventitious roots. Some bulbs have thick fleshy "contractile roots" that serve to pull the bulb to the proper depth in the soil. Bulbs can be classified as either tunicate or nontunicate (scaly). Tunicate bulbs, such as tulips and daffodils, have a dry papery outer membrane that serves to protect the fleshy inner scales. Scales are attached to the basal plate and arranged in concentric layers. Nontunicate bulbs have no dry papery outer membrane, making them more susceptible to mechanical damage and drying out. Like tunicate bulbs, nontunicate bulbs have scales attached to the basal plant; however, the scales overlap giving a bumpy scale-like appearance. Lilies are an example of nontunicate bulbs.

BULB PROPAGATION

Offsets are miniature bulbs (bulblets) formed naturally underground on bulbs and can be used to efficiently propagate many bulbous plants. Offsets will produce vegetative growth and adventitious roots following removal from the mother plant and planting in soil. The most important ornamental bulbous plant in the world, the tulip, is propagated by bulb offsets. Unfortunately, some important bulbous plants, such as hyacinth (*Hyacinthus*) and lily (*Lilium*), do not produce offsets fast enough for commercial production and, so, must be propagated using different methods.

Scooping (see also chapter 30) is a method of bulb propagation that involves the removal of the basal plate and growing point of a mature bulb and is commonly used to propagate hyacinths. It is important to treat the scooped bulb with a fungicide to prevent rotting or decay. Following scooping and fungicide treatment, bulbs are placed in a dark, dry environment at 21°C (70°F) for a few weeks for wound tissue (callus) formation on the cut surface. After callus has formed, bulbs are exposed to increased temperature of 30°C (85°F) and high relative humidity (~85%) to facilitate bulblet formation on the callus tissue. A scooped bulb might produce 30 to 60 bulblets, depending on the species and condition of the mother bulb.

Scoring (see also chapter 30) is another bulb propagation method similar to scooping. For scoring, three to six vertical cuts are made through the basal plate and growing point. The cuts may either be straight or made wedge-shaped by removing sections of the bulb to expose greater wounded surface area. Scored bulbs are thereafter treated as with scooped bulbs, including a fungicide treatment. Scored bulbs typically produce 15 to 25 bulblets.

Coring and sectioning are two additional propagation methods that are similar to scooping and scoring. Coring involves the removal of a cone-shaped section of the bulb through the center of the basal plate and the entire growing point. Sectioning differs from scoring only in that the cuts are made vertically through the entire bulb, resulting in 6 to 12 or more individual wedge-shaped sections. Both methods follow the same protocol described for scooping and scoring, but typically result in relatively fewer bulblets formed. With sectioning, there is a tradeoff in terms of the number of bulblets produced (often, a greater total number of bulblets are produced as a bulb is further sectioned) versus the weight of individual bulblets (larger sections yield fewer, but larger, bulblets).

Bulb cuttings begin by first sectioning the bulb as described above; however, each section is then divided by bulb scales into groups of two to four each containing a portion of the basal plate. When two scale pieces are used, the resulting cutting is commonly referred to as a "twinscale." The bulb cuttings are placed vertically in sand or vermiculite in a warm, moist environment. Bulblets form between the scales from the basal plate in two to three weeks.

Individual scales of nontunicate bulbs can be easily separated and used as a propagule for new bulb formation. Bulbs to be used for scaling are harvested following flowering, and care must be taken to prevent the bulbs from drying out. Although all the scales present could be used, inner scales are not as productive, and the mother bulb will be diminished or destroyed. Easter lily is an important bulbous crop that is commercially propagated by scaling. Following removal of scales, they are treated with a fungicide and

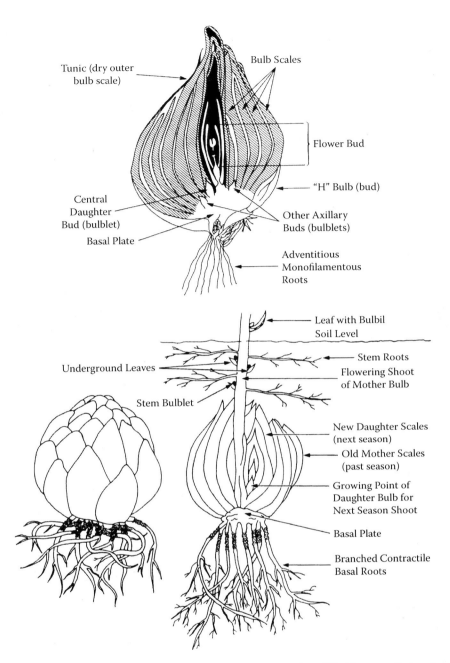

FIGURE 29.1 A true bulb is comprised of a compressed stem, or basal plate, and modified leaves called scales serve as the primary storage tissue. Examples include tulip, lily, allium (e.g., onion), Dutch iris, hyacinth, daffodils, and others. A bulb is referred to as tunicated if the outermost scale is dried out and papery, serving as protection for the fleshy scales beneath. Tulip bulbs are tunicated. Bulbs that lack the papery outer covering, such as lily bulbs, are described as nontunicated. (From De Hertogh, A.A. and M. Le Nard (Eds.) 1993. *The Physiology of Flower Bulbs*. Elsevier, Amsterdam, The Netherlands. With permission.)

placed in a rooting medium either in trays basal side down or in sealed plastic bags. Three to five new bulblets will form at the base of each scale within two to three months.

CORMS

A corm, like a bulb, is a compressed stem with a basal plate formed from an underground swollen stem surrounded by dry leaves or scales (Figure 29.2). Corms can be tunicate or nontunicate; persistent dry leaf bases protect the fleshy scales by forming a tunic (cover), often over the entire corm. Unlike true bulbs, the thickened stem serves as the primary storage site, and scales are attached at nodes, which, along with internodes, are distinguishable.

Most corms undergo annual replacement where a new corm is formed each year. One or more new corms usually begin development following flowering on top of the old corm at the base of the current season's stem. The new

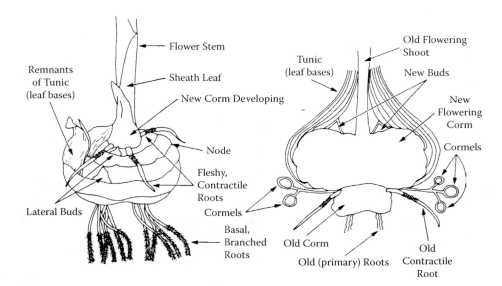

FIGURE 29.2 Like a true bulb, the corm is a modified stem with a basal plate, but the primary storage tissue is the stem tissue itself, rather than leaf tissue, so corms are frequently described as "solid bulbs." These organs also may be tunicated or nontunicated, and they have nodes from which meristems originate. Gladiolus, freesia, crocus, and ixia are some examples. (From De Hertogh, A.A. and M. Le Nard (Eds.) 1993. *The Physiology of Flower Bulbs*. Elsevier, Amsterdam, The Netherlands. With permission.)

corm develops fleshy contractile roots that will move the new corm down to the depth of the current season's corm. The mother corm has a fibrous root system at the base. In addition to new corms, miniature corms, or cormels, often form between the new and old corms.

CORM PROPAGATION

Propagation of corms is usually accomplished by harvesting new corms and cormels. Fortunately, modifying environmental conditions can increase the number of new corms and cormels produced. Corm propagation usually begins in the field. Mother corms are planted and allowed to develop vegetatively and eventually flower. Proper fertilization and moisture control will encourage the development of large corms and an increased number of new corms and cormels. Corms are lifted only after the tops have withered or have been killed by frost. Lifted corms are first clean of dirt, placed in slatted trays or on screens for air circulation, and stored for 12 to 24 hours at 30°C (87°F). This curing treatment eases separation of new corms and cormels from the mother corm. New corms and cormels are graded for size, and any diseased corms are discarded. The new corms should be treated with an appropriate fungicide and then stored at 30°C (87°F) for one week and then stored in a cool (5°C/40°F) moist environment until planting time the following season.

TUBERS

Tubers, like corms, are thickened underground stems that often develop at the tip of stolons or rhizomes and serve as storage organs (Figure 29.3). Unlike corms, however, tubers have no basal plate, since they do not originate from the base of a stem. In potatoes (*Solanum tuberosum*) and many other tuberous plants, many tubers may form on a single plant. Tubers are anatomically the same as stems, having internodes and nodes from which "eyes" develop, containing one or more shoot buds. Tuberous plants produce tubers each season. The tubers then serve as an overwintering storage site, producing new roots and shoots during the following season. The new shoots use the reserves from the tuber for initial growth and produce new tubers for the following year.

TUBER PROPAGATION

Many tuberous plants are propagated asexually by division of the tuber. Tubers are divided into sections, each containing one or more eyes. Tubers can be planted directly; however, the freshly cut surface will be highly susceptible to disease organisms. To avoid this problem, cut sections are often stored in a warm (20°C/68°F) moist environment to allow the cut areas to heal. A few tuberous plants will produce tubercles that form from buds in the leaf axils above ground. Tubercles are easily removed from the stems and can be stored overwinter and planted like a seed in the spring. *Begonia grandis* ssp. *evansiana* and *Dioscorea bulbifera* are examples of tubercle-producing plants.

TUBEROUS ROOTS AND ENLARGED HYPOCOTYLS

Tuberous roots develop during seed germination with the enlargement of the developing radicle and from sec-

Storage Organs

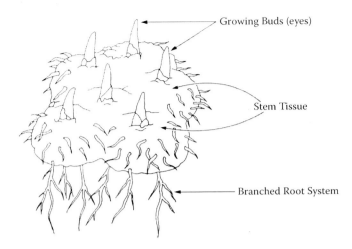

FIGURE 29.3 A tuber is a thickened underground stem, and the stem tissue serves as the primary storage tissue, but, unlike a corm, it has no basal plate. Meristems occur on the tuber and are commonly called "eyes" on a potato. Caladium is the most common floricultural tuber. (From De Hertogh, A.A. and M. Le Nard (Eds.) 1993. *The Physiology of Flower Bulbs*. Elsevier, Amsterdam, The Netherlands. With permission.)

ondary roots attached to the crown (Figure 29.4). They differ from "true" tubers in that they develop from roots instead of stems and do not have the anatomical features of a stem, such as nodes and internodes. Most tuberous roots do not develop adventitious shoots except from the proximal end where a small portion of the crown is attached. However, some tuberous roots, such as sweet potato (*Ipomoea batatas*), have the capacity to produce adventitious shoots under the right environmental conditions. Tuberous roots serve as the overwintering portion of the plant and provide energy for growth the following season. The tuberous root declines as reserves are used up and new tuberous roots develop. After a few years of growth, plants may have numerous tuberous roots attached to the same crown.

The hypocotyl is the area of a germinating seedling between the cotyledons and the root radicle. In plants such as cyclamen (*Cyclamen persicum*) and gloxinia (*Sinningia speciosa*), the hypocotyl enlarges to becoming a storage organ called an enlarged hypocotyl or tuberous stem (Figure 29.5). Tuberous roots are usually biennial, whereas enlarged hypocotyls are perennial. Leaves and stems emerge from the crown of the enlarged hypocotyls and roots from the base. Cool night temperatures promote carbohydrate storage in the hypocotyl, increasing overwintering capacity and vegetative and reproductive growth. This is particularly important in food crops that develop enlarged hypocotyls, such as beets (*Beta vulgaris*), where both the enlarged hypocotyls and the leaves are consumed.

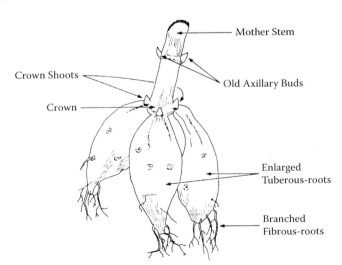

FIGURE 29.4 A tuberous root consists of enlarged fleshy root tissue. Roots are fleshy because they are the primary storage tissue. Growth arises from buds at the top (crown) of the root mass. Examples include dahlia, anemone, and ranunculus. (From De Hertogh, A.A. and M. Le Nard (Eds.) 1993. *The Physiology of Flower Bulbs*. Elsevier, Amsterdam, The Netherlands. With permission.)

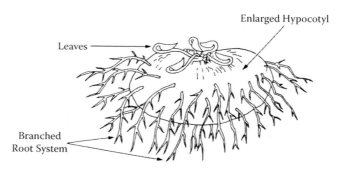

FIGURE 29.5 The hypocotyl of a seedling is the portion of the stem below the cotyledon and above the roots. In some plants, the hypocotyl enlarges, becoming a fleshy storage site as the plant develops. Cyclamen and gloxinia are examples of plants with enlarged hypocotyls. Also, "tuberous" begonias are not actually tuberous, but have enlarged hypocotyls. (From De Hertogh, A.A. and M. Le Nard (Eds.) 1993. *The Physiology of Flower Bulbs*. Elsevier, Amsterdam, The Netherlands. With permission.)

TUBEROUS ROOTS AND ENLARGED HYPOCOTYL PROPAGATION

Tuberous roots and enlarged hypocotyls can be propagated by division. Propagation of tuberous roots by division is accomplished by separating individual tuberous roots from the crown. Care must be taken to ensure each tuberous root has a portion of the attached crown and has at least one bud present. To divide enlarged hypocotyls, the hypocotyl is sectioned much like a tuber, with each section having at least one bud. The cut sections of the hypocotyl will be susceptible to disease organisms and should be allowed to heal before planting. Fungicide treatment may also be beneficial.

Shoot cuttings from plants having enlarged hypocotyls can be rooted similarly to those from other herbaceous and woody plants. Some evidence suggests that including a portion of the hypocotyl may enhance propagation success. Leaf and bud cuttings can also be accomplished by taking a single leaf and corresponding bud from the hypocotyl. The leaf and bud cutting is placed in a suitable rooting medium and moist environment to encourage bud break and eventually adventitious root development. Opportunities for shoot cuttings from tuberous roots are limited. However, plants such as sweet potatoes can be propagated by taking stem cuttings from the current season's vines and treating as with other cuttings. Unlike most tuberous roots, sweet potato tubers will produce adventitious shoots that can be rooted. To promote adventitious shoot development in sweet potato (slips), tuberous roots are placed in an open bed and covered with sand or other suitable medium. If outdoors, beds should be covered to maintain a warm moist environment. The beds should be ventilated and kept moist after adventitious shoots emerge. Adding additional medium at the base of the new shoots (i.e., healing) will promote adventitious root development. When new shoots are about eight in. tall and have three to five nodes, they can be removed from the tuber by cutting below the new root system.

RHIZOMES

Rhizomes are another type of specialized stem that grows horizontally and serves as a storage organ (Figure 29.6). Rhizomes may grow at the soil surface as with irises (*Iris* sp.) or below ground as with ginger (*Zingiber officinale*). Rhizomes that grow on the surface tend to be large and thick, while many underground rhizomes are more slender and rapidly spread. Many garden weeds, such as Johnsongrass (*Sorghum halepense*) and quackgrass (*Elytrigia repens*), have very fast-growing slender rhizomes, making them difficult to control using mechanical cultivation. When these underground rhizomes are severed with a shovel or hoe, new adventitious shoots and roots develop, and growth continues. Many rhizomes are segmented in appearance, indicating the nodes and internodes present on the stem. Also, most plants having large rhizomes form crowns and spread slowly, whereas those with slender rhizome are crownless. Most, but not all, plants with rhizomes are monocots, and new leaves develop from the nodes, initially sheath the rhizome, and continue growth vertically away from the rhizome.

RHIZOME PROPAGATION

Crown-forming rhizomes, such as iris and asparagus (*Asparagus officinalis*), can be divided during the dormant season. Lift rhizomes from the ground or container and divide with a shovel or sharp knife. As with other divisions, care must be taken to ensure the inclusion of at least one bud. Divisions can be replanted immediately or placed in containers filled with an appropriate medium and stored in a warm moist environment. Irises and gingers may also be divided after flowering in the summer. Rhizomes are lifted and divided into segments containing at least one culm. Culms are shoots arising from the rhizome and are often hollow, except at the nodes. Culms are then cut back, and the rhizomes are replanted.

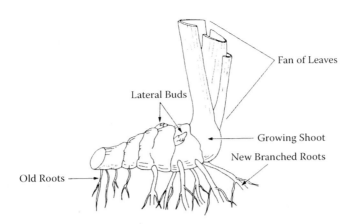

FIGURE 29.6 A rhizome is a modified stem, and the stem tissue itself is the primary storage tissue. The rhizome, however, is unique in that it grows horizontally through soil. Some irises have rhizomes, as well as lily-of-the-valley, calla lily, oxalis, and achimenes. (From De Hertogh, A.A. and M. Le Nard (Eds.) 1993. *The Physiology of Flower Bulbs*. Elsevier, Amsterdam, The Netherlands. With permission.)

Some rhizomatous plants, such as bamboo, can be propagated by culm cuttings. Culms with lateral branches present are cut from the rhizome and either sectioned (single or double node with branch) or used whole. The cuttings are placed in sand beds, and the culm, but not the culm branch, is covered. Auxin treatment may hasten adventitious root develop.

REFERENCES AND SUGGESTED READINGS

Bryan, J.E. 1989. *Bulbs*. 3 vols. Timber Press, Portland, OR. 218 pp.

De Hertogh, A.A. and M. Le Nard (Eds.) 1993. *The Physiology of Flower Bulbs*. Elsevier, Amsterdam, The Netherlands. 812 pp.

Hamrick, D. (Ed.) 2003. *Ball Redbook: Crop Production*. 17th ed. vol. 2. Ball Publishing, Batavia, IL. 724 pp.

Hartmann, H.T., D.E. Kester, F.T. Davies, Jr., and R.L. Geneve. 2002. *Plant Propagation: Principles and Practices*. 7th ed. Prentice Hall, Inc., Upper Saddle River, NJ. 880 pp.

Huxley, A., M. Griffiths, and M. Levy (Eds.) 1992. *The New Royal Horticulture Society Dictionary of Gardening*. vol. 1. MacMillan Press, London.

Toogood, A.R. (Ed.) 1999. *American Horticulture Society Plant Propagation*. DK Publishing, New York. 320 pp.

30 Propagating Selected Flower Bulb Species

William B. Miller, Jeffrey A. Adkins, and John E. Preece

Most bulbous plants can be propagated asexually through a variety of means, including offset division, basal cuttage (scooping or scoring), scaling, or division (chapter 29). Such asexual methods are used to maintain clonal characteristics, which is important because most bulbous, tuberous, cormous, or rhizomatous ornamental plants are sold with a cultivar designation. Unless the objective is breeding of new cultivars, where the sexual phase and seed production is necessary, with rare exceptions, almost all ornamental bulbous plants are propagated vegetatively.

Ornamental bulbs are colorful additions to the spring or summer garden. Spring-flowering bulbs, such as tulip, daffodil, and hyacinth, are available and planted in the fall, whereas summer-flowering bulbs, such as lily, begonia, and gladiolus, are typically planted in the spring. By definition, bulbs planted in the fall and surviving the local winter are "hardy." In the case of the summer-flowering bulbs, many are in fact substantially winter hardy (e.g., most lilies), whereas others are very noncold hardy and must be planted after all frost is past, or even started indoors (e.g., tuberous begonia).

The following laboratory exercises assume a fall course offering, as the hyacinths would be commercially available from bulb suppliers or local garden centers. The exercises on lily, dahlia, or potato can, in principle, be done at any time of the year. The lily exercise shows the adventitious formation of bulblets at the base of lily bulb scales in relation to different environmental conditions. Recall that a bulb is a compressed stem, and that the bulb's scales are actually leaves modified to store large quantities of starch and other carbohydrates. Bulblets that form from the scales have a small basal plate of stem tissue from which new scales grow at nodes. Adventitious roots are attached to each basal plate, not the mother scale. The influence of new leaf growth from the bulblets on the subsequent growth of the bulblets will also become evident between the two treatments. The hyacinth, potato, dahlia, and garden experiments are mainly descriptive and manipulative in nature and allow students practice with different forms of asexual propagation of flower bulbs.

EXERCISES

EXPERIMENT 1: EFFECT OF THE PROPAGATION ENVIRONMENT ON LILY BULBLET FORMATION AND GROWTH

Materials

The following items will be sufficient for individual students or teams:

- One healthy lily bulb
- Clean potting medium
- Sphagnum peat:perlite:vermiculite (2:1:1, by volume) with nutrients and wetting agent provides a suitable environment for bulblet formation. Commercially available mixes or handmade blends work well, if there is good wettability of the medium
- One 4- to 6-in. azalea or standard pot or a 3 × 5 × 8 in. market pack or similar container. The container should be new or properly disinfested of fungal pathogens. Drainage holes are essential
- One new 1-qt, clear plastic bag. This can have a zippered top, or can be closed with a twist tie

Follow the instructions presented in Procedure 30.1 to complete this experiment.

Anticipated Results

Almost all lily bulb scales will form bulblets, regardless of whether they are in the plastic bag or pot or market pack. The bulblets will sprout and grow faster in the pot or market pack in full sun. The bulblets in the plastic bag will grow slowly and will sprout leaves that become etiolated because of low light. Callus (looks like small bumps on the surface) will form within the first two weeks. The bumps grow into the bulblets and can be counted. Root growth will be greater on bulblets in full sunlight. The bulblets should have well-formed, rooted, and sprouted bulblets within five weeks, and certainly within six to eight weeks.

	Procedure 30.1 Propagating Lilies from Scale Leaves
Step	Instructions
1	Fill the pot or market pack with the potting medium, then water. Add enough moist (not wet) potting medium to the bag so that the medium is "three fingers" deep (measure by holding three fingers up to the back of the bag).
2	Select one healthy lily bulb. Unlike an onion or hyacinth, there is no papery covering on this bulb. Discard the dried and damaged scales from the outside of the nontunicate bulb. Keep removing scales until only white, plump scales remain. Break off at least 10 of the largest, whitest scales and divide them into two uniform groups.
3	Plant all of the scales in one of these groups in the medium in your pot or in a market pack. Plant them so that the pointed end is up and the end where the scale was attached to the mother bulb is within the medium—the lower halves of the scales should be within the medium.
4	In the plastic bag, mix all of the scales from your other group into the medium with either your fingers or by shaking the bag. Close the bag using the zipper or twist-tie. Water thoroughly the scales in the pot or market pack, then place in full sun. This is typically done on a greenhouse bench, but a bright window or other protected environment is suitable. The container must be watered when it dries.
5	Place the plastic bag on a bench or table in the classroom. Add water only if necessary. Some bags may not require water for a year. Room temperature and the light that is on when the students are present are sufficient for growth and formation of bulblets.
6	Data should be collected once each week on number of bulblets, bulblet diameter, rooting, and sprouting.

Questions

- Why do the upper half of the bulb scales that are exposed to full sunlight become green, often with pink, red, or purple?
- If you tried different colored cultivars, is the scale color in the light-exposed area similar to the flower color? Why or why not?
- What are the implications of chlorophyll formation in the scale portions exposed to sunlight on the propagation of bulblets?
- What are the facilities requirements, labor requirements, and overall expenses for each method of bulblet propagation using scales? Which method would you use if propagating lilies commercially?
- Other than passing direct genetic information to the bulblets, mother scale pieces may also transmit pathogens (disease causing agents). How can this be reduced or avoided?
- What other methods can be used to propagate lily?

EXPERIMENT 2: THE EFFECTS OF PLANTING METHOD ON PROPAGATION OF HYACINTHS BY SCALING

Hyacinths (*Hyacinthus orientalis*) produce beautiful, fragrant flowers and are used in gardens and forced in containers. Commercially, hyacinths are propagated by removal or cutting through the basal plate of stem tissue at the base of the bulb. The basal plate can be scored or scooped out, or bulbs can be cored vertically to produce bulblets (chapter 29). To be successful with these techniques, the proper facilities must be available to allow for proper humidity, temperature, and ventilation control. These "basal cuttage" operations are of prime importance, since experience in The Netherlands has shown that each hyacinth cultivar requires a slightly different treatment (e.g., depth or width of scooping, or postcuttage incubation temperature). Commercially, scooping or scoring is done in early to mid-July.

For the purposes of these exercises, the best quality hyacinth bulbs (>18 cm in circumference) are available in late summer and autumn. Sometimes, they may be available in the spring, but are of lower quality, and cultivar selection will be limited. The cultivar 'L'Innocence' will produce bulblets reliably. If the experiments are conducted in the spring, it will be necessary to store the bulbs. Hyacinth bulbs can be difficult to store for five to six months without rotting. If the bulbs are to be stored for more than two months, they should be soaked for one to two h in 0.63 g/L (6 g/gal) Truban 30% WP fungicide (30% ethridiazole, a.i.). They should then be placed in open containers of bone-dry sphagnum moss at 4°C.

To be successful in scaling hyacinths, cuts must be made in the proper area at the base of the scale. Scales are modified leaves that store food and nutrients for the plant during its dormant season. Each hyacinth scale is more or less globe-shaped, and each completely surrounds the inner scales. The outermost scales have dried into the papery tunic or covering. Each scale is attached to the basal plate (stem tissue) at a node, and the cells at the base of each scale leaf have the competence to dedifferentiate and form bulblets. This ability decreases and then is lost with increasing distance from the basal plate. The basal plate also cannot form adventitious bulblets. Therefore, the cut at the bottom of the scale should be within 5 mm of the basal plate and should include no basal plate tissue. This will ensure that the scale piece contains competent cells for producing callus and adventitious bulblets.

This laboratory exercise will give students experience with vegetative propagation of a tunicate bulb. Although the focus of the experiment is on scaling, other propagation techniques, such as scoring and scooping, can be practiced as well. This will still allow for all of the bulb scale tissue that is necessary for the scaling exercise.

Materials

The following items will be sufficient for individuals or teams of students.

- Three healthy hyacinth bulbs
- Pasteurized potting medium. A medium containing sphagnum peat:perlite and vermiculite (2:1:1, by volume) is recommended. Commercially available or handmade media can yield good results, if they have good wettability
- One 6-in. azalea or standard pot, 3 × 5 × 8 in. market pack, or other similar container will work well. Whatever is used, it must have drainage holes and should either be new or properly disinfested
- A tray with ca. 1/2 in. of sand or dry sphagnum moss in it to help "hold" bulbs upright after scooping or scoring
- Sharp knife
- Greenhouse or bright, sunny windows or other protected, sunny location
- A temperature of 25°C/20°C ± 3°C should be maintained under a natural photoperiod

Follow instructions outlined in Procedure 30.2 to complete the primary and optional experiments.

Anticipated Results

Almost all hyacinth scales will form bulblets. There may be slightly more on the horizontal pieces. The bulblets will require at least six to eight weeks to form roots, and perhaps longer for green photosynthetic leaves to appear. Initially, creamy to light brown-colored callus at the cut base of the scale pieces will be seen. The callus will gradually form "bumps" that will grow into bulblets. There are similarities and differences between bulblet formation from hyacinth bulb scale pieces and lily bulb scales. This makes an interesting comparison if done at the same time as the propagating lily bulbs experiment.

Questions

- How does hyacinth propagate itself in nature? Is that method more or less efficient than scoring, scooping, or scaling?
- Which technique, scoring, scooping, or scaling, results in the most bulblets per bulb?
- How many years does it take from the formation of a new bulblet to the time that it is 18 to 19 cm in circumference and ready to be stored?
- Why does the mother scale piece change color when exposed to sunlight?
- Is the change in color good or bad for production?

	Procedure 30.2 Propagating Hyacinth Bulbs (Including Some Optional Experiment)
Step	Instructions
	Sectioning
1	Cutting from the top down (longitudinally), quarter the bulb. Flake apart the scale pieces with your fingers. Select at least 10 uniform, plump, large, healthy scale pieces.
2	Plant at least five scale pieces vertically, so that the upper half of the scale piece is above the surface of the medium. Plant the remaining scale pieces horizontally, so that the curved (convex) side is up, and the cut bottom is within the surface 2 to 3 mm of the medium.
3	Dust the horizontally planted scale pieces lightly with medium, leaving the curved portion exposed to the sunlight.
4	Water and place in the greenhouse, sunny window, or full sunlight in another protected environment.
5	Data should be collected once each week on callus formation, number of bulblets, bulblet diameter, rooting, and sprouting.
	Scoring
1	Holding the bulb so that the basal plate is facing you, cut completely across and through the basal plate. The basal plate is shaped conically within the bulb, so cut deeper in the center than toward the edges of the basal plate.
2	Angle the knife slightly and make a second parallel cut about 2 mm away to remove a thin wedge. Turn the bulb 90°, and make a second wedge cut so that the basal plate looks like a "+" sign. The cuts must be sufficiently deep to wound the bottom 5 mm of the scales above the cut. If the cut bottoms of scales are not visible, continue cutting until they are. It is the wounded and exposed portion of the bases of the bulb scales that will produce bulblets. Follow procedure in Step 2, below, in the "Scooping" section.
	Scooping
1	Hold the scored basal plate toward you and remove the entire basal plate. This can be done by angling the knife and cutting so that the concentric rings of the bases of the scales are visible. A sharpened spoon or melon baller could also be used for scooping. Care must be taken to remove all of the basal plate tissue. Keep cutting if you cannot see the concentric rings at the base of the bulb. If cuts are too deep, all competent cells will be removed, and the remaining scale pieces will either not produce bulblets or do poorly. Scooped bulbs are processed like scored bulbs for bulblet production.
2	For either scooped or scored bulbs, place them vertically, with the cut base plates up, into trays with clean sand (this helps hold them upright). Exposure to 70°F is needed to heal the cuts; then, if possible gradually increase the temperature to ca. 85°F. If this is not possible, use room temperature, although the process will go more slowly.
3	Make observations on time to see callus and on the appearance of any bulblets. This takes a long time (two and a half to three months) and may not be possible within the length of a typical propagation class. In commercial practice, hyacinth propagation is done in July, soon after the bulbs are lifted. Intact scooped or scored bulbs, with numerous bulblets attached, are planted in fields in October for overwintering and growth the following year.

EXPERIMENT 3: DIVISION OF DAHLIA ROOTS AND CUTTING OF A POTATO TUBER

Materials

The following items will be sufficient for individual students or teams:

- One or two healthy tuberous root clumps of *Dahlia*. These may be purchased in the spring or may be lifted from a garden in the fall before the first frost
- A medium-sized, healthy potato
- Clean, good-quality potting mix
- Two 6-in. diameter plastic pots
- Sharp knife

Follow instructions outlined in Procedure 30.3 to complete the primary and optional experiments.

Propagating Selected Flower Bulb Species

Procedure 30.3
Division of Dahlia and Potato

Step	Instructions
	Dahlia
1	Observe the dahlia clump and identify the plump storage roots and also the stem tissue (the remnants of the prior growing season's stem is also usually prominent) to which they are joined. Closer examination should reveal several small, pointed buds associated with the stem tissue.
2	Carefully cut two to four roots from the clump without any stem or bud tissue being present. Cut two to four other roots in such a manner that stem tissue, including at least one bud, is present per root.
3	Plant the roots or each treatment into separate pots. Orient such that the stem (apical) end is about 1 cm above the surface of the mix. Label the pots.
4	Make observations at weekly or biweekly intervals. Make notes of presence or absence of shoots, and of shoot length. Continue observations for six to eight weeks.
5	At the conclusion of the experiment, remove the plants, and remove soil from them. Observe and make notes on root growth on all roots.
	Potato
1	Take the potatoes and make a cut across the length of each tuber. Save two of the halves. With the remaining halves, cut in half again, so you now have four quarters. Save two of these, then cut each quarter again, so you have eighths. By this time, be sure you have at least one eye per piece—the smaller the pieces, the more important this is. Continue cutting if possible. The result will be a range of potato pieces, from halves through perhaps 16ths of a whole tuber.
2	Store the pieces at about 70°F for two to three days to allow wound healing.
3	Plant such that the eyes are at or slightly under the soil surface.
4	Make observations at weekly or biweekly intervals. Make notes of the presence or absence of shoots, and of shoot length. Continue observations for six to eight weeks.
5	At the conclusion of the experiment, remove the plants and remove the soil from them. Observe and make notes on root growth and on the condition of the originally planted potato pieces.

Anticipated Results

The underground portion of a dahlia is a clump of tuberous roots, which are attached to stem tissue at their apical ends. When roots are separated from the stem tissue (the "crown"), the absence of a viable bud will result in no growth and a failed propagation effort. When one or more buds are present on the removed root, a shoot will grow, and eventually adventitious roots will form. Roots will also form on the original tuberous root. A potato is a tuber, and the entire volume of the potato is stem tissue. The "eyes" of a potato are buds, arranged on a spiral around the tuber. When cutting the potato, one invariably obtains one to several buds per chunk of tissue, and successful propagation results.

Questions

- Why is more care needed in dividing dahlia tuberous roots than in cutting a potato?
- With the dahlia, is there a difference in visual root growth at the end of the experiment? Why or why not?

EXPERIMENT 4: OBSERVATION OF NATURAL PROPAGATION STRUCTURES IN BULBOUS PLANTS

The following exercise assumes the availability of a garden containing examples of bulbous plants, such as tulip, daffodil, hyacinth, lily, gladiolus, tuberous begonia, canna, and iris. Permission to dig plants and observe their underground structures is also assumed. While not an experiment per se, this is a great observational exercise that allows students to gain significant experience with the underground storage organs of many ornamental bulbous plants.

Materials

The following items will be sufficient for a student team or the entire class:

- One garden containing bulbous plants
- One shovel

Follow instructions outlined in Procedure 30.4 to complete exercise.

	Procedure 30.4
	Observation of Natural Propagation of Storage Organs
Step	Instructions
	Dahlia
1	In the garden area where you have permission to dig, work the shovel into the ground and lift up the soil clump containing the bulbous plant of interest. Shake off the majority of the soil to expose the bulb/tuber clump. Wash in a stream of water, if desired. Try to lift a variety of species. Of particular interest are daffodil (usually these show prominently offset bulbs, which may be detached from the parent bulb), tulip (also have offset bulbs), lily (often, established bulbs will have multiple growing points, or "noses," which can be separated; lilies also commonly have stem bulblets that form on the stem under the soil line), gladiolus (in the fall right before frost, the new corm can be clearly seen growing on top of the last years' corm, and also you can usually see numerous cormels attached between the old and new corm), canna (a good example of a rhizomatous plant), etc.
2	Observations can be made about the overall structure of the below-ground parts. If there are sufficient quantities available, the parts may be sectioned or dissected and internal components identified.

Part XII

Micropropagation

31 Micropropagation

Michael E. Kane, Philip Kauth, and Scott Stewart

CHAPTER 31 CONCEPTS

- Shoot meristem cells retain the embryonic capacity for unlimited division.

- Isolated smaller meristem explants require more complex culture media for survival.

- Meristem and meristem tip culture are methods for disease eradication.

- Shoot culture provides a means to multiply periclinal chimeras.

- Cytokinins disrupt apical dominance and enhance axillary shoot production.

- Increased auxin concentrations increase percent rooting and root number, but decrease root elongation.

- Negative carryover effects of auxins used for Stage III rooting may affect *ex vitro* survival and growth of plantlets.

- The application of shoot organogenesis or nonzygotic embryogenesis for commercial micropropagation has been limited in many species.

- For nonzygotic embryogenesis to be a viable propagation method, delivery systems must be developed through which nonzygotic embryos are processed and function as synthetic seeds that survive handling, planting, and further develop into vigorous plants.

Micropropagation is defined as aseptic asexual plant propagation on a defined culture medium in culture vessels under controlled conditions of light and temperature. Since plants are propagated in culture vessels, the procedure is often referred to as *in vitro* (Latin: "in glass") propagation. In reality, numerous types of culture vessels constructed of plastic are often used. The commercial application of micropropagation became established in the mid-1960s with orchids in France. Since this time, commercial micropropagation has developed into a worldwide industry producing more than 500 million plants yearly. By 2001, there were approximately 90 commercial micropropagation laboratories in the United States. Requirements for specialized equipment, highly trained technicians, and high labor costs make the technology very expensive. Consequently, commercial micropropagation is limited to crops generating high unit prices including ornamental plants and food crops, such as potato.

Depending on the species and cultural conditions, micropropagation can be achieved by the following methods:

1. Node culture
2. Enhanced axillary shoot proliferation
3. *De novo* formation of adventitious shoots through shoot organogenesis
4. Nonzygotic embryogenesis

Currently, the most frequently used micropropagation method for commercial production, shoot culture, relies on enhanced axillary shoot proliferation from cultured meristems (Figure 31.5a). This method often provides acceptable levels of genetic stability and is more readily attainable for many plant species. Given the importance of shoot culture, our review of micropropagation concepts will focus on this procedure.

SHOOT APICAL MERISTEMS

Given that all four micropropagation methods rely on the formation of shoot apical meristems, it is important to briefly review the general structure of shoot meristems. Shoot growth in mature plants is restricted to specialized regions, which exhibit little differentiation and in which

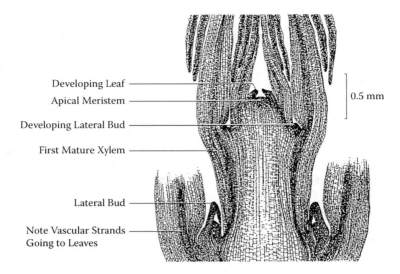

FIGURE 31.1 Diagrammatic representation of a dicotyledonous shoot tip. The shoot tip comprises the apical meristem, subtending leaf primordia and lateral buds.

the cells retain the embryonic capacity for unlimited division. These regions, called apical meristems, are located in the apices of the main and lateral buds of the plant (Figure 31.1). Cells derived from these apical meristems subsequently undergo differentiation to form the mature tissues of the plant body. Due to their highly organized structure, apical meristems tend to be genetically stable.

Significant differences exist in the shape and size of shoot apices and associated layers between different taxonomic plant groups. A typical dicotyledon shoot apical meristem consists of a layered dome of actively dividing cells located at the extreme tip of a shoot and measures about 0.1 to 0.2 mm in diameter and 0.2 to 0.3 mm in length. The apical meristem has no connection to the vascular system of the stem. Below the apical meristem, localized areas of cell division and elongation represent sites of newly developing leaf primordia (Figure 31.1). Lateral buds, each containing an apical meristem, develop within the axils of the subtending leaves. In the intact plant, outgrowth of the lateral buds is usually inhibited by apical dominance of the terminal shoot tip. Organized shoot growth from plant apical meristems is potentially unlimited and said to be indeterminate. However, shoot apical meristems may become committed to the formation of determinate organs, such as flowers.

IN VITRO CULTURE OF SHOOT MERISTEMS

The recognized potential for unlimited shoot growth prompted early and largely unsuccessful attempts to aseptically culture isolated shoot meristems in the 1920s. By the mid-1940s, sustained growth and maintenance of cultured shoot meristems through repeated subculture was achieved for several species. However, in 1946, Dr. Ernest Ball (at that time a professor at North Carolina State University) provided the first detailed procedures for the isolation and production of plants from cultured shoot meristem tips and the successful transfer of rooted plantlets into soil. Ball is often called the "Father of Micropropagation" because his shoot tip culture procedure is the one most commonly used by commercial micropropagation laboratories today. Although these studies demonstrated the feasibility of regenerating shoots from cultured shoot tips, the procedures typically yielded unbranched shoots.

Several important discoveries facilitated application of *in vitro* culture techniques for large-scale clonal propagation from shoot meristems. The discovery that virus-eradicated plants could be generated from cultured excised meristems led to the widespread application of the procedure also for routine fungal and bacterial pathogen eradication. Demonstration of the rapid production of orchids from cultured shoot tips in 1960 supported the possibility of rapid clonal propagation in other crops. It should be noted that *in vitro* propagation in many orchids does not occur via axillary shoot proliferation, but rather the cultured meristems become disorganized and form spheroid protocorm-like bodies that are actually nonzygotic embryos.

The final discovery critical for production of plants from cultured shoot meristems was the elucidation of the role of cytokinins in disrupting apical dominance by Wickson and Thimann in 1958. This finding was eventually applied to enhance axillary shoot production *in vitro*. Application of this method was further expedited by development of improved culture media, such as Murashige and Skoog medium, which supported propagation of a large number of plant species.

MERISTEM AND MERISTEM TIP CULTURE

Although not directly used for propagation, meristem and meristem tip culture will be briefly described

Micropropagation

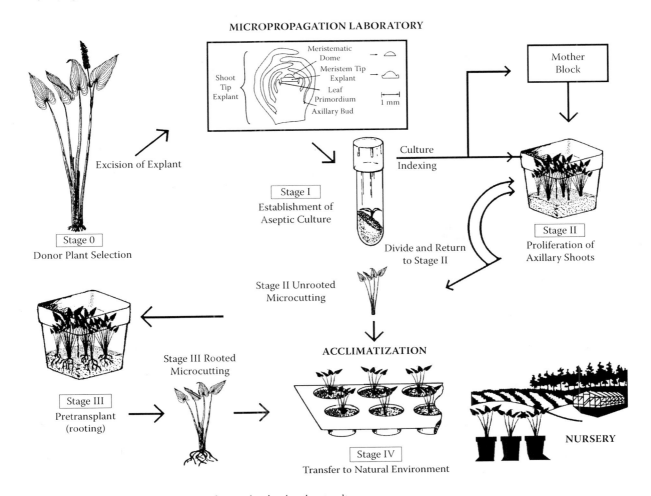

FIGURE 31.2 Micropropagation stages for production by shoot culture.

because these procedures are used to generate pathogen-eradicated shoots that subsequently may serve as propagules for *in vitro* propagation. Culture of the apical meristematic dome alone (Figure 31.2) from either terminal or lateral buds, for the purpose of pathogen elimination, is termed "meristem culture." Culture of larger explants consisting of the apical meristematic dome plus several subtending leaf primordia is termed "meristem tip culture" (Figure 31.2). Disease eradication by meristem culture is possible because the pathogens do not infect the meristematic cells due to the lack of vascular development in the extreme shoot tip. In reality, true meristem culture is rarely used because isolated apical meristems of many species exhibit both low survival rates and increased chance of genetic variability following callus formation and adventitious shoot formation. Caution should be taken when interpreting much of the early published literature of successful "meristem" culture because, in many instances, meristem tip or even larger shoot tip explants were actually used. The terms "meristemming" and "mericloning" commonly used in the orchid literature are equally ambiguous.

SHOOT AND NODE CULTURE

Although not the most efficient procedure, propagation from axillary shoots has proved to be a reliable method for the micropropagation of a large number of species. Depending on the species, two methods are used: shoot or node culture. Both methods rely on stimulation of lateral shoot growth following disruption of apical dominance. Shoot culture refers to the *in vitro* propagation by repeated enhanced formation of axillary shoots from shoot tip or lateral bud explants cultured on medium supplemented with growth regulators, typically a cytokinin. The axillary shoots produced are either subdivided into shoot tips and nodal segments that serve as secondary explants for further proliferation or are treated as microcuttings for rooting. In some species modified storage organs, such as miniaturized tubers or corms (Figure 31.3), develop from axillary shoots or rhizomes under inductive culture conditions and may serve as the propagules for either direct planting or long-term storage.

When specific pathogen-eradicated donor plants are used, or when pathogen elimination is not a concern, rela-

FIGURE 31.3 In some species, such as *Sagittaria latifolia* L., *in vitro* multiplication may occur through production of shoots (left) or corms (right), depending on culture conditions of light and temperature. Scale bar = 1 cm.

tively larger (1 to 20 mm long) shoot tip or lateral bud primary explants (Figure 31.2) can be used for culture establishment and subsequent shoot culture. Advantages of using larger shoot tips include greater survival, more rapid growth responses, and the presence of more axillary buds. However, these larger explants are more difficult to completely surface sterilize and can potentially harbor undetected latent or systemic microbial infection. Compared to other micropropagation methods, shoot cultures:

1. Provide reliable rates and consistency of multiplication following culture stabilization
2. Are less susceptible to genetic variation
3. May provide for clonal propagation of periclinal chimeras

Node culture, a simplified form of shoot culture, is another method for production from preexisting meristems. Numerous plants, such as potato (*Solanum tuberosum* L.), do not respond well to cytokinin stimulation of axillary shoot proliferation observed in the micropropagation of other crops. Axillary shoot growth is promoted by the culture of either intact shoots positioned horizontally on the medium (*in vitro* layering) or single or multiple node segments. Typically, single elongated unbranched shoots, comprised of multiple nodes, are rapidly produced. These shoots (microcuttings) are either rooted or acclimatized to *ex vitro* conditions, or repeatedly subdivided into nodal cuttings to initiate additional cultures. Node culture is the simplest method, and it is associated with the least genetic variation among the plants produced.

MICROPROPAGATION STAGES

In 1974, Dr. Toshio Murashige described three basic stages (I through III) for successful micropropagation. Recognition of contamination problems often observed after inoculation of primary explants prompted inclusion of a Stage 0. This additional stage describes specific cultural practices that maintain the hygiene of stock plants and decrease the contamination frequency during initial establishment of primary explants. As a result of our increased awareness of the requirements for successful micropropagation, five stages (Stages 0 through IV) are currently recognized. These stages not only describe the procedural steps in the micropropagation process, but also represent points at which the cultural environment is altered. This system has been adopted by most commercial and research laboratories, as it simplifies production scheduling, accounting, and cost analysis. Requirements for completion of each stage will depend on the plant material and specific method used. A diagrammatic representation of the micropropagation stages for propagation by shoot culture is provided in Figure 31.2.

STAGE 0: DONOR PLANT SELECTION AND PREPARATION

Explant quality and subsequent responsiveness *in vitro* is significantly influenced by the phytosanitary and physiological conditions of the donor plant. Prior to culture establishment, careful attention is given to the selection and maintenance of the stock plants used as the source of explants. Stock plants are maintained in clean controlled conditions that allow active growth but reduce the probability of disease. Maintenance of specific pathogen tested stock plants under conditions of relatively lower humidity, use of drip irrigation, and antibiotic sprays have proved effective in reducing the contamination potential of candidate explants. Such practices also allow excision of relatively larger and more responsive explants, often without increased risks of contamination.

Numerous practices are also employed to increase explant responsiveness *in vitro* by modifying the physi-

Micropropagation

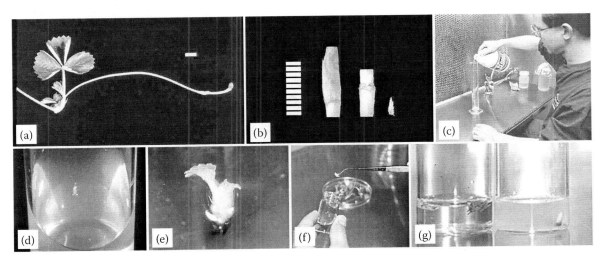

FIGURE 31.4 Establishment of strawberry (*Fragaria* x *ananassa*) shoot cultures. (a) Strawberry stolons serve as the source of axillary and apical shoot meristems for establishment of Stage I cultures. Bar = 1 cm. (b) Excised stolon nodes with bracts covering lateral buds (left) are surface sterilized and rinsed in sterile water. The nodal bract is then removed to expose the lateral bud (center). The lateral shoot tip (right) is removed, using a sterile scalpel and forceps, usually with the aid of a dissecting microscope placed in a laminar flow transfer hood. Bar = 1 cm. (c) Diluted bleach (sodium hypochlorite) is often used as the surface sterilizing agent. (d) Excised shoot tip explants are inoculated onto a Stage I medium that promotes establishment. (e) Enlargement of strawberry shoot tip after four weeks culture. (f) Following establishment, small tissue samples are cut from the established shoot culture and inoculated into an indexing medium to screen for the presence of microbial contaminants. The shoot culture is transferred to fresh Stage I medium. (g) Clouding of inoculated indexing medium (right) is a positive indication of the presence of cultivable contaminants in the tissue sample.

ological status of the stock plant. These include the following: (1) trimming to stimulate lateral shoot growth, (2) pretreatment sprays containing cytokinins or gibberellic acid, and (3) use of forcing solutions containing 2% sucrose and 200 mg/L 8-hydroxyquinoline citrate for induction of bud break and delivery of growth regulators to target explant tissues. Currently, information on the effects of other factors, such as stock plant nutrition, light, and temperature treatments on the subsequent *in vitro* performance of meristem explants, is limited.

STAGE I: ESTABLISHMENT OF ASEPTIC CULTURES

Initiation and aseptic establishment of pathogen-eradicated and responsive terminal or lateral shoot meristems explants is the goal of this stage. The primary explants obtained from the stock plants may consist of surface-sterilized shoot apical meristems or meristem tips for pathogen elimination or shoot tips from terminal or lateral buds (Figure 31.2). This can be demonstrated using strawberry (*Fragaria* x *ananassa* Duch.) as an example (Figure 31.4a).

Rapid clonal production of specific pathogen-eradicated plants is a fundamental goal of the micropropagation process. However, it is desirable to establish and maintain plant cultures that are also free of nonpathogenic microbial contaminants. The surfaces of plants are naturally populated with a diverse microflora consisting of bacteria, fungi, yeast, and other organisms. A primary objective of Stage I is the elimination of this microflora and the subsequent physiological adjustment of the explants to *in vitro* culture. This is usually accomplished through surface disinfecting explants with alcohol and/or sodium hypochlorite prior to culture inoculation (Figure 31.4c). The transfer of surface-sterilized explants into culture vessels onto sterile medium is performed in a laminar airflow transfer hood (Figure 31.10a), which provides for a particulate-free work area. Tissues are cut and handled using scalpels and forceps (Figure 31.4f) sterilized by flame or dry heating such as that achieved using glass bead sterilizers (Figure 31.10a)

Once explants become establish and grow *in vitro* (Figure 31.4e), it is essential that cultures be indexed (screened) for the presence of common microbial contaminants. Bacteria and fungal contaminants often persist within cultured tissues that visually appear contaminant free. Contaminated cultures may exhibit no symptoms, variable growth, regeneration, reduced shoot proliferation, rooting, or poor survival. Many culture contaminants are not pathogenic to plants under natural conditions, but become pathogenic *in vitro* due to the release of toxic secondary metabolites into the medium. Consequently, it is essential that Stage I cultures be indexed for the presence of internal microbial contaminants prior to serving as sources of shoot tip or nodal explants for Stage II multiplication. Since secondary culture contamination can occur as a result of poor aseptic technique or contaminant vectors, such as mites, cultures should be routinely reindexed. Indexing for microbial contaminants is usually accomplished by inoculating tissue sections or intact shoots into enriched nutrient medium that will promote the visible growth of bacteria, filamentous fungi, yeast, or other contaminants (Figure 31.4f,g).

FACTORS AFFECTING CULTURE ESTABLISHMENT

Many factors may affect successful Stage I establishment of meristem explants: explantation time, position of the explant on the stem, explant size, and polyphenol oxidation. Time of explantation can significantly affect explant response *in vitro*. In deciduous woody perennials, shoot tip explants collected at various times during the spring growth flush of the donor plants may vary in their ability for shoot proliferation. Shoot tips collected during or at the end of the period of most rapid shoot elongation on the donor plants exhibit weak proliferation potential. Explants collected before or after this period are capable of strong shoot proliferation *in vitro*. Conversely, the best results are obtained with herbaceous perennials that form storage organs, such as tubers or corms, when explants are excised at the end of dormancy and after sprouting.

Explants also exhibit different capacities for establishment *in vitro* depending on their location on the donor plant. For example, survival and growth of terminal bud explants are typically greater than lateral bud explants. Often similar lateral meristem explants from the top and bottom of a single shoot may respond differently *in vitro*. In woody plants exhibiting phasic development, juvenile explants are typically more responsive than those obtained from the often nonresponsive mature tissues of the same plant. Sources of juvenile explants include root suckers, basal parts of mature trees, stump sprouts, and lateral shoots produced on heavily pruned plants.

The excision of primary explants from donor plants often promotes the release of polyphenols and stimulates polyphenol oxidase activity within the damaged tissues. The polyphenol oxidation products often blacken the explant tissue and medium. Accumulation of these polyphenol oxidation products can eventually kill the explants. Procedures used to decrease tissue browning include use of liquid medium with frequent transfer, adding antioxidants, such as ascorbic or citric acids, polyvinylpyrrolidone (PVP), activated charcoal, or culture in reduced light or darkness.

Clearly there is no one universal culture medium for establishment of all species; however, modifications to the Murashige and Skoog medium formulation are most often used. Cytokinins and auxins are most frequently added to Stage I media to enhance explant survival and shoot development. The types and levels of growth regulators used in Stage I media are dependent on the species, genotype, and explant size.

Knowledge of the specific sites of hormone biosynthesis in intact plants provides insight into the relationship between explant size and dependence on exogenous growth regulators in the medium. Endogenous cytokinins and auxins are synthesized primarily in root tips and leaf primordia, respectively. Consequently, smaller explants, especially cultured apical meristem domes, exhibit greater dependence on medium supplementation with exogenous cytokinin and auxin for maximum shoot survival and development. Larger shoot tip explants usually do not require the addition of auxin in Stage I medium for establishment. Rapid adventitious rooting of shoot tip explants often provides a primary endogenous cytokinin source. Most Stage I media consist of mineral salts, sucrose, and vitamins, supplemented with at least a cytokinin and solidified with a gelling agent such as agar or gellan gum (Figure 31.4d). The most frequently used cytokinins are N^6-benzyladenine (BA), kinetin (Kin), and N^6-(2-isopentenyl)-adenine (2-iP). Due to its low cost and high effectiveness, the cytokinin BA is most widely used. Substituted urea compounds, such as thidiazuron, exhibit strong cytokinin-like activity and have been used to facilitate the shoot culture of recalcitrant woody species like maples (*Acer* spp.).

Several types of auxins are also used in Stage I media. The naturally occurring auxin indole-3-acetic acid (IAA) is the least active, whereas the stronger and more stable compounds α-naphthalene acetic acid (NAA), a synthetic auxin, and indole-3-butyric acid (IBA), a naturally-occurring auxin, are more often used. Stage I medium plant growth regulator (PGR) levels and combinations that promote explant establishment and shoot growth but limit formation of callus and adventitious shoot formation should be selected.

A commonly held misconception is that primary explants exhibit immediate and predictable growth responses following inoculation. For many species, particularly herbaceous and woody perennials, consistency in growth rate and shoot multiplication is achieved only after multiple subculture on Stage I medium. Physiological stabilization may require from 3 to 24 months and four to six subcultures. Failure to allow culture stabilization, before transfer to a Stage II medium containing higher cytokinin levels, may result in diminished shoot multiplication rates or production of undesirable basal callus and adventitious shoots. With some species, the time required for stabilization can be reduced by initial culture in liquid medium.

In many commercial laboratories, stabilized cultures, verified as being specific pathogen tested and free of cultivable contaminants, are often maintained on media that limit shoot production to maintain genetic stability. These cultures, called mother blocks, serve as sources of shoot tips or nodal segments for initiation of new Stage II cultures (Figure 31.10c).

STAGE II: PROLIFERATION OF AXILLARY SHOOTS

Stage II propagation is characterized by repeated enhanced formation of axillary shoots from shoot tips or

FIGURE 31.5 Shoot culture of strawberry (*Fragaria* x *ananassa*). (a) Stage II culture consisting of multiple axillary shoots produced from a single shoot after four weeks culture. The Stage II medium typically contains a cytokinin to promote axillary shoot production. (b) Stage II shoot clusters are separated into single unrooted microcuttings. Bar = 1 cm. (c) Microcuttings are inoculated onto a Stage III rooting medium, usually containing an auxin to promote adventitious rooting. (d) Microcutting rooting after four weeks. (e) Rooted microcuttings are acclimatized *ex vitro* to the lower humidity and higher light conditions in a greenhouse.

lateral buds cultured on a medium supplemented with a relatively higher cytokinin level to disrupt apical dominance of the shoot tip (Figure 31.5a). A subculture interval of four weeks with a three- to eightfold increase in shoot number is common for many crops propagated by shoot culture (Figure 31.5a). Given these multiplication rates, conservatively, more than 4.3×10^7 shoots could be produced yearly from a single starting explant.

Stage II cultures are routinely subdivided into smaller clusters, individual shoot tips, or nodal segments that serve as propagules for further proliferation (Figure 31.5b). Additionally, axillary shoot clusters may be harvested as individual unrooted Stage II microcuttings or multiple shoot clusters for *ex vitro* rooting and acclimatization (Figure 31.2; Figure 31.5b). Clearly, Stage II represents one of the most costly stages in the production process.

Both source and orientation of explants can affect Stage II axillary shoot proliferation. Subcultures inoculated with explants that had been shoot apices in the previous subculture often exhibit higher multiplication rates than lateral bud explants. In some species, inverting shoot explants in the medium can double or triple the number of axillary shoots produced on vertically oriented explants per culture period.

The number of Stage II subcultures possible before new cultures from the mother block are required depends on the species or cultivar and its inherent ability to maintain acceptable multiplication rates while exhibiting minimal genetic variation. Some species can be maintained with monthly subculture from 8 to 48 months in Stage II. In contrast, as few as three subcultures may only be possible before the frequency of off types increases to unacceptable levels, such as in the fern genus *Nephrolepsis*.

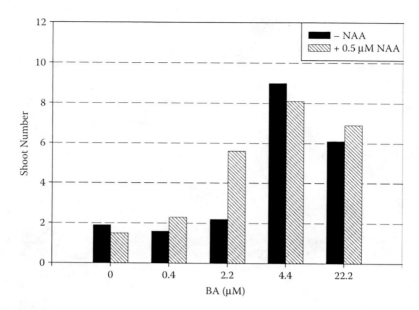

FIGURE 31.6 Effect of benzyladenine (BA) concentration on Stage II axillary shoot proliferation from two-node explants of *Aronia arbutifolia* L., a woody plant, after 28-day culture in presence and absence of 0.5 µM NAA.

Increased production of off types is often attributed to production of adventitious shoots, particularly through an intermediary callus stage. Stage II cultures, originally regenerating from axillary shoots, often begin producing adventitious shoots (see Figure 31.11) at the base of axillary shoot clusters after a number of subcultures on the same medium. These so-called mixed cultures can develop without any morphological differences being apparent. Selecting only terminal shoots of axillary origin for subculture, instead of shoot bases, decreases the frequency of off types including the segregation of periclinal chimeras (chapter 16).

Selection of Stage II cytokinin type and concentration is made on the basis of shoot multiplication rate, shoot length, and frequency of genetic variation. Although shoot proliferation is enhanced at higher cytokinin concentrations, the shoots produced are usually smaller and may exhibit symptoms of hyperhydricity (i.e., plants appear water-soaked). Depending on the species, exogenous auxins may or may not enhance cytokinin-induced axillary shoot proliferation. This is exemplified with the woody shrub *Aronia arbutifolia* L. (Figure 31.6). Addition of auxin in the medium often mitigates the inhibitory effect of cytokinin on shoot elongation, thus increasing the number of usable shoots of sufficient length for rooting (Figure 31.7). This benefit must be weighed against the increased chance of callus formation. Similarly, shoot elongation in Stage II cultures may be achieved by adding gibberellic acid to the medium.

The possibility of adverse carryover effects on rooting of plantlets and survivability in Stage IV should be evaluated when selecting a Stage II cytokinin. For example, with some tropical foliage plants species, the use of BA in Stage II can significantly reduce Stage IV plantlet survival and rooting to as low as 10%. Use of Kin or 2-iP instead of BA yields survival rates in excess of 90%. In some species the adverse effect of BA on Stage IV survival and rooting has been attributed to production of an inhibitory BA metabolite, which can be reduced by substituting with *meta*-topolin, a BA analogue.

STAGE III: PRETRANSPLANT (ROOTING)

This step is characterized by preparation of Stage II shoots (microcuttings) or shoot clusters for successful transfer to soil (Figure 31.2; Figure 31.5b). The process may involve the following:

1. Elongation of shoots prior to rooting
2. Rooting of individual shoots (microcuttings) or shoot clumps
3. Fulfilling dormancy requirements of storage organs *in vitro* by cold treatment
4. Prehardening cultures *in vitro* to increase survival

Where possible, commercial laboratories have developed procedures to transfer Stage II microcuttings to soil, thus bypassing Stage III rooting (Figure 31.2).

There are several reasons for eliminating Stage III rooting in commercial production. Estimated costs for Stage III range from 35 to 75% of the total production costs. This reflects the significant input of labor and supplies required to complete Stage III rooting. Considerable cost savings can be achieved if Stage III is eliminated. Furthermore, it has been observed that *in vitro*–formed

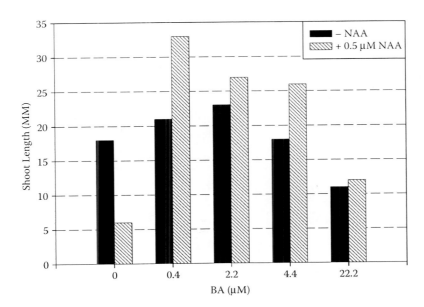

FIGURE 31.7 Inclusion of auxin may reduce the inhibitory effect of BA on axillary shoot elongation. Data shown for shoots generated from two-node explants of *Aronia arbutifolia* after 28-day culture in presence and absence of 0.5 µM NAA.

root systems are often nonfunctional and die following transplanting. This results in a delay in transplant growth prior to production of new adventitious roots.

For various reasons, however, it may not always be feasible to transplant Stage II microcuttings directly to soil. Given the aforementioned limitations of Stage III rooting, sometimes Stage III culture is used solely to elongate Stage II shoot clusters prior to separation and rooting *ex vitro*. Elongated shoots may be further pretreated in an aqueous auxin solution prior to transplanting. Often Stage III rooting of herbaceous plants can be achieved on medium in the absence of auxins (Figure 31.5d). However, with many woody species, the addition of an auxin (IBA or NAA) in Stage III medium is required to enhance adventitious rooting (Figure 31.8) and plant performance *ex vitro*. Optimal auxin concentration is determined based upon percent rooting, root number, and length (Figure 31.8). Roots should not be allowed to elongate excessively, to help prevent root damage during transplanting. Care must be taken when selecting an auxin. For example, compared to IBA, use of NAA for Stage III rooting has been shown to decrease *ex vitro* survival rates or suppress posttransplant growth (Figure 31.9).

STAGE IV: TRANSFER TO NATURAL ENVIRONMENT

The ultimate success of shoot or node culture depends on the ability to transfer and reestablish vigorously growing plants from *in vitro* to greenhouse conditions (Figure 31.2; Figure 31.5e; Figure 31.10h). This involves acclimatizing or hardening-off plantlets to conditions of significantly lower relative humidity and higher light intensity. Even when acclimatization procedures are carefully followed, poor survival rates are frequently encountered. Micropropagated plants are often difficult to transplant because of a heterotrophic mode of nutrition *in vitro* and poor control of water loss.

Plants cultured *in vitro* in the presence of sucrose and under conditions of limited light and gas exchange exhibit no or extremely reduced capacities for photosynthesis. Reduced photosynthetic activity is associated with low photosynthetic enzyme activity. During acclimatization, there is a need for plants to rapidly transition from the heterotrophic to photoautotrophic state for survival. Unfortunately, this transition is not immediate. For example, in cauliflower, no net increase in CO_2 uptake is achieved until 14 days after transplantation. This occurs only following development of new leaves, since leaves produced *in vitro*, in the presence of sucrose, never develop photosynthetic competency under greenhouse conditions. Interestingly, before senescencing, these older leaves function as "lifeboats" by supplying stored carbohydrate to the developing and photosynthetically competent new leaves. This is not the rule with all micropropagated plants, since the leaves of some species become photosynthetic and persist after acclimatization.

A variety of anatomical and physiological features, characteristic of plants produced *in vitro* under 100% relative humidity, contribute to the limited capacity of micropropagated plants to regulate water loss immediately following transplanting. These features include reductions in leaf epicuticular wax, poorly differentiated mesophyll, abnormal stomata function, and poor vascular connection between shoots and roots. Genotypic differences in capacity for acclimatization have been reported in species.

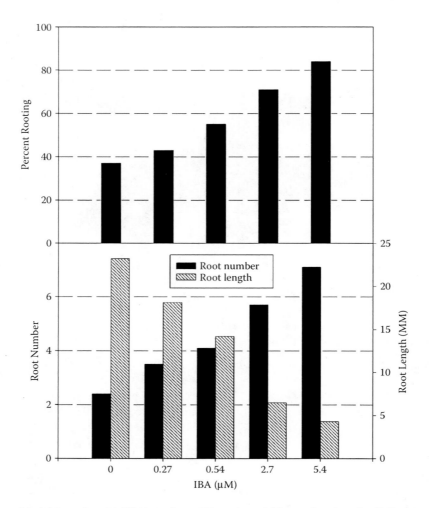

FIGURE 31.8 Effects of indolebutyric acid (IBA) on Stage III rooting of 10-mm *Aronia arbutifolia* microcuttings after 28-day culture. Increased IBA concentrations enhance percent rooting, but inhibit root elongation.

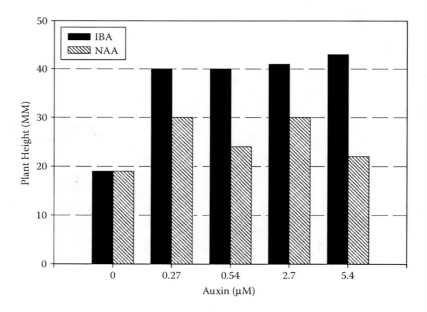

FIGURE 31.9 Comparative effects of Stage III rooting of *Aronia arbutifolia* microcuttings with IBA or NAA on posttransplant growth *ex vitro* after 28 days.

Micropropagation

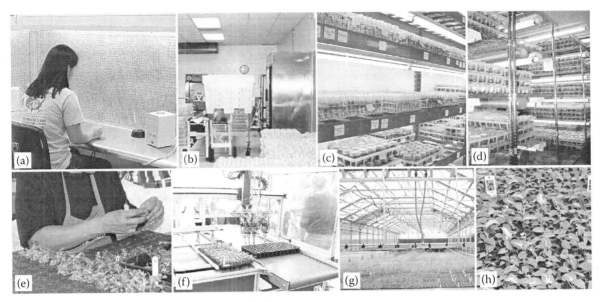

FIGURE 31.10 Commercial micropropagation laboratory. (a) To maintain sterility, culture transfers are performed in a laminar flow clean bench, which provides a particulate-free work area. (b) Culture medium preparation room with autoclave for medium sterilization. (c) Mother block room where indexed stabilized cultures are maintained to ensure a source of genetically identical cultures. (d) Production culture room maintained under controlled light and temperature conditions. (e) Stage II unrooted or Stage III rooted microcuttings are transferred to growing medium in plug trays before being placed under greenhouse conditions. (f) Robotic plug transfer equipment is used to increase efficiency and decrease labor costs. (g) Stage IV plants acclimatized in a large production greenhouse with computer-controlled environmental and irrigation/fertilization systems. (h) Acclimatized micropropagated plantlets.

To overcome these limitations, plantlets are transplanted into a well-drained "sterile" growing medium and maintained initially at high relative humidity and reduced light (40 to 160 $\mu mol \cdot m^{-2} sec^{-1}$) at 20°C to 27°C. Relative humidity may be maintained with humidity tents, single tray propagation domes, intermittent misting, or fog systems. However, use of intermittent mist often results in slow plantlet growth following water logging of the medium and excessive leaching of nutrients. Transplants are acclimatized by gradually lowering the relative humidity over a one- to four-week period. Plants are gradually moved to higher light intensities to promote vigorous growth. During this stage, container size and growing medium type can have a profound effect on the quality of the plants produced. Labor costs for commercial laboratories are reduced through adaptation of automation, including robotics for planting (Figure 31.10f), greenhouse environmental control, and fertilization and pesticide application (Figure 31.10g).

OTHER MICROPROPAGATION METHODS

Besides propagation by proliferation of preexisting shoot meristems though shoot culture, plants can be produced in vitro via two additional development pathways: shoot organogenesis (formation of adventitious shoots from tissues without preexisting meristem) or nonzygotic embryogenesis (formation of embryos from cells other than zygotes). When formed, adventitious shoot buds consist of an apical shoot but no root meristem (Figure 31.11c). Adventitious shoots may develop vascular connections with the vasculature of the explant from which it arises or produce adventitious roots at the base (Figure 31.11e). Nonzygotic embryos, being bipolar, possess both shoot and root meristems and develop from nonzygotic cells without any vascular connection to the original explant (Figure 31.12a). Nonzygotic embryos develop through similar developmental stages as zygotic embryos. Most reports of nonzygotic embryogenesis involve development from somatic (vegetative) cells. This method is often termed somatic embryogenesis. Adventitious shoots may form directly on the explants (direct shoot organogenesis; Figure 31.11b) or indirectly on an intermediary callus stage (indirect shoot organogenesis; Figure 31.11f). Similarly, nonzygotic embryos may develop directly or indirectly from the explant (Figure 31.12a,d). Both adventitious shoots and nonzygotic embryos typically arise from single cells.

Given the single-cell origin of adventitious shoots and nonzygotic embryogenesis, the potential for high plant multiplication rates from a single explant can be high. As many as 2600 shoot meristems have been observed on a single leaf explant cultured for four weeks. Depending on the species, genotype, and culture conditions, indirect shoot organogenesis and nonzygotic embryogenesis can result in an increased frequency of genetic variability in the plants produced. This variation, called somaclonal variation, can be a limitation to using shoot organogenesis or nonzygotic embryogenesis for commercial micropropagation of some species.

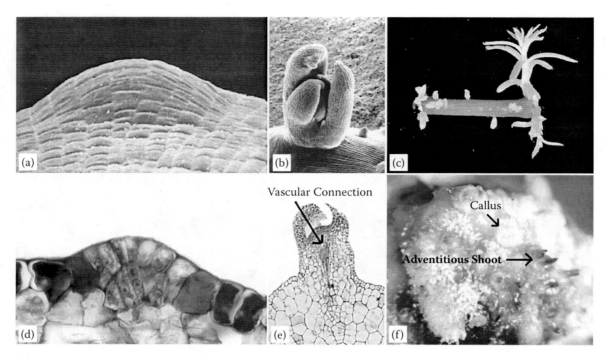

FIGURE 31.11 Micropropagation by shoot organogenesis. (a,d) Epidermal cells of explants divide to form meristematic domes when cultured on medium containing cytokinin and auxin. Each meristem typically arises from a single cell and may form directly on the explant. (b,e) Meristems develop into adventitious shoots consisting of an apical meristem and leaves. The vascular system of the shoot may connect with the vascular system of the explant. (e) This is an example of direct shoot organogenesis. (c) Adventitious shoots develop to a size sufficient for rooting. (f) Indirect shoot organogenesis occurs when adventitious shoots form from cells on an intermediary callus that forms on the explant. The frequency of genetic variability is often greater in plants produced from callus.

Many potential advantages exist for producing plants via nonzygotic embryogenesis. These include high rates of clonal multiplication, presence of both shoot and root meristems, and the adaptation of automated liquid culture/bioreactor technology for production using embryogenic cell suspension cultures (Figure 31.12e). One limitation has been the absence of a protective seed coat, which makes the processing and handling of nonzygotic embryos problematic (Figure 31.12b). For nonzygotic embryogenesis to be a viable propagation method, delivery systems must be developed through which nonzygotic embryos are processed and function as synthetic seeds that survive handling, planting, and further develop into vigorous plants. One technique involves encapsulation of nonzygotic embryos in an alginate gel to create an artificial seed coat (Figure 31.12f). To date, the application of synthetic seed technology as a commercially viable plant propagation method has been limited by high costs.

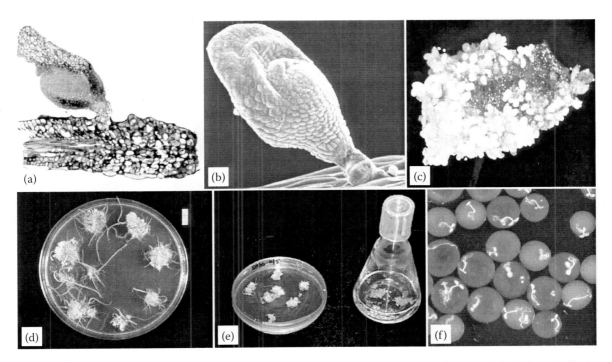

FIGURE 31.12 Micropropagation by nonzygotic embryogenesis in orchardgrass (*Dactylis glomerata* L.). (a) Longitudinal section of a nonzygotic embryo. Note the absence of a vascular connection between the nonzygotic embryo and the explant. (Used with permission of D.J. Gray.) (b) Scanning electron micrograph depicting direct nonzygotic embryogenesis. The absence of a seed coat can make handling of these embryos problematic. (Used with permission of D.J. Gray.) (c) Formation of multiple nonzygotic embryos on a leaf explant. (Used with permission of D.J. Gray.) (d) Indirect formation of "germinated" nonzygotic embryos on embryogenic callus. (e) Embryogenic callus cultures (left) can be used to establish highly efficient embryogenic cell suspensions (right). The use of bioreactors to maintain cell cultures offers significant potential for increased propagation efficiency. For example, more than 60,000 carrot nonzygotic embryos per liter can be generated in this manner. (f) Synthetic seeds are produced by encapsulating nonzygotic embryos in alginate beads. Each bead functions as an artificial seed coat, which increases survival and improves handling and storage.

LITERATURE CITED AND SUGGESTED READINGS

Aitken-Christie, J., T. Kozai, and M.A.L. Smith. 1993. *Automation and Environmental Control in Plant Tissue Culture.* Kluwer Academic Publishers, Dordrecht, The Netherlands.

De Klerk, G.J. 2002. Rooting of microcuttings: theory and practice. *In Vitro Cell. Dev. Biol.-Plant.* 38: 415–422.

Gaspar, T., C. Kevers, C. Penel, H. Greppin, D.M. Reid, and T. Thorpe. 1996. Plant hormones and plant growth regulators in plant tissue culture. *In Vitro Cell. Dev. Biol.-Plant.* 32: 272–289.

George, E.F. 1993. *Plant Propagation by Tissue Culture. Part 1. The Technology.* Exegetics, Ltd., London.

Gray, D. 2005. Propagation from non-meristematic tissues: nonzygotic embryogenesis, in *Plant Development and Biotechnology.* Trigiano, R.N. and D.J. Gray (Eds.) CRC Press, Boca Raton, FL, pp. 187–200.

Jain, S.M. and G.J. De Klerk. 1998. Somaclonal variation in breeding and propagation of ornamental crops. *Plant Tissue Culture Biotech.* 4: 63–75.

Jenks, M.A., M.E. Kane, and D.B. McConnell. 2000. Shoot organogenesis from petiole explants in the aquatic plant *Nymphoides indica. Plant Cell Tissue Organ Culture* 63: 1–8.

Kane, M.E., G.L. Davis, D.B. McConnell, and J.A. Gargiulo. 1999. *In vitro* propagation of *Cryptocoryne wendtii. Aquat. Bot.* 63: 197–202.

Kane, M.E. 2000. Micropropagation of potato by node culture and microtuber production, in *Plant Tissue Culture Concepts and Laboratory Exercises,* 2nd ed. Trigiano, R.N. and D.J. Gray (Eds.) CRC Press, Boca Raton, FL, pp. 103–109.

Kane, M.E. 2000. Culture indexing for bacterial and fungal contaminants, in *Plant Tissue Culture Concepts and Laboratory Exercises,* 2nd ed. Trigiano, R.N. and D.J. Gray (Eds.). CRC Press, Boca Raton, FL, pp. 427–431.

Lloyd, G. and B. McCown. 1980. Commercially-feasible micropropagation of Mountain laurel, *Kalmia latifolia*, by use of shoot-tip culture. *Int. Plant Prop. Soc. Proc.* 30: 421–427.

Murashige, T. and F. Skoog. 1962. A revised medium for rapid growth and bioassays with tobacco tissue cultures. *Physiol. Plant.* 15: 473–497.

Preece, J.E. and E.G. Sutter. 1991. Acclimatization of micropropagated plants to greenhouse and field, in *Micropropagation Technology and Application.* Debergh P.C. and R.H. Zimmerman (Eds.) Kluwer Academic Publishers, Boston, pp. 71–93.

Schwartz, O.J., A.R. Sharma, and R. M. Beaty. 2005. Propagation from non-meristematic tissues: organogenesis, in *Plant Development and Biotechnology.* Trigiano, R.N. and D.J. Gray (Eds.) CRC Press, Boca Raton, FL, pp. 159–172.

Styer, D.J. and C.K. Chin. 1984. Meristem and shoot-tip culture for propagation, pathogen elimination, and germplasm preservation. *Hort. Rev.* 5: 221–277.

Valero Aracama, C., M.E. Kane, S.B. Wilson, J.C. Vu, J. Anderson, and N.L. Philman. 2006. Photosynthetic and carbohydrate status of easy- and difficult-to-acclimatize sea oats (*Uniola paniculata* L.) genotypes during *in vitro* culture and *ex vitro* acclimatization. *In Vitro Cell. Dev. Biol.-Plant.* 42: 572–583.

Valero Aracama, C., S.B. Wilson, M.E. Kane, and N.L. Philman. 2007. Influence of *in vitro* growth conditions on *in vitro* and *ex vitro* photosynthetic rates of easy- and difficult-to-acclimatize sea oats (*Uniola paniculata* L.) genotypes. *In Vitro Cell. Dev. Biol.-Plant.* 43: 237–246.

Werbrouck, S.P.O., M. Strnad, H.A. Van Onckelen, and P.C. Debergh. 1996. *Meta*-topolin, an alternative to benzyladenine in tissue culture? *Physiol. Plant.* 98: 291–297.

Wickson, M. and K.V. Thimann. 1958. The antagonism of auxin and kinetin in apical dominance. *Physiol. Plant.* 11: 62–74.

32 Getting Started with Tissue Culture: Media Preparation, Sterile Technique, and Laboratory Equipment*

Caula A. Beyl

CHAPTER 32 CONCEPTS

- A tissue laboratory needs adequate physical space for work and storage and equipment, such as an autoclave, a distilled water source, balances, refrigerators, various laboratory instruments, culture vessels, and flow hoods, to name a few items.

- There are many types of growth media available, and the type of basal culture medium selected depends on the species to be cultured. The growth of the plant in culture is also affected by the selection of plant growth regulators (PGR) and environmental (cultural) conditions.

- There are about 20 different components in a tissue culture medium, including inorganic mineral elements, various organic compounds, PGRs, and support substances (e.g., agar or filter paper).

- Plant growth regulators are typically expressed in media as mg/L or µM. When comparing the effects of several PGRs on tissues in culture, prepare media using µM concentrations since an equal number of molecules of the various PGRs will be present in each of the media.

A plant tissue culture laboratory has several functional areas, whether it is designed for teaching or research and no matter what its size or how elaborate it is. It has some elements similar to a well-run kitchen and other elements that more closely resemble an operating room. There are areas devoted to preliminary handling of plant tissue destined for culture, media preparation, sterilization of media and tools, a sterile transfer hood or "clean room" for aseptic manipulations, a culture growth room, and an area devoted to washing and cleaning glassware, and tools. The following chapter will serve as an introduction to what goes into setting up a tissue culture laboratory, what supplies and equipment are necessary, some basics concerning making stock solutions, calculating molar concentrations, making tissue culture media, preparing a transfer (sterile) hood, and culturing various cells, tissues, and organs.

EQUIPMENT AND SUPPLIES FOR A TISSUE CULTURE TEACHING LABORATORY

Ideally, there should be enough bench area to allow for both preparation of media and storage space for chemicals and glassware. In addition to the usual glassware and instrumentation found in laboratories, a tissue culture laboratory needs an assortment of glassware, which may include graduated measuring cylinders, wide-necked Erlenmeyer flasks, medium bottles, test tubes with caps, petri dishes, volumetric flasks, beakers, and a range of pipettes. In general, glassware should be able to withstand repeated autoclaving. Baby food jars are inexpensive alternative tissue culture containers well suited for teaching. Ample quantities can be obtained by preceding the recycling truck on its pickup day (provided you are not embarrassed by the practice). Some tissue culture laboratories find presteril-

---

* Slightly modified from Beyl, C.A. 2005. Getting started with tissue culture—media preparation, sterile technique and laboratory equipment, in *Plant Development and Biotechnology*. Trigiano, R.N. and D.J. Gray (Eds.) CRC Press, Boca Raton, FL, pp. 19–37.

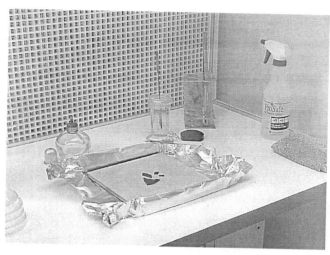

FIGURE 32.1 A typical layout of materials in the hood, showing placement of the sterile tile work surface, an alcohol lamp, spray bottle containing 80% ethanol, a cloth for wiping down the hood, and two different kinds of tool holders—a glass staining (Coplin) jar and a metal rack for holding test tubes.

ized disposable culture containers and plastic petri dishes to be convenient, but the cost may be prohibitive for others on a tight budget. There are also reusable plastic containers available, but their longevity and resistance to wear, heat, and chemicals vary considerably.

It is also good to stock metal or wooden racks to support culture tubes both for cooling and later during their time in the culture room, metal trays (such as cafeteria trays) and carts for transport of cultures, stoppers and various closures, nonabsorbent cotton, cheese cloth, foam plugs, metal or plastic caps, aluminum foil, and Parafilm™ and plastic wrap.

To teach tissue culture effectively, some equipment is necessary, such as a pH meter, balances (one analytical to four decimal places and one to two decimal places), Bunsen burners, alcohol lamps, electric sterilizing devices or heating beads, several hot plates with magnetic stirrers, a microwave oven for rapid melting of large volumes of agar medium, a compound microscope and hemocytometer for cell counting, a low-speed centrifuge, stereomicroscopes (ideally with fiber-optic light sources), large (10- or 25-L) plastic carboys to store high-quality (purity) water, a fume hood, an autoclave (or, at the very least, a pressure cooker), and a refrigerator to store media, stock solutions, plant growth regulators (PGRs), etc. A dishwasher is useful, but a large sink with drying racks, pipette and acid baths, and a forced air oven for drying glassware will also work. Also, deionized distilled water for the final rinsing of glassware is needed. Aseptic manipulations and transfers are done in multistation laminar-flow hoods (one for each pair of students).

Equipment used in the sterile transfer hood usually includes a spray bottle containing 70% ethanol, spatulas (useful for transferring callus clumps), forceps (short, long, and fine-tipped), scalpel handles (#3), disposable scalpel blades (#10 and #11), a rack for holding sterile tools, a pipette bulb or pump, Bunsen burner, alcohol lamp, or other sterilizing device, and a sterile surface for cutting explants (see below). If necessary for the experimental design, uniform-sized leaf explants can be obtained using a sterile cork borer.

There are a number of options for providing a sterile surface for cutting explants. An autoclaved stack of paper towels wrapped in aluminum foil is effective, and as each layer becomes messy, it can be peeled off and the next layer beneath it used (Figure 32.1). Others prefer reusable surfaces, such as ceramic tiles (local tile retailers are quite generous and will donate samples), metal commercial ashtrays, or glass petri dishes (100 × 15 mm). Sterile plastic petri dishes also can be used, but the cost may outweigh the advantages. A container is needed to hold the alcohol used for flaming instruments, if flame sterilization is used. An ideal container for this purpose is a slide staining Coplin jar with a small wad of cheesecloth at the bottom to prevent breakage of the glass when tools are dropped in. It has the advantage that it is heavier glass, and because the base is flared, it is not prone to tipping over. Other containers can also serve the same purpose, such as test tubes in a rack or placed in a flask or beaker to prevent them from spilling. Plastic containers, which can catch fire and melt, never should be used to hold alcohol.

WATER

High-quality water is a required ingredient of plant tissue culture media. Ordinary tap water contains cations, anions, particulates of various kinds, microorganisms, and gases that make it unsuitable for use in tissue culture media. Various methods are used to treat water, including filtration through activated carbon to remove organics and chlorine, deionization or demineralization by passing water through exchange resins to remove dissolved ionized impurities, and distillation that eliminates most ionic and particulate impurities, volatile and nonvolatile chemicals, organic matter, and microorganisms. The process of reverse osmosis, which removes 99% of the dissolved ionized impurities, uses a semipermeable membrane through which a portion of the water is forced under pressure, and the remainder containing the concentrated impurities is rejected. The most universally reliable method of water purification for tissue culture use is a deionization treatment followed by one or two glass distillations, although simple deionization alone is sometimes successfully used. In some cases, newer reverse osmosis purifying equipment (Milli-RO™, Millipore™; RO pure™, Barnstead™; Bion™, Pierce™), combined with cartridge ion exchange, adsorption, and membrane filtering equipment has replaced the traditional glass distillation of water.

THE CULTURE ROOM

After the explants are plated on the tissue culture medium under the sterile transfer hood, they are moved to the culture room. It can be as simple as a room with shelves equipped with lights or as complex as a room with intricate climate control. Most culture rooms tend to be rather simple consisting of cool white fluorescent lights mounted to shine on each shelf. Adjustable shelves are an asset and allow for differently sized tissue culture containers and for moving the light closer to the containers to achieve higher light intensities. Putting the lights on timers allows for photoperiod manipulation. Some cultures grow equally as well in dark or light. Temperatures of 26°C to 28°C are usually optimum. Heat buildup can be a problem if the room is small, so adequate air conditioning is required. Good air flow also helps to reduce condensation occurring inside petri dishes or other vessels. Some laboratories purchase incubators designed for plant tissue culture. If a liquid medium is used, the culture room should be equipped with a rotary or reciprocal shaker to provide sufficient oxygenation. The optimum temperature, light, and shaker conditions vary depending on the plant species being cultured.

CHARACTERISTICS OF SOME OF THE MORE COMMON TISSUE CULTURE MEDIA

The type of tissue culture medium selected depends on the species to be cultured. Some species are sensitive to high salts or have different requirements for PGRs. The age of the plant also has an effect. For example, juvenile tissues generally regenerate roots more readily than adult tissues. The type of organ cultured is important (roots require thiamine). Each desired cultural effect has its own unique requirements, such as auxin (see below) for induction of adventitious roots and altering the cytokinin:auxin ratio for initiation and development of adventitious shoots.

Development of culture medium formulations was a result of systematic trial and experimentation. Table 32.1 provides a comparison of the composition of several of the most commonly used plant tissue culture media with respect to their components in mg/L and molar units. Murashige and Skoog (1962) medium (MS) is the most suitable and most commonly used basic tissue culture medium for plant regeneration from tissues and callus. It was developed for tobacco, based primarily on the mineral analysis of tobacco tissue. This is a "high salt" medium due to its content of K and N-based salts, such as NO_3^- and NH_4^+. Linsmaier and Skoog medium (1965) is basically Murashige and Skoog (1962) medium with respect to its inorganic portion, but only inositol and thiamine HCl are retained among the organic components. To counteract salt sensitivity of some woody species, Lloyd and McCown (1980) developed the woody plant medium (WPM).

Gamborg's B5 medium (Gamborg et al., 1968) was devised for soybean callus culture and has lesser amounts of nitrate and particularly ammonium salts than MS medium. Although B5 was originally developed for the purpose of obtaining callus or for use with suspension culture, it also works well as a basal medium for whole-plant regeneration. Schenk and Hildebrandt (1972) developed SH medium for the callus culture of monocots and dicots. White's medium (1963), which was designed for the tissue culture of tomato roots, has a lower concentration of salts than MS medium. Nitsch's medium (Nitsch and Nitsch, 1969) was developed for anther culture and contains a salt concentration intermediate between those of MS and White's media.

Many companies sell packaged prepared mixtures of the better-known media recipes. These are easy to make because they merely involve dissolving the packaged mix in a specified volume of water. These can be purchased as the salts, the vitamins, or the entire mix with or without PGRs, agar, and sucrose. These are convenient, less prone to individual error, and make keeping stock solutions unnecessary. However, they are more expensive than making media from scratch.

COMPONENTS OF THE TISSUE CULTURE MEDIUM

Growth and development of explants *in vitro* is a product of its genetics, surrounding environment, and components of the tissue culture medium, the last of which is easiest to manipulate to our own ends. Tissue culture medium consists of 95% water, macro- and micronutrients, PGRs, vitamins, sugars (because plants *in vitro* are often not pho-

TABLE 32.1
Composition of Five Commonly Used Tissue Culture Media in Milligrams per Liter and Molar Concentrations

Compounds	Murashige and Skoog	Gamborg B-5	WPM	Nitsch and Nitsch	Schenk and Hildebrandt	White
Macronutrients in mg/L (mM)						
NH_4NO_3	1650 (20.6)	—	400 (5.0)	720	—	—
$NH_4H_2PO_4$	—	—	—	—	300 (2.6)	—
NH_4SO_4	—	134 (1.0)	—	—	—	—
$CaCl_2 \cdot 2H_2O$	440 (3.0)	150 (1.0)	96 (0.7)	166 (1.1)	151 (1.0)	—
$Ca(NO_3)_2 \cdot 4H_2O$	—	—	556 (2.4)	—	—	288 (1.2)
$MgSO_4 \cdot 7H_2O$	370 (1.5)	250 (1.0)	370 (1.5)	185 (0.75)	400 (1.6)	737 (3.0)
KCl	—	—	—	—	—	65 (0.9)
KNO_3	1900 (18.8)	2500 (24.8)	—	950 (9.4)	2500 (24.8)	80 (0.8)
K_2SO_4	—	—	990	—	—	—
KH_2PO_4	170 (1.3)	—	170 (1.3)	68 (0.5)	—	—
NaH_2PO_4	—	130.5 (0.9)	—	—	—	16.5 (0.12)
Na_2SO_4	—	—	—	—	—	200 (1.4)
Micronutrients in mg/L (µM)						
H_3BO_3	6.2 (100)	3.0 (49)	6.2 (100)	10 (162)	5 (80)	1.5 (25)
$CoCl_2 \cdot 6H_2O$	0.025 (0.1)	0.025 (0.1)	—	—	0.1 (0.4)	—
$CuSO_4 \cdot 5H_2O$	0.025 (0.1)	0.025 (0.1)	0.25 (1)	0.025 (0.1)	0.2 (0.08)	0.01 (0.04)
Na_2EDTA	37.3 (100)	37.3 (100)	37.3 (100)	37.3 (100)	20.1 (54)	—
$Fe_2(SO_4)_3$	—	—	—	—	—	2.5 (6.2)
$FeSO_4 \cdot 7H_2O$	27.8 (100)	27.8 (100)	27.8 (100)	27.8 (100)	15 (54)	—
$MnSO_4 \cdot H_2O$	16.9 (100)	10.0 (59)	22.3 (132)	18.9 (112)	10.0 (59)	5.04 (30)
KI	0.83 (5)	0.75 (5)	—	—	0.1 (0.6)	0.75 (5)
$NaMoO_3$	—	—	—	—	—	0.001 (0.001)
$Na_2MoO_4 \cdot 2H_2O$	0.25 (1)	0.25 (1)	0.25 (1)	0.25 (1)	0.1 (0.4)	—
$ZnSO_4 \cdot 7H_2O$	8.6 (30)	2.0 (7.0)	8.6 (30)	10 (35)	1 (3)	2.67 (9)
Organics in mg/L (µM)						
Myo-inositol	100 (550)	100 (550)	100 (550)	100 (550)	1000 (550)	—
Glycine	2.0 (26.6)	—	2.0 (26.6)	2.0 (26.6)	—	3.0 (40)
Nicotinic Acid	0.5 (4.1)	1.0 (8.2)	0.5 (4.1)	5 (40.6)	5.0 (41)	0.5 (4.1)
Pyridoxine HCl	0.5 (2.4)	0.1 (0.45)	0.5 (2.4)	0.5 (2.4)	0.5 (2.4)	0.1 (0.45)
Thiamine HCl	0.1 (0.3)	10.0 (30)	1.0 (3.0)	0.5 (1.5)	5.0 (14.8)	0.1 (0.3)
Biotin	—	—	—	0.05 (0.13)	—	—
Folic Acid			0.50			

tosynthetically competent), and sometimes various other simple-to-complex organic materials. All in all, about 20 different components are usually needed.

INORGANIC MINERAL ELEMENTS

Just as a plant growing *in vivo* requires many different elements from either soil or fertilizers, the plant tissue growing *in vitro* requires a combination of macro- and micronutrients. The choice of macro- and microsalts and their concentrations is species dependent. MS medium is very popular because most plants react to it favorably; however, it may not necessarily result in the optimum growth and development for every species because the salt content is so high.

The macronutrients are required in millimolar (mM) quantities in most plant basal media. Nitrogen (N) is usually supplied in the form of ammonium (NH_4^+) and nitrate (NO_3^-) ions, although sometimes more complex organic sources, such as urea, amino acids like glutamine, or casein hydrolysate, which is a complex mixture of amino acids and ammonium, are used too. Although most plants prefer NO_3^- to NH_4^+, the right balance of the two ions for optimum *in vitro* growth and development for the selected species may differ.

In addition to nitrogen, potassium, magnesium, calcium, phosphorus, and sulfur are provided in the medium as different components referred to as the macrosalts. $MgSO_4$ provides both magnesium and sulfur; $NH_4H_2PO_4$, KH_2PO_4, or NaH_2PO_4 provide phosphorus; $CaCl_2 \cdot 2H_2O$ or $Ca(NO_3)_2 \cdot 4H_2O$ provide calcium; and KCl, KNO_3, or KH_2PO_4 provide potassium. Chloride is provided by KCl and/or $CaCl_2 \cdot 2H_2O$.

Microsalts typically include boron (H_3BO_3), cobalt ($CoCl_2 \cdot 6H_2O$), iron (complex of $FeSO_4 \cdot 7H_2O$ and Na_2EDTA or rarely as $Fe_2(SO_4)_3$), manganese ($MnSO_4 \cdot H_2O$), molybdenum ($NaMoO_3$), copper ($CuSO_4 \cdot 5H_2O$), and zinc ($ZnSO_4 \cdot 7H_2O$). Microsalts are needed in much lower (micromolar; µM) concentrations than the macronutrients. Some media may contain very small amounts of iodide (KI), but sufficient quantities of many of the trace elements inadvertently may be provided because reagent grade chemicals contain inorganic contaminants.

Organic Compounds

Sugar is a very important part of any nutrient medium and its addition is essential for *in vitro* growth and development of the culture. Most plant cultures are unable to photosynthesize effectively for a variety of reasons, including insufficiently organized cellular and tissue development, lack of chlorophyll, limited gas exchange and CO_2 in the tissue culture vessels, and less than optimum environmental conditions, such as low light. A concentration of 20 to 60 g/L sucrose (a disaccharide made up of glucose and fructose) is the most often used carbon or energy source because this sugar is also synthesized and transported naturally by the plant. Other mono- or disaccharides and sugar alcohols, such as glucose, fructose, sorbitol, and maltose, may be used. The sugar concentration chosen is dependent on the type and age of the explant in culture. For example, very young embryos require a relatively high sugar concentration (>3%). For mulberry buds *in vitro*, fructose was found to be better than sucrose, glucose, maltose, raffinose, or lactose (Coffin et al., 1976). For apple, sorbitol and sucrose supported callus initiation and growth equally as well, but sorbitol was better for peach after the fourth subculture (Oka and Ohyama, 1982).

Sugar (sucrose) that is bought from the supermarket is usually adequate, but be careful to get pure cane sugar, as corn sugar is primarily fructose. Raw cane sugar is purified and, according to the manufacturer's analysis, consists of 99.94% sucrose, 0.02% water, and 0.04% other material (inorganic elements and also raffinose, fructose, and glucose). Nutrient salts contribute approximately 20% to 50% to the osmotic potential of the medium, and sucrose is responsible for the remainder. The contribution of sucrose to the osmotic potential increases as it is hydrolyzed into glucose and fructose during autoclaving. This may be an important consideration when performing osmotically sensitive procedures, such as protoplast isolation and culture.

Vitamins are organic substances that are parts of enzymes or cofactors for essential metabolic functions. Of the vitamins, only thiamine (vitamin B_1 at 0.1 to 5.0 mg/L) is essential in culture, as it is involved in carbohydrate metabolism and the biosynthesis of some amino acids. It is usually added to tissue culture media as thiamine hydrochloride. Nicotinic acid, also known as niacin, vitamin B_3, or vitamin PP, forms part of a respiratory coenzyme and is used at concentrations between 0.1 and 5 mg/L. MS medium contains thiamine HCl as well as two other vitamins, nicotinic acid and pyridoxine (vitamin B_6) in the HCl form. Pyridoxine is an important coenzyme in many metabolic reactions and is used in media at concentrations of 0.1 to 1.0 mg/L. Biotin (vitamin H) is commonly added to tissue culture media at 0.01 to 1.0 mg/L. Other vitamins that are sometimes used are folic acid (vitamin M; 0.1 ot 0.5 mg/L), riboflavin (vitamin B_2; 0.1 to 10 mg/L), ascorbic acid (vitamin C; 1 to 100 mg/L), pantothenic acid (vitamin B_5; 0.5 to 2.5 mg/L), tocopherol (vitamin E; 1 to 50 mg/L) and para-aminobenzoic acid (0.5 to 1.0 mg/L).

Inositol is sometimes characterized as one of the B complex vitamin group, but it is really a sugar alcohol involved in the synthesis of phospholipids, cell wall pectins, and membrane systems in cell cytoplasm. It is added to tissue culture media at a concentration of about 0.1 to 1.0 g/L and has been demonstrated to be necessary for some monocots, dicots, and gymnosperms.

In addition, other amino acids are sometimes used in tissue culture media. These include L-glutamine, asparagine, serine, and proline, which are used as sources of reduced organic nitrogen, especially for inducing and maintaining somatic embryogenesis. Glycine, the simplest amino acid, is a common additive, since it is essential in purine synthesis and is a part of the porphyrin ring structure of chlorophyll.

Complex organic compounds are a group of undefined supplements, such as casein hydrolysate, coconut milk (the liquid endosperm of the coconut), orange juice, tomato juice, grape juice, pineapple juice, sap from birch, banana puree, etc. These compounds are often used when no other combination of known defined components produces the desired growth or development. However, the composition of these supplements is basically unknown and may also vary from lot to lot, causing variable responses. For example, the composition of coconut milk (used at a dilution of 50 to 150 mL/L), a natural source of the PGR, zeatin (see below), not only differs between young and old coconuts, but also between coconuts of the same age.

Some complex organic compounds are used as organic sources of nitrogen, such as casein hydrolysate, a mixture of about 20 different amino acids, and ammonium (0.1

to 1.0 g/L), peptone (0.25 to 3.0 g/L), tryptone (0.25 to 2.0 g/L), and malt extract (0.5 to 1.0 g/L). These mixtures are very complex and contain vitamins as well as amino acids. Yeast extract (0.25 to 2.0 g/L) is used because of the high concentration and quality of B vitamins.

Polyamines, particularly putrescine and spermidine, are sometimes beneficial for somatic embryogenesis. Polyamines are also cofactors for adventitious root formation. Putrescine is capable of synchronizing the embryogenic process of carrot.

Activated charcoal is useful for absorption of the brown or black pigments and oxidized phenolic compounds. It is incorporated into the medium at concentrations of 0.2% to 3.0% (w/v). It is also useful for absorbing other organic compounds, including PGRs, such as auxins and cytokinins, and vitamins, iron, and zinc chelates (Nissen and Sutter, 1990). Carryover effects of PGRs are minimized by adding activated charcoal when transferring explants to media without PGRs. Another feature of using activated charcoal is that it changes the light environment by darkening the medium so it can help with root formation and growth. It may also promote somatic embryogenesis and enhance growth and organogenesis of woody species.

Leached pigments and oxidized polyphenolic compounds and tannins can greatly inhibit growth and development. These are formed by some explants as a result of wounding. If charcoal does not reduce the inhibitory effects of polyphenols, addition of polyvinylpyrrolidone (PVP, 250 to 1000 mg/L), or antioxidants, such as citric acid, ascorbic acid, or thiourea, can be tested.

PLANT GROWTH REGULATORS (PGRs)

PGRs (chapter 14) exert dramatic effects at low concentrations (0.001 to 10 µM). They regulate the initiation and development of shoots and roots on explants and embryos on semisolid or in liquid medium cultures. They also stimulate cell division and expansion. Sometimes a tissue or an explant is autotrophic and can produce its own supply of PGRs. Usually PGRs must be supplied in the medium for growth and development of the culture.

The most important classes of the PGRs used in tissue culture are the auxins and cytokinins. The relative effects of auxin and cytokinin ratio on morphogenesis of cultured tissues were demonstrated by Skoog and Miller (1957) and still serve as the basis for plant tissue culture manipulations today. Some of the PGRs used are hormones (naturally synthesized by higher plants), and others are synthetic compounds. PGRs exert dramatic effects, depending on the concentrations used, the target tissue, and their inherent activity, even though they are used in very low concentrations in the media (from 0.1 to 100 µM). The concentrations of PGRs are typically reported in mg/L or in µM units of concentration. Comparisons of PGRs based on their molar concentrations is more useful because the molar concentration is a reflection of the actual number of molecules of the PGR per unit volume (Table 32.3).

Auxins play a role in many developmental processes, including cell elongation and swelling of tissue, apical dominance, adventitious root formation, and somatic embryogenesis. Generally, when the concentration of auxin is low, root initiation is favored, and when the concentration is high, callus formation occurs. The most common synthetic auxins used are 1-naphthaleneacetic acid (NAA), 2,4-dichlorophenoxyacetic acid (2,4-D), and 4-amino-3,5,6-trichloro-2-pyridinecarboxylic acid (picloram). Naturally occurring indoleacetic acid (IAA) and indolebutyric acid (IBA) are also used. IBA was once considered synthetic, but has also been found to occur naturally in many plants including olive and tobacco (Epstein and Ludwig-Müller, 1993). Both IBA and IAA are photosensitive, so stock solutions must be stored in the dark. IAA is also easily broken down by enzymes (peroxidases and IAA oxidase). IAA is the weakest auxin and is typically used at concentrations between 0.01 and 10 mg/L. The relatively more active auxins such as IBA, NAA, 2,4-D, and picloram are used at concentrations ranging from 0.001 to 10 mg/L. Picloram and 2,4-D are examples of auxins used primarily to induce and regulate somatic embryogenesis.

Cytokinins promote cell division and stimulate initiation and growth of shoots *in vitro*. The cytokinins most commonly used are zeatin, dihydrozeatin, kinetin, benzyladenine, thidiazuron, and 2iP. In higher concentrations (1 to 10 mg/L) they induce adventitious shoot formation, but inhibit root formation They promote axillary shoot formation by opposing apical dominance regulated by auxins. Benzyladenine has significantly stronger cytokinin activity than the naturally occurring zeatin. However, a concentration of 0.05 to 0.1 µM thidiazuron, a diphenyl substituted urea, is more active than 4 to 10 µM BA, but thidiazuron may inhibit root formation, causing difficulties in plant regeneration. Adenine (used at concentrations of 2 to 120 mg/L) is occasionally added to tissue culture media and acts as a weak cytokinin by promoting shoot formation.

Gibberellins are less commonly used in plant tissue culture. Of the many gibberellins thus far described, GA_3 is the most often used, but it is very heat sensitive (after autoclaving, 90% of the biological activity is lost). Typically it is filter sterilized and added to autoclaved medium after it has cooled. Gibberellins help to stimulate elongation of internodes and have proved to be necessary for meristem growth for some species.

Abscisic acid is not normally considered an important PGR for tissue culture except for somatic embryogenesis and in the culture of some woody plants. For example, it promotes maturation and germination of somatic embryos of caraway (Ammirato, 1974) and spruce (Roberts et al., 1990).

Organ and callus cultures are able to produce the gaseous PGR, ethylene. Since culture vessels are almost

> **Procedure 32.1**
> Converting Molar Solutions to mg/L and mg/L to Molar Solutions Using Conversion Factors
>
> **How to determine how many mg/liter are needed for a 1.0 molar concentration**
>
> First, look up the molecular weight of the plant growth regulator. In this example, we will use kinetin. The molecular weight is 215.2, so a one molar solution will consist of 215.2 g per liter (L) of solution. By using conversion factors and crossing out terms, you cannot go wrong.
>
> $$1 \text{ molar solution} = \frac{1 \text{ mole}}{\text{liter solution}} \times \frac{215.2 \text{ grams}}{1 \text{ mole}} = 215.2 \text{ grams/L}$$
>
> To see what a 1.0 mM solution of kinetin would consist of, multiply the grams necessary for a 1 molar solution by 10^{-3}, which would give you 215.2×10^{-3} g/L or 215.2 mg/L.
>
> $$1 \text{ mmolar sol'n} = \frac{1 \text{ mmole}}{\text{liter solution}} \times \frac{215.2 \text{ mg}}{1 \text{ mmole}} = 215.2 \text{ mg/L}$$
>
> To see what a 1.0 μM solution of kinetin would consist of, multiply the grams necessary for a 1 molar solution by 10^{-6}, which would give you 215.2×10^{-6} g/L or 215.2 μg/L.
>
> $$1 \text{ μmolar sol'n} = \frac{1 \text{ μmole}}{\text{liter solution}} \times \frac{215.2 \text{ mg}}{1 \text{ μmole}} = 215.2 \text{ μg/liter or } 0.215 \text{ mg/L}$$
>
> **How to determine the molar concentration of a solution that is in mg/L**
>
> Let us assume you are given a 10 mg/L solution of indolebutyric acid (IBA), and you wish to know its molarity. First, look up the molecular weight of the plant growth regulator. In this example, using IBA, the molecular weight of IBA is 203.2. Again, using conversion factors and crossing out terms:
>
> $$\frac{10 \text{ mg IBA}}{\text{liter of sol'n}} \times \frac{1 \text{ gram}}{1000 \text{ mg}} \times \frac{1 \text{ mole}}{203.2 \text{ grams}} = \frac{0.0000492 \text{ moles}}{\text{liter of sol'n}} = 49.2 \text{ μM IBA}$$
>
> *Note:* remember 1 mole = 10^3 mmol = 10^6 μmol.

entirely closed, ethylene can sometimes accumulate. Many plastic containers also contribute to ethylene content in the vessel. There are contrasting reports in the literature concerning the role played by ethylene *in vitro*. It appears to influence embryogenesis and organ formation in some gymnosperms. Sometimes *in vitro* growth can be promoted by ethylene. At other times, addition of ethylene inhibitors results in better initiation or growth. For example, ethylene inhibitors, particularly silver nitrate, are used to enhance embryogenic culture initiation in corn. High levels of 2,4-D can induce ethylene formation.

AGAR AND ALTERNATIVE CULTURE SUPPORT SYSTEMS

Agar is used to solidify tissue culture media into a gel. It enables the explant to be placed in precise contact with the medium (for example on the surface or embedded), but remain aerated. Agar is a high-molecular-weight polysaccharide that can bind water and is derived from seaweed. It is added to the medium in concentrations ranging from 0.5% to 1.0% (w/v). High concentrations of agar result in a harder medium. If a lower concentration of agar is used (0.4%) or if the pH is low, the medium will be too soft and will not gel properly. The consistency of the agar can also influence the growth. If it is too hard, plant growth is reduced. If it is too soft, hyperhydric plants may be the result (Singha, 1982). To gel properly, a medium with 0.6% agar must have a pH above 4.8. Sometimes activated charcoal in the medium will interfere with gelling. Typical tissue culture agar melts easily at ~65°C and solidifies at ~45°C.

Agar also contains organic and inorganic contaminants, the amount of which varies between brands. Organic acids, phenolic compounds, and long-chain fatty acids are common contaminants. A manufacturer's analysis shows that Difco Bacto agar also contains (amounts

TABLE 32.2
Macro- and Micronutrient 100× Stock Solutions for Murashige-Skoog (MS) Medium (1962)[a]

Stock (100×)	Component	Amount
Nitrate	NH_4NO_3	165.0 g
	KNO_3	190.0 g
Sulfate	$MgSO_4 \cdot 7H_2O$	37.0 g
	$MnSO_4 \cdot H_2O$	1.7 g
	$ZnSO_4 \cdot 7H_2O$	0.86 g
	$CuSO_4 \cdot 5H_2O$	2.50 mg[b]
Halide	$CaCl_2 \cdot 2H_2O$	44.0 g
	KI	83.0 mg
	$CoCl_2 \cdot 6H_2O$	5.0 mg[c]
PBMo	KH_2PO_4	17.0 g
	H_3BO_3	620.0 mg
	Na_2MoO_4	25.0 mg
NaFeEDTA[d]	$FeSO_4 \cdot 7H_2O$	2.78 g
	Na_2EDTA	3.74 g

Note: The number of grams or milligrams indicated in the amount column should be added to 1000 mL of deionized distilled water to make one liter of the appropriate stock solution. For each liter of the medium made, 10 mL of each stock solution will be used.

[a] To make 100× stock solutions for any of the media listed in Table 32.1, multiply the amount of chemical listed in the table by 100, and dissolve in one liter of deionized, distilled water.

[b] Because this amount is too small to weight conveniently, dissolve 25 mg of $CuSO_4 \cdot 5H_2O$ in 100 mL of deionized distilled water, then add 10 mL of this solution to the sulfate stock.

[c] Because this amount is too small to weight conveniently, dissolve 25 mg of $CoCl_2 \cdot 6H_2O$ in 100 mL of deionized distilled water, then add 10 mL of this solution to the halide stock.

[d] Mix the $FeSO_4 \cdot 7H_2O$ and NaEDTA together, and heat gently until the solution becomes orange. Store in an amber bottle or protect from light.

---

in ppm): 0.0 to 0.5 cadmium, 0.0 to 0.1 chromium, 0.5 to 1.5 copper, 1.5 to 5.0 iron, 0.0 to 0.5 lead, 210.0 to 430.0 magnesium, 0.1 to 0.5 manganese, and 5.0 to 10.0 zinc. Generally, relatively pure, plant tissue culture-tested types of agar should be used. Poor-quality agar can interfere or inhibit the growth of cultures.

Agarose is often used when culturing protoplasts or single cells. Agarose is a purified extract of agar that leaves behind agaropectin and sulphate groups. Since its gel strength is greater, less is used to create a suitable support or suspending medium.

Gellan gums like Gelrite™ and Phytagel™ are alternative gelling agents. They are made from a polysaccharide produced by a bacterium. Rather than being translucent (like agar), they are clear, so it is much easier to detect contamination, but they cannot be reliquefied by heating and gelled again, and the concentration of divalent cations like calcium and magnesium must be within a restricted range or gelling will not occur.

Mechanical supports, such as filter paper bridges or polyethylene rafts do not rely on a gelling agent. They can be used with liquid media, which then circulates better, but keeps the explant at the medium surface so that it remains oxygenated. The types of support systems that have been used are as varied as the imagination and include rock wool, cheesecloth, pieces of foam, and glass beads.

STEPS IN THE PREPARATION OF TISSUE CULTURE MEDIUM

The first step in making tissue culture medium is to assemble needed glassware; for example, a one-liter beaker, one-liter volumetric flask, stirring bar, balance, pipettes, and the various stock solutions (Table 32.2 and Procedure 32.2). The plethora of units used to measure concentration may be confusing when first encountered, so a description of the most common units and what they mean is given

Getting Started with Tissue Culture: Media Preparation, Sterile Technique, and Laboratory Equipment

---

Procedure 32.2
Making Stock Solutions for Vitamins and Plant Growth Regulators

Step	Instructions and Comments
1	To make a stock solution for nicotinic acid, look at how much is required for 1 L of medium. In this example, we are going to assume we are making MS medium, so we will need 0.5 mg.
2	It would be convenient to be able to add the 0.5 mg of nicotinic acid by adding a volume of the nicotinic acid stock that corresponded to 1 mL. If 0.5 mg of nicotinic acid must be in 1 mL of the stock solution, then (multiplying by 100) use 50 mg for the 100 mL of stock solution. One mL may be dispensed into 1.5 mL Eppendorf tubes and frozen (−20°C).
3	Now prepare a PGR stock—IBA for a rooting medium. If 1.0 mg/L of IBA is require for rooting medium, then our solution must contain 1 mg in each mL of the IBA stock. We must weigh 100 mg of IBA for the 100 mL of IBA stock solution. IBA is not very soluble in water, so first dissolve it in a small amount (~1 mL) of a solvent such as 95% ethanol, 100% propanol, or 1 N KOH. Swirl it to dissolve, and then slowly add the remainder of the water to a final volume of 100 mL. Label the stock solution, and add 1 mL for every liter of medium to be made. Store in a brown bottle in the refrigerator (4°C). Note: Many growth regulators need special handling to dissolve. Indoleacetic acid, indolebutyric acid, naphthaleneacetic acid, 2,4-D, benzyladenine, 2-iP, and zeatin can be dissolved in either 95% ethanol, 100% propanol, or 1 N KOH. Kinetin and ABA are best dissolved in 1 N KOH, but thidiazuron will not dissolve in either alcohol or base, so a small amount of dimethylsulphoxide (DMSO) must be used. By using only 1 mL of each stock, only very small amounts of the solvent are added to the medium, which will minimize toxic effects.
4	Now let us make a stock solution of IBA, but this time we want to have a 5 µM solution of IBA in the medium. Look up the molecular weight of IBA in Table 32.3 and find that it is 203.2. Using the same rationale in the two examples above, we wish to have 5 µmol delivered to the medium by using 1 mL of IBA stock solution. So to make 100 mL of stock solution, you must have 500 µmol or 0.5 mmol in the 100 mL of stock solution. If one mole is 203.2 grams, one mmol is 203.2 mg, so 0.5 mmol IBA × 203.2 mg/1 mmol = 101.6 mg of IBA needed for 100 mL of IBA stock solution. To prepare the solution, weigh out the IBA, dissolve it in 1 N KOH as described above, and label the stock solution and store in the refrigerator (4°C) in a brown bottle. Add 1 mL of the IBA stock solution to deliver 5 µmol for every liter of the medium.

---

below. Once familiar with these, you can confidently proceed to the section on making stock solutions.

UNITS OF CONCENTRATION CLARIFIED

Concentrations of any substance can be given in several ways. The following list gives some of the methods of indicating the concentration commonly found in literature on tissue culture:

- Percentage based upon volume % (v/v): Used for coconut milk, tomato juice, and orange juice. For example, if 100 mL of a 5% (v/v) coconut milk was desired, 5 mL of coconut milk would be diluted to 100 mL with water.
- Percentage based upon weight % (w/v): This is often used to express concentrations of agar or sugar. For example, to make a 1% (w/v) agar solution, dissolve 10 g of agar in 1 L of nutrient medium.
- Molar solution: A mole (mol) is the same number of grams as the molecular weight (Avogadro's number of molecules), so a 1 molar solution represents one mole of the substance in 1 L of solution, and 0.01 M represents 0.01 times the molecular weight in 1 L. A millimolar (mM) solution is 0.001 mole/L, and a micromolar (µM) solution is 0.000001 moles/L. Substances like plant growth regulators are active at micromolar concentrations. Molar concentration is used to accurately compare relative reactivity among different compounds. For example, a 1 µM concentration of IAA would contain the same number of molecules as a 1 µM concentration of kinetin, although the same could not be said for units based upon weight.
- Milligrams per liter (mg/L): Although not an accurate means of comparing substances molecularly, this is simpler to calculate and use, since it is a direct weight. Such direct measurement is commonly used for macronutrients and some-

TABLE 32.3
Plant Growth Regulators, Their Molecular Weights, Conversions of mg/L Concentrations into µM equivalents, and Conversion of µM Concentrations into mg/L Equivalents

Plant Growth Regulator	Abbreviation	M.W.	mg/L Equivalents for These µM Concentrations				µM Equivalents for These mg/L Concentrations			
			0.1	1.0	10.0	100.0	0.1	0.5	1.0	10.0
Abscisic acid	ABA	264.3	0.0264	0.264	2.64	26.4	0.38	1.89	3.78	37.8
Benzyladenine	BA	225.2	0.0225	0.225	2.25	22.5	0.44	2.22	4.44	44.4
Dihydrozeatin	DHZ	220.3	0.0220	0.220	2.20	22.0	0.45	2.27	4.53	45.3
Gibberellic acid	GA_3	346.4	0.0346	0.346	3.46	34.6	0.29	1.44	2.89	28.9
Indoleacetic acid	IAA	175.2	0.0175	0.175	1.75	17.5	0.57	2.85	5.71	57.1
Indolebutyric acid	IBA	203.2	0.0203	0.203	2.03	20.3	0.49	2.46	4.90	49.0
Potassium salt of IBA	K-IBA	241.3	0.0241	0.241	2.41	24.1	0.41	2.07	4.14	41.4
Kinetin	KIN	215.2	0.0215	0.215	2.15	21.5	0.46	2.32	4.65	46.7
Naphthaleneacetic acid	NAA	186.2	0.0186	0.186	1.86	18.6	0.54	2.69	5.37	53.7
Picloram	PIC	241.5	0.0242	0.242	2.42	24.2	0.41	2.07	4.14	41.4
Thidiazuron	TDZ	220.3	0.0220	0.220	2.20	22.0	0.45	2.27	4.54	45.4
Zeatin	ZEA	219.2	0.0219	0.219	2.19	21.9	0.46	2.28	4.56	45.6
2-Isopentenyl adenine	2IP	203.3	0.0203	0.203	2.03	20.3	0.49	2.46	4.92	49.2
2,4-Dichlorophenoxy acetic acid	2,4-D	221.04	0.0221	0.221	2.21	22.1	0.45	2.26	4.52	45.2

times with PGRs. One mg/L means placing 1 mg of the desired substance in a final volume of 1 L of solution. One mg is 10^{-3} g.

- Microgram per liter (µg/L): This is used with micronutrients and also sometimes with growth regulators. It means placing 1 µg of substance in 1 L of solution. One µg = 0.001 or 10^{-3} mg = 0.000001 or 10^{-6} g.
- Parts per million (ppm): Sometimes media components are expressed in ppm, which means 1 part per million or 1 mg/L.

Instructions for making media can be found in Procedure 32.3. These instructions describe MS medium preparation, but will work just as effectively for any other media that you choose. Merely follow the same steps, and substitute the macro- and micronutrient stocks that you have made for the desired medium. Omit the agar to produce a liquid medium for use in suspension culture. PGRs can be customized for the medium of your choice, whether it is intended for initiation of callus, shoots, roots, or some other purpose.

Adjusting the pH of the medium is an essential step. Plant cells in culture prefer a slightly acidic pH, generally between 5.3 and 5.8. When pH values are lower than 4.5 or higher than 7.0, growth and development *in vitro* generally is greatly inhibited. This is probably due to several factors including PGRs, such as IAA and gibberellic acid, becoming less stable, phosphate and ion salts precipitating, vitamin B_1 and pantothenic acid becoming less stable, reduced uptake of ammonium ions, and changes in the consistency of the agar (the agar becomes liquefied at lower pH). Adjusting the pH is the last step before adding and dissolving the agar then distributing it into the culture vessels and autoclaving. If the pH is not what it should be, it can be adjusted using KOH to raise pH or HCl to lower it (0.1 to 1.0 N), depending on if the pH is too low or too high. While NaOH can be used, it can lead to an undesirable increase in sodium ions. The pH of a culture medium generally drops by 0.3 to 0.5 units after it is autoclaved and then changes throughout the period of culture, due to both oxidation and the differential uptake and secretion of substances by growing tissue.

STOCK SOLUTIONS OF THE MINERAL SALTS

Mineral salts can be prepared as stock solutions 10 to 1000 times the concentration specified in the medium. Mineral salts are often grouped into two stock solutions, one for macroelements and one for microelements, but unless these are kept relatively dilute (10×), precipitation can occur. In order to produce more concentrated solutions, the preferred method is to group the compounds by the ions they contain, such as nitrate, sulfate, halide, P, B, and Mo (phosphorus, boron, molybdenum, respectively) and iron and make them up as 100× stocks. Table 32.2 lists the stock solutions for MS medium made at 100× final concentration, which means that 10 mL of

Getting Started with Tissue Culture: Media Preparation, Sterile Technique, and Laboratory Equipment

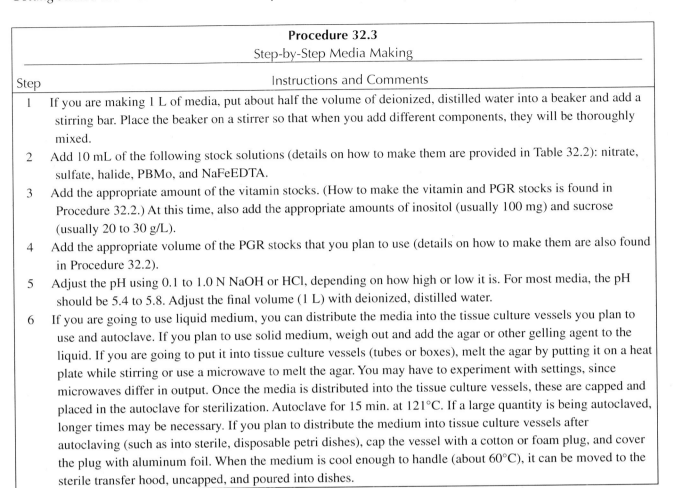

Procedure 32.3
Step-by-Step Media Making

Step	Instructions and Comments
1	If you are making 1 L of media, put about half the volume of deionized, distilled water into a beaker and add a stirring bar. Place the beaker on a stirrer so that when you add different components, they will be thoroughly mixed.
2	Add 10 mL of the following stock solutions (details on how to make them are provided in Table 32.2): nitrate, sulfate, halide, PBMo, and NaFeEDTA.
3	Add the appropriate amount of the vitamin stocks. (How to make the vitamin and PGR stocks is found in Procedure 32.2.) At this time, also add the appropriate amounts of inositol (usually 100 mg) and sucrose (usually 20 to 30 g/L).
4	Add the appropriate volume of the PGR stocks that you plan to use (details on how to make them are also found in Procedure 32.2).
5	Adjust the pH using 0.1 to 1.0 N NaOH or HCl, depending on how high or low it is. For most media, the pH should be 5.4 to 5.8. Adjust the final volume (1 L) with deionized, distilled water.
6	If you are going to use liquid medium, you can distribute the media into the tissue culture vessels you plan to use and autoclave. If you plan to use solid medium, weigh out and add the agar or other gelling agent to the liquid. If you are going to put it into tissue culture vessels (tubes or boxes), melt the agar by putting it on a heat plate while stirring or use a microwave to melt the agar. You may have to experiment with settings, since microwaves differ in output. Once the media is distributed into the tissue culture vessels, these are capped and placed in the autoclave for sterilization. Autoclave for 15 min. at 121°C. If a large quantity is being autoclaved, longer times may be necessary. If you plan to distribute the medium into tissue culture vessels after autoclaving (such as into sterile, disposable petri dishes), cap the vessel with a cotton or foam plug, and cover the plug with aluminum foil. When the medium is cool enough to handle (about 60°C), it can be moved to the sterile transfer hood, uncapped, and poured into dishes.

each stock is used to make one liter of medium. Some of the stock solutions require extra steps to get the components into solution (for example, NaFeEDTA stock), or may require making a serial dilution to obtain the amount of a trace component for the stock (sulfate and halide stocks).

Sometimes the amount of a particular component needed for a tissue culture medium is extremely small, so that it is difficult to weight out the amount even for the 100× stock solution. Because such small quantities of a substance cannot be weighed accurately, a serial dilution technique is used. The following example illustrates how serial dilutions can be used to obtain the correct amount of a component (in this case $CuSO_4 \cdot 5H_2O$) of the medium for its appropriate stock solution.

The stock solution calls for 2.5 mg of $CuSO_4 \cdot 5H_2O$ as a part of the sulfate stock of MS medium. Make an initial stock solution by placing 25 mg of $CuSO_4 \cdot 5H_2O$ in 100 mL of deionized/distilled water. After mixing thoroughly, use 10 mL of this solution, which will contain the desired 2.5 mg, and place it into the sulfate stock. This procedure required only one serial dilution, but any component can be subjected to one or more dilutions to obtain the desired amount.

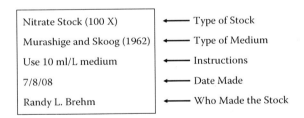

FIGURE 32.2

Once a stock solution is made, it should be labeled as in Figure 32.2 to avoid error and inadvertently keeping a stock solution too long.

MAKING STOCK SOLUTIONS OF THE PGRS AND VITAMINS

Vitamins and PGR stock solutions can be made up in concentrations 100× to 1000× of that required in the medium. Determine the desired amount for one liter of medium, the volume of stock solution needed to deliver that dosage of vitamin or PGR, and the volume of stock you wish to make. Procedure 32.2 gives examples of how to make vitamin and PGR stock solutions. Many of the PGRs require special handling to get them into solution. You will also find this information in Procedure 32.2.

STERILIZING EQUIPMENT AND MEDIA

Tissue culture media, in addition to providing an ideal medium for growth of plant cells, also are an ideal substratum for growth of bacteria and fungi. So it is necessary to sterilize the media, culture vessels, tools, and instruments and surface disinfest the explants as well. The most commonly used means of sterilizing equipment and media is by autoclaving at 121°C with a pressure of 15 psi for 15 min or longer for large volumes. Glassware and instruments are usually wrapped in heavy-duty aluminum foil or put in autoclave bags. Media (even those that contain agar) are in a liquid form in the autoclave, requiring a slow exhaust cycle to prevent the media from boiling over when pressure is reduced. Media should be sterilized in tissue culture vessels with some kind of closure, such as caps or plugs made of nonabsorbent cotton covered by aluminum foil. This way they do not become contaminated when they are removed from the autoclave for cooling. Use of racks that tilt the tissue culture tubes during cooling can give a slanted surface to the agar medium. These can be purchased or made from scratch using a little ingenuity.

Some components of the medium may be heat labile or altered by the heat so that they become inactive. These are usually added to the media after it has been autoclaved, but before the media has solidified. It is filtered through a bacteria-proof membrane (0.22 μm) filter and added to the sterilized medium after it has cooled enough not to harm the heat-labile compound (less than 60°C) and then thoroughly mixed before distributing it into the culture vessels. A rule of thumb is to add the filtered material at a point when the culture flask is just cool enough to be handled without burning one's hands. Some filters are available presterilized and fit on the end of a syringe for volumes ranging from 1 to 200 mL. More elaborate disposable assemblies are also available.

PREPARING THE STERILE TRANSFER HOOD

Successfully transferring explants to the sterile tissue culture medium is done in a laminar flow transfer hood. A transfer hood is equipped with positive-pressure ventilation and a bacteria-proof high-efficiency particulate air (HEPA) filter. The laminar flow hoods come in two basic types. Generally, air is forced into the cabinet through a dust filter and a HEPA filter, and then it is directed either downward (vertical flow unit) or outward (horizontal flow unit) over the working surface at a uniform rate. The constant flow of bacteria- and fungal spore-free filtered air prevents nonfiltered air and particles from settling on the working area that must be kept clean and disinfected. The simplest transfer cabinet is an enclosed plastic box or shield with an UV light and no airflow. A glove box can also be used, but both of these low-cost, low-technology options are not convenient for large numbers of transfers.

In the transfer hood, you should have ready some standard tools for use, such as a scalpel (with a sharp blade), long-handled forceps, and sometimes a spatula. Occasionally, fine-pointed forceps, scissors, razor blades, or cork borers are needed for preparing explants.

Many people prefer doing sterile manipulations on the surface of a presterilized disposable petri dish. Other alternatives that have worked well are stacks of standard laboratory-grade paper towels. These can be wrapped in aluminum foil and autoclaved. When the top sheet of the stack is used, it can be peeled off and discarded, leaving the clean one beneath it exposed to act as the next working surface. Another alternative is to use ceramic tiles. These can also be wrapped in aluminum foil and autoclaved, but tiles with very slick or very rough surfaces should be avoided.

In Procedure 32.3 is a suggested protocol to follow for preparing the sterile transfer hood. It also contains tips for keeping your work surface clean, eliminating contamination, and avoiding burns when flaming your instruments.

STORAGE OF CULTURE MEDIA

Once culture media has been made, distributed into tissue culture vessels, and sterilized by autoclaving, it can be stored for up to one month, provided it is kept sealed to prevent excessive evaporation of water from the medium. It should also be placed in a dark, cool place to minimize degradation of light-labile components, such as IAA. Storage at 4°C prolongs the time that media can be stored, but condensate may form in the container and encourage contamination. By making media five to seven days in advance, you allow time to check for any unwanted microbial contamination before explants are transferred onto the medium. However, media for certain sensitive species or operations or media that contains particularly unstable ingredients must be used fresh and cannot be stored.

SURFACE DISINFESTING PLANT TISSUES

Just as the media, instruments, and tools must be sterilized, so must the plant tissue be disinfested before it is placed on culture medium. Many different materials have been used to surface disinfest explants, but the most commonly used are 0.5% to 1% (v/v) sodium hypochlorite (commercial bleach contains 5.25% sodium hypochlorite), 70% ethanol, or 10% hydrogen peroxide. Others include using a 7% saturated solution of calcium hypochlorite, 1% solution of bromine water, 0.2% mercuric chloride solution, and 1% silver nitrate solution. If these more rigorous techniques are used, precautions should be taken to mini-

Procedure 32.4
Getting under the Hood

Step	Instructions and Comments
1	Turn on the transfer hood so that positive air pressure is maintained. This ensures that all of the air passing over the work surface is sterile. You should feel air flowing against your face at the opening. Make sure there are no drafts, such as open windows or air conditioning vents, that may interfere with the air flow coming out of the hood.
2	Use a spray bottle filled with 70% (v/v) ethanol to spray down the interior of the hood. Do not spray the HEPA filter. This is more effective than absolute alcohol for sterilizing surfaces, perhaps because 70% (v/v) ethanol denatures DNA. You can also use a piece of cheesecloth saturated with 70% (v/v) ethanol to help to distribute the ethanol more uniformly. Allow it to dry. To maintain the cleanliness of the interior, anything that is now placed inside the hood must be sprayed with 70% (v/v) ethanol. This includes the alcohol lamp, the slide staining jar filled with 80% to 95% (v/v) ethanol for flaming the instruments, a rack for the tools, racks of the tissue culture vessels containing medium, and all of the presterilized wrapped bundles containing your working surface (tiles, paper towels, ashtrays, etc.) and your tools (forceps, spatula, scalpel, etc.)
3	When you are ready to begin, remove jewelry and wash your hands thoroughly with soap and water, and then just before placing then in the hood, spray them down with 70% ethanol.
4	You may now open your work surface by peeling back the heavy duty aluminum foil exposing the surface of the tile (or other alternative). Never block the air flow across the surface coming from the filter unit. Also do not pass your hands across the surface of the tile. Talking while you are in the unit also compromises sterility. If you must talk, turn your head to one side. If you have long hair, fasten it so that it does not dangle onto your work surface.
5	Keep any open sterile containers as far back in the hood as you can. When you open containers that have been sterilized, keep your fingers away from the opening. If you open a glass container or vessel, as general rule, pass the opening through the flame. This creates a warm updraft from the vessel helping to prevent contamination from entering it.
6	Flaming instruments can be hazardous if you forget that ethanol is flammable. When you are flame sterilizing an instrument, for example, forceps, dip them into the jar containing the 80% ethanol, and when you lift them out, keep the tip of the instrument at an angle downward (away from your fingers) so that any excess alcohol does not run onto your hand (Figure 32.1). Then pass the tip through the flame of the alcohol burner, and hold the instrument parallel to the work surface. When the flame has consumed the alcohol, let the tool cool on the rack until you are ready to use it. Never, never place a hot tool back into the jar containing 80% ethanol. I know of an experienced scientist who momentarily forgot this simple rule and the resulting fire singed his hand and the hair off his forearm.

mize health and safety risks, especially with the heavy metal-containing solutions.

The type of disinfectant used, the concentration, and the amount of exposure time (1 to 30 min) vary depending on the sensitivity of the tissue and how difficult it is to disinfest. Woody or field-grown plants are sometimes very difficult to disinfest and may benefit from being placed in a beaker with cheesecloth over the top and placed under running water overnight. In some cases, employing a two-step protocol (70% ethanol followed by bleach) or adding a wetting agent, such as Tween 20 or detergent, helps to increase the effectiveness. In any case, the final step before trimming the explant and placing onto sterile medium is to rinse it several times in sterile, distilled water to eliminate the residue of the disinfesting agent. Laboratory exercises presented throughout the book will contain detailed directions for various procedures to disinfest tissue.

FINAL WORD

Tissue culture is much like good cooking. There are simple recipes, and then there is "haute cuisine." Cooking is also a very rewarding activity, particularly when the end result is delicious. By following the procedures outlined in this chapter and in the succeeding chapters of the book, you should find that, with care and attention to detail, you will be a "chef extraordinaire," and your tissue culture ventures will be successful.

LITERATURE CITED

Ammirato, P.V. 1974. The effects of abscisic acid on the development of somatic embryos from cells of caraway (*Carum carvi* L.). *Bot. Gaz.* 135: 328–337.

Coffin, R., C.D. Taper, and C. Chong. 1976. Sorbitol and sucrose as carbon source for callus culture of some species of the Rosaceae. *Can. J. Bot.* 54: 547–551.

Epstein, E. and J. Ludwig-Müller. 1993. Indole-3-butyric acid in plants: occurrence, synthesis, metabolism, and transport. *Physiol. Plant.* 88: 382–389.

Gamborg, O.L., R.A. Miller, and K. Ojima. 1968. Nutrient requirements of suspension cultures of soybean root cells. *Exp. Cell Res.* 50: 151–158.

Linsmaier, E.M. and F. Skoog. 1965. Organic growth factor requirements of tobacco tissue cultures. *Physiol. Plant.* 18: 100–127.

Lloyd, G. and B. McCown. 1980. Commercially feasible micropropagation of mountain laurel, *Kalmia latifolia*, by use of shoot tip culture. *Intl. Plant Prop. Soc. Proc.* 30: 421–427.

Murashige, T. and F. Skoog. 1962. A revised medium for rapid growth and bio-assays with tobacco tissue cultures. *Physiol. Plant.* 15: 473–497.

Nissen, S.J. and E.G. Sutter. 1990. Stability of IAA and IBA in nutrient medium to several tissue culture procedures. *HortScience.* 25: 800–802.

Nitsch, J.P. and C. Nitsch. 1969. Haploid plants from pollen grains. *Science.* 163: 85–87.

Oka, S. and K. Ohyama. 1982. Sugar utilization in mulberry (*Morus alba* L.) bud culture, in *Plant Tissue Culture*. Fujiwara, A. (Ed.), Proc. 5th Int. Cong. Plant Tiss. Cell Cult., Jap. Assoc. Plant Tissue Culture, Tokyo, Japan, pp. 67–68.

Roberts, D.R., B.C.S. Sutton, and B.S. Flinn. 1990. Synchronous and high frequency germination of interior spruce somatic embryos following partial drying at high relative humidity. *Can. J. Bot.* 68: 1086–1090.

Schenk, R.V. and A.C. Hildebrandt. 1972. Medium and techniques for induction and growth of monocotyledonous plant cell cultures. *Can. J. Bot.* 50: 199–204.

Singha, S. 1982. Influence of agar concentration on *in vitro* shoot proliferation of *Malus* sp. 'Almey' and *Pyrus communis* 'Seckel.' *J. Amer. Soc. Hort. Sci.* 107: 657–660.

Skoog, F. and C.O. Miller. 1957. Chemical regulation of growth and organ formation in plant tissues cultured *in vitro*. *Symp. Soc. Exp. Biol.* 11: 118–131.

White, P.R. 1963. *The Cultivation of Plant and Animal Cells.* 2nd ed. Ronald Press Co., New York.

33 Micropropagation of Mint (*Mentha* spp.)

Sherry L. Kitto

Mint oils, from both peppermint (*Mentha x piperita* L.) and spearmints (*M. spicata* L. and *M. cardiaca* G.), are important flavoring agents for pharmaceuticals, foods, and cosmetics. In 1999, peppermint oil, harvested from roughly 100,000 acres, was the largest-volume essential oil exported from the United States. Infestations with pests, such as weeds, insects, or pathogens, greatly impact yields of this vegetatively propagated and, therefore, clonal crop. While there is a great deal of interest in improving mints that would be tolerant of or resistant to such pests, crop improvement through breeding is challenging mostly because of sterility. However, improvement of commercial mint cultivars may be realized in the near future as a result of genetic engineering or transformation. The release of a new, improved mint cultivar for commercial distribution would require the development of a quick, economical, and vegetative propagation protocol.

The purpose of micropropagation is to produce the highest number of clonally identical, quality plants in the shortest timeframe. The objective of this laboratory exercise is to examine some variables that impact the rapidity with which mint can be micropropagated. This exercise will examine the following classic stages of a micropropagation system: establishment of sterile cultures, proliferation of shoots, rooting of microcuttings, and acclimatization/reestablishment of plants outside the laboratory environment (chapter 31). Each experiment of this multiweek laboratory is described in detail; however, experimental protocols can be modified to discover the "best" system for micropropagating mint under various physical and environmental conditions. Where applicable, the exercises that follow generally describe some variables that can be explored in each stage of micropropagation. Pick no more than four variables to examine prior to beginning each week's work. It is a good idea to decide on the table format to be used to report the results, so that the correct data can be collected. If possible, use at least 10 explants for each treatment.

EXERCISES

EXPERIMENT 1: ESTABLISHMENT OF CLEAN/STERILE CULTURES

Sterile cultures can be established using seeds or vegetative material. Although commercial mints are not propagated using seeds, for the purposes of this exercise, purchasing seeds may be the easiest way to obtain mint germplasm. Addresses for sellers of mint seeds can be found on the Internet. Vegetative material may be obtained from a friend's yard, or commonly in the United States, mint shoots can be found in the produce section of a local grocery store. If you decide to use vegetative material, be sure to select healthy, actively growing, aboveground shoots. Actively growing plant material is usually less contaminated by windborne pests. If you are unable to find actively growing shoots, a rootstock can be dug, potted, and placed in a greenhouse or on a window ledge to encourage new shoot growth for this exercise. There are advantages to using each type of initial explant: seeds are more tolerant to the surface disinfestation protocols, and therefore are easier to establish as aseptic (or axenic) cultures; however, starting with vegetative material more accurately mimics what would actually occur in a commercial micropropagation facility working with a new cultivar of mint.

Aseptic cultures need to be established for use in the proliferation phase of the laboratory exercise. The procedures for seed germination (Table 33.1) and surface disinfestation of the vegetative shoots (Table 33.2) can be started the same day. Within a couple of weeks, seedling shoots will be available for subculture, and the initial identification of clean vegetative cultures should be possible. Keep in mind that there might be "clean" cultures that are false negatives and may still be contaminated. Be sure to observe the cultures carefully for contamination. Holding cultures up to the light and looking through the bottom of the container and gelled medium help to visualize some bacterial contaminants. Avoid turning the cultures upside

TABLE 33.1
Disinfestation Protocols

Explant	Disinfestation Protocol	N[a]	Contamination (%)	Mean No. of Axillary Shoots[b]	Mean No. Nodes/Shoot[b]	No. Roots[b]
Seeds	20% bleach					
Vegetative	10% bleach					
Vegetative	20% bleach					

[a] Initial number cultured
[b] Data based on number of clean cultures

TABLE 33.2
Contamination and Establishment

Shoot Source	N[a]	Contamination (%)	Mean No. of Axillary Shoots[b]	Mean No. of Nodes/Shoot[b]	General Appearance[b]	No. Roots[b]
Seedling						
Vegetative						

[a] Initial number of shoots cultured
[b] Data based on number of clean cultures

down. Contaminated cultures should be discarded (disposable containers) or autoclaved (reusable containers). Do not open contaminated cultures.

Materials

The following materials will be required for the experiment:

- Seeds and/or vegetative shoots of mint
- Sterile glass or plastic petri dishes containing moist filter paper
- Culture tubes (25 × 150 mm) containing sterile water
- Establishment medium: MS medium supplemented with 3% sucrose and 0.6% Phytagar or other suitable brand of agar, 15 to 25 mL per 25 × 150-mm culture tube
- Liquid establishment medium: MS medium supplemented with 3% sucrose, 10 mL/culture tube
- Muslin or cheesecloth, rubber bands, labels, ethanol, commercial bleach, Tween 20 or some other surfactant, Parafilm
- Dissecting microscope

Follow the protocols listed in Procedure 33.1 and Procedure 33.2 to complete the experiments.

Anticipated Results

Do not be surprised if many of the seeds do not germinate, or if browned-out leaves appear on the vegetative shoots. As long as a shoot's growing point is green, you should be able to establish a viable culture. There is a fine line between "cleaning up" and killing vegetative plant tissue (bleach is phytotoxic), so some damage (browning) is to be expected. The amount of browning, resulting from the sensitivity of the vegetative tissues to the bleach, is dependent on how hardened the initial material is. Mints are fairly malleable and successful establishment of clean, vigorous shoots should be fairly easy.

Questions

- What was the most common type (bacterium or fungus) of contamination?
- Why are seeds easier to surface disinfest than vegetative shoots?
- Why do some cultures initially appear to be clean, but later become contaminated?
- What are some alternative procedures that might be tried to establish clean cultures?

EXPERIMENT 2: ESTABLISHMENT OF PROLIFERATING SHOOT CULTURES

The objective of shoot proliferation is to maximize axillary shoot (Figure 33.1) production between successive

Micropropagation of Mint (Mentha spp.)

	Procedure 33.1
	Mint Seed Germination
Step	Instructions and Comments
1	Scarify 20 seeds using fine grit sandpaper.
2	Place 20 scarified and 20 nonscarified seeds in separate 5 cm × 5 cm squares of muslin or cheesecloth (place paper and pencil treatment identification labels in with each group of seeds). Secure with a rubber band, and surface disinfest by dipping in 70% ethanol for 1 min, followed by 20% bleach + 0.1% Tween 20 solution for 20 min. Rinse three times with sterile water, 10 min each.
3	Culture five seeds on presterilized, moist filter paper in glass or plastic petri dishes. It is important that the filter paper be moist. Seal the dishes with Parafilm™, and label with the date and your name and the plant ID. Species = _____.
4	Place seeds in dark for germination. Seeds should germinate in a week or two. Examine the seeds using a dissecting microscope for obvious contamination. Do not open dishes that contain contamination.
5	Keep clear records on the number of seeds germinated, number of contaminated seeds, and type of contamination (i.e., bacterium or fungus). Use Table 33.1 as a model for reporting results. Using the results presented in Table 33.1, write a brief summary of this exercise.
6	Go to shoot proliferation protocols (Procedure 33.3) to continue.

	Procedure 33.2
	Surface Disinfestation of Vegetative Shoots
Step	Instructions and Comments
1	Collect shoots that visually appear to be free from contaminants.
2	Remove larger leaves from the shoots and cut 30 nodal/tip explants that are ca. 2.5 to 4.0 cm. Leave enough plant material, petioles and stem, so that the bleach-killed bases can be trimmed later.
3	Surface disinfest 1/2 of the mint shoots for 20 min. in 20% bleach + 0.1% Tween 20; rinse three times with sterile water for 5 min. each.
4	Trim bleach-damaged ends off cuttings, and culture on establishment medium. Be sure to label the containers with the treatment, date, your name, and the plant ID. Species = _____.
5	Surface disinfest the other 1/2 of the mint shoots for 10 min. in 10% bleach + 0.1% Tween 20 solution; rinse three times with sterile water for 5 min. each.
6	Culture the shoots in liquid establishment medium and place on a shaker (100 rpm) overnight. Surface disinfest for 10 min by adding an equal volume of 20% bleach + 0.1% Tween 20 solution and shaking. Rinse three times with sterile water for 5 min. each.
7	Treat cuttings as described in Steps 5 and 6.
8	Collect data weekly and, after one month, compare the two surface disinfestation protocols for percentage of clean cultures. Count the number of shoots and number of nodes for each culture/treatment. Use Table 33.1 as a model for reporting results. Using the results presented in Table 33.1, write a brief summary of this exercise.
9	Go to shoot proliferation protocols (Procedure 33.3) to continue.

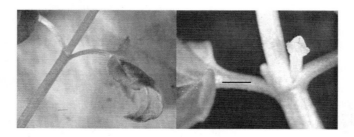

FIGURE 33.1 Leaf axils just after subculture (left) and after a couple of weeks (right); note the axillary shoots in the leaf axils on the right. (Bar = 1 mm.)

subcultures. Aseptic, proliferating shoot cultures need to be established to generate shoots for use in the maintenance of proliferating cultures experiment.

Materials

The following items will be needed for this experiment:

- Aseptically grown mint seedlings
- Axenic vegetative shoot cultures
- Shoot Proliferation Medium: MS medium supplemented with 3% sucrose, 5 µM BA and 0.6% Phytagar; 15 to 25 mL per 25 × 150-mm culture tube

Follow the protocols listed in Procedure 33.3 to complete this phase of the experiment.

Anticipated Results

If you initiated shoot cultures from both seeds and vegetative plant material, you may notice that the seedling-derived shoot cultures are less vigorous. Remember one reason that the vegetative material from the grocery store is being clonally maintained and cultivated is because it is vigorous. Response variation among explants that appeared to be initially identical is normal. A few cultures may become contaminated due to false negatives in the sterile-culture establishment stage.

Questions

- With respect to efficiency, what is the best size for an explant for shoot proliferation, one node or two nodes?
- Do seedling shoots proliferate differently than vegetative shoots?

EXPERIMENT 3: MAINTENANCE AND PERIODIC SUBCULTURE OF PROLIFERATING CULTURES

Once aseptic shoot cultures are established, they must be maintained. It is important to develop a protocol to maximize production of axillary shoots or microcuttings. Mint, having opposite leaves, has two axillary buds at each node; however, some plants can be massed up very effectively from single-node explants. In this experiment, it will be important to know how many axillary shoots are generated as well as the number of nodes present on each shoot. Experiments that can be attempted here include (1) comparing proliferation from shoot-tip cuttings (Figure 33.2), nodal cuttings (two axillary buds) (Figure 33.2), and split-node cuttings (one axillary bud); or (2) determining if there is a gradient of proliferation based on node position on a shoot (how do nodal explants collected nearer the tip proliferate compared to nodal explants located nearer the base?). Orientation of the explants can also be explored; stem explants may be placed in the medium vertically, horizontally, or even

Procedure 33.3 Establishment of Proliferating Shoot Cultures of Mint	
Step	Instructions and Comments
1	Excise six shoots (three with one node and three with two nodes) each from the surface disinfestation cultures containing the seedlings and/or the vegetative shoots, and culture on shoot proliferation medium. Be sure to label the cultures with explant source, date, your name, and plant ID. Species = _____.
2	Observe weekly, and keep clear records for each explant source on the number of contaminated cultures and the type of contamination (i.e., bacterium or fungus).
3	At the end of four weeks, collect these additional data: number of shoots generated, presence of callus, general appearance of the shoots (i.e., green, yellow, brown), and root production. Use Table 33.2 as a model for reporting results. Using the results presented in Table 33.2, write a brief summary of this exercise.
4	Go to maintenance of proliferating cultures protocols (Procedure 33.4) to continue.

Micropropagation of Mint (*Mentha* spp.)

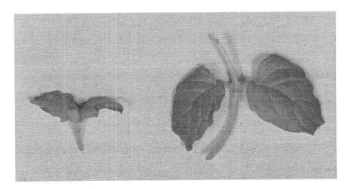

FIGURE 33.2 Shoot-tip cutting (left) and nodal cutting (right) may be used to maintain proliferation cultures or may be rooted. (Bar = 1 mm.)

FIGURE 33.3 Microcutting that has rooted *in vitro* on growth regulator-free medium. (Bar = 0.3 cm.)

upside down. Shoots that result during the establishment of proliferating shoot cultures can be rooted or recultured to maintain shoot proliferation cultures.

Select the treatments you are interested in comparing. Set up at least 10 containers for each treatment with one explant per container. After four weeks, using sterile technique, collect shoot and/or node production data for each shoot. Repeat the experiment once, and collect the data. Present the numerical and observational data in table format (e.g., Table 33.3).

Materials

The following items are needed for the experiment:

- Clean shoot cultures
- Shoot Proliferation Medium: MS medium supplemented with 3% sucrose, 4 µM BA, and 0.6% Phytagar, 15 to 25 mL per 25 × 150-mm culture tube

Follow the instructions outlined in Procedure 33.4 to complete this experiment.

Anticipated Results

Seedling-derived cultures may remain less vigorous after repeated subcultures. Shoot-tip cuttings may grow taller, while nodal cuttings may produce more axillary shoots. Commercially, the goal is to produce the highest number

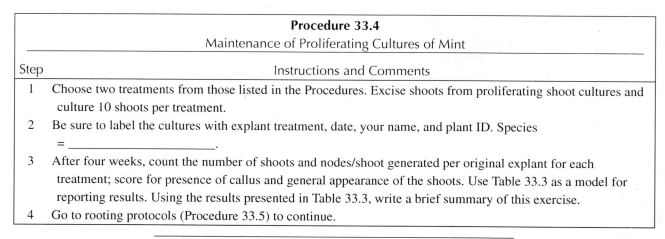

of explants, whether for subculture or rooting. Split-node explants may be less responsive due to damage while cutting (the sharper the blade, the more practiced the operator, the quicker the cut, all result in less damage).

Questions

- Why is it important to look for callus production?
- What are the differences between axillary and adventitious shoots? Is one type of shoot better than the other?
- Which treatment was the best for the production of shoots?
- Which treatment "looked" the best (greenest, least amount of browning or yellowing, least amount of callus, etc.)?

EXPERIMENT 4: ROOTING AND ACCLIMATIZATION

Mint cuttings root fairly rapidly and easily without the use of rooting compounds (Figure 33.3); therefore, rooting environment, the use of mist, and/or humidity domes, will be examined. It is critical that the microcuttings not dry out at any point. They need to be kept in a high-humidity environment until they have had the chance to initiate roots and acclimate to the *in situ* environment. Microcuttings will be placed into a soilless greenhouse medium, such as a 2 sphagnum peat:1 perlite:1 vermiculite medium like Promix BX or Redi-Earth®. They are then placed under the humidity domes or under mist for one week before being transferred to the greenhouse bench. If humidity domes are used, they must be removed prior to placement on the greenhouse bench. Microcuttings may be placed into individual cells or planted as a group into a larger container. Be sure to label them correctly. Select a container with drainage holes and a soilless greenhouse medium that allows for good drainage.

It is not necessary to plant a node in the rooting medium for mint microshoots to root. Place the microcuttings in the moist soilless greenhouse medium. Half of the containers will be placed in a flat with a humidity dome. The other half of the containers will be placed in a flat that is set under mist. You will need to work fairly quickly so that the microcuttings do not dry out, use the water bottle mister frequently.

Materials

The following materials will be needed for this experiment:

- Microcuttings from proliferating shoot cultures
- Water bottle with mist attachment
- Pencil, scalpel or sharp knife
- Greenhouse containers, flats, humidity domes
- Greenhouse mist system
- Soilless greenhouse medium such as Promix BX or Redi-Earth®

Micropropagation of Mint (Mentha spp.)

	Procedure 33.5
	Rooting of Mint Microcuttings
Step	Instructions and Comments
1	In the laboratory, using forceps, take hold of the base of a proliferating shoot clump and remove it from the container. Working with one container at a time, wash the agar away from the shoot clump using warm tap water and gentle rubbing. Place washed shoot clumps between moist paper towels. It is critical that the shoot clumps not dry out. Repeat until all the shoot clumps have been cleaned.
2	Again, working with one clump at a time, use a scalpel to cut individual shoots away from a clump. Be careful not to cut yourself. Place these microcuttings between moist paper towels. It is critical that the microcuttings not dry out. Collect 40 microcuttings.
3	Wrap the moist towel-wrapped microcuttings in plastic wrap, a plastic zipper bag, or aluminum foil for travel to the greenhouse.
4	In the greenhouse, fill four containers with soilless greenhouse medium and thoroughly wet.
5	Using a small dibble (a pencil works well), make a shallow hole. Remove one microcutting from the paper towels (be sure to keep the remaining microcuttings covered with the moist towels), place in the hole, and using the dibble, carefully backfill with the medium and firm the medium around the microcutting with your thumb and forefinger. The microcutting should remain upright when stuck properly. Mist the microcutting.
6	Repeat Step 5 until 10 of the microcuttings are planted. Remember to mist all of the microcuttings with water repeatedly until the flats have been placed under mist.
7	Repeat Steps 5 and 6 for the other three containers. Label each container with treatment (plus or minus dome), date, your name, and plant ID. Species = _____. Place each container in a flat. Mist the inside of the humidity domes, and place over the two appropriate flats. Put all four flats under mist that provides a fine mist over the flats every five min.
9	After two weeks, remove 1/2 of the cuttings from each container and count and measure the length of the roots. Four weeks later, remove the remaining cuttings and collect data. Keep a record of survival.
10	Using the collected data, calculate, for each two-week interval, the percentage of survival, mean number of roots, and mean length of roots. Use Table 33.4 as a model for reporting results. Using the results presented in Table 33.4, write a brief summary of this exercise.

TABLE 33.4
Data for Rooting and Acclimatization

		Week 2			Week 4		
Treatment	N[a]	Survival (%)	Mean No. Roots[b]	Mean Length of Roots[b]	Survival (%)	Mean No. Roots[b]	Mean Length of Roots[b]
+ Humidity dome							
− Humidity dome							

[a] Initial number of shoots planted
[b] Data based on five shoots

Follow the instructions outlined in Procedure 33.5 to complete this phase of the experiment.

Anticipated Results

Mint microcuttings are fairly easy to root and reestablish.

Questions

- How would you determine whether your protocols are cost effective?
- If mint plants are planted in 50- to 75-cm (20- to 30-in.) rows with plants 15 cm (4 to 6 in.) apart in the row, how many plants will you need to plant

for each hectare (acre)? How long will it take you, using your protocol, to produce the required number of plants required for a hectare (acre)?
- If 10 stock cultures, each with 10 shoots, are maintained and some shoots have to be held back for the "mother" stock plants, how far in advance would you have to start production to guarantee spring delivery of 10,000 plants?

SUMMARY

Prepare a laboratory report in a scientific manuscript format that includes the following sections: abstract, introduction, materials and methods, results, and discussion.

SUGGESTED READINGS

Lee, J. 2001. Making a mint of money: essential oils' export success. *AgExporter.* Feb. 13: 20 USDA, FAS, Washington, D.C.

Murashige, T. and F. Skoog. 1962. A revised medium for rapid growth and bioassays with tobacco tissue cultures. *Physiol. Plant.* 15: 473–497.

Tucker, A.O. 1992. The truth about mints. *The Herb Companion.* August/September, pp. 51–52.

USDA National Agricultural Statistics Service. 2002. www.nass.usda.gov/

Van Eck, J.M. and S.L. Kitto. 1990. Callus initiation and regeneration in *Mentha. HortScience.* 25: 804–806.

Veronese, P., X. Li, X. Niu, S.C. Weller, R.A. Bressan, and P.M. Hasegawa. 2001. Bioengineering mint crop improvement. *Plant Cell. Tiss. Org. Cult.* 64: 133–144.

34 Micropropagation of Tropical Root and Tuber Crops

Leopold M. Nyochembeng

CHAPTER 34 CONCEPTS

- A tuberous root is an enlarged fleshy secondary root containing carbohydrate reserves (often starch).

- Examples of tropical root and tuber crops are cassava, yam, cocoyam, potato, and sweet potato.

- Cassava is a woody perennial shrub with palmate leaves that bears multiple nodes containing dormant buds on the stem and branches and produces an edible root.

- A tuber is a swollen subapical portion of a modified stem that develops underground as a storage organ. Tubers contain large amounts of starch reserves serving as the major sink for the products of photosynthesis.

- Edible tubers, such as yam, cocoyam, and potato, are important sources of carbohydrates, vitamins, and minerals.

- Yam is a tuber-producing annual or perennial climbing monocotyledonous vine. It produces a stem tuber. Yams are a different species than sweet potatoes, and yam tubers have white flesh.

- Cocoyam is an Aroid that produces underground starchy edible tubers known as cormels from a swollen cormous stem base.

- A yam minisett is a piece of yam derived from the main tuber and weighing 30 to 40 g that is used as "seed" for yam production.

- Microtuberization is the development of miniature tuberous propagules from shoots in tissue culture.

Root and tubers crops are essentially geophytic plants (chapter 29), whose survival rests on the development of fleshy subterranean organs with or without dormant buds. These food storage organs allow the plant to survive during adverse climatic conditions and support renewed growth when environmental conditions become favorable again. For these important food crops adapted to the wet–dry cycle of tropical and subtropical regions (cassava, yam, cocoyam, sweet potato, potato, and taro) and warm–cold cycle of temperate climates (sweet potato and potato), survival means stashing away large amounts of carbohydrates and water during their annual growth cycle. These root and tuber crops are tropical perennials, cultivated mainly as annuals. The subterranean environment is ideal for storing carbohydrates because its moisture content is relatively stable, and temperature fluctuations are not as extreme as in the aboveground environment. After attaining physiological maturity, the tuberous structures undergo variable durations of dormancy prior to sprouting.

Botanically, a tuber is a subapical swelling of a genetically programmed modification of a stem that serves as a subterranean storage structure. Examples include yam, cocoyam, taro, and potato. Tuberous roots are enlarged swellings of fleshy secondary roots that contain significant carbohydrate reserves, such as cassava and sweet potato. A distinguishing feature of tubers and tuberous roots is that their storage organs are generally bulky, due to high starch and moisture content. They also store poorly after harvest if not properly conditioned, thus some form of postharvest processing is required to increase their shelf life. Root and tuber crops are conventionally vegetatively propagated either by stem cuttings (cassava and sweet potato), whole root or root fragments (sweet potato), and tubers (yam, cocoyam, taro, and potato). This method of propagation

FIGURE 34.1 Cassava plant growing from a potted stem cutting. Note the prominent nodes.

offers some advantages to root and tuber crops in that it makes them easy and convenient to propagate and also helps maintain clonal fidelity in the propagated populations with little or no genetic variability. Seed propagation is rare for several reasons including unpredictable flowering, poor seed set, genetic heterozygosity, and variable germination rate and seedling vigor.

Root and tuber crops are important food resources, serving as staple food crops in the humid regions of many developing countries. The foliage may also be used for food or feed. Many roots and tubers serve as sources of important raw materials for various industries in developed economies. Cassava starch, for example, is useful in the paper and textiles industry. It is also used for alcohol production in South America. In Africa, cassava flour is an important substitute for wheat flour.

CASSAVA

Cassava (*Manihot esculenta* Crantz) is a perennial woody shrub (Figure 34.1) with an edible root, which grows in tropical and subtropical areas of the world. It is also known in other regions as yuca, manioc, manihot, tapioca, and mandioca. A unique feature of cassava is its good drought tolerance and the ability to thrive in soils with low fertility. Cassava roots (Figure 34.2) can be stored in the ground for up to two years or more depending on the cultivar, which provides much flexibility in harvesting the tuberous roots and cuttings for propagation.

Cassava Plant Propagation

Cassava is propagated conventionally from hardwood stem cuttings consisting of two to six nodes (10 to 20 cm long) depending on the portion of the stem and cultivar used. The cassava plant produces seed, but seed propagation is not usually used. Recently, a propagation technique that employs ministems was developed at the International Institute of Tropical Agriculture (IITA), Nigeria. This technique involves cutting healthy cassava stems into several short pieces, each with at least oné node. If cuttings are taken from hardwood stems, one to two nodes are used,

FIGURE 34.2 Fresh cassava roots harvested at maturity. (Photo courtesy of Dr. K.N. Konan, Alabama A&M University.)

Micropropagation of Tropical Root and Tuber Crops

from semihardwood, 4 to 6 nodes, and from the soft terminal portion of shoots, 6 to 10 nodes. Cuttings are then dipped into a fungicide suspension and placed loosely in perforated transparent polyethylene bags under shade. The cuttings sprout three to five days later and are then transplanted into the field. Ministem propagation increases the in-field multiplication ratio of cassava severalfold compared to traditional hardwood cutting propagation.

IN VITRO PROPAGATION

In vitro procedures have been applied extensively to cassava for rapid micropropagation, meristem culture for generation of pathogen-free plants for international germplasm distribution, embryo culture, somatic embryogenesis, and plant transformation for development of transgenic cassava. Several kinds of explants and regeneration systems exist for micropropagating cassava. Most commonly used systems include apical meristems in meristem tip culture, somatic embryogenesis from cotyledonary explants, organogenesis and shoot production from stem callus, and rapid micropropagation via multiple shoot production from nodal explants and axillary bud-derived meristems.

MERISTEM CULTURE

Shoot meristem culture involves excising shoot tips consisting of the meristematic dome plus two or more subtending leaf primordia and culturing these under well-defined *in vitro* conditions. Introduced over three decades ago, meristem culture has been used to eliminate several viral pathogens in important crop plants. For example, African cassava mosaic virus-free cassava has been obtained via meristem culture (chapter 11). The technique gained prominence in micropropagation with the observation that plants regenerated via meristem culture maintained clonal fidelity in addition to lacking viral disease symptoms that were present in parental clones.

The following two experiments will demonstrate cassava shoot meristem culture and rapid shoot multiplication in cassava using axillary shoot buds.

EXPERIMENT 1: SHOOT MERISTEM CULTURE

Materials

The following materials will be needed for each student team:

- Actively growing cassava plants
- Scissors
- Liquid chlorine bleach (containing 5.25% NaClO)
- Whatman No. 1 filter paper
- Stereo dissecting microscope
- Scalpel and forceps
- Growth chamber
- pH meter
- Stir plate and stir bar
- Balance
- Laminar air-flow bench
- Murashige and Skoog (1962) (MS) basal salt medium plus B5 vitamins (Gamborg et al., 1968) and glassware

Follow the protocols listed in Procedure 34.1 to complete the experiment.

Anticipated Results

The growth and morphogenic response of cassava meristems in culture is influenced by the growth factors in the medium as well as the cultivar. Some combinations of growth regulators should work better than others in inducing callus, shoots, and roots.

Questions

- Why is meristem culture used for the production of disease-free plants for germplasm distribution? Is meristem culture a viable option for the production of virus-free plants from virus-infected ones?
- What factors should be considered in the use of meristem culture as a tool for micropropagation?

EXPERIMENT 2: RAPID MULTIPLICATION FROM CASSAVA AXILLARY SHOOT BUDS

Materials

The following materials will be needed by each team of students:

- Tissue-cultured cassava shoots
- Scalpel and forceps
- Tissue culture glassware
- Growth chamber, tissue culture room, or lab bench equipped with lights
- pH meter
- Stirring plate and stirring bar
- Balance
- Laminar air-flow bench
- Magenta GA7 containers or baby food jars
- Ziploc bags
- Marker pens and plastic labels
- Vermiculite
- Perlite
- Plastic pots (10 × 10 cm)
- Murashige and Skoog (MS) basal salt medium (Murashige and Skoog, 1962)

Follow the instructions outlined in Procedure 34.2 to complete this experiment.

	Procedure 34.1
	Shoot Meristem Culture of Cassava
Steps	Instructions and Comments
1	Prepare two to three node cuttings per group; place in moist 1 sphagnum peat:1 perlite by volume or sawdust in the greenhouse to force shoots. Shoots will be ready in about two weeks.
2	Detach (24 per group) shoot tips about 2 cm long, place in a beaker, and rinse under running tap water for one hour, followed by two rinses in distilled water.
3	Within the sterile transfer hood, disinfest with 100 mL 10% liquid chlorine bleach (v/v, 0.525% NaClO) containing one drop of Tween 20 (a wetting agent) for 15 min. Rinse four times in sterile distilled water.
4	Place shoot tips under a dissecting microscope and, using a sterile scalpel and forceps, dissect meristems approximately 0.5 to 1 mm long at a magnification of 4×. Place the meristems in fresh petri dishes containing a moist sterile Whatman # 1 filter paper, and cover. Students should work in groups of two to three and take turns in dissecting the meristem tips. Care should be taken to minimize the time that meristems are exposed on the dissecting scope to avoid dehydration and microbial contamination.
5	Transfer the meristem tips onto agar solidified Murashige and Skoog (MS:1962) basal medium in 60 × 15 mm petri dishes containing 100 mg/L inositol, 80 mg/L adenine, 0.58 µM GA_3, B5 vitamins, 0.7% agar and supplemented with benzyladenine (BA) (0, 0.44, and 1.32 µM) and naphthalene acetic acid (NAA) (0 and 0.05 µM). Culture two meristem tips per petri dish, seal with Parafilm™, and label the dishes with your name, date, and concentrations of growth regulators used. Make four replications per growth regulator treatment combination.
6	Incubate cultures at 26°C to 28°C in a 16-h photoperiod at 90 µmol·m^{-2}·sec^{-1} and observe weekly for up to six weeks. Record your observations on viability, callus formation, root formation, number and length of roots, shoot development, and number and length of shoots.
7	Analyze your data using ANOVA (analysis of variance), and compare treatment means where necessary using Tukey's HSD at 5% probability.

Anticipated Results

Shoot and plantlet development will depend on the morphogenic potential of the cassava cultivar used and its response to BA treatment. Rapid shoot multiplication can be achieved from shoot or axillary bud-derived explants of cassava; however, the rates of shoot multiplication may be quite variable with different cultivars or clones.

Questions

- How would you improve the efficiency of *in vitro* rapid shoot multiplication of cassava?
- What is the relative advantage in using axillary bud-derived meristems for rapid multiplication of cassava compared to meristem tip culture?
- What characteristics does a tissue-cultured plantlet have that makes acclimatization necessary?

YAMS

Yams (*Dioscorea* spp.) are tuber-producing annual or perennial climbing monocotyledonous plants that should not be confused with sweet potato (*Ipomoea batatas* (L.) Lam.). The stem is a vine, which may bear aerial tubers (tubercles) in some species (*D. bulbifera* L.). In most edible species, the tubers (Figure 34.3) are subterranean and can grow over 40 cm long and 15 cm wide. Over 95% of the world's edible yams (36 million tons) are cultivated in sub-Saharan Africa, within the region commonly known as the yam belt that stretches across West Africa. Other minor producers are found in the West Indies, South and Central America, and parts of Asia. Of the 600 yam species grown worldwide, only three species are most popular: white yam (*D. rotundata* Poir), yellow yam (*D. cayenensis* Lam.), and water yam (*D. alata* L), with white yam being the most predominant cultivated species. Yam is an important cash crop that competes with cassava as the preferred staple food crop and plays an important sociocultural role within some indigenous communities in West Africa.

PROPAGATION OF YAMS

Traditionally, yams are propagated from the seed yams, which are made up of relatively smaller tubers (200 to 1000 g) selected from the current harvest for use in the next planting season. With this method, a significant portion of the harvest has to be set aside as planting material.

Procedure 34.2
Rapid Multiplication from Cassava Axillary Shoot Buds

Steps	Instructions and Comments
1	Prepare single-node explants from shoots and plantlets of preexisting axenic tissue cultures of cassava under aseptic conditions. Axenic cultures are advantageous for use because additional disinfestation is not required.
2	Place two explants in MS medium containing 0.7% agar, 100 mg/L inositol, and supplemented with 0, 22.2, 44.4, and 66.6 µM BA in Magenta GA7 containers. For each treatment make 10 replications.
3	Incubate cultures at 26°C to 28°C in a 16-h photoperiod at 90 $\mu mol \cdot m^{-2} \cdot sec^{-1}$ for five days.
4	Excise axillary shoot buds from each stock sulture and reculture in the same formulation, but fresh medium (first subculture) and incubate as above. Monitor shoot development in each treatment.
5	Transfer axillary bud-derived regenerated shoots (second subculture) onto fresh MS basal medium containing 3% sucrose and without BA, and incubate as above to root and develop as plantlets.
6	Remove well-rooted plantlets, and wash the agar off the roots with distilled water. Caution: *In vitro*-derived roots of cassava plantlets are very brittle, so care must be exercised in rinsing the agar off the roots. Place plantlets in pots (10 × 10 cm) containing a mixture of 1:1 vermiculite:perlite (by volume), and cover the newly potted plantlet with an inverted Ziploc bag to retain high levels of humidity. Systematically widen the opening of the bag (once every five days) until the cover is completely removed in about two weeks. This process is called acclimatization and prepares the plantlet for survival in the harsher *ex vitro* environment. The Ziploc bag is placed over the plantlets in the lab and then continued in the shaded greenhouse.
7	Record the number of shoots, roots, and plantlets produced per nodal explant for different concentrations of BA. Determine the survival rate for the acclimatized plantlets.
8	Analyze BA concentration effect on shoot, root, and plantlet formation, including the number and lengths of roots and shoots formed, using ANOVA. Share results with other groups of students.

FIGURE 34.3 Dormant tubers of white yam.

Over two decades ago, a new technology was introduced called the minisett technique for seed yam production.

YAM PROPAGATION BY MINISETT PRODUCTION

The minisett technology for yam propagation is a semi-rapid multiplication method for seed yam production in major yam species. The technology was developed at the National Root Crops Research Institute (NRCRI), Nigeria, tested and improved at the International Institute of Tropical Agriculture (IITA), and adopted by yam growers in West Africa. Minisetts are obtained by sectioning a yam tuber weighing 1.2 to 1.5 kg into 45 to 50 pieces, each weighing about 25 to 30 g. The sections can be dipped in a fungicide solution and cured (air-dried) under shade for a day and then planted in a sprouting medium, such as garden soil or moist sawdust. Sprouting occurs five weeks after planting, and the sprouted minisetts are planted in a nursery bed for seed yam production. Seed yams weighing an average of 100 g are produced. Under proper manage-

	Procedure 34.3 Effect of Portion of Tuber (Head, Middle, Tail, or Mixed) on the Sprouting Frequency of Minisetts
Steps	Instructions and Comments
1	Cut a healthy yam tuber weighing about 2 kg into three equal cylindrical sections, and label sections as head (A), middle (B), and tail (C). The head is the section closest (proximal end) to the stem, and the tail section comes from the distal end. Cut each section into similarly sized disks from which small wedge-like pieces (30–40 g) can be obtained by cutting longitudinally. These pieces constitute the minisetts. Use another tuber to obtain minisetts for the control treatment. For the control, blend equal numbers of minisetts obtained from all three sections (D). About 50 minisetts can be derived from one 2-kg tuber.
2	Spread the minisetts on a shaded flat surface, and allow to air dry for one to two h.
3	Plant the minisetts with the "skin" portion down in trays (50 cm × 36 cm × 10 cm) containing sawdust in four rows with four minisetts per row. The rows should be approximately 10 cm apart. Make sure minisetts are completely covered by the sawdust.
4	Prepare enough minisetts for a randomized complete block design with four replications.
5	Place the minisetts in a greenhouse or nursery at 26°C–28°C, and irrigate periodically using an overhead fine mist sprinkler to avoid drying.
6	Monitor sprouting of minisetts that begins in three to four weeks and continues up to seven weeks after planting. Sprouting involving root and shoot emergence takes place on the "skin" of the minisett after dormancy is broken.
7	Count and record data on the number of sprouted minisetts, including number of roots or shoots produced per minisett in each treatment and the root or shoot length.
8	Analyze your data statistically using ANOVA, and run mean separation where necessary. Compare your results with those obtained by other student groups.

ment and production practices, seed yams weighing up to 300 g have been obtained. Still, there is a great potential for higher yields. Larger seed yams are desirable for large yam tuber production. Therefore, the goal is to optimize those factors (e.g., mulching, sanitation) that influence good seed yam production.

IN VITRO PROPAGATION

Tissue culture regeneration experiments in yam have been geared toward shoot production and clonal multiplication, germplasm preservation and distribution, microtuber induction, and subsequent field production of seed yams. Micropropagation of yams can be achieved through the development of multiple axillary shoots or microtuber induction. Microtuberization is a useful method for rapidly propagating yams for seed yam production. It also offers tremendous opportunity for storage and safe international movement and distribution of pathogen-free germplasm and clonal yam propagules (chapter 11). Microtuberization, similar to other tissue culture procedures in yam, is influenced by several growth factors, including all the major classes of endogenous plant growth regulators, sucrose concentration, and environmental factors, such as photoperiod.

The following experiments will illustrate sprouting in minisetts and disinfestation procedures for the establishment of aseptic cultures of nodal vine cuttings.

EXPERIMENT 3: EFFECT OF PORTION OF TUBER (HEAD, MIDDLE, TAIL, OR MIXED) ON THE SPROUTING FREQUENCY OF MINISETTS

Minisetts obtained from various portions of the yam tuber may sprout at different rates. The purpose of this experiment is to evaluate the response of minisetts taken from various sections of the yam tuber on sprouting frequency.

Materials

The following materials will be needed for each student or team of students:

- Two large yam tubers each weighing 1.5 to 2 kg (large yams used for eating are "ware" yams, and the smaller ones less than 1 kg used for planting are "seed" yams)
- Sharp nursery knife
- Planting trays
- Sawdust or other porous potting mix (1:1 perlite: vermiculite)
- Overhead misting system and bench space in greenhouse
- Labels and permanent marker

Follow the instructions in Procedure 34.3 to complete this experiment.

Microproagation of Tropical Root and Tuber Crops

Anticipated Results

Unequal sprouting among minisetts taken from the different sections of yam tuber will occur. The most sprouting will be obtained in minisetts derived from the middle sections of the yam (treatment B) and minisetts taken from all sections of the tuber (treatment D) that served as control.

Questions

- What accounts for the differences in sprouting of minisetts taken from various sections of the tuber?
- Is it possible to achieve uniform sprouting with the minisett technology? Why?
- How would you attain uniform sprouting in minisetts?
- How would you enhance the rate of sprouting in minisetts?

EXPERIMENT 4: EFFECTS OF DISINFESTING AGENTS ON CONTAMINATION IN YAM TISSUE CULTURES

Yam tubers sprouted in potting soil or other nonsterilized commercial propagation mixes in the greenhouse or nursery are often difficult to disinfest because of the presence of microorganisms in the shoots and axillary buds. The objective of this experiment is to compare two methods of explant disinfestation, including a post rinse dip of explants in dilute aqueous solutions (0.1% or 0.2% [w/v]) of calcium hypochlorite, to enhance success in the aseptic culture establishment.

Materials

The following items will be required for each group of three to four students:

- Access to sprouted shoots of yam (*D. rotundata* or *D. cayenensis*)
- Scissors
- Cheesecloth
- Liquid chlorine bleach
- Calcium hypochlorite—To prepare 8% (w/v) solution, add 8 g in 100 mL distilled water and stir for 10 to 15 min. Filter through Whatman No. 1 using funnel and gravity
- 95% ethanol
- Baby food jars
- Whatman No. 1 filter paper
- Large funnel
- Growth chamber, tissue culture room, or lab bench with lights
- pH meter
- Stirring plate and stirring bar
- Balance
- Laminar air-flow bench
- Murashige and Skoog (MS) basal salt medium and glassware

Follow the instructions in Procedure 34.4 to complete this experiment.

Anticipated Results

Treatments A and B should result in a low contamination rate. Subtreatments I, II, or III may or may not enhance this result significantly.

Questions

- Which agent was most effective for disinfesting nodal yam cuttings? Why?
- What factors would you consider in selecting a chemical agent for disinfesting your explants?
- What were the predominate contaminants—fungi or bacteria?

COCOYAM

Cocoyam (*Xanthosoma sagittifolium* L. Schott) (Figure 34.4) is a monoecious herbaceous perennial monocot in the Araceae, which grows in the humid tropics and subtropics. It is known under a variety of names, including macabo, tannia, yautia, malanga, or maffafa, depending on the region where it is cultivated. It is mainly grown as an annual over an 8- to 12-month production cycle. The enlarged underground storage organ extending from the base of the crown is a modified stem with compressed internodes referred to as a corm. Toward the end of the first annual growth cycle, it becomes enveloped in senesced or shreddy vestiges of the petioles that, when removed, reveal rings of small dormant axillary buds. When maintained in the ground for up to two or three years without cultivation, this structure may grow longer and larger, becoming a prominent stem with several rings of small dormant buds interspersed with larger secondary buds. The primary, small dormant shoot buds need special treatments, such as tissue culture conditions, in order to grow into shoots. The corms produce small tubers called cormels (Figure 34.5) that are detached at maturity and used for food or propagation. Cocoyams are conventionally propagated by fragmenting the corms or cormels into 30- to 40-g pieces with large buds and planting. Propagation by seed is not practical because of the absence or scarcity of flowering.

IN VITRO PROPAGATION

Conventional methods of propagating cocoyams are not adequate to meet demand for good-quality seedstock and also contribute to dissemination of pathogens. *In vitro* techniques have been considered and applied successfully for

	Procedure 34.4
	Effects of Disinfesting Agents on Contamination in Yam Tissue Cultures
Steps	Instructions and Comments
1	Collect nodal yam cuttings (3 cm tall), divide into three groups (A, B, and C), and place in distilled water. Trim off the lamina (leaf blades), leaving the petioles attached, and rinse twice with distilled water containing a few drops of Tween 20.
2	Surface disinfest explants of two groups (A and B) of cuttings in 8% aqueous calcium hypochlorite solution with one drop of Tween 20 for 10 min. After one rinse in sterile distilled water, place explants in 6% calcium hypochlorite solution plus one drop of Tween 20 for 5 min, and then rinse three times with sterile distilled water. Disinfest the third group (C) of explants in a 1:1 Clorox (5.25% NaOCl):ethanol (v/v) solution for 7 min., and rinse three times with sterile distilled water.
3	Cut off the bleached "burnt" ends of the nodal stem cutting and the petioles to expose healthy tissue.
4	Instantly dip explants of group B in either 0.1% (I) or 0.2% (II) calcium hypochlorite (0.1%–0.1 g Ca(OCl)2 in 100 mL distilled water; 0.2%—0.2 g in 100 mL distilled water), or in sterile distilled water (III) prior to culture. Sterile distilled water serves as the control.
5	Culture all explants in baby food jars containing agar solidified MS medium supplemented with 10 µM indole acetic acid (IAA) and 10 µM kinetin with one explant per jar. Make seven replications (jars) per treatment.
6	Incubate cultures at 26°C–27°C with 60 µmol·m^{-2}·sec^{-1} of fluorescent light for up to three weeks.
7	After three weeks, determine explant viability and the number of contaminated explants in each treatment, and compare treatments A and C using a t-test.
8	Analyze the results of treatments I, II, and III using the parameters stated in Step 7. Share and compare your results with those of other student groups.

FIGURE 34.4 Tissue culture-derived cocoyam plants in a greenhouse.

rapid micropropagation via shoot tip culture and axillary bud enhancement. Regenerating cocoyams adventitiously has contributed to a tremendous increase in multiplication rate of cocoyam and provided opportunities for biotechnological applications.

SHOOT TIP CULTURE AND AXILLARY BUD MULTIPLICATION

The following experiment will demonstrate rapid multiplication of cocoyam via shoot tip culture and axillary bud enhancement. This experiment is to be conducted by groups of four to five students.

EXPERIMENT 5: THE EFFECTS OF KINETIN AND NAA ON SHOOT MULTIPLICATION IN COCOYAM

Materials

The following materials are required for each group of students:

- Cocoyam shoots

Micropropagation of Tropical Root and Tuber Crops

FIGURE 34.5 Cocoyam cormels.

- Liquid chlorine bleach
- 95% ethanol
- Balance
- Growth chamber, tissue culture room, or lab bench with lights
- Petri dishes
- pH meter
- Stir plate and stir bar
- Laminar air-flow bench
- Kinetin and NAA stocks
- Test tubes with caps (150 × 20 mm)
- B5 basal salt medium (Gamborg et al., 1968) and glassware

Follow the instructions in Procedure 34.5 to complete this experiment.

Anticipated Results

Properly dissected shoot tips should float on the surface of the liquid medium. Explants with large amounts of tissue from the original corm at the base may sink in the medium

Procedure 34.5	
The Effects of Kinetin and NAA on Shoot Multiplication in Cocoyam	
Steps	Instructions and Comments
1	Collect one- to two-week-old cocoyam shoot buds from sprouted tuber fragments in the greenhouse, and wash under running tap water to remove growing medium. Three weeks before the laboratory exercise is planned, cocoyam tubers should be cut into 40-g fragments bearing dormant buds, allowed to air dry, and placed in a mixture of 1:1 vermiculite:perlite (by volume) in plastic flats (50 × 36 × 10 cm) to encourage sprouting. The trays are misted for 30 seconds every 2 h.
2	With the aid of a scalpel blade, remove several whorls of leaves to obtain shoot tips 5 to 6 mm long, leaving a thin layer of tuber tissue at the base. Place in a beaker containing distilled water.
3	Disinfest the shoot tips in 15% liquid chlorine bleach (0.79% NaClO) solution containing one drop of Tween 20 for 20 min. with constant agitation followed by four rinses in sterile distilled water under the laminar airflow hood. Because of their small size, the shoot tips should be disinfested in a small container such as a petri dish.
4	Transfer explants into B5 basal liquid medium (Gamborg et al., 1968) containing kinetin (0, 0.46, 4.65, and 13.9 μM) and NAA (0, 0.54, and 1.62 μM) in 150 × 20 mm culture tubes. Place one explant in each culture tube, cap, and seal with Parafilm. A completely randomized design with five replications should be used.
5	Incubate and monitor cultures at 25°C to 27°C and photon flux of 50 $\mu mol \cdot m^{-2} \cdot sec^{-1}$, 16-h photoperiod, for up to 6 weeks. Note and remove contaminated cultures from the growth chamber.
6	Collect data on contamination, viability, number and length of shoots and roots, and plantlet formation, and perform the ANOVA procedure. Perform treatment means separation using LSD or Tukey's HSD at 5% probability. Share data with other student groups and report your findings.

and, thus, fail to grow. Bulging and expansion of explants should occur in 7 to 10 days in the medium containing the highest concentration of kinetin. Multiple shoots may be formed in media containing relatively high concentrations of kinetin in combination with NAA.

Questions

- How would you improve explant development and morphogenesis when using liquid media?
- How would you increase the frequency of shoot production and plantlet regeneration?

LITERATURE CITED AND SUGGESTED READINGS

Alizadeh, S., S.H. Mantell, and A.M. Viana. 1998. *In vitro* shoot culture and microtuber induction in the steroid yam *Dioscorea compisita* Hemsl. *Plant Cell, Tiss. Organ Cult.* 53: 107–112.

Gamborg, O.L., R.A. Miller, and K. Ojima. 1968. Nutrient requirement of suspension cultures of soybean root cells. *Exp. Cell. Res.* 50: 157–158.

Hartmann, H.T., D.E. Kester, F.T. Davis, Jr., and R.L. Geneve. 2002. *Plant Propagation: Principles and Practices,* 7th ed. Prentice Hall, Inc., Upper Saddle River, New Jersey. 880 pp.

Konan, N.K., C. Schöpke., R. Cárcamo., R.N. Beachy, and C. Fauquet. 1997. An efficient mass propagation system for cassava (*Manihot esculenta* Crantz) based on nodal explants and axillary bud-derived meristems. *Plant Cell Rep.* 16: 444–449.

Murashige, T. and F. Skoog. 1962. A revised medium for rapid growth and bioassays with tobacco tissue cultures. *Physiol. Plant.* 15: 473–497.

Nyochembeng, L.M. and S. Garton. 1998. Plant regeneration from cocoyam callus derived from shoot tips and petioles. *Plant Cell, Tiss. Organ Cult.* 53: 127–134.

Okoli, O.O., A.F.K. Kissiedu, and J.A. Otoo. 1992. Production of seed yams by the minisett technique: Effect of mulch, stakes and plant population. Proc. 4th Symposium, Intl. Soc. Trop. Root Crops—Africa Branch (ISTRC-AB). pp. 277–279.

Otoo, J.A. 1992. Substitutes for chemicals, sawdust and plastic mulch in improved seed yam production. Proc. 4th Symposium, Intl. Soc. Trop. Root Crops—Africa Branch (ISTRC-AB). pp. 281–284.

Watanabe, K.N. 2002. Challenges in biotechnology for abiotic stress tolerance on roots and tubers. Japan International Research Center for Agricultural Sciences (JIRCAS) Working Report. pp. 75–83.

Zok, S., Sama A, and L. Nyochembeng. 1992. Elimination of yam culture contamination using double surface sterilization. Proc. 4th Symposium, Intl. Soc. Trop. Root Crops—Africa Branch (ISTRC-AB). pp. 291–293.

35 Micropropagation of Woody Plants

Robert R. Tripepi

Micropropagation is the production of plants from plant parts cultured *in vitro*. Many *in vitro* techniques can be used to produce large numbers of plants. Although micropropagation refers to many tissue culture techniques, axillary shoot proliferation is the procedure most often used during commercial micropropagation to reproduce true-to-type plants. Axillary shoot production is preferred over shoot organogenesis (adventitious shoot production) because the former technique reduces the chances for producing off-type plants. As you read this chapter and complete the lab exercises, keep in mind that micropropagation and axillary shoot proliferation are used synonymously.

Micropropagation (axillary shoot proliferation) involves several phases during the production of plants; these phases are referred to as stages of micropropagation (chapter 31). Stage 0 is the growing and care of stock plants from which explants are harvested. Stage I of axillary shoot proliferation involves the establishment of shoot tip or nodal explants in culture. Stage II entails multiplication (proliferation) of axillary shoots from the established explants (propagules). Stage III involves preparing the plant parts (usually shoots) for transfer to the outside (*ex vitro*) environment. This stage usually involves rooting microshoots. Finally, Stage IV involves transferring and acclimatizing plantlets from *in vitro* conditions to a greenhouse or other controlled environment facility.

Many species of woody plants, including temperate hardwoods and conifers as well as tropical plants, can be micropropagated. Some woody species are relatively easy to establish in micropropagation, whereas other species may survive for only several subcultures before all microshoots die. For this reason, micropropagation is a commercially useful tissue culture technique for only amenable species.

Explant establishment is the first objective of this laboratory exercise. Establishment depends on the physiological state of the stock plant, components in the culture medium, and environmental conditions under which the explants are grown. European birch (*Betula pendula* Roth) is a good candidate for micropropagation because it is an important landscape species in North America and a prominent timber species in northern European countries. This species can be micropropagated for commercial and research purposes. It also multiplies easily (Stage II of micropropagation), and microshoots easily form roots *in vitro* (Stage III). Finally, a modest amount of care is needed to acclimatize rooted microshoots to the environment outside the tissue culture vessel (Stage IV). Alternatively, birch microcuttings can be taken from Stage II shoot cultures and rooted directly in potting mix held in a protected environment. This laboratory exercise will provide experiences in all stages of micropropagation of European birch, and the procedures can be completed in 16 weeks. Although, European birch is easy to micropropagate, the use of two other woody species will be highlighted for comparative purposes.

The objectives of this part of the exercises are the following:

1. To demonstrate methods used to make and disinfect nodal explants of a deciduous tree species
2. To provide experience establishing nodal explants, subculturing, and multiplying microshoots, rooting microshoots *in vitro*, and acclimatizing the resulting plantlets to the ambient environment

PROCEDURES

The culture medium used throughout this laboratory exercise is Woody Plant Medium (WPM) basal salts (Lloyd and McCown, 1980) supplemented with various organic components (e.g., vitamins, sugar alcohol; Table 35.1). The pH of the medium is adjusted to 5.2 before sterilization, and about 25 mL of medium are poured into each 250-mL baby food jar.

STAGE I: EXPLANT ESTABLISHMENT

Stem nodes used as explants (Figure 35.1) in this stage of micropropagation must be freed from microbial contamination or explant establishment on the medium will be impossible. The concentrations of plant growth regulators (PGRs) used in the establishment medium can also have significant effects on explant establishment. During Stage I of micropropagation, the effects of different concentrations of auxin and cytokinin growth regulators on explant axillary bud break and subsequent growth will be determined.

TABLE 35.1
Woody Plant Medium (WPM) Used as the Culture Medium for Three Stages of European Birch Micropropagation

Ingredient	Stage I Establishment Medium	Stage II Proliferation Medium	Stage III Rooting Medium
WPM salts	1.15 g	1.15 g	1.15 g
Sucrose	10 g	10 g	10 g
WPM organics stock solution[a]	10 mL	10 mL	10 mL
Plant growth regulators	See Table 35.2	See Table 35.2	NAA stock solution[b] 0.5 mL
Adjust volume	Bring up to 500 mL	Bring up to 500 mL	Bring up to 500 mL
Adjust pH of medium	5.2	5.2	5.2
Agar	3.25 g	3.25 g	3.25 g

Note: Each separate column is used to make 500 mL of medium, which is enough for 20 jars of medium. To make more or less culture medium, the amounts of each ingredient need to be adjusted accordingly.

[a] WPM organics stock solution is made by mixing (on a grams per liter basis) the following organic compounds: myo-inositol 5; glycine 0.1; nicotinic acid 0.025; pyridoxine HCl 0.025; and thiamin HCl 0.05. Dissolve each component, starting with myo-inositol, before adding the next one. These compounds can be purchased from various vendors (Sigma-Aldrich Company, Caisson Laboratories, and *Phyto*Technology Laboratories).

[b] A 1 mM naphthaleneacetic acid (NAA) stock solution is made by dissolving 0.0186 g of naphthaleneacetic acid in 10 to 15 drops of 1 N NaOH (40 g NaOH per liter of water) and then bringing the solution up to 100 mL.

FIGURE 35.1 Single-node explants of birch stems being established during Stage I of micropropagation.

Materials

The following materials will be needed to complete this stage of micropropagation by individuals or groups of students:

- Ten percent bleach (0.6% w/v NaOCl—commercial product) plus two drops of Tween 20 per 200 mL of bleach solution
- Seventy percent (v/v) ethanol solution (aqueous)
- Spray bottle with 250 mL capacity
- Three autoclaved baby food jars filled 2/3 full of sterile distilled water
- Several European birch stems, enough to contain 50 nodes when cut into explants
- Razor blades or pruning shears
- Scalpels and forceps (two each per work station or hood)
- Sixteen baby food jars, four of each containing medium with different combinations of cytokinin (benzyladenine [BA]) and auxin (naphthaleneacetic acid [NAA])—see Table 35.2

TABLE 35.2
Plant Growth Regulators Used during Stage I and II for European Birch Micropropagation Treatments

Treatment	Stage I Benzyladenine Stock Solution[a] (mL)	Stage I Naphthaleneacetic Acid Stock Solution[b] (mL)	Stage II Benzyladenine Stock Solution (mL)	Stage II Naphthaleneacetic Acid Stock Solution (mL)
A	0	0	2.2	0
B	1.1	0	4.4	0
C	2.2	0	6.6	0
D	2.2	0.5	—	—

Note: The stated volumes of stock solutions are added to 500 mL of medium, which is enough to for 20 jars of medium. To make more or less culture medium, the amounts of each stock solution need to be adjusted accordingly.

[a] A 1 mM benzyladenine stock solution is made by dissolving 0.0225 g of benzyladenine in 10 to 15 drops of 1 N NaOH (40 g NaOH per liter of water) and then bringing the solution up to 100 mL.

[b] A 1 mM naphthaleneacetic acid stock solution is made by dissolving 0.0186 g of naphthaleneacetic acid in 10 to 15 drops of 1N NaOH and then bringing the solution up to 100 mL.

- Sterile paper towels (wrapped in aluminum foil and sterilized)

Follow the protocol provided in Procedure 35.1 to complete this experiment.

Anticipated Results

The axillary buds will begin to grow within four to seven days after being placed on the medium if stems were taken at the proper stage of development. Contamination will be present on the medium either if the explant surface sterilization procedure (using the bleach solution) was inadequate or if aseptic techniques (handling procedures) were poor. The buds should grow from nodal explants on all four combinations of plant growth regulators, but the combinations of cytokinin and auxin will affect shoot growth differently. These differences should be noted.

Questions

- Why must you avoid contaminating any of the sterile water rinses used in Stage I?
- Why were you asked to remove (cut off) 2 mm of tissue from the ends of the single-node explants after completing the last sterile rinse in Stage I?
- How many days passed before the buds started to expand?
- Was cytokinin necessary for bud elongation?
- What were the effects, if any, of the changes in cytokinin concentrations on the explants?
- Was auxin necessary for bud expansion, or did this growth regulator have any effect on the explants?

STAGE II: SHOOT MULTIPLICATION

Axillary buds that have grown to form stems are removed from the original stem (single-node explant) and placed on a multiplication medium to induce more axillary shoots to grow. Therefore, the new growth from the original explant now becomes a propagule that is divided and put back into culture (i.e., subcultured) to produce additional axillary shoots (for axillary shoot proliferation) (Figure 35.2). The concentrations of PGRs in the medium during this stage can also have significant effects on the number of axillary buds that grow (proliferate) (Figure 35.3). During Stage II of micropropagation, the effects of different concentrations of a cytokinin growth regulator on the number of axillary shoots that grow from the subcultured propagules will be determined.

Materials

The following materials will be needed to complete this stage of micropropagation by individuals or groups of students.

- Birch cultures from Procedure 35.1
- Seventy percent (v/v) ethanol solution (aqueous)
- Spray bottle with a 250 mL capacity
- Scalpels and forceps
- Twelve baby food jars, three of each different combination of cytokinin (benzyladenine) and auxin (naphthaleneacetic acid)—see Table 35.2
- 15-cm-long ruler that is surface disinfested with 70% ethanol (a flat ruler without raised letters or markings works best)
- Sterile paper towels

Procedure 35.1
Establishing Single-Node Birch Stem Explants in Stage I of Micropropagation

Step	Instructions and Comments
1	Make a 10% (v/v) bleach solution (20 mL liquid chlorine bleach + 180 mL water + 2 drops Tween 20).
2	Be sure to have three baby food jars filled about 2/3 full of distilled water, and sterilize these jars in an autoclave or pressure cooker for 30 min. These will be used for rinsing.
3	European birch stock plants will be used for micropropagation. Please be careful as you cut branches off the plants with razor blades, scalpels, or pruning shears. Note the condition of the leaves and stems (stem appearance—woody, semiwoody, etc.; leaf color; disease or tissue damage present). Succulent small stems as well as older woody branches should be avoided. The best stems to use are those that are still green and pliable (softwood to semiwoody).
4	Make stem sections from the branches you cut, with each section containing one node (axillary bud). Cut the stem about 1 cm above the bud and 2 cm below the bud, if possible. Some adjustments for these cutting distances from the bud may be necessary so that as many explants as possible can be made from the plant material available. These stem sections are called single-node explants.
5	Remove the leaf blade and petiole, being careful to avoid tearing or damaging the bud or stem. Cut the petiole if necessary. Make 50 nodal sections for each group (typically two to three students).
6	Place the stem sections into the 10% bleach solution (in a baby food jar) for 20 min. The baby food jar used to hold the bleach solution does not need to be sterile (the bleach solution will surface sterilize it). Agitate every few minutes. Spray the jar with 70% ethanol before moving it into the laminar flow (clean) hood.
7	Spray 70% ethanol on the outside of each capped baby food jar that contains sterile water before placing jars in the laminar flow hood. In the hood, rinse the stem sections in three changes of sterile distilled water in baby food jars for 1 to 2 min. per rinse. Use sterile forceps to transfer stem sections. Forceps and scalpels can be sterilized in a Bacti-Cinerator® or other appropriate equipment that will heat these tools to high temperatures to kill microbes on tool surfaces. Forceps and scalpels can be surface sterilized by using 70% ethanol and an alcohol lamp or gas burner to flame these tools, but a fire can easily be started when flaming tools and using alcohol. Therefore, use equipment that lacks flames (e.g., Bacti-Cinerator) to maintain laboratory safety.
8	Sterilize forceps between rinses. Leave the stem sections in the rinse water 1 to 2 min. before transferring them to another rinse jar. Rinse the stem sections three times.
9	After the third rinse, remove three stem sections at a time, and place them on a sterile paper towel. Make two cuts on each stem section to remove about 2 to 3 mm from each end of the stem section. With sterile forceps, place the stem sections in an upright position into one of the jars of culture medium. The base of the explant should be submerged 3 to 5 mm deep into the agar. Be sure to keep the proper orientation of the stem sections—proximal end down (right side up). Put three stem sections in each jar, and use four jars of each medium (treatment). The four treatments are listed in Table 35.2.
10	After the three stem sections are in place, cap the jar before removing more stem sections from the last rinse of sterile distilled water. Label all jars, and put these jars on a lighted shelf (25 to 40 $\mu mol \cdot m^{-2} \cdot sec^{-1}$) in a culture room. The jars can be placed about 35 cm (14 in.) under fluorescent shop lights fitted with cool white lamps to provide a 16-h photoperiod in a room at 25°C. These Stage I explants will be allowed to grow four weeks before subculturing.
11	Observe the explants once every few days for contamination, and then observe at weekly intervals. Record the following data: percentage of contaminated stem explants at Stage I, percentage of stem sections with expanded/elongated buds per treatment, the number of shoots that grew or formed from each single-node explant per treatment, and the length of each elongated shoot (when subculturing it). Calculate means for the number of shoots formed per explant and the lengths of elongated shoots for each treatment. Measure elongated shoots under aseptic conditions.

FIGURE 35.2 European birch shoot culture with proliferating microshoots on multiplication medium during Stage II.

FIGURE 35.3 Axillary shoot proliferation from two stem propagules of European birch during Stage II of micropropagation. Note the number of axillary shoots that have formed.

Follow the outline provided in Procedure 35.2 to complete this experiment.

Anticipated Results

The shoots will continue to elongate, but the cytokinin concentrations will affect the length of shoot growth and the number of axillary buds that grow on each stem (propagule). These differences should be noted because the results will provide an idea of how much cytokinin should be used in the proliferation medium if this birch species were to be micropropagated commercially.

Questions

- How did the cytokinin concentration affect the number of axillary buds that grew (proliferated)?
- Did the original establishment (Stage I) medium have any effects on the number of new shoots formed? How do you know?
- How did the cytokinin concentration affect the increase in stem length of the propagules (stems removed from Stage I explants)?
- Would the number of axillary buds that proliferated increase or decrease if the Stage I propagules were oriented horizontally (placed on their sides) rather than vertically? Why?

STAGE III: TRANSFER PREPARATION AND ROOTING THE MICROSHOOTS

A critical part of micropropagation is to induce roots to form on microshoots. Root formation on microshoots is typically the limiting factor in micropropagation rather than shoot proliferation from axillary buds. Microshoots can be rooted either *in vitro* or in the regular environment (in which case microshoots are now called microcuttings). *In vitro*-rooted shoots can be placed on a medium supplemented with auxin or one that lacks cytokinin.

	Procedure 35.2
	Multiplying Birch Stems Obtained from Stage I Nodal Explants
Step	Instructions and Comments
1	About four weeks after making the Stage I explants, prepare to subculture the expanded birch shoots. Be sure to record the data requested in Step 11 from Procedure 35.1.
2	Remove the jars containing the microshoots from the culture room shelf. Spray 70% ethanol on the outside of these jars and those that contain the multiplication media before transferring the jars to the laminar flow hood.
3	Open one jar that contains the Stage I explants, remove one stem explant with sterile forceps, and place it on a sterile paper towel. All of the following procedures should be completed with only sterile forceps and sterile scalpels touching the plant tissues. Cut off the expanded shoot at the point where it arises from the old (original explant) stem, and place the new shoot beside the surface sterilized ruler on a sterile paper towel. Avoid touching the shoot with the ruler to reduce chances of contamination. Be sure to measure the length of the new (expanded) shoot while in the hood.
4	Place the shoot on a new medium (A, B, or C). This shoot and all others should be oriented vertically when placed on new medium. Repeat this procedure with the next two shoots in the jar, and then go on to the next jars to subculture the other shoots. Put four shoots in each jar of new medium. Since 12 shoots are on each Stage I medium (A, B, C, or D), put four shoots into each jar of each multiplication (Stage II) medium (new A, B, or C). See Figure 35.4 for a diagram that shows how the stems should be transferred from Stage I to Stage II media. Be sure to note the original establishment medium on the new jar. You will need to use four jars of each multiplication medium.
5	After completing this work, the jars containing the shoots should be placed in the culture room under the fluorescent lights. Let the shoots grow for about four weeks before transferring microshoots to a rooting medium (Stage III).
6	Observe the explants at least twice a week, looking for contamination, axillary bud break, and stem growth. Record the following data: the number of shoots formed on each propagule per Stage II medium, the mean number of shoots that formed from each propagule per medium, and the length of the original propagule (stem) after four weeks on multiplication medium. Determine this stem length when preparing the shoots for Stage III, and calculate a mean length for each treatment. Make measurements under aseptic conditions within the laminar flow hood. Also, be sure to note the relative amount of callus formed on stems in each treatment.

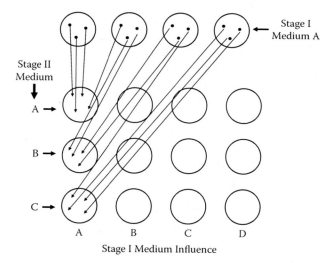

FIGURE 35.4 Schematic diagram for moving stems from one jar of Stage I medium (A) to three different media used in Stage II of micropropagation. Note that the original treatment should also be listed on the Stage II treatment jar.

Basal mineral concentration is sometimes reduced to half strength to promote rooting. Addition of activated charcoal is sometimes helpful for root formation on microshoots.

In this experiment, axillary buds that have broken on birch microshoots and subsequently grown to form stems are counted and removed from the original explant (propagule). Stems that are longer than 1.5 cm can be rooted to form plantlets (Figure 35.5), which will later be transferred to the *ex vitro* environment. The concentrations of cytokinin used during Stage II may have an effect on the ability of the microshoots to form roots. During Stage III of micropropagation, proliferated microshoots will be cut from Stage II shoot cultures and rooted individually on a culture medium that contains an auxin.

Materials

The following materials will be needed to complete this stage of micropropagation by individuals or groups of students:

Micropropagation of Woody Plants

FIGURE 35.5 European birch microshoots rooted during Stage III of micropropagation.

- Stage II shoot cultures of European birch
- Seventy percent (v/v) ethanol solution (aqueous)
- Spray bottle with a 250 mL capacity
- Scalpels and forceps
- Twelve baby food jars of rooting medium containing 1 µM NAA, but lacking a cytokinin—see Table 35.1
- A 15-cm long ruler that is surface sterilized with 70% ethanol (a flat ruler without raised letters or markings works best)
- Sterile paper towels

Follow the procedures outlined in Procedure 35.3 to complete this exercise.

Anticipated Results

Many or most of the microshoots should form roots, depending on their health and vigor. The cytokinin concentrations used in Stage II may have an effect on the percentage of shoots that form roots. Proliferated microshoots on Stage II medium that contained higher cytokinin concentrations may also grow to similar lengths as those on lower cytokinin concentrations. Since Stage II European birch microshoots often root readily when transferred to potting mix, nonrooted shoots can also be used in Stage IV of this experiment.

Rooting Experiment for Birch Microshoots

European birch microshoots readily respond to different concentrations of auxin in the culture medium. Higher concentrations may induce more roots to form, but root and shoot growth can be affected adversely by auxin concentrations that are too high. The question is what concentration is too high? An alternative experiment to try is to use WPM supplemented with different levels of NAA to determine the effects of this plant growth regulator on root and shoot growth of Stage III plantlets.

For this alternative experiment, the culture media should be supplemented with 0, 0.1, 1.0, or 5.0 µM NAA (0, 0.05, 0.5, or 2.5 mL of 1 mM NAA stock solution per 500 mL of medium, respectively). Follow Procedure 35.3 to prepare Stage II microshoots for Stage III. Cut all microshoots to a consistent length (e.g., 2.5 cm), if possible. Place equal numbers of microshoots in each jar containing one of the four rooting media. The microshoots should form roots in about two weeks, but count roots after three to four weeks to allow time for root initiation and growth to be completed. Note the numbers of roots formed and their thickness as well as overall root lengths. In addition, measure the length of the plantlet stems (rooted microshoots).

Questions

- Did the multiplication treatment used in Stage II affect the percentage of rooting and the number of roots formed per shoot? If so, how?
- Did roots form directly on the stem of each shoot or on callus? From what tissue(s) did the root appear to originate?
- What conclusions can you draw about the effects of NAA concentrations on root and shoot growth of the plantlets?
- How would you be able to tell if the NAA concentration used in Stage III was too high for the microshoots? What type of plant symptoms would you expect to see?
- What concentration of NAA would you recommend to use when rooting European birch microshoots?

STAGE IV: TRANSFERRING AND ACCLIMATIZING BIRCH PLANTLETS AND MICROSHOOTS

Rooted microshoots (plantlets) and nonrooted shoots can be transferred to a simple vessel or container to con-

	Procedure 35.3
	Rooting Birch Microshoots Obtained from Stage II Shoot Cultures
Step	Instructions and Comments
1	About four weeks after making the Stage II explants, prepare to excise the birch microshoots. Be sure to record the data requested in Step 6 from Procedure 35.2.
2	Remove the jars containing the microshoots from the culture room shelf. Spray 70% ethanol on the outside of these jars and those that contain the rooting medium before placing the jars in the laminar flow hood.
3	Open one jar that contains the Stage II explants, remove one microshoot, which should contain one or several shoots, with sterile forceps, and place it on a sterile paper towel. All of the following procedures should be completed with only sterile forceps and sterile scalpels touching the plant tissues. Measure the shoot length before cutting off any callus that has formed at the base of the shoot. Make the measurements while under the hood, and be sure to maintain aseptic techniques.
4	Use a scalpel to remove any callus at the base of the shoot. Cut off proliferated axillary shoots at their points of attachment on the original propagules (stems), and place the shoots (also referred to as microshoots) vertically in a jar that contains rooting medium. This medium contains 1 μM NAA, but lacks a cytokinin. Although any shoot about 1 cm long may root, use shoots longer than 1.5 cm in this part of the lab exercise for ease of handling. When placing the shoot in the jar, 3 to 5 mm of the basal end of the stem should be inserted into the medium. Repeat Steps 3 and 4 until all jars contain at least five or six microshoots from the same Stage II proliferation treatment. If extra shoots are available, transfer only the most vigorous shoots to Stage III medium. Be sure to note the Stage II proliferation medium on the jar that contains the rooting medium. You will need to use four jars of rooting medium for each Stage II multiplication medium.
5	After completing this work, the jars containing the shoots should be placed in the culture room under the fluorescent lights. Let the shoots grow for two to four weeks before transferring microshoots to the *ex vitro* environment (Stage IV).
6	Observe the explants at least twice a week, looking for contamination, adventitious root initiation, and stem growth. Record the following data: the percentage of shoots that formed at least one root, the number of roots per shoot only for shoots that formed roots, and the number of root initials per shoot. All these data should be grouped for each Stage II medium. To determine if the cytokinin concentrations in the multiplication media affected rooting, calculate a mean number of roots per shoot (ONLY for responding shoots) for each Stage II treatment. A mean for the number of root initials per responding stem can also be calculated. When counting roots, rooted stems can be handled under nonaseptic conditions, but they should be held in a capped vessel, since they can readily dry out. Take these data before transferring the rooted microshoots and those that lack roots to Stage IV potting mix.

tinue rooting and promote acclimatization of the shoots. Although a greenhouse mist bench covered with shade cloth can be used to acclimatize the shoots, other less elaborate environments can be used to help the birch plantlets and nonrooted shoots successfully make the transition from *in vitro* to *ex vitro* conditions. Since European birch shoots require only a modest amount of care during the acclimatization phase, healthy plantlets can make the transition to an environment with higher light and lower humidity in about two weeks. The amount of new growth formed in a short period of time depends on the vigor of the plantlets or nonrooted microshoots when transplanted as well as the acclimatization conditions.

Materials

The following materials will be needed to complete this stage of micropropagation by individuals or groups of students:

- Stage II shoot cultures
- Pasteurized potting mix
- Clear polystyrene (plastic) clamshell container (typically a donut, cookie, or salad container)
- Small-diameter cork borer or dissecting probe

Follow the protocol provided in Procedure 35.4 to complete this experiment.

Procedure 35.4
Transferring Stage III Plantlets to *Ex Vitro* Conditions

Step	Instructions and Comments
1	Between two to four weeks after beginning Stage III, roots will form on the birch microshoots. Be sure to record the data requested in Step 6 from Procedure 35.3.
2	Obtain a clear polystyrene (plastic) clamshell container. The type of container suggested for this stage of micropropagation has hinged lid and often is used to hold donuts or cookies. Used food containers can be washed and used rather than purchasing new containers. Use a small-diameter cork borer (≤3 mm) or a dissecting probe to punch several holes in the bottom of the container. These holes will provide drainage for the vessel. Next, moisten enough pasteurized potting mix (any good brand should work) to fill the bottom of the food container to a depth of 4 cm. The potting mix should be moist NOT wet; if the mix is too wet, the rooted plantlets can become severely stressed and die.
3	Remove the jars containing the rooted microshoots from the culture room shelf and place them on a classroom table or other suitable area for working with potting mix. Avoid transferring the shoots in the laminar flow hood or culture room.
4	Carefully remove the plantlets from the jar. You may find it easier to remove the plantlets and agar all at once and separate the plantlets outside the jar. Remove each plantlet from the agar, and gently remove any agar adhering to the roots by pulling or rinsing off the material with tap water. Be sure to record the data requested in Step 6 from Procedure 35.3 before planting the rooted plantlets. Use your finger or a pencil to poke a small hole in the moistened potting mix, place the plantlet base and roots in the hole, and gently cover the base with the mix. Repeat this procedure for all shoots, noting the multiplication (Stage II) treatment used on the shoots. To keep track of the multiplication treatment used, keep plantlets receiving the same treatment in the same row in the container. Also transplant nonrooted shoots into the container.
5	The clamshell container with the microshoots can be placed back on a culture room shelf, but insects and potting mix may contaminate the room. An alternative choice would be to place the container about 30 cm below a 1.3-m (4-ft) fluorescent shop light. The container can also be placed on a shaded greenhouse bench to keep potting mix out of the culture room. If the plantlets are acclimatized on a greenhouse bench, make sure the container is kept under shade cloth rather than direct sunlight. After one or two weeks, check the plantlets, and count the numbers that appear sick, wilted, or dead. Also note if the plantlets have resumed shoot and root growth. Root growth can be checked by placing a finger under the plantlet and gently lifting the shoot out of the potting mix.
6	To help the plants acclimatize to the environment, the lid of the clamshell container can be opened for one to several h a day. The lid should be held open for longer periods of time each day over a four-day period. After the plants become acclimatized to ambient humidity conditions, they can be transplanted to pots and grown in a greenhouse or outdoors. Bright light should be avoided for the first week after transplanting.

Anticipated Results

Most of the birch plantlets should acclimatize to the conditions in the clamshell container. Even microshoots that lacked roots may start rooting in the potting mix. Rooted plantlets that have successfully made the transition will form new leaves and roots that are fully functional. In fact, rooted plantlets may grow about 2 to 3 cm within three weeks of being transferred out of the baby food jars. Once the plantlets have formed new leaves and roots (one to three weeks) and been acclimatized to lower humidity, the plants can be transplanted into pots and grown in a greenhouse or outdoors.

Questions

- Did the multiplication treatment affect the ability of rooted microshoots to be transplanted successfully?
- Did nonrooted shoots die? Were any starting to form roots?
- Did the new leaves that formed after the plantlets started to grow in the clamshell container appear different than those formed under *in vitro* conditions? If so, how did they differ?

TABLE 35.3
Tissue Culture Media Used for Stage I of Micropropagation of 'Kardinal' Rose and American Elm Plants

Ingredient	Rose Micropropagation Medium	American Elm Micropropagation Medium
Salt formulation[a]	MS 2.15 g	DKW 2.6 g
Sucrose	15 g	15 g
Organics stock solution	MS organics stock[b] 5 mL	DKW organics stock[c] 5 mL
Plant growth regulators	See Table 35.4	See Table 35.4
Adjust volume	Bring up to 500 mL	Bring up to 500 mL
Adjust pH of medium	5.6	5.7
Agar	3.25 g	3.25 g

Note: Each separate column is used to make 500 mL of medium. To make more or less culture medium, the amounts of each ingredient need to be adjusted accordingly.

[a] Abbreviations for salt formulations: MS is Murashige and Skoog medium, whereas DKW is Driver-Kuniyuki Walnut medium. These can be purchased from vendors such as Sigma-Aldrich Company, Caisson Laboratories, and *Phyto*Technology Laboratories.

[b] MS organics stock solution is made by mixing (on a g/L basis) the following organic compounds: myo-inositol 10; glycine 0.2; nicotinic acid 0.05; pyridoxine HCl 0.05; and thiamin HCl 0.01. Dissolve each component, starting with myo-inositol, before adding the next one.

[c] DKW organics stock solution is made by mixing (on a g/L basis) the following organic compounds: myo-inositol 10; glycine 0.2; nicotinic acid 0.1; and thiamin HCl 0.2. (Note: pyridoxine HCl is excluded from this medium.) Dissolve each component, starting with myo-inositol, before adding the next one.

ALTERNATIVE SPECIES FOR MICROPROPAGATION

European birch plants are often easy to establish *in vitro*, and they usually form only one shoot per explant. Other plant species, however, respond differently to plant growth regulators used in the establishment medium, which will affect the number of shoots that form on Stage I explants. Alternative species to micropropagate are 'Kardinal' rose (*Rosa* x *hybrida*) or American elm (*Ulmus americana* L.).

Stems from 'Kardinal' rose and American elm plants growing outdoors or in a greenhouse can be used for explants. Greenhouse plants are preferred because they should have fewer microorganisms on their surfaces, making them easier to surface sterilize compared to outdoor-grown plants. Procedure 35.1 can be used for making single-node explants from rose or elm stems. The differences for these two species will be the tissue culture media and plant growth regulator concentrations used (Tables 35.3 and 35.4).

Surface-sterilized stem explants from 'Kardinal' rose plants should be placed on Murashige and Skoog (MS) medium, whereas American elm stem explants should be cultured on Driver-Kuniyuki Walnut (DKW) medium (Table 35.3). Higher concentrations of BA can be used for establishing rose stem explants compared to elm stem explants (Table 35.4). After 4 weeks, be sure to record the data described in Step 11 of Procedure 35.1. Pay particular attention to the number of buds that expanded per explant and the length of the new shoots.

Stages II, III, and IV for rose and elm microshoots can be completed as described in Procedures 35.2, 35.3, and 35.4, respectively. The media formulations described in Table 35.3 can be used for these species for Stages II and III, but the plant growth regulators need to be changed. Benzyladenine will be the only plant growth regulator needed for Stage II for rose and elm multiplication (Table 35.5). Rose microshoots taken from Stage II can be placed on half-strength MS medium supplemented with 0.5 µM NAA (0.25 mL of 1 mM NAA stock solution per 500 mL of medium) to promote rooting during Stage III. Elm microshoots are somewhat more difficult to root *in vitro* compared to rose and birch microshoots. An NAA concentration of 1 µM (0.5 mL of 1 mM NAA stock solution per 500 mL of medium) in DKW medium should help promote root initiation. Elm microshoots taken from Stage II can also be rooted directly in a clear polystyrene (plastic) clamshell container, bypassing Stage III, but this process can take a long time, and the percentage of successfully rooted microshoots may be low. Follow directions in Procedure 35.4 for acclimatization of rooted rose (Figure 35.6) and elm microshoots.

TABLE 35.4
Plant Growth Regulators Used during Stage I for 'Kardinal' Rose and American Elm Micropropagation Treatments

Treatment	'Kardinal' Rose Stage I Benzyladenine Stock Solution[a] (mL)	'Kardinal' Rose Stage I Naphthaleneacetic Acid Stock Solution[b] (mL)	American Elm Stage I Benzyladenine Stock Solution (mL)	American Elm Stage I Naphthaleneacetic Acid Stock Solution (mL)
A	0	0	0	0
B	1.8	0	0.8	0
C	1.8	0.5	1.5	0
D	3.5	1	1.5	0.5

Note: The stated volumes of stock solutions are added to 500 mL of medium, which is enough to for 20 jars of medium. To make more or less culture medium, the amounts of each stock solution need to be adjusted accordingly.

[a] A 1 mM benzyladenine stock solution is made by dissolving 0.0225 g of benzyladenine in 10 to 15 drops of 1 N NaOH (40 g NaOH per liter of water) and then bringing the solution up to 100 mL.

[b] A 1 mM naphthaleneacetic acid stock solution is made by dissolving 0.0186 g of naphthaleneacetic acid in 10 to 15 drops of 1N NaOH and then bringing the solution up to 100 mL.

TABLE 35.5
Amount of Benzyladenine Stock Solution Used during Stage II for Rose and Elm Micropropagation Treatments

Treatment	Rose Micropropagation Stage II Benzyladenine Stock Solution[a] (mL)	Elm Micropropagation Stage II Benzyladenine Stock Solution (mL)
A	1.8	0.8
B	3.5	1.5
C	7	3

Note: The stated volumes of BA stock solution are added to 500 mL of medium, which is enough for 20 jars of medium. To make more or less culture medium, the amount of stock solution needs to be adjusted accordingly.

[a] A 1 mM benzyladenine stock solution is made by dissolving 0.0225 g of benzyladenine in 10 to 15 drops of 1 N NaOH (40 g NaOH per liter of water) and then bringing the solution up to 100 mL.

FIGURE 35.6 Rose plantlets that have successfully made the transition from Stage IV plantlets to the ambient environment. Note the new (glossy) leaves on the rooted microshoots, indicating the plantlets are actively growing.

LITERATURE CITED AND SUGGESTED READING

Lloyd, G. and B. McCown. 1980. Commercially-feasible micropropagation of mountain laurel, *Kalmia latifolia*, by shoot-tip culture. *Comb. Proc. Intl. Plant Prop. Soc.* 30: 421–427.

McGranahan, G.H., J.A. Criver, and W. Tulecke. 1987. Tissue culture of *Juglans*, in *Cell and Tissue Culture in Forestry*, vol 3. Bonga, J.M. and D.J. Durzan (Eds.) Martinus Nijoff Publishers, Boston, pp. 261–271.

Murashige, T. and F. Skoog. 1962. A revised medium for rapid growth and bioassays with tobacco tissue cultures. *Physiol. Plant.* 15: 473–497.

Pierik, R.L.M. 1997. In Vitro *Culture of Higher Plants*. 4th ed. Kluwer Academic Publishers, Dordrecht, The Netherlands. 348 pp.

Preece, J.E. 1997. Axillary shoot proliferation, in *Biotechnology of Ornamental Plants*. Geneve, R.L., J.E. Preece, and S.A. Merkle (Eds.) CAB International, Wallingford, U.K., pp. 35–43.

Part XIII

Seed Production and Propagation

36 Sexual Reproduction in Angiosperms

Robert N. Trigiano, Renee A. Follum, and Caula A. Beyl

CHAPTER 36 CONCEPTS

- Most plants exhibit alternation of generations. In this case, the plant generations are the sporophytic or diploid (2n) phase that produces spores and the gametophytic or haploid (n) phase that produces gametes. The gametophyte is more conspicuous in less advanced plants, such as some algae and mosses (bryophytes), whereas the sporophyte is the conspicuous and dominant phase in more advanced plants, such as ferns, gymnosperms, and angiosperms.

- Sexual reproduction is a major source of genetic diversity in populations and individuals.

- The four parts of flowers are sepals, petals, anthers, and pistils. Flowers of some species are composed of all four parts, whereas flowers of other species lack one of more of the parts.

- The mature functioning female gametophyte of angiosperms is typically composed of eight nuclei and the mature male gametophyte has only three nuclei.

- Angiosperms exhibit double fertilization, in which one sperm (n) unites with the egg (n) to form the zygote (2n), and the other sperm fuses with two polar nuclei (either n and/or 3n) to become the primary endosperm nucleus (3 or 5n). Water is not necessary for fertilization.

- Zygotic embryogenesis for monocotyledonous and dicotyledonous plants typically follows a predetermined developmental pattern, although there are many variations between species. The development of somatic embryos in tissue culture generally follows the same pattern as their zygotic counterparts, but often exhibits more plasticity and may manifest some "abnormalities."

Many horticultural crops are produced exclusively via vegetative or asexual means. Asexual reproduction using grafting, rooted cuttings, and various other methods assure clonal fidelity of a cultivar or an elite germ line. Phenotypic and genetic variation in these crops is undesirable, especially if the consumer is expecting to purchase a plant that they have seen growing somewhere else or pictured in a catalog. For example, all *Cornus florida* L. (flowering dogwood) 'Cherokee Sunset' plants have yellow variegated leaves and more or less pink-red bracts subtending the floral disk and are produced exclusively by a grafting technique (chapter 27). There is almost no variation in plants produced using this technique, and the technique is capable of making millions of exact genetic copies of 'Cherokee Sunset.' In the end, the consumer receives the desired product with all of the expected characteristics.

Other horticultural plants, especially many bedding species and vegetable crops, for example, are propagated mainly by seeds. This is a sexual process that entails meiosis in both the male and female cells to form haploid (n) gametes and recombination of chromosomes and restoration of the diploid (2n) number by fusion of gametes. As will be discussed later in more detail, sexual reproduction in angiosperms involves pollination in which pollen or the male gametophyte lands on a receptive female surface, usually the stigma of the ovary. Eventually, a process called double fertilization takes place, in which one of the male sperm fuses with female egg nucleus to create the zygote and the other sperm unites with two other nuclei within the female gametophyte to form the primary endosperm nucleus.

Sexual reproduction would seem then to increase genetic variability, so how do seed producers maintain the characteristics of various cultivars? After all, when we plant a certain variety of petunia or tomato, we have expectations of growing a uniform crop. If the horticultural traits of a plant or a population of plants are the same and can be "selfed" (some plants can mate with themselves or other plants in the population that are essentially identical) and sexual compatibility (chapter 37) is not a problem, then producing seed that yields uniform plants is not difficult. Although oversimplified, seed producers allow the plants to mate exclusively with each other and then collect the seeds.

New cultivar development often involves sexual reproduction. An example of breeding a new cultivar is incorporating disease resistance into an existing, standard, susceptible cultivar. Sexual reproduction can be used to "transfer" the gene (s) for resistance from one cultivar to the other, while still maintaining the characteristics of the nonresistant cultivar. In this case, the goal of the breeder is to transfer just that trait, while maintaining most or all of the other characteristics of the other plant. Repeated (recurrent) selection for disease resistance linked to other horticultural characteristics and backcrossing to the original cultivar with progeny that has the desired trait will eventually produce seeds and plants with 99.5% or more of the characteristics of the original plant, plus the new desired trait. For most practical purposes, all of the plants in the population are now considered identical, at least for the horticultural traits of the cultivar and selected trait of disease resistance, although there probably are some minuscule differences. Using various breeding techniques, the trait for resistance has now become fixed into the standard cultivar, so seed that will produce plants with resistance to the disease and have the horticultural characteristics of the original parent can be grown on a large scale for sale. There are many breeding schemes and strategies, and some of these are discussed in chapter 37.

In the above example, sexually compatible plants were used, but there are many instances where plants cannot mate with themselves or with very similar other genotypes in the population. This reproductive limitation is called sexual incompatibility and is found commonly in many woody ornamental plants, such as flowering dogwood. Sexual incompatibility can be manifested as the failure of pollen to germinate and form germ tubes, or even the poor (slow) growth of pollen tubes, which as explained later in the chapter, are the tubes by which sperm nuclei are brought to the egg. Selfing may be prevented by incompatibility genetics, either one allele (gametophytic) or two alleles (sporophytic). Generally, self-incompatibility or obligate outcrossing encourages and favors genetic diversity by forcing mating of different genotypes. In the case of flowering dogwood, each and every seed is unique, and the resultant plants would possess traits that are intermediate between both of the parents. The details of sexual incompatibility are provided in chapter 37, "Breeding Horticultural Plants."

There are other mechanisms that plants have to effectively reduce selfing or even prevent hybridization between closely related species. Detailed descriptions of these methods are beyond the scope of this brief chapter (see chapter 37), but are worth mentioning in the context of sexual reproduction. Timing of maturity of the sexual floral organs may play an important role in the ability to self. The pollen (male gametophyte) in the perfect flower (having both functional male and female floral organs) may not be mature when the stigma (the receptive surface of the female floral organ–see Figure 36.4) is receptive. In other words, pollination may occur in this instance, but the pollen will not germinate and, therefore, cannot complete fertilization. The converse may be true as well. The stigma may be receptive either before or after the anther (part of the male floral organ—see Figure 36.4) dehisces, and pollination/fertilization is thwarted. This forces individual plants to mate with other plants, often genetically very different, and not themselves. Another simple solution plants utilize to eliminate selfing is to have male and female floral organs on separate plants (dioecious = two houses). In this case, there is no alternative except to cross with another plant.

So, sexual reproduction generally increases genetic variability in plants and can be used to transfer traits (genes) from one plant (cultivar) to another plant (cultivar). We have briefly outlined only a few of the important propagation techniques and ideas involving sexual reproduction, and it is not our intention to describe additional aspects of plant breeding, population genetics, etc. in this chapter—see chapter 37 for details. Suffice to say that sexual reproduction in plants is very important to agronomic and horticultural endeavors. Our goal in this chapter is to describe the biological principles behind sexual reproduction. To this end, we will begin with the concept of alternation of generations, explore flower structure and development of gametophytes, and finally discuss embryos and seeds. We will also discuss somatic embryogenesis—a special case of embryo development.

ALTERNATION OF GENERATIONS

The concept of alternation of generations is central to understanding sexual reproduction in angiosperms as well as other organisms. In this case, generations do not refer to parents and children, but rather to the nuclear condition or ploidy number of cells. Ploidy level is the number of copies of chromosomes in the nucleus. Remember that most of the blueprints, the genetic code (DNA), for all the structures and life processes for plants are contained in genes found on chromosomes located in the nucleus, whereas a small portion of coding DNA is situated in chloroplasts and mitochondria. For all organisms that reproduce sexually (fusion of male and female gametes), including the angiosperms, parts of the life cycles are represented by haploid (n) cells (gametes that only have one-half the number of chromosomes), whereas the other portions are manifested as diploid (2n) cells (having the full complement of chromosomes). It is important to mention here that plants maybe contain single or multiple sets of chromosomes (polyploidy). For example, flowering dogwood haploid (n) cells contain one set of 11 chromosomes and, therefore, can be represented by the equation: $n = x = 11$, where $x = $ a set of 11 chromosomes. Therefore, the diploid (2n) cells of flowering dogwood all contain 22 chromosomes or $2n = 2x = 22$. In contrast, other plants,

Sexual Reproduction in Angiosperms

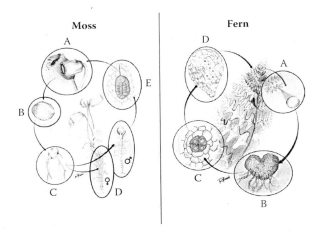

FIGURE 36.1 Alternation of generations. **Moss.** The most notable phase of moss is the green "leafy" gametophyte (n), the plant that produces gametes. The gametophytic phase begins with the formation of n meiospores (product of meiosis) in the sporophyte capsule (A). Meiospores are released from the capsule (B) and germinate (C). The gametophyte grows by typical cell division and differentiation and ultimately produces male and female gametangia containing gametes (D). Sperm from the male gametangia swim in film of water to the female gametangia and fuse with the egg to produce a 2n zygote (E). The zygote develops and produces the mature sporophyte, which remains attached to the living gametophytic phase (A). **Fern.** The most conspicuous phase of ferns is the sporophyte, which is green and can be very large. Meiosis occurs in sporangia and heralds the beginning of the gametophytic phase of the life cycle (A). The haploid meiospore germinates and grows into a very small and inconspicuous heart-shaped gametophyte containing both female and male gametangia (B). Fertilization requires water, and sperm and egg fuse, creating the zygote and restoring the diploid (2n) or sporophytic phase (C). The zygote grows by mitotic divisions and differentiates into the mature sporophyte, which we typically think of as a fern (D).

such as the common chrysanthemum, have multiple sets of chromosomes. The haploid gametes of this plant each contain three sets of nine chromosomes or expressed in the above equation form: $n = 3x = 27$. Each diploid cell contains six sets, each consisting of nine chromosomes or $2n = 6x = 54$.

The life cycle phase of plants that produces gametes (n) is termed the gametophytic. For example, the green, "leafy" moss growing on a log or a rock is the gametophytic phase. Often individual plants produce either sperm (male) or eggs (female) gametes (Figure 36.1A). All of the cells of the gametophyte are haploid. Using this same example, the green (turning brown with age) hooked stalk with a terminal capsule growing from the leafy gametophyte of mosses is the sporophyte or sporophytic phase, which will eventually produce spores. All of the cells of the sporophyte are diploid. The gametophyte is the most conspicuous phase of many "lower" plants, such as algae and mosses, whereas the gametophyte is reduced and is much less conspicuous in higher plants, such as ferns and gymnosperms (Figure 36.1B). In the case of ferns, we typically see the green, leafy phase, and seldom observe the much-reduced in size gametophytic phase, which typically contains both male and female structures.

The angiosperms represent the epitome of gametophyte reduction. In this group, the female and male gametophytes are represented only by a few cells, and will be discussed later on in this chapter. The sporophyte of angiosperms is typically what you see in the forests, fields, and landscapes, whereas the gametophyte is very small, and for the most part, we are unaware of its physical existence except perhaps as a coating of "dust" on a window or car. However, many of us are allergic (runny noses, itchy eyes, etc.) to the unseen male gametophyte of angiosperms—pollen.

MITOSIS AND MEIOSIS ESSENTIAL CELLULAR PROCESSES IN REPRODUCTION

So, how do alternation and development of generations occur in angiosperms? The driving mechanisms for the alternation of generations are meiosis, which reduces the ploidy number to haploid from diploid, and sexual fusion of gametes to form the zygote, which restores the diploid number. To understand the maintenance and reduction of ploidy levels, we need to consider two cellular processes that are related—mitosis (Figure 36.2) and meiosis (Figure 36.3).

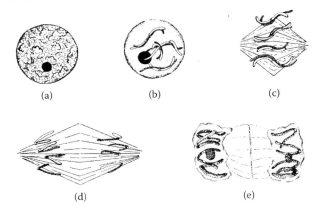

FIGURE 36.2 Idealized representation of mitosis. (a) Interphase nucleus. In this phase chromosomes were replicated and remain uncondensed. Nucleoli are prominent. (b) Prophase nucleus. In this example, the set of four sister chromatids (two copies of each chromosome) have condensed (thickened and shortened). In late prophase, the nuclear membrane and the nucleolus disappear. (c) Metaphase. The sister chromatids have aligned on the equatorial (metaphase) plate and spindle fibers attached to centromeres of the chromatids. The individual sister chromatids (now chromosomes) begin to move toward the two poles. (d) Anaphase (late). The chromosomes have migrated to the poles. (e) Telophase. The chromosomes become long, threadlike (uncondensed) molecules, the nuclear membrane and nucleoli are reformed, and the wall between the two daughter cells is formed.

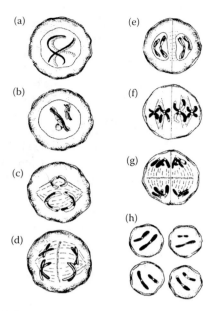

FIGURE 36.3 Idealized representation of meiosis. 2n = 4. (a) Interphase. Replication of chromosomes (sister chromatids). (b) Prophase I. Condensation of two sets of homologous chromosomes (sister chromatids: black and white represents chromosomes from the two parents). Crossing-over events (mixed black and white) occur during this phase. (c) Early anaphase I. Homologous chromosomes paired on the metaphase plate and one-half of the homologous pair begins to migrate toward each of the poles. This has effectively reduced the ploidy of the cell to haploid or, in this case, n = 2. This is the beginning of the gametophytic phase. (d) and (e) Late anaphase I and telophase I. (f) Metaphase II. Two sets of sister chromatids line up on the equatorial plate as in mitosis, and an individual sister chromatid (chromosome) migrates to the pole (f) in anaphase II. (g) Late telophase II. (h) Four haploid (n) nuclei are created.

Let's consider mitosis first because it is less complicated then meiosis. This process is integral in cell division and allows cells to maintain the ploidy level of the original cell, regardless of whether it was haploid or diploid. In other words, the number of chromosomes or sets of chromosomes will be the same in the daughter cells resulting from a mitotic division as it was in the mother cell. Sometimes a mitotic division is referred to as an "equal division." The process of mitosis (followed by cytokinesis—division of the cytoplasm) allows an organism to grow by adding new cells. For this equal division to happen, duplication of the chromosomes must occur sometime during the cell cycle. The cell cycle can be conveniently organized into the following five phases: interphase, prophase, metaphase, anaphase and telophase. These phases are continuous with one another in the cell, but are demarked by more or less recognizable characteristics of the nucleus. The nucleus of the cell remains in interphase for the majority of the cell cycle (Figure 36.2a). The nucleus appears granular after staining with an appropriate stain, and the nucleolus is apparent as a more densely staining sphere within the nucleus. At this time, the DNA molecules that chromosomes are made of are very thin and long and cannot be recognized as separate structures within the nucleus. The chromosomes are replicated or copied during this time. The template (the original DNA strand) and the replicated chromosome copy do not separate, but stay joined and together are now called sister chromatids instead of chromosomes. Prophase (Figure 36.2b) is marked by the sister chromatids shortening and thickening. This process is also referred to as condensation. The end of prophase is marked by the disappearance of the nucleolus and nuclear membrane. The next phase is metaphase, in which the sister chromatids line up on the equatorial (metaphase) plate of the cell, and spindle fibers emanating from the poles attach to the centromeres (Figure 36.2c). These fibers are made of microtubules and contract, pulling one of each of the sister chromatids toward the opposite poles. Individual sister chromatids are now termed chromosomes. The migration of the chromosomes signals the beginning of anaphase (Figure 36.2d). Telophase (Figure 36.2e) occurs when the chromosomes reach the poles and start to become thin and threadlike, just as they were before prophase. They continue this process until they are no longer discernable as individual entities. The nucleoli reappear, and nuclear membranes are established around each of the daughter nuclei. Finally, a cell plate forms perpendicular to the spindle fibers and between the two nuclei. This will divide the cytoplasm and eventually become the cell walls of the daughter cells (not shown). One final thought. Sometimes nuclear division is not followed by cytokinesis (free nuclear division), and the cell becomes multinucleated as endosperm (see double fertilization). So, true cell division that results in two equal cells requires nuclear division followed by cytokinesis.

Now let's consider meiosis (Figure 36.3), which is considerably more complex than mitosis. Meiosis basically reduces the ploidy number by half and is essential to the formation of gametophytes and gametes. A diploid cell undergoing meiosis will produce four haploid cells, of which one or more will be functional. Meiosis may be thought of as a reduction division (diploid to haploid) followed by an equal division, similar to mitosis. As in mitosis, the chromosomes are replicated during interphase (Figure 36.3a) and are now in the form of sister chromatids. The chromosomes thicken and become visible in prophase as discussed in mitosis. However, at the end of this phase and unlike in mitosis, chromosomes are paired. These are homologous chromosomes, and one chromosome is derived from the male parent and the other from the female parent (Figure 36.3b). As mentioned before, each chromosome consists of sister chromosomes, so that each homologous pair actually consists of four sister chromatids. Another phenomenon happens in this stage as well. Segments from sister chromatids of homologous pair of chromosomes are exchanged. In other

Sexual Reproduction in Angiosperms

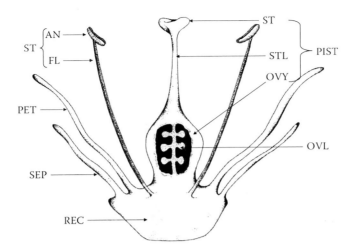

FIGURE 36.4 Idealized diagram of a perfect flower. The four parts of a flower are sepals (SEP), petals (PET), stamens (ST), and pistil(s) (PIST). In this case, all parts are mounted on a receptacle (REC). Stamens are composed of filaments (FL) that support the anthers (AN), which contain pollen. A pistil has three parts—a stigma (ST), a knob-like structure were pollination occurs, a style (STL), and an ovary (OVY), which contains ovules (OVL).

words, a portion from the male-derived chromosome is "transferred" to the corresponding site on the original female chromosome and vice versa. Crossing over represents a rearrangement of genes and contributes to the genetic diversity of the species. The homologous pairs of chromosomes line up on the equator in metaphase I (the "I" refers to the first division of meiosis), and spindle fibers are formed. However, instead of the sister chromosomes migrating to opposite poles as in mitosis, one-half of the homologous pair of chromosomes, (each consisting of two sister chromatids) moves toward each pole during anaphase I (Figure 36.3c). Cytokinesis occurs at the end of telophase I, and each of the daughter nuclei contains one-half of the number of chromosomes (Figure 36.3d,e). The second (II) stage of meiosis begins as metaphase II in the daughter cells, where the sister chromatids align on the equatorial plate (Figure 36.3f). Just like in mitosis, spindle fibers form and attach to the centromeres. During anaphase II (Figure 36.3g), the sister chromatids (now called chromosomes) separate, and one of each is drawn to the opposite pole (telophase II). Cytokinesis follows and meiosis results in four haploid cells, each with one-half of the diploid complement of chromosomes (Figure 36.3h).

FLOWERS

The angiosperms have evolved specialize structures called flowers to accomplish sexual reproduction. Flowers come in many shapes, colors, and arrangements (inflorescences), and depending on the species of plants, may appear at anytime during the growing season. Flowers may be borne singly as shown in Figure 36.5, or many flowers may be clustered on a common axis (inflorescences). There are many types of inflorescences, and these arrangements are sometimes distinguishing characteristics of certain groups or families of plants. For those of you that have additional interest in this area, please consult any one of a number of basic botany books listed in the references. Some species of plants produce flowers that may lack one or more basic parts (incomplete), whereas others have all of the structures (complete). Flowers may contain both male and female structures (bisexual), whereas other flowers contain only female (pistillate) or male (staminate) parts. Individual plants may have both male and female flowers on the same plant (monoecious—literally one house), whereas other plants may only possess flowers that are either male or female (dioecious—literally two houses). We could go on to describe many other differences between flowers of different species, some very obvious and others very subtle, but one unifying function of flowers among the very diverse species of angiosperms is that meiosis and fusion of gametes to form the zygote (sexual reproduction) always occurs here—within the flowers.

It is not our intention to explore all of the different variations of flower morphology, so we will restrict our discussion to the very simple case of a perfect flower—a flower that has all four basic parts with a superior ovary (Figure 36.4 to Figure 36.6). The four basic building units of flowers, working from the outside to interior, are the sepals, green leaf-like appendages at the base; the petals, which are often pigmented; the stamens, consisting of filaments (stalk) and anthers; and pistils made up of stigma, style, and ovary. All of the floral components in this case are mounted on a receptacle (Figure 36.4 and Figure 36.6a). The male portions of the flower are the stamens (Figure 36.4), which contain many cells (microspore mother cells in the anther of the stamen) that will undergo meiosis and eventually form the male gametophytes or pollen. The female part of the flower is the pistil. The ovary contains one to many ovules or unripened seeds

FIGURE 36.5 Morphology of a dicot flower. The perfect flower of this *Oxalis* species has five sepals (SEP), which are hidden from view in the open flowers by the five petals (PET), five stamens (ST), and one pistil (PIS). Compare to Figure 36.4 and Figure 36.6.

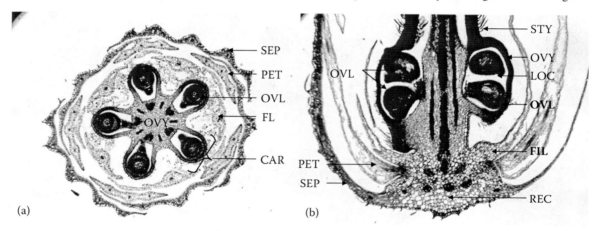

FIGURE 36.6 Flower anatomy. (a) Cross section through a *Geranium* species (Stork's-bill) flower. Progressing from the outside inward, sepals (SEP), petals (PET), and filaments (FIL). Occupying the center of the flower is the ovary (OVY), which is composed of five carpels (CAR) in which two ovules (OVL) are borne. Note: Only a single ovule can be seen in each locule (space) with this section. (b) Longitudinal section through the Stork's-bill flower seen in (a). In this view, two carpels (CAR) and locules (LOC) of the ovary (OVY) are visible, but each contains two ovules (OVL), of which only one will mature. Note the trichomes (hairs) on the style (STY) portion of the pistil. The sepals (SEP), petals (PET), and filaments (FL) are inserted below the ovary on the receptacle (REC). (Adapted from Trigiano, R.N and D.J. Gray. 2005. *Plant Biotechnology and Development*. CRC Press, Boca Raton, FL. With permission.)

(Figure 36.4 and Figure 36.6a,b). Each ovule contains one megaspore mother cell, which through the process of meiosis will give rise to the female gametophyte.

FEMALE AND MALE GAMETOGENESIS

Gametogenesis is the development of either the female (Figure 36.7) or male (Figure 36.8) gametophyte or haploid (n) generation of the life cycle. The female gametophyte develops from a single megaspore mother cell located in the ovule (Figure 36.7a). The megaspore mother cell, which is 2n, divides by meiosis and produces four haploid megaspore nuclei (Figure 36.7b). There are a number of species-specific developmental scenarios that can happen after meiosis, but we will use the *Polygonium* model to illustrate embryo sac or female gametophyte formation. Typically in this model, three of the nuclei from meiosis degenerate (Figure 36.7c), and the remaining megaspore divides via mitosis three times to produce an eight-celled (nucleated) female gametophyte bounded by the nucellus. The nucellus is sporophytic tissue and is not formed from the megaspore mother cell. The solitary egg (n) or female gamete, flanked on either side by synergids (n), is located at the micropylar end of the ovule (Figure 36.7a,d). The micropyle is the opening formed by the immature integuments (later in development they become the seed coat or testa) of the ovule. The polar nuclei, both haploid, are located in more or less the center

Sexual Reproduction in Angiosperms

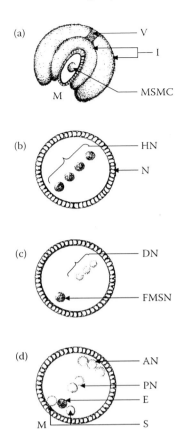

FIGURE 36.7 Idealized diagrammatic representation of female gametophyte development. (a) A single diploid (2n) megaspore mother cell (MSMC) resides within each ovule contained in the ovary. I = integuments (may be one to several); M = micropyle; V = vascular tissue at the opposite end of the ovule. (b) The MSMC undergoes meiosis (see Figure 36.3) to produce four haploid (n) nuclei (HN) within the nucellus (N). (c) In this example, three of the HN degenerate (DN), leaving one functional megaspore nucleus (FMSN). (d) The FMSN will divide via mitosis three times, and the eight haploid nuclei migrate to form the complete female gametophyte or embryo sac consisting of an egg (E), two synergids (S), which lie on either side of the egg at the micropylar end, two polar nuclei (PN) in the center, and three antipodals (AN) at the distal portion of the embryo sac. M = micropylar end of the ovule.

of the embryo sac. Lastly, there are three antipodal nuclei (n) located at the end opposite of the micropyle.

Male gametogenesis occurs from hundreds of microspore mother cells, which are all 2n, located in the anthers (Figure 36.8a,b). Each microspore mother cell undergoes meiosis to form four (a tetrad) microspores, which are haploid (Figure 36.8c). These microspores while in the anther divide via mitosis once again to produce binucleated pollen grains (Figure 36.8d). Each grain has a tube nucleus and a generative nucleus, which will give rise to two sperm nuclei—the male gametes (Figure 36.8e). The male gametophyte usually matures (produces sperm nuclei) only after dehiscence of the anther, which liberates the pollen and allows subsequent pollination.

POLLINATION AND DOUBLE FERTILIZATION

Now that the male and female gametophytes have been produced, how do the two gametes come into close proximity? The act of pollen grain(s) being transferred from the anthers to the stigma of a flower is called pollination. Pollen can be delivered to the stigma of the pistil either via wind currents or by a vector. Those plant species that rely on wind dissemination usually produce copious amounts of light, dry pollen. Some examples of plants depending primarily on wind are some trees, such as oaks and maples, corn and wheat, and perhaps the most notorious of them all, ragweed, a cause of "hay fever" in people during the fall months. Pollen may be vectored to the stigmatic surface of the pistil by numerous types of insects, including honeybees, bumblebees, moths and beetles, other insects, and vertebrates, such as bats and birds. In general, those species relying on vectors for pollination produce more conservative amounts of heavy, sticky pollen than by those depending on wind pollination.

Pollination may occur between pollen and the stigma in the same flower (self-pollination) or plant or may happen between two different plants (cross-pollination). The surface of the stigma is typically covered with exudates (usually some type of sugar) that help the pollen grain adhere to the tissue and provide physical and physiological cues for germination. Remember, pollination does not mean fertilization—they are two different processes. After pollination occurs, the pollen grain will germinate, producing a pollen tube, which grows through the style toward the ovules in the ovary. Shortly after germination, the generative cell divides, and two sperm nuclei are produced. There are a number of factors that will affect germination of pollen and the growth rate of pollen tubes. For example, in some plants genetic incompatibility factors exist that either prevent germination of the pollen or retard pollen tube growth. These factors control mating and thus direct recombination and genetic variability. Assuming that all physical, environmental, and genetic factors are favorable for pollen germination, pollen tubes grow through the length of the style to the micropylar end of the ovule in the ovary and deliver two nonflagellated sperm (Figure 36.9a,b). Since the sperm are not required to swim, free water is not necessary for fertilization in the angiosperms and represents a significant advancement over most other plants, which require free water for fertilization. Another process that is unique to the life cycle of angiosperms is double fertilization. Double fertilization entails one of the sperm nuclei uniting with the egg to form the zygote, which is the first cell of the 2n sporophytic

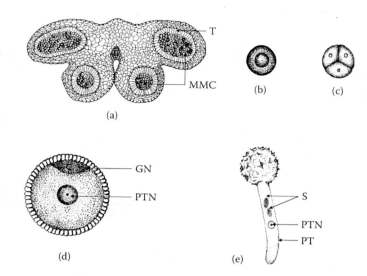

FIGURE 36.8 Idealize diagrammatic representation of male gametophyte development. (a) Many diploid (2n) microspore mother cells (MMC) located in the anther portion of the stamens are surrounded by the tapetum (T), the cell layer derived from the sporophyte that surrounds the MMC and later the pollen. (b) A single MMC undergoes meiotic division to produce four haploid (n) microspores, which typically are found in a group called a tetrad shown in (c). (d) After the microspores separate (while still in the anther), the nucleus divides via mitosis to produce a generative nucleus (GN) and a pollen tube nucleus (PTN). Hereafter, the binucleated cell is termed pollen. (e) After pollination, the pollen grain germinates, and the generative nucleus divides to produce two sperm nuclei that are haploid.

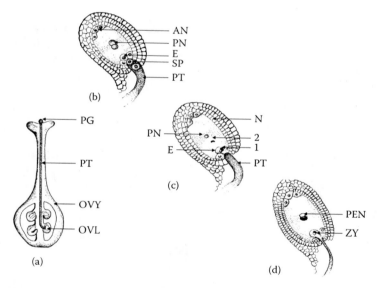

FIGURE 36.9 Idealized diagrammatic representation of pollination, pollen tube growth, and double fertilization. (a) Pollen grain (PG) germinates on the stigma of the pistil and grows via a pollen tube (PT) toward an ovule (OVL) containing a mature female gametophyte. OVY = Ovary. (b) The pollen tube (PT) grows through the micropylar end of the ovule toward the egg (E). AN = antipodal nuclei; PN = polar nuclei; SP = sperm nuclei. (c) Double fertilization occurs when one sperm nucleus (1) fuses with the egg and the other sperm nucleus (2) unites with the polar nuclei. N = nucellus. (d) Double fertilization results in the zygote (ZY), which is diploid (2n), and the primary endosperm nucleus (PEN), which in our example is triploid or 3n.

phase, and the other sperm fusing with both of the polar nuclei to form the primary endosperm nucleus, which is triploid (3n) in our *Polygonium* example (Figure 36.9c,d). Note that endosperm can be pentaploid (5n) in some species of plants. The primary endosperm nucleus divides repeatedly to form endosperm, which functions as food and provides other physiological and physical factors for the developing sporophyte and/or will be absorbed into the cotyledons of the embryo.

ZYGOTIC EMBRYOGENESIS

In this section, we will describe only zygotic embryogenesis in dicotyledonous plants, and our example will be red-

Sexual Reproduction in Angiosperms

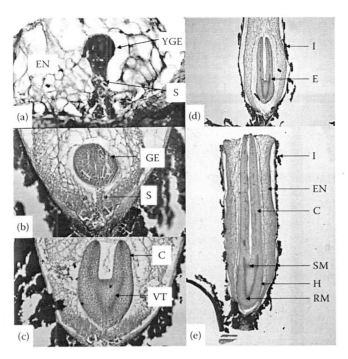

FIGURE 36.10 Embryo development in *Cercis canadensis* (redbud), a dicotyledonous plant. (a) A longitudinal section through a very young globular-stage embryo (YGE). Note the suspensor (S) and the cellular endosperm (EN). (b) Longitudinal section through a late globular-stage embryo (GE) with characteristic radial symmetry. S = suspensor. (c) A median longitudinal section through a bipolar, late heart-shaped embryo exhibiting bilateral symmetry. The shoot and root meristems (arrows) are not well differentiated at this stage. C = cotyledon; VT = vascular tissue development. (Figure 36.10b,c modified and adapted with permission from Jain, S.M., P.K. Gupta, and R.J. Newton (Eds.). 1999. *Somatic Embryogenesis in Woody Plants,* Kluwer Academic Publishers, Dordrecht, The Netherlands.) (d) A median longitudinal section through an early torpedo-stage embryo (E). Meristems, cotyledons, and vascular tissue are well differentiated. I = integuments. (e) A median longitudinal section through a cotyledonary-stage embryo with distinct cotyledons (C), a shoot meristem (SM) and a less conspicuous root meristem (RM). H = hypocotyl; I = integuments, and EN = endosperm. Compare to a mature monocotyledonous embryo shown in Figure 36.11a. (Figure 36.10e modified and adapted with permission from Trigiano, R.N. and D.J. Gray (Eds.). 2005. *Plant Biotechnology and Development.* CRC Press, Boca Raton, FL.)

bud (*Cercis canadensis* L.). Zygotic embryogenesis begins with the first mitotic division of the zygote to form the embryo proper (the cell that will form the embryo) and the suspensor. The cells of the suspensor continue to divide and push the developing embryo into the endosperm. The suspensor may help transfer nutrients and water to the developing embryo, but does not have any vascular tissue (xylem and phloem). The mature suspensor may be a long single file of cells (uniseriate) as in flowering dogwood (*Cornus florida* L.) or may have multiple files of cells (multiseriate) as in redbud (Figure 36.10a,b). The cell that becomes the embryo proper divides via mitosis many times and produces a more or less round (isodiametric) ball of cells with radial symmetry (Figure 36.10a,b). This is the globular stage of embryogenesis, and at this point there is very little morphological/physiological specialization evident. The protoderm, the cells on the perimeter of the embryo that will form the epidermis, may be distinguishable from other cells in this stage. With continued divisions and development, the distil portion of the globular embryo begins to flatten (Figure 36.10b), and the embryo develops two distinct poles or is bipolar (Figure 36.10c). The symmetry of the embryo changes from radial to bilateral. The radial symmetric embryo may be sectioned in half in any number of planes, and each section is the mirror image of the other. The bisymmetrical embryo can only be sectioned in one longitudinal plane to produce mirror halves. This phase of development is termed the cordate or heart-shaped stage. The shoot apical meristem or the area that will become the apical meristem begins to differentiate; the root apical meristem remains obscure and relatively undifferentiated. Cotyledons are initiated, and vascular tissue in the hypocotyl (below where the cotyledons join the long axis of the embryo) starts to form (Figure 36.10c). The embryo continues to grow through the torpedo stage (Figure 36.10d) and cotyledonary stage (Figure 36.10e). At this time, the ovule containing the mature embryo becomes a seed, and the integuments harden and form the seed coat. Further maturation usually includes accumulation of storage food materials and loss of moisture in preparation for dormancy.

Seeds may or may not be able to germinate immediately. Seeds often serve as overwintering or survival structures for many annual plants and, therefore, may remain dormant for extended periods of time. Dormancy is also found in seeds from perennial plants. Dormancy assures that seeds will not germinate until favorable conditions for growth are present. Many species require that specific conditions be met to break dormancy and encourage germination. For example in the laboratory, a prolonged (four months) cold–moist treatment of redbud and flowering dogwood seeds is essential for germination. One might easily equate this treatment to what the winter environment is like in soil. In fact, nursery producers growing these two woody species plant the seeds in the fall to ensure germination in the spring.

Remember that young ovules were located in the ovary of the flower. In many species, the ovary and/or the receptacle also have continued to develop and have now formed a conspicuous fruit. Seeds produced inside the fruit are characteristic of only the angiosperms—no other group of plants forms fruit. Fruit morphology and composition is extremely variable, ranging from the simple pod of legumes (redbud is a legume), to the aggregate fruit of a blackberry (*Rubus* spp.), to the kernel of corn (*Zea mays* L.). For additional information on fruit types, consult any number of introductory botany textbooks. Fruit, besides being good to eat, serves to protect and aid in dispersal of seeds.

SOMATIC EMBRYOGENESIS— A SPECIAL CASE

There is a special case of embryogenesis called somatic (nonzygotic or adventive) embryogenesis in plants that neither has reliance on fusion of gametes and formation of endosperm nor production of seeds and fruit (Table 36.1). It is a natural phenomenon, also called apospory, that occurs most notably in citrus seeds from the nucellus, which is sporophytic tissue. However, the process is relatively common in tissue culture (*in vitro*) of plants and is used to clone specific genotypes as well as to regenerate plants from single cells that have been modified using genetic engineering techniques (i.e., insertion of genes). Somatic embryos form entire plants much like zygotic embryos by first producing a root and then the shoot.

Somatic embryogenesis takes advantage of the totipotency of some plant cells. Totipotency is the word that describes the ability of a single cell to reproduce the entire organism and, in our case, the entire plant. Not all cells in the plant are naturally totipotent, but with certain treatments, such as exposure to specific growth regulators [synthetic auxins (2,4-D, dicamba, etc.) and cytokinins (e.g., benzyladenine)], some cells can be induced to a less differentiated state that is capable of forming somatic embryos. Somatic embryos are derived from single cells and generally follow the same ontological or developmental sequences that were discussed for their zygotic counterparts. Typically, somatic embryos are aberrant, which may be an expression of the plasticity of the genome. Probably some of these abnormalities are directly linked to environmental/nutritional/physiological differences between the petri dish and the confines of the ovule in which the zygotic embryo develops. However, when compared side by side, the morphology of zygotic embryos (Figure 36.11a,c) is remarkably similar to the corresponding somatic embryos (Figure 36.11b,c). Of course, somatic embryos do not have seed coats and have sometimes been called "naked embryos."

We have attempted to describe the general processes and structures involved in sexual reproduction and embryo development in angiosperms in this chapter. As the reader, you should be aware that there are many species-specific variations on the general themes explored here and that, although we have tried to be accurate, many details have been omitted. We sincerely hope that the preceding pages have whetted your appetite for a greater understanding not only of angiosperm sexual reproduction, but also for sexual reproduction in the other broad groupings of plants. Some of the gymnosperms (conifers etc.), ferns, and lower

TABLE 36.1
Comparison of Zygotic and Somatic Embryogenesis

Zygotic Embryo(genesis)	Somatic Embryo(genesis)
The zygote (2n) results from union or fusion of two haploid (n) cells, sperm and egg.	Formed from a single cell. This cell may be haploid or diploid.
Sexual recombination—genetic traits derived from both parents or, if selfed, rearrangement of parental traits. Results in increased genetic diversity.	Generally thought to be a clone (identical copy) of the original plant.
Endosperm (either 3n or 5n) is associated with the embryo and is the result of the fusion of a sperm with polar nuclei.	No endosperm associated with somatic embryo.
Follows a prescribed series of developmental events. Abnormalities are not common.	Generally follows the same developmental sequence as zygotic embryos, but may often manifest some abnormalities.
Embryo within seed coat (seeds) and seeds contained in fruit.	No seed coat (seed) or fruit—a "naked" embryo.

Sexual Reproduction in Angiosperms

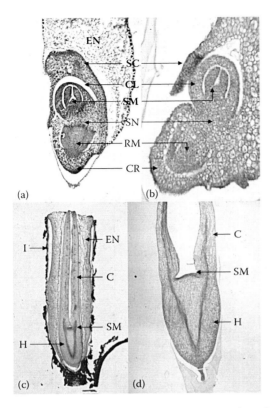

FIGURE 36.11 Comparison of zygotic and somatic embryogenesis in monocotyledonous and dicotyledonous plants. (a) Median longitudinal section through a mature zygotic embryo of *Dactylis glomerata* (orchardgrass) showing endosperm (EN), the scutellum (SC) or cotyledon coleoptile (CL), shoot meristem (SM), the scutellar notch (node) (SN), root meristem (RM), and coleorhizza (CR). Compare to Figure 36.11b. (Modified and reprinted with permission from McDaniel, J.K., B.V. Conger, and E.T. Graham. 1982. *Protoplasma* 110: 121–128.) (b) Median longitudinal section through a morphologically mature somatic embryo of *Dactylis glomerata* (orchardgrass). The somatic embryo is morphologically very similar to its zygotic counterpart shown in Figure 36.11a, except for the lack of endosperm. (d) and (d) Median longitudinal sections through a morphologically mature zygotic embryo (c) and a morphologically mature somatic embryo (d) of *Cercis canadensis*. The two embryos are very similar except that the hypocotyl (H) of the somatic embryo is swollen, and the meristem of the somatic embryo is malformed. The somatic embryo also lacks endosperm (EN). (Figure 36.11d modified and adapted from Jain, S.M., P.K. Gupta, and R.J. Newton (Eds.) 1999. *Somatic Embryogenesis in Woody Plants*, Kluwer Academic Publishers, Dordrecht, The Netherlands. With permission.)

plants complete sexual reproduction in unique and interesting ways. There are many good introductory botany texts that provide fantastic information on this subject, and if you have the opportunity, we encourage you to further your formal education by enrolling in additional horticultural and botany courses, such as plant morphology/survey and anatomy courses.

LITERATURE CITED

McDaniel, J.K., B.V. Conger, and E.T. Graham. 1982. A histological study of tissue proliferation, embryogenesis, and organogenesis from tissue cultures of *Dactylis glomerata* L. *Protoplasma*. 110: 121–128.

Trigiano, R.N., L.G. Buckley, and S.A. Merkle. 1999. Somatic embryogenesis in woody legumes, in *Somatic Embryogenesis in Woody Plants*, vol. 4. Jain, S.M., P.K. Gupta, and R.J. Newton (Eds). Kluwer Academic Publishers, Dordrecht, The Netherlands, pp. 189–208.

Trigiano, R.N. and D.J. Gray. 2005. A brief introduction to plant anatomy, in *Plant Biotechnology and Development*. Trigiano, R.N. and D.J. Gray (Eds.) CRC Press, Boca Raton, FL, pp. 73–85.

Wilson, C.L., W.E. Loomis, and T.A. Steeves. 1971. *Botany*, 5th ed. Holt, Rinehart and Winston, New York. 752 pp.

OTHER EXCELLENT SOURCES OF INFORMATION

Hartmann, H.T., D.E. Kester, F.T. Davies, Jr., and R.L. Geneve (Eds.) 2002. *Plant Propagation: Principles and Practices*, 7th ed. Prentice Hall, Inc. Upper Saddle River, NJ. 880 pp.

Toogood, A. (Ed.) 1999. *The American Horticultural Society Plant Propagation*. DK Publishing, New York. 320 pp.

37 Breeding Horticultural Plants

Timothy A. Rinehart and Sandra M. Reed

CHAPTER 37 CONCEPTS

- Plant breeding involves selection of plants with combinations of improved traits that are inherited in a predictable manner.

- Collecting, understanding, and incorporating genetic variation into a horticultural breeding program are critical to success.

- Clearly defined goals help plant breeders choose an appropriate plant breeding strategy to avoid problems, such as inbreeding depression.

- Forced hybridization between species or genera often requires laboratory methods, such as embryo rescue, to assist the growth of hybrid progeny.

- Manipulating chromosome numbers is an effective strategy to change plant morphology and/or overcome fertility barriers.

- Mutation breeding and biotechnology can be used to breed new cultivars from clonal material without sexual reproduction.

Modern genetics originated when Gregor Mendel (1822–1884) discovered that hereditary traits are determined by elementary units transmitted between generations in a uniform and predictable fashion. Each genetic unit is called a gene and must adhere to two principles. First, it must be inherited between generations such that each descendent has a physical copy of the material. Second, it must provide information regarding structure, function, or other biological attribute. Thus, there are two ways to think about genes: one is to identify the specific genetic material, or DNA, that is transferred between generations (genotype), and the other is to focus on the way that a specific piece of DNA manifests itself (phenotype), or the biological characters that are inherited. Both concepts are critical to successful breeding of horticultural plants, and in this chapter we will demonstrate how these lines of investigation are linked.

Mendel's experiments with plant hybrids effectively disproved the notion that heredity was a blending process and the dilution of parental characteristics. He also cast serious doubt on the idea of pangenesis, that physical traits acquired during a parent's lifetime could be inherited in their progeny. Several biologists performed experiments with plant hybrids, comparing similarities and dissimilarities between offspring of different generations and parental combinations. Kolreuter (1733–1806), Gartner (1772–1850), Naudin (1815–1899), and Darwin (1809–1882) each observed that hybridization often produces progeny with considerable diversity compared to parental stock. However, it was Mendel who paid attention to the numerical ratio in which the different parental characteristics appeared in offspring. It was Mendel who demonstrated that the appearance of traits in offspring followed specific mathematical rules, which he deduced by counting the diverse kinds of offspring produced from a particular set of crosses. Although Mendel published the basic groundwork for modern genetics in 1866, his experiments were not recognized until the 1900s after it was shown that chromosomes are the hereditary material passed from one generation to the next (Olby, 1966).

Modern genetic technology can clearly visualize chromosomes and has confirmed that DNA is inherited as discrete, physical units. Statistical tools predict the behavior of DNA from one generation to the next, and gene expression assays can confirm the relationship between genotype and phenotype. However, the basic principles of inheritance remain unchanged from Mendel's day. Plant breeding relies heavily on identifying plants that exhibit desirable traits from among the morphological and physiological variation that exists in natural populations. These plants are used as parents to produce seedlings, some of which inherit the desirable traits. Individuals with improved combinations of traits are selected by plant breeders and become the parents for the next genera-

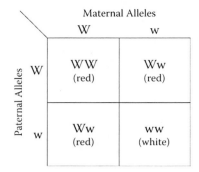

FIGURE 37.1 Punnett square depicts the possible allele combinations found in the progeny of two parents that are heterozygous for a gene controlling flower color. As shown in the diagram, 75% of the offspring are expected to produce red flowers because the homozygous (WW) and heterozygous (Ww) allele combinations produce the red phenotype. The remaining 25% are expected to produce white flowers because they contain two copies of the recessive (w) allele.

	Possible Maternal Allele Combinations			
	DW	Dw	dW	dw
DW	DDWW (tall) (red)	DDWw (tall) (pink)	DdWW (tall) (red)	DdWw (tall) (pink)
Dw	DDWw (tall) (pink)	DDww (tall) (white)	DdWw (tall) (pink)	Ddww (tall) (white)
dW	DdWW (tall) (red)	DdWw (tall) (pink)	ddWW (dwarf) (red)	ddWw (dwarf) (pink)
dw	DdWw (tall) (pink)	Ddww (tall) (white)	ddWw (dwarf) (pink)	ddww (dwarf) (white)

(Possible Paternal Allele Combinations along the rows)

FIGURE 37.2 Punnett square depicts the possible allele combinations in progeny from two parents that are heterozygous for two different genes. In this case, flower color is controlled by a gene exhibiting incomplete dominance, with heterozygous (Ww) progeny producing intermediate, or pink, flowers. The gene for plant height is completely dominant, and heterozygous progeny (Dd) are indistinguishable from progeny homozygous for the dominant trait (DD). As diagrammed, a plant breeder would expect to observe white-flowered dwarf plants in 1/16 of the progeny, or 6.25% of the population. The percentage of plants expected to be homozygous recessive for all alleles is $(1 \div 4^n) \times 100$, where n = number of gene pairs.

tion, until a superior plant is chosen for mass production. Selecting plants with improved traits is a central theme of modern plant breeding, and we will discuss breeding strategies later in this chapter. We will also touch on how biotechnology can accelerate the evaluation and selection process, but first we need to explore the underlying theory of genetic inheritance.

We already know that sexual reproduction is intrinsic to the transfer of genes from one generation to the next. We also covered mitosis, meiosis, and possible differences in chromosome numbers (ploidy) in the previous chapter (chapter 36). In order to discuss Mendel's Laws of Genetics, we will assume that the plants in each example are diploid, which means they contain two copies of each gene. The paternal copy of the gene came from the pollen and the maternal copy from the egg. When fused, they reconstitute the diploid genome in the progeny. This splitting and uniting of genetic material during sexual reproduction is the basis for segregation, which is Mendel's First Law of Genetics. Simply put, the mother and father plants contain two copies of each gene, called alleles, but only contribute one copy to each progeny. This is because each member of an allelic pair separates, or segregates, from the other during meiosis and gamete formation. The unique combination of one allele from your mother (maternal contribution) and one from your father (paternal contribution) is a significant source of genetic variation.

We can visualize the segregation of alleles using some examples. In their simplest form, alleles are either dominant or recessive. Dominant alleles (denoted by uppercase letters) can be observed in the plant even when there is only one copy. Recessive alleles are only observed in the phenotype when the dominant allele is not present. In a diploid plant, there must be two copies of the recessive allele to see the recessive phenotype. Figure 37.1 is a Punnett square depicting all the possible allele combinations of a gene for flower color when you cross two heterozygous parents. Segregation of genotypes is different than the segregation of phenotypes because of the dominant and recessive allele combinations. You can follow this gene transfer forward an infinite number of generations, with each generation segregating for the maternal or paternal contribution of the previous parents.

The second predictable source of genetic variation is Mendel's Law of Independent Assortment. This law states that the segregation of one gene is independent of the segregation of a second gene. It can also be visualized by Punnett square (Figure 37.2). As more desirable traits are selected, the number of progeny to be evaluated increases due to the smaller ratio of progeny expected to contain rare combinations of alleles.

There are exceptions to independent assortment. Genes that are physically located next to each other on a chromosome may experience linkage disequilibrium. The close physical connection between genes reduces the likelihood of recombination during meiosis and increases the probability that both paternal or both maternal alleles will be found in individual progeny. Genes on different chromosomes are unlinked. Linkage distorts the ratio between observed and expected number of progeny with a particular allele combination. The more tightly linked the genes, the more the ratio is skewed from the expected 50% probability that two genes in any given individual hybrid came from different parents. Genetic linkage maps are based on the recombination frequencies observed between genes on the same chromosome and reveal the order of genes along a chromosome. Linkage between genes often results in unwanted or unrelated traits being linked to desirable traits. Breeding strategies can break the linkage between genes and/or make use of the linkage between genes to indirectly select for traits.

All of the genetic traits that we have talked about so far have been qualitative. That is, the phenotype is controlled by a single or few genes, each usually having a dominant and recessive phenotype. We have also explored the mathematical repercussions of selecting for more than one qualitative trait at a time. However, some traits are controlled by multiple genes, each contributing unequally to the observed phenotype. Thus, the addition or subtraction of a particular allele does not have the dramatic, qualitative effect seen in previous examples. Instead, allele combinations display incremental, or quantitative, differences. Quantitative phenotypes are often measured on a continuous scale where progeny display a distribution of phenotypes based on the allele combinations present in the population. The frequency of a particular phenotype depends on the number of genes involved and the allele combinations present in the parental genomes (Figure 37.3). Breeding strategies for quantitative traits generally focus on increasing the frequency of favorable alleles and moving the distribution of phenotypes toward one end of the curve.

Selecting improved plants, particularly for traits having aesthetic value, is both a science and an art. Natural selection occurs over evolutionary time and reflects the effects of pressure from environmental conditions and competition for resources on plant reproduction. Natural selection can be disruptive, stabilizing, or directed, depending on the trait. Selection during plant breeding, however, is directional and acts on only a few traits at a time. Directed selection is accelerated during plant breeding by applying selection over multiple generations, typically with much smaller populations than are found in the wild, and by forcing the hybridization of selected plants. The results are plants with exaggerated traits in combinations that would not typically be selected for in natural environments. Some traits, such as increased yield or number of flowers, are easily quantified, and improvement, no matter how incremental, can be objectively measured. Other traits are not easily quantified. For example, insect resistance, disease resistance, and drought tolerance may be hard to objectively measure because they involve environmental interactions or require the fortuitous occurrence of disease or insects.

Some traits are selected based entirely on aesthetics. Flower and fruit color and shape, plant growth habit, bark characteristics, and many other horticultural traits have aesthetic components based on consumer appeal. An "eye

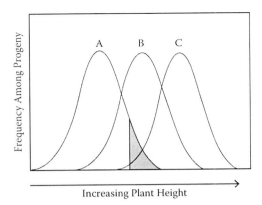

FIGURE 37.3 Graph of plant height from three seedling populations created from three different sets of parents. The distribution of plant heights from populations A, B, and C overlap, but the mean plant height for population C is larger than that of A or B, presumably due to increased number of favorable alleles affecting plant height. Breeding for increased plant height with population A would be inefficient even if the tallest plants on the most extreme right side of the curve (shaded area) were selected as parents for subsequent generations.

for plants" is often considered one of the most important qualities of a horticultural plant breeder, since breeding necessarily produces hundreds to thousands of seedlings, out of which only a few are expected to become new cultivars. It is not surprising that commercial plant breeding efforts often include test marketing. Some successful ornamental plant introductions are based almost entirely on novelty rather than production value. Historically, the emphasis on consumer appeal has tainted the accomplishments of some plant breeders, such as Luther Burbank (1849–1926), who was often accused of exaggerating his claims. Modern plant breeders rely less on intuition and more on science, but a keen sense of observation will always be critical for successful breeding.

Clear objectives, whether based on production or consumer interests, should be defined before initiating a breeding program because unfocused breeding can result in an exponential increase in the numbers of plants. Crossing every conceivable plant and evaluating the tens or hundreds of thousands of progeny produced year after year is simply not physically possible or financially feasible. Luther Burbank once wrote, "Plant breeding to be successful must be conducted like architecture. Definite plans must be carefully laid for the proposed creation; suitable materials selected with judgment, and these must be securely placed in their proper order and position. No occupation requires more accuracy, foresight and skill than does scientific plant breeding."

Germplasm resources play an important role in plant breeding. Without genetic variation there is little chance that traditional breeding approaches will improve traits. The "gene pool," or total number and variety of genes and alleles in a sexually reproducing population, determines what alleles are available for transmission to the next generation. Natural populations are the most basic source of germplasm, but horticultural breeding often includes previously selected plants and commercially available cultivars. These cultivars often contain allele combinations not predicted by Hardy-Weinberg equilibrium equations, which are used to estimate the frequency of alleles in natural populations (Weinberg, 1908). Hardy-Weinberg is a mathematical model for the gene pool and is based on the following five factors: population size, migration between populations, mutation within a population, differences in fertility, and mating systems for a specific crop. We can calculate theoretical allele frequencies using Hardy-Weinberg equations, but for our purposes, it is sufficient to understand that allele frequencies are generally balanced for large, random mating populations in nature where the effects of mutation, migration, and selection are minimal. In the altered framework of plant breeding, however, the directed evolution of small breeding populations is mainly due to the parents, the type of cross, and the artificial selection of desired traits. Therefore, careful consideration must be given to the selection of parents because it has a profound effect on the results. Crossing and selection schemes cannot be successful if the desired alleles are not present in the first place.

Two types of variation are observed when evaluating germplasm. Environmental variation is based on nutrients, moisture, light, temperature, and other factors and can be seen even among plants with identical genotypes, or individuals that have been clonally propagated. The expression of some genes is more affected by environment than others, but typically plant populations are heterogeneous with regard to plant height, vigor, and flowering, as well as their response to abiotic and biotic stress. Environmental conditions may even create situations where inferior genotypes outperform superior genotypes. This can be demonstrated by growing two different cultivars under the same conditions, but only applying fertilizer to one cultivar. The cultivar with fertilization may outperform the unfertilized cultivar even if it is known to be inferior under identical environmental conditions. To offset the confounding effects of environment, plants are evaluated under uniform conditions that approximate production or landscape use and are evaluated in multiple locations and/or over several years. Factors such as irrigation, fertilization, and photoperiod are often manipulated during plant evaluation to simulate real-world conditions. Environmental variation is not heritable, but it can significantly impact the selection of heritable traits.

Genetic variation is heritable and can be detected at the molecular level using biotechnology. Plant breeding is designed to take advantage of genetic variation, but in order to do so it must first be separated from environmental variation. Statistical tools were developed to estimate the relative importance of genotypic (G) and environmental (E) effects on phenotype (P), or $P = G + E$. Plant breeders measure the environmental effects that cause phenotypes to deviate from expected genotypes using a statistical tool called variance. Phenotypic variance (pe), genotypic variance (ge), and environmental variance (ee) all play a role, especially in analyzing quantitative traits where genotypic variance is expected to be significantly larger than with qualitative traits that have essentially no genotypic variance. The ratio of genotypic to phenotypic variance, called heritability, represents the proportion of trait variation that is actually due to DNA. A plant breeder's ability to make progress increases as this ratio increases.

There is one more critical factor that needs attention before discussing breeding strategies. In natural populations, plants are either self-pollinating or cross-pollinating. While these naturally occurring reproductive systems sometimes can be artificially changed (self-pollinating species can be forced to outbreed and vice versa), the underlying genetic structure that these reproductive systems impose cannot be changed. This "genetic architecture," along with dominance of traits, gene interactions, and heritability, plays an important role in the decision to use a breeding strategy. For example, some breeding strategies include the creation of pure lines, which are the products of prolonged self-fertilization, the most extreme form of inbreeding. With each generation of self-fertilization, heterozygosity decreases by 50%. The coefficient of inbreeding (F), which can be plugged into the Hardy-Weinberg equilibrium equations, suggests that plants approach 100% homozygosity in just eight generations of self-fertilization. Full-sib and half-sib mating schemes also significantly increase homozygosity over time (Figure 37.4). Thus, crops that are self-incompatible are not good candidates for breeding strategies that require pure lines, and crops that outbreed in nature may suffer from inbreeding depression when forced to self-fertilize. Inbreeding depression varies among outcrossing species, but in its extreme form it can be lethal after just a few generations. In other outcrossing species the consequences of inbreeding is observed as reduced vigor, typically most pronounced in the first four generations. This is due to increasing homozygosity of deleterious recessive alleles whose effects are mitigated when heterozygous. For plants that regularly self-fertilize in the wild, inbreeding can be desirable because it increases the genetic diversity among individuals in a population, theoretically making selection of desirable traits easier.

BREEDING SELF-POLLINATED PLANTS

While several breeding strategies have been developed for self-pollinated plants, all involve the development of pure, or highly homozygous, lines. Most begin with controlled pollinations between two highly homozygous plants, followed by self-pollination to produce the F_2 population.

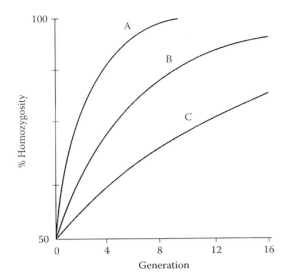

FIGURE 37.4 Effects of inbreeding on the reduction of heterozygosity. A is self-fertilization. B is full-sib mating, and C is half-sib mating. Full-sib mating is the hybridization between progeny from a single cross (siblings), while half-sib mating is the crossing of plants that only share one parent. Over time, all three crossing schemes result in high levels of homozygosity.

Parents are often chosen because they have complementary traits. For example, if the goal of a breeding program is to develop a high-yielding cultivar that matures early, an early-season, low-yielding cultivar may be hybridized to a cultivar that matures later but has high yields. Alternatively, parents that possess the same desirable trait may be hybridized if it is thought that they possess different genes for the trait of interest. As an example, two cultivars from diverse genetic backgrounds, both of which mature fairly early in the season, may be hybridized in an effort to breed for even earlier maturity.

In the classical pedigree method, superior F_2 plants are selected and self-pollinated. From the F_3 to F_5 or F_6 generation, both individual plant performance and performance of families (e.g., all F_3 plants derived from a single F_2 plant) are evaluated. Superior plants from the best-performing families are advanced to the next generation. In early generations, individual plant performance plays a greater role than family performance in determining which plants are self-pollinated and advanced to the next generation. As heterozygosity declines, selection is based more on family than on single plant performance. Following the F_5 or F_6 generation, selection is based solely on family performance. Because the pedigree method is extremely labor-intensive—requiring detailed recordkeeping and extensive evaluations of families in early generations when heterozygosity is high and heritability low—it has mostly been replaced by other breeding methods.

Most of today's breeding schemes for self-pollinated plants are based on the bulk population method. In this breeding scheme, selection in early generations (F_2 to F_5 or F_6) is based solely on single plant performance. Since superior plants are self-pollinated and their seed bulked to produce the next generation, no progeny rows or pedigree records are required in early generations. Selection among family lines is delayed until homozygosity, and therefore heritability, is high. Various bulk population methods have been proposed, most of which differ in terms of number of plants grown and stringency of selection in the early generations.

In the single-seed-descent breeding scheme, no selection is practiced during early generations. Only a single selfed-seed of each individual plant is carried forward to the next generation. Because only a single seed is required and no selection is practiced, during early generations plants can be grown under less than ideal conditions (e.g., on a greenhouse bench), allowing two or even three generations to be grown in a year. Advanced generations produced by the single-seed-descent method are expected to be a general representation of a random sample of the pure lines that any given F_2 hybrid is capable of producing. This increases the chances of a superior, true-breeding individual being present in advanced generations. In contrast, the sample of pure lines developed from bulk population methods will usually be a selected sample, and thus a superior one, assuming that early plant selection was effective.

Backcross breeding is a modified form of inbreeding and is generally applied to situations where existing cultivars are only deficient in one or two traits. Backcross breeding involves repeated crossing back to one original parent, called the recurrent parent, to produce progeny containing the desirable traits. When the trait of interest is controlled by a recessive gene, each backcross will be followed by self-pollination. The recurrent parent should not contain other deleterious traits because the resulting progeny will likely also contain those inferior traits. The parent containing the trait missing from the recurrent parent is called the donor parent. Traits that are qualitative, dominant, and produce a readily observed phenotype are easiest to introgress from the donor parent. Transferring quantitative traits through backcross breeding is usually not productive. Depending on the crop and trait, backcross breeding usually takes at least five generations because the proportion of donor genes is reduced by 50% with each backcross generation.

Because pure lines of self-pollinated crops are true breeding, seed from a pure line cultivar can be saved and used to generate the next season's crop. For seed companies, which have often invested considerable resources into developing a cultivar, it is not advantageous to develop and market pure lines. Instead, F_1 hybrids are increasingly being developed in self-pollinated crops. In addition to providing a continual revenue stream to the developer of the cultivar, F_1 hybrids provide a convenient way of combining multiple desirable traits, such as resistance to a large number of diseases, into a single cultivar.

BREEDING CROSS-POLLINATED CROPS

The oldest method of breeding cross-pollinated crops is mass selection. The performance of individual plants within a heterogeneous population is used to select those plants whose progeny will constitute the next generation. Generation after generation, seed is collected from only those plants having the desired trait(s). Assuming that selection is effective, the end result of mass selection is an increase in the frequency of desirable genes in the population. The major drawbacks of this method are that there is no control, or testing, of the plant that served as pollen donor and that it is only effective for traits that have high heritability. A variation of phenotypic mass selection that was developed with corn is called ear-to-row selection. Ears from selected parents are planted in separate rows and only those progeny rows that are visually superior are continued in the breeding program. Adding progeny testing to phenotypic selection improves the effectiveness of plant breeding efforts, but generally only for traits with a high heritability. For many cross-pollinated crops, hybrid varieties have been found to be superior to those developed through mass selection.

Hybrid varieties are produced by hybridizing two or more inbred lines. As mentioned earlier, inbreeding in cross-pollinated crops is typically deleterious due to inbreeding depression. However, inbreeding depression can often be reversed through the effects of heterosis, otherwise known as hybrid vigor. In these cases, elite breeding lines can display the ill effects of inbreeding depression as long as the final hybridization prior to cultivar selection is made between lines with distinct genetic backgrounds. In addition to uniting the horticultural traits found in the elite breeding lines, mixing the genetic backgrounds can restore vigor and produce offspring that exhibit improvement in traits beyond what either parent displays. Hybrid vigor is presumably due to the new combination of hundreds of unselected genes with small effects. The few genes responsible for the inbreeding phenotype are reshuffled when the genetic backgrounds are mixed.

Several recurrent selection schemes have been developed for cross-pollinated plants and are used to concentrate genes for quantitative characters without incurring a marked loss in genetic variability. All recurrent selection schemes involve repeated (or recurrent) cycles of (1) self-pollinating plants from a heterozygous source, (2) saving self-seed from only those plants that are judged to be superior, and (3) intercrossing the self-progeny from the previous cycle. In simple recurrent selection, selection is based on phenotype of individual plants or their selfed progeny. The other three types of recurrent selection—recurrent selection for general combining ability (GCA), recurrent selection for specific combining ability (SCA), and reciprocal recurrent selection—differ from simple recurrent selection in that they all require some type of test cross to measure combining ability. Combining ability is the measure a plant's potential to transfer traits. GCA refers to the additive actions of genes when considering the behavior of one parent in a series of hybrid combinations. Additive effects are subject to selection and, therefore, GCA is a good measure of whether a particular parent will produce improved progeny when crossed to any other parents (Jenkins, 1940). SCA refers to specific hybrid combinations that produce better or poorer progeny than expected based on the GCA. Thus, SCA is a good measure of nonadditive genetic effects (Hull, 1945). Reciprocal recurrent selection provides for simultaneous selection for both GCA and SCA. Backcross breeding, as described for self-pollinated crops, is sometimes used for improving a specific trait in an elite inbred line. For more details related to the different recurrent selection schemes, see Allard (1999) or another plant breeding text.

BREEDING CLONALLY PROPAGATED PLANTS

Cultivars of many horticultural species—such as fruit (grapes, berries, fruit trees), ornamental (annuals, perennials, woody trees and shrubs), and root (potato, sweet potato) crops—are reproduced using vegetative propagation. Because nearly all clonally propagated species are perennial outcrossers in nature, they are highly heterozygous. Breeding these crops usually involves hybridizing superior clones to produce populations of genetically variable individuals. Superior plants within the populations may be (1) selected and propagated for replicated evaluations; (2) used to produce F_2 progeny via self, full-, or half-sib pollinations (especially important if dealing with recessive traits); or (3) hybridized to superior plants from other populations. Because selections made early in the breeding program are based on single-plant observations, they are most effective for traits with high heritability. Later in the breeding cycle, when sufficient numbers of clones have been obtained through vegetative propagation of the superior plants, selection for traits of lower heritability becomes more effective.

Breeding clonally propagated plants is in some ways easier and faster than for those crops typically produced from seed. Often only one or two generations are required to produce populations showing the maximum genetic variability, and once a superior plant is selected, its genetic makeup becomes immediately "fixed" via vegetative reproduction. However, many clonally produced crops have long life cycles, and it may not be possible to select for some of the most important traits until the plants have reached a mature stage. Also, many clonally propagated species are polyploid, which complicates breeding and genetic studies.

OTHER BREEDING CONSIDERATIONS

Linkage

Getting rid of unwanted genetic baggage requires large populations. Genes for desirable traits are sometimes physically located next to genes responsible for undesirable phenotypes. This linkage means that selecting for the desired trait has a high probability of dragging along an undesirable gene. Recombination events can separate tightly linked genes, but the frequency of recombination is proportional to the distance between the genes. The closer the physical association, the less likely they are to be split apart during meiosis. Genetic linkage maps are based on the empirically determined rate at which meiotic recombination separates genes and describe the order and distance between genes on a chromosome. Breaking an undesirable genetic linkage is influenced by the genetic distance between genes and chosen breeding strategy. Recombination between tightly linked genes is rare and proportional larger populations must be generated and screened to identify plants with separation between undesirable linkage groups. The chances of breaking genetic linkage can be increased by backcross breeding. The repeated use of the recurrent parent generates a new population each cycle that may contain individuals where tightly linked genes from the donor parent are separated.

The good news about genetic linkage is that indirect selection can be used to select for traits that are difficult to evaluate. For example, a particular type of insect resistance may be tightly linked to a unique leaf shape due to proximity of the genes despite the fact that there is no physiological or mechanistic association between the two traits. Direct screening for insect resistance may be very difficult, either in getting the insects or analyzing the results of feeding. On the other hand, leaf shape is easily evaluated, and hybrids showing the correct leaf shape will have a higher probability of having the desired insect resistance. These linkages are rare but easily identified when recording the segregation of multiple traits in a single population. Gene interactions are more readily uncovered in crops that have genetic linkage maps.

Non-Mendelian Inheritance

Extrachromosomal inheritance may affect selection of some traits. Nuclear DNA is evenly contributed by maternal and paternal sources at fertilization because meiosis evenly divides genomes during gametogenesis. However, only the egg contains cytoplasm, so plastid DNA from chloroplasts and mitochondria is contributed exclusively from maternal sources. This maternal bias influences some traits, particularly those that involve chlorophyll, such as variegation. Most horticultural breeding strategies mitigate possible maternal effects by including reciprocal crossing schemes. Phenotype differences between reciprocal populations are evidence of maternal inheritance.

Other factors affecting inheritance include transposable elements which readily occur in plants. Transposable elements, also known as jumping genes, are small, mobile regions of DNA that can increase the rate of mutation during plant breeding by jumping into genes (McClintock, 1950). Some transposable elements can also jump back out of genes causing them to revert to their wild-type state, or nonmutated phenotype. Phenotypes affected by transposable element movement are not inherited in a Mendelian fashion, and traits are often unstable during vegetative propagation. Traits displaying non-Mendelian inheritance and instability may also be under the influence of other epigenetic phenomenon, such as chromatin remodeling, paramutation, imprinting, position effects, or sex-chromosome inactivation (for review, see Chandler and Stam, 2004).

Wide Hybridization, Incompatibility, and Polyploidization

Intergeneric or interspecific hybrids are important tools in breeding strategy, since they broaden the genetic base, introduce new genes and gene combinations, and can bridge gaps between species not sexually compatible. Backcross breeding is often used to transfer desirable traits from one species into another after a wide hybrid is successfully produced. Fertility issues in wide hybrids often stem from ploidy differences because the chromosome number can vary between different genera, species, and sometimes even between subspecies or landraces. Differences in chromosome number do not necessarily prevent fertilization, but zygotes are often aborted or show abnormal growth as the two different sets of haploid chromosome attempt to establish new gene expression patterns. Wide hybrids often display unstable phenotypes. Even if embryos grow into flowering plants, they may not produce viable gametes if chromosomes are unable to recognize and pair with their diploid partners during meiosis. Sterile hybrids are a breeding dead end.

One method used to recover fertility in wide hybrids is the deliberate doubling of chromosome number, thus providing a diploid partner for every chromosome before gamete formation. Mitotic inhibitors, such colchicine and oryzalin, are known to arrest the mitotic cell cycle after DNA is replicated, but before the cell divides (Nakasone and Kamemoto, 1961). Plant tissues treated with these chemicals contain higher proportions of cells with doubled genomes, and some of these cells may go on to form sexual reproduction structures and gametes. Chromosome manipulation is also an effective strategy to change plant phenotype without sexual reproduction. Tetraploid plants generally display shorter internode lengths, larger flowers, increased leaf and stem thickness, and other exaggerated morphology when compared with diploid starting material. Because they are clonally produced, ploidy manipulation does not usually affect flower color, disease or insect resistance, or other environmental adaptations present in the starting material.

Incompatibility is often encountered when making wide hybridizations. Prezygotic incompatibility is associated with mechanisms that keep sperm and egg from ever fusing. Chemicals secreted on the stigma may prevent germination of pollen grains once deposited. Pollen tube growth may be inhibited before reaching the ovule. Even if the pollen cells reach the egg, the embryo sac may deteriorate before fertilization can occur. Late-acting, or postzygotic, incompatibility includes embryo lethality, halted growth, and aberrant development. Tissue culture techniques, such as embryo rescue, have been widely used to overcome postzygotic incompatibility.

It is worth noting that the spontaneous generation of unreduced gametes, usually pollen containing the diploid complement of genes rather than the haploid number expected, occurs naturally in some horticultural crops. Plants produced by fertilization with an unreduced gamete are usually triploid, sometimes tetraploid if both gametes are unreduced. Triploids are generally sterile but not always. While sterility in breeding lines is ruinous, sterility in released cultivars can be beneficial. Terminal crosses between diploid and tetraploid breeding lines to produce sterile triploid forms is one strategy to overcome plant invasiveness. Triploidy has also been used as a way of developing seedless fruit crops.

MUTAGENESIS BREEDING

Mutagenesis breeding is another strategy that does not necessarily involve sexual reproduction. Spontaneous mutant phenotypes, called sport mutations, generally retain all the characteristics of the original germplasm except for a single trait. For example, four variegated *H. macrophylla* cultivars are all naturally occurring sport mutations of a very old, nonvariegated cultivar (Reed and Rinehart, 2007). While these *H. macrophylla* sports are easily distinguished by their different patterns of variegation, they are indistinguishable at the genetic level. Mutations can be induced. Ethyl methane sulfonate (EMS) is a chemical mutagen that produces single-nucleotide changes, typically affecting only a few genes in the exposed plant. Conversely, mutations caused by x-ray and gamma ray sources can cause chromosomal rearrangements, breaks, and aneuploidy (loss of a chromosome) as well as single-nucleotide changes. Because these events are random, mutation breeding typically involves screening large numbers of plants for novel phenotypes. Seed or clonal tissue may be mutagenized, and the process is relatively straightforward. If the crop is clonally propagated, any stable sport mutant is a potential new cultivar. Mutation breeding often starts with a proven cultivar that shows superior ornamental features in an effort to expand the range of flower or leaf color and/or shape, variegation, and sometimes environmental tolerance. Improvements in disease and insect resistance or other complex horticultural traits are generally not found in mutation populations.

BIOTECHNOLOGY

Molecular markers, which are generally random regions of DNA that show variation, can also be indirectly associated with phenotypic traits. Selection for plants containing molecular markers linked to desirable traits has a probability of containing the gene for the desired trait that is proportional to the genetic distance between the two. When using multiple, tightly linked molecular markers, plant breeders can indirectly select for genes with relative confidence. While this type of biotechnology, called marker-assisted selection (MAS), is commonly associated with agronomic crops, it is becoming more feasible to use DNA-based tools in plant breeding of horticultural crops. MAS is typically associated with breeding for disease and insect resistance, especially where multiple forms of resistance to a single pest need to be incorporated into a single genotype (Dudley, 1993). Without indirect selection, one gene for pest resistance would hide the effects of additional pest resistance genes making it impossible to select for both genes in a single population. Using molecular markers, multiple genes can be indirectly selected regardless of phenotype. In this manner, agronomic crops, such as barley can contain multiple genes for disease resistance and are less likely to succumb to new pathogen variants when they arise (Hayes et al., 2001; Castro et al., 2003).

Molecular markers have been extensively analyzed for association with quantitative traits, or traits controlled by multiple genes, and those that show linkage are called quantitative trait loci (QTL). QTL markers can be genetically mapped based on the degree of cosegregation with the desired phenotype and are often associated with the percentage that they contribute to the desired phenotype (Lynch and Walsh, 1998). This detailed information allows plant breeders to select for genes that represent minor contributions to a quantitative trait. The impact of these genes would not be observed in traditional plant evaluations, but when assembled together in the final cultivar they can contribute significantly to the final phenotype. For example, a gene that adds 10% more height might not be selected for in a population that is 50% shorter than the rest, but it might still boost plant height 10% when combined with the tallest phenotypes. QTL markers are often associated with agronomic traits such as increased yield, but could also be applied to horticultural plant breeding of high-value crops.

Biotechnology has opened new doors in plant breeding, such as gene transfer and the creation of genetically modified organisms. Gene transfer requires tissue culture or plant regeneration systems, which have not been developed for all horticultural crops. Foreign DNA with proven usefulness must also be available. A few successful examples of genetically modified horticultural crops exist, including blue roses developed by Florigene, Inc. Given the legal complications associated with genetically modified organisms and the biotechnology used to create them, it remains to be seen whether these methods will show widespread benefit. Suffice to say that the traditional horticultural breeding methods described in this chapter will remain relevant and useful in the near future, as they have for the past 100 years.

LITERATURE CITED

Allard, R.W. 1999. *Principles of Plant Breeding,* 2nd ed. John Wiley & Sons, New York.

Castro, A., X. Chen, P. Hayes, and M. Johnston. 2003. Pyramiding quantitative trait locus (QTL) alleles determining resistance to barley stripe rust. *Crop Science.* 43: 651–659.

Chandler V.L. and M. Stam. 2004. Chromatin conversations: Mechanisms and implications of paramutation. *Nat. Rev. Genet.* 5: 532–544.

Dudley, S.W. 1993. Molecular markers in plant improvement: Manipulation of genes affecting quantitative traits. *Crop Science.* 33: 660–668.

Hayes, P., A. Castro, A. Corey, T. Filichkin, C. Rossi, J. Sandoval, M. Vales, H. Vivar, and J. von Zitzewitz. 2001. Collaborative stripe rust resistance gene mapping and deployment efforts, in *Breeding Barley in the New Millennium: Proceeding of an International Symposium,* Vivar H.E. and A. McNab (Eds.) Ciudad Obregon, Sonora, Mexico, D.F. CIMMYT. pp. 47–60.

Hull, H. 1945. Recurrent selection for specific combining ability. *J. Am. Soc. Agron.* 37: 134–145.

Jenkins, M. 1940. The segregation of genes affecting yield of grain in maize. *J. Am. Soc. Agron.* 32: 56–63.

Lynch, M. and B. Walsh. 1998. *Genetics and Analysis of Quantitative Traits.* Sinaue Assoc., Inc., Sunderland, MA.

McClintock, B. 1950. The origin and behavior of mutable loci in maize. *Proc. Natl. Acad. Sci. U.S.A.* 36: 344–355.

Nakasone H. and H. Kamemoto. 1961. Artificial induction of polyploidy in orchids by the use of colchicine. *Hawaii Agric. Exp. Stn. Tech. Bull.* 42.

Olby, R. 1966. *Origins of Mendelism,* 2nd ed. University of Chicago Press, Chicago.

Reed, S.M. and T.A. Rinehart. 2007. Simple sequence repeat marker analysis of genetic relationships within *Hydrangea macrophylla. J. Am. Soc. Hort. Sci.* 132(3): 341–351.

Weinberg, W. (1908). Über den Nachweis der Vererbung beim Menschen. *Jahreshefte des Vereins für vaterländische Naturkunde in Württemberg.* 64: 368–382.

38 Seed Production, Processing, and Analysis

J. Kim Pittcock

CHAPTER 38 CONCEPTS

- Pure seed is the percentage of seed of the desired species that actually exists in the seed lot.

- Inert matter is percentage of nonviable material in a seed package. This may include broken seeds, rocks, sand, trash, or any material that is not a complete seed.

- Other crop seed is the percentage of seeds found in the seed package that are not the species or variety on the label.

- Weed seed is the percentage of noncrop seed found in the seed package. Most labels will include the name and percentage of each type of weed seed included.

- Noxious weed seed is the percentage of weed seeds found that are on the state or federal noxious weed list. Noxious weeds are the weeds that a state has determined to be objectionable.

Modern day farming and horticulture is dependent on seeds and seed production. Seeds are self-contained units, which have the ability to regenerate the species. For the continued use of superior or pure seed lines, they must be produced in sufficient quantities to meet the needs of the market. After the plants have produced the seed, the seeds must also be properly handled and stored to continue this production cycle.

SEED PRODUCTION

Seed production can occur in regions with commercially produced crops, or seeds may be produced in isolated areas away from production centers. For seed crops that are self-pollinated or when pollination can be controlled, the crops are not isolated.

Seed production in isolated regions is necessary when the plants are open pollinated or cross-pollinated, or when hybrid seed is being produced. Climatic factors, such as low relative humidity or lack of rainfall during harvest, are critical site location factors for high-quality seed, especially when disease is an issue. Isolated areas are also important when the crop encounters extreme pest problems in the production regions.

Normal production practices are required for quality seed production. Initial field selection is critical, and field history needs to be considered. Seed production of a cereal grain should not be done in a field that was previously planted with a similar grain, as volunteer grains could contaminate the seed lot at harvest. Optimum soil fertility and proper moisture levels must be maintained. Additionally, proper insect, disease, and weed control need to be performed to produce the highest quality seed. Production fields and storage facilities are monitored by the certifying agency to guarantee that the crop meets all the certification standards.

Harvesting of agronomic crops is done mainly after the seeds are mature and have reached the specific moisture content required for the species. Basic harvesting methods are combining/threshing, stripping, and windrowing.

Combining/threshing involves removing the seed heads and any plant material in close proximity. Stripping is the process of removing only the mature seed from the plant, without injury to the plant and any future seed. Windrowing involves harvesting the upper portion of the plant before the seed has reached the proper moisture content. The seed and plant are allowed to dry, and then the seed is removed for cleaning and storage.

After the seed is harvested, it may need to be conditioned by hulling (cereal crops), debearding (grasses), or scalping and rough cleaning. Air-cleaning machines are used to remove trash and to grade the seeds. Additional conditioning is required on grain seed. They are finished by sizing for length and thickness and then are density graded.

If the seed is not dried in the field before harvest, then it must be mechanically dried or naturally air-dried after harvest. Seeds are normally dried to 7% moisture content or below and then stored under low temperatures (0°C to 4°C) and low relative humidity (below 25%). If the temperatures and relative humidity are elevated, the longevity of the seeds will be reduced. To maintain the maximum life of the seed, each factor must be carefully regulated.

Harvesting of vegetable and flower seed is more labor intensive than most agronomic crops. Vegetables seed production can be divided into the following two categories: wet-seeded (tomato [*Lycopersicon esculentum* var. *esculentum* Mil.], melons, cucurbits) and dry-seeded (onion [*Allium spp.* L.], legumes, cole crops).

Wet-seeded crops are either hand harvested or machine harvested by removing the entire plant. The fruit may be sorted to select only the superior ones. Fruits are crushed or macerated to pulverize the tissue, so the seeds may be extracted. This gelatinous material is fermented, which allows the mature seeds to sink to the bottom of the tanks. After extraction, the seeds are immediately washed in water troughs. Drying is accomplished in batch dryers, which expose the seed to heated air. A few crops, such as cucumber, can be air-dried in the sun. The seeds are separated from inert matter by an air-screen cleaner. Immature seeds are separated from mature seeds by gravity tables. Seeds are stored under cool, dry conditions at approximately 6% moisture content.

Dry-seeded vegetables have the potential for seed loss due to shattering when the seed is ripe. Immediately after the seed is ready, the entire green plant is harvested and placed in windrows. The plant and fruit are allowed to air-dry. The seed is hand harvested or combined before it shatters. Dry-seeded vegetable seed lots must be pre-screened because they normally contain more inert material and undesirable appendages. The appendages are removed by processes, such as de-tailing or debearding. Conditioning and storage conditions are similar to wet-seeded vegetables.

SEED PROCESSING

SEED TREATMENTS

To assist the grower, many enhancements have been added to the seed and/or coat to make seeding easier or to ensure higher seedling survivability rates. Many bedding plants and vegetable seeds are extremely small, therefore, they are pelleted with inert materials, such as clay, that more than double the size of the seed. Larger seed allows for easier mechanical and hand seeding.

Another seed treatment is the addition of a product that will improve the health or assist the seedling in growth. These include fungicides, insecticides, and micronutrients, such as zinc. Chemical pretreatments are common on cotton (*Gossypium hirsutum* L.), corn (*Zea mays* L.), and rice (*Oryza sativa* L.). A polymer-based or plastic film coating can also be applied to seed. This allows the chemicals to adhere better and also lessens the chance that the pretreatment will rub off during handling. The coating also greatly decreases the possible chemical exposure to workers.

Priming seeds is the newest pretreatment process and is gaining popularity, mainly in the vegetable industry. The seeds are hydrated to allow seed germination to begin, but the germination process is stopped before radicle emergence (chapter 14). The seed can be planted or redried and stored. Primed seeds will develop plants faster when sown and are less prone to disease.

Mechanized treatments remove any unwanted part of the reproductive structure, such as wings, barbs, fuzz, or chaff from the seed. These treatments include debearding, bobtailing, defuzzing, and de-tailing. This allows the seed to more easily be separated for mechanical or hand planting.

SEED CLASSIFICATION

Certified seed is an important aspect of the production of genetically pure varieties. Many agronomic crops, except hybrid varieties, such as corn, have been certified. New interest has occurred in the horticulture industry to participate in certification programs for sod production and tree seed breeding programs. The use of certified seed ensures that producers are obtaining varietally pure seed that is of the highest quality. The four seed classes are breeder seed, foundation seed, registered seed, and certified seed. Breeder seed is directly controlled by the originator or plant breeder and is used for seed increases. It is designated by a white tag and labeled "Breeder Seed." Foundation seed is the first generation of breeder seed and is controlled by the foundation producer. White tags are also used for the foundation class. Registered seed is the progeny of foundation seed and controlled by registered producers. It is also used for seed increases and carries purple tags. Certified seed is commonly available to the farmer and in large quantities. Blue tags represent certified seed.

SEED ANALYSIS

The quality of seed produced from a plant is the determining factor in a crop's future success or failure. Seed quality involves genetic factors, physical purity, germination, and health. By law, seed analyses (germination percentage and purity information) must be conducted, with the results displayed on the tag or seed packet. Numerous seed tests can be performed and normally fall into the following three basic categories: germination testing, vigor testing, and biochemical testing. The most common laboratory analyses performed are the germination test, tetrazolium test (viability), and purity tests.

The germination test is a basic seed analysis test and determines the actual percentage of normal seedlings that develop. Controlled environmental conditions of moisture, humidity, and light are combined with a specific time frame for each species of seed tested. Cereal crops are tested for 12 to 14 days, grasses require 21 days, and many woody plant seeds and some wildflowers must be evaluated for six weeks. Seeds are observed at regular intervals with newly germinating seedlings removed. A weakness to this test is that all seeds are given the opportunity to germinate under optimum conditions during the time frame, but there is no distinction between strong seedlings and weak seedlings. The weak seeds may never germinate under normal growing conditions, so this measure of germination percentage may be overestimated.

The tetrazolium (TTZ) test is a biochemical vigor test that gives a rapid estimation of the potential germination percentage by estimating viability. Seeds are soaked in a tetrazolium chloride (2,3,5-triphenyl tetrazolium chloride) solution and staining of various tissues occurs. Respirating cells will absorb the solution at varying amounts and will reveal various staining intensity patterns. Trained and experienced technicians evaluate the staining patterns to determine if the seed is viable.

Purity tests are conducted to establish the percentages of pure seed, inert matter, other crop seed, weed seed, and noxious weed seed. Pure Live Seed (PLS) is becoming a more commonly used category in purity testing. PLS is a combination of the pure seed percentage and the germination percentage. This calculation gives the percentage of usable seed in the lot. Many other tests can be included under purity tests depending on individual state laws or for a specific plant species. For most horticultural crops, the seed lot will be cleaned, and the seed package will contain only pure seed.

Vigor testing is more specific to individual species or categories of plants. These tests include the warm germination test, cold test, and the accelerated aging (AA) stress test. The warm germination test uses sand as the growing medium instead of germination paper and is used by large seed companies because it provides a more realistic germination percentage due to the seed being grown in a natural environment. The cold test is a stress vigor test and is used to determine the germination percentage of seeds planted at low temperatures and high moisture contents. This test mimics environmental field conditions in early spring and is primarily used for corn, soybean (*Glycine max* Merr.), and sorghum (*Sorghum bicolor* (L.) Moench.). The accelerated aging (AA) test is designed to simulate aging (harvest problems and storage) of a seed lot. It also provides an estimate of the germination percentage that will be obtained under less-than-ideal conditions. Seeds are exposed to high temperatures and high humidity prior to a standard germination test. Accelerated aging testing is commonly performed on soybeans and corn.

Another beneficial test that can be conducted in the field or laboratory is the mechanical seed damage test or bleach test. The percent of damaged seeds from harvesting or processing can be determined quickly, and adjustments can be made to the equipment to minimize future damage.

EXERCISES

EXPERIMENT 1: GERMINATION TEST

This exercise is designed to provide students the necessary tools to determine the actual germination percentage of a seed lot. To determine the germination percent, each species of seed chosen may need to be placed at different temperatures, although most seeds germinate between 21°C and 23°C. The germination paper towels (commonly called rag dolls in the industry) must be kept moist throughout the experiment. After a seed has germinated and been recorded, it is normally removed. After 10 to 14 days, the majority of agronomic species will have reached maximum germination percent. Recorded data will be both the number of seeds that germinate to produce healthy, normal seedlings and the number of ungerminated seeds.

Materials

Items needed for each student or group of students include the following:

- Ten of each seed provided (corn, bean (*Phaseolus vulgaris* L.), garden pea (*Pisum sativum* L.), tomato, marigold (*Tagetes patula*. L.), etc.)
- Distilled water
- Germination paper
- Rubber bands
- Markers
- Sealable plastic container
- Germination or growth chamber

Follow the methods in Procedure 38.1 to complete this experiment.

Anticipated Results

Germination should have begun on the majority of agronomic crop seeds by the first observation date. Vegetable and flower seedlings may not begin germinating until the second or third observation date. Most seeds from herbaceous plant material should be concluded in 21 days or less.

Most fresh commercial seed should have germination percentages above 85%. Variations that can be added to the experiment are to conduct germination tests on old seed lots and the fresh seed lots to illustrate the reduction in germination percentages as seeds age.

	Procedure 38.1
	Germination Test
Step	Instructions and Comments
1	Select 10 seeds randomly from a seed lot.
2	Place the seeds along the center line of the germination paper towel.
3	Roll the paper towel into a cylindrical tube. Place a rubber band on each end, and label each towel.
4	Thoroughly soak the germination paper with distilled water, and place the towels in a sealed container in the germination/growth chamber at the appropriate temperature (as indicated by the instructor).
5	Check the seeds at three-day intervals until maximum germination has been recorded (10 to 14 days or longer if determined by the instructor). Remoisten the towels if needed. Remove seeds that have germinated once they have been counted.
6	Calculate the final germination percent [(number of germinated seed / total number of seeds) × 100].

Questions

- If a seed lot has a germination percent of 95%, why would a grower obtain only 80%?
- Why would a grower need to know the germination percentage for a seed lot?

EXPERIMENT 2: TETRAZOLIUM TESTS

This exercise is designed to provide students the necessary tools to determine the potential for seed germination percentage within a seed lot. This is a rapid test in comparison to a standard germination test. This test depends on tetrazolium chloride (2,3,5-triphenyl tetrazolium chloride), which produces a chemical reaction when applied to living tissue. A red coloration indicates cellular respiratory activity. The intensity of the red staining is correlated with the rates of respiration of different tissue types. Dark red staining indicates damaged or deteriorated tissues, and no staining is indicative of nonrespiring or dead tissue. The potential seed germination percentage is calculated by the proportion of seeds with stained embryos and essential germination structures. Potential germination percentage may be higher than the percent obtained from a standard germination test. Dormant and diseased seeds may stain, but may not germinate and/or produce healthy seedlings due to additional internal or external factors affecting their successful germination.

Materials

Items needed for each student or group of students include the following:

- Ten of each seed provided (bean, corn, garden pea, wheat (*Triticum aestivum* L.), etc.)
- Distilled water
- 10-cm diameter petri dishes (3 per group)
- Scalpel or razor blade
- 0.1% tetrazolium chloride solution. (To prepare 0.1% tetrazolium solution, mix 0.1mg of 2,3,5-triphenyl tetrazolium chloride in 99 mL of distilled water. Adjust the pH to 7.0.)
- Marker
- Paper towels
- Dark chamber (box or inside cabinet)
- Microscope or hand lens

Follow the methods in Procedure 38.2 to complete this experiment.

Note: Most seeds require 1 to 1.5 h soaking in the tetrazolium solution, but soybeans require 3 to 4 h at room temperature to complete the test. The soaking time may be decreased by placing the seeds at higher temperatures (ca. 35°C).

Anticipated Results

Students may have a difficult time evaluating the plant structures and staining patterns unless examples are provided as a guide. The AOSA (Association of Official Seed Analysts) and ISTA (International Seed Testing Association) have good examples on their Web sites. The embryo structures will stain a darker red than cotyledon tissue. Radicle tips will normally have a redder staining intensity than other embryo structures.

Questions

- Why is the tetrazolium test only a measure of potential seed germination percent and not actual germination percent?
- Why do different structures of an embryo have varying stain intensities?

EXPERIMENT 3: MECHANICAL SEED DAMAGE TEST (BLEACH TEST)

This exercise provides a quick and easy method of determining the damage percentage seed lots suffered from harvesting, threshing, or seed conditioning. This test is

Seed Production, Processing, and Analysis

	Procedure 38.2
	Tetrazolium Test for Viability
Step	Instructions and Comments
1	Select 10 seeds (of each species provided) that have been soaked overnight in distilled water. There will be total of 30 seeds.
2	Cut the seeds longitudinally through the embryo with a scalpel or razor blade.
3	In a petri dish pour 0.1% tetrazolium solution (2,3,5-triphenyl tetrazolium chloride) to cover the bottom.
4	Place the seeds, cut side down in the tetrazolium solution. Soak in a dark location for 1 to 1.5 h.
5	Remove seeds from tetrazolium solution when they are light pink. Rinse and cover the seeds with distilled water until each seed can be evaluated.
6	Place the seed on paper towels for evaluation of red staining. Categorize each seed's germination potential by observing the embryo and other essential germination structures. Small seeds or seed structures may need to be observed with a hands lens or under a dissection microscope. (Variations of red stained areas are normal respiring tissue, dead tissue will not be stained, dark red stained areas indicate damaged or deteriorated tissue.)
7	Determine the potential for seed germination percentage of the seed lot [(number of seed with viable germination structures / total number of seeds) × 100].

specifically used for soybean and dry edible beans. An advantage of this test is that it can be conducted quickly under field conditions or in the laboratory. Each seed crop has an optimum moisture content for harvesting and, if seeds are allowed to drop below this level before harvesting, the seed coats are susceptible to cracking. Combines and threshers are also a potential threat to cause damage to the seed when they are not in proper adjustment. If damage is noted by this test conducted while in the field, harvesting equipment can immediately be adjusted. Damaged seeds are a major concern when the seed is sold for "whole seed markets," such as for seed production.

Materials

Items needed for each student or group of students include the following:

- One hundred soybean or edible bean seeds (eliminate the severely damaged)
- Household bleach solution (sodium hypochlorite—mix one part bleach to five parts water)
- 10-cm diameter petri dishes
- Paper towels

Follow the methods in Procedure 38.3 to complete this experiment.

Anticipated Results

Seeds that do not swell have sustained very little or no damage. Wrinkled seeds (seed coat appears blistered) are only moderately damaged and will normally germinate the subsequent season. Swollen seeds that have absorbed large quantities of bleach are considered damaged and normally do not have the ability to germinate. When the swollen seed percentage is 10% or above, harvest or handling equipment adjustments must be made to lower the quantity of damaged seed.

From commercial bags of soybeans, our results have averaged approximately 50% to 60% of the seeds not swollen, 30% to 40% with blistering, and the remainder swollen.

	Procedure 38.3
	Mechanical Seed Damage Test (Bleach Test)
Step	Instructions and Comments
1	Randomly select 100 seeds from the seed lot.
2	Place the seeds in a petri dish containing a bleach solution. (Caution: bleach may damage clothes.)
3	Soak seeds for 15 min. (If blistering and swollen seeds are not observed, continue the soak for a total of 30 min.)
4	Pour off the bleach solution and place the seeds on a paper towel for evaluation.
5	Count the number of swollen seeds.
6	Calculate the percent damaged seed [(number of swollen seeds / total number of seeds) × 100].

Question

- If a mechanical damage test is performed on a trailer load of soybean seed and the percentage of swollen seed is 35%, what can this load of seed be used for?

SUGGESTED READINGS

Acquaah, G. 2002. *Horticulture: Principles and Practices.* Prentice-Hall, Inc., Upper Saddle River, NJ.

AOSA. 1999. *Rules for Testing Seeds.* Association of Official Seed Analysts, Lincoln, NE.

Baskin, C. Seed Facts, Information Sheet 1062. Mississippi State Cooperative Extension Service. Mississippi State University.

Bassett, M.J. 1986. *Breeding Vegetable Crops.* Avi Publishing, Westport, CT.

George, R.A.T. 1985. *Vegetable Seed Production.* Longman Press, Essex, U.K.

ISTA. 1999. International Rules for Seed Testing. *Seed Sci. Technol. (Supplement)* 27: 1–333.

McDonald, M.B. and L.O. Copeland. 1989. *Seed Science and Technology Laboratory.* Iowa State University Press, Ames, IA.

McDonald, M.B. and L.O. Copeland. 1997. *Seed Production: Principles and Practices.* Chapman and Hall, New York.

Thompson, J.R. 1979. *An Introduction to Seed Technology.* Halstead Press, New York.

39 Environmental Factors Affecting Seed Germination

Emily E. Hoover

The effects of environmental factors are crucial to seed germination and continued seedling growth. These factors include water, oxygen, temperature, light, and nutrients. A constant supply of water must be available for uptake by the seed for germination to occur. However, a balance between water and oxygen in the propagation medium is also essential. In a waterlogged medium, oxygen is not present in high enough concentrations for germination and seedling growth to occur. Monitoring the moisture level and, thus, the oxygen level in the medium will boost success.

The optimum temperature for seed germination and seedling growth differs with plant species, although the range of 24°C to 26°C is used as a starting place for many plants. Plants native to warmer climates may have higher optimum temperatures for germination than plants that are native to cooler, temperate regions. Plant species can also be grouped according to whether light is required for germination. Some species require light for germination, whereas germination is inhibited by light in other species.

Although germination can occur without added nutrients, nutrients present in low to moderate concentrations at time of germination will result in more rapid seedling growth. Care should be taken to avoid excessive salt levels, which will injure new actively growing root tissue.

Seeds differ greatly when it comes to the environmental requirements for germination. Some seeds will germinate immediately without special treatment, whereas others may require manipulation and/or specific environmental conditions for germination to occur. For example, seeds of most annuals, including vegetable seeds, will germinate quickly when placed in a favorable environment (moisture, temperature, aeration, and sometimes light). Such seeds are said to be quiescent (ecodormancy). Seeds that are viable, but fail to germinate when exposed to conditions favorable for germination, are said to be dormant (endodormancy). The seeds of many woody plants and some herbaceous perennials exhibit some type of dormancy. Seed dormancy may be caused by a variety of factors and is an evolutionary characteristic important to the survival of many plant species. Overcoming dormancy, which may be physically and/or physiologically induced, is often a challenge when propagating plants from seed. Pretreatments, such as scarification (breaking seed coats) and stratification (cool, moist conditions), are often critical to successful germination when propagating plants from seed (chapter 41). In addition, providing conditions that are optimum for the germination of specific seeds can often determine success.

FACILITIES

The first experiment is conducted in two phases. During the first phase, seeds will be stratified in a 4°C cooler. After stratification, the seeds will be grown in full sun. In temperate climates during the academic year, this requires a greenhouse. Mist will not be used, and growing containers should be hand watered. A temperature of 25/20°C ± 3°C (day/night) should be maintained under a natural photoperiod in the greenhouse. In the second experiment, a growth chamber will be needed with light cycles of 14 hours of light with cool white fluorescent bulbs.

EXERCISES

EXPERIMENT 1: STRATIFICATION AND GIBBERELLIC ACID EFFECT ON *MALUS DOMESTICA* (APPLE) SEED GERMINATION

Apple seeds demonstrate the importance of stratification on seed germination. Apples are readily available; you can also ask students to bring their own. When planning, you can assume six to eight seeds per fruit. We have also found that stratifying seeds more than three weeks does not appreciably alter the results. If the seeds are to germinate, three weeks is ample time.

Materials

Each student or team of students will require the following items for the exercise:

- Apples (*Malus domestica* L.—Rosaceae or rose family) for seed extraction
- Razor blades
- Permanent marking pen
- Labels
- Pots
- Moistened growing medium

	Procedure 39.1
	Factors Affecting Germination of Apple Seeds.
Step	Instructions or Comments

1	There are a total of four treatments, with five seeds used for each treatment. The treatments are the following:
	  1. Sow seeds directly into growing medium.
	  2. Place into stratifying conditions.
	  3. Soak seeds in a solution of 400 ppm gibberellic acid for 60 min.; then sow seeds directly into growing medium.
	  4. Soak seeds in a solution of 400 ppm gibberellic acid for 60 min.; place into stratifying conditions.

2	Steps 2 through 4 are to be completed during the first lab period. Make labels for treatments two and four, which will go into stratifying conditions for six weeks. Fill pots with moist media.
3	Extract 20 seeds from the apples, keeping them moist AT ALL TIMES. Each treatment will be replicated five times (5 seeds per treatment) for a total of 20 seeds.
4	Sow the seeds in a pot; then place pot in a plastic bag, and secure the opening so growing medium does not dry out. Place pots into the cooler (4°C) for stratification (remember to label all pots with your name).
5	Steps 5 through 8 are to be completed 42 days after stratification of the apple seeds begins. Remove the pots from the cooler and place in greenhouse 42 days (6 weeks) after the seeds were placed into conditions of stratification.
6	Make labels for treatments one and three.
7	Extract 20 seeds from the apples, keeping them moist AT ALL TIMES. Each treatment will be replicated five times (5 seeds per treatment) for a total of 20 seeds.
8	Treat the seeds as described in Step 1. Then sow the seeds in a pot; place pot in the greenhouse with the two treatments from stratification.
9	Data and observations that may be taken for each treatment include (but are not limited to) number of seeds rotted, germination percentage (number of seeds germinated/total number of seeds), and time to emergence.

Follow the instructions and comments listed in Procedure 39.1 to complete the experiment.

Anticipated Results

Although apple is supposed to germinate after stratification, we have found that gibberellic acid can replace the stratification requirement. Planting freshly harvested seed, which has not been allowed to dry out, usually yields a germination percentage of less than 20%.

Questions

Compare each of the treatments with respect to germination percentage.

- What other environmental parameters could be tested?
- How effective were each of the treatments?
- What other treatments might be used to overcome dormancy requirements?
- What recommendations would you give to someone who wanted to know about overcoming dormancy of woody species?

EXPERIMENT 2: LIGHT EFFECTS ON SEED GERMINATION

Seeds of native plant species can be collected by students during certain times of the year. There are also many nurseries that sell species native to the area.

Materials

Each student or team of students will require the following items for the exercise:

- Seeds from native plant species with unknown germination requirements
- Permanent marking pen
- Labels
- Petri dishes with moistened filter paper or sand, depending on seed size
- Aluminum foil

Follow the instructions and comments listed in Procedure 39.2 to complete the experiment.

	Procedure 39. 2
	Light Effects on Germination of Native Plant Species
Step	Instructions or Comments
1	There are a total of three treatments with 10 seeds used for each. The treatments are the following:
	1. Light treatment of 14 h for 3 weeks (light control).
	2. Light treatment of 14 h for 1 week.
	3. Dark treatment for 3 weeks (dark control) by covering petri dish in 2 layers of aluminum foil.
2	Seed viability should be determined.
3	Label each treatment and your name on the bottom of each petri dish. Place 10 seeds per dish, and place in growth chamber.
4	At the end of one week, cover the dishes from treatment two and place back in growth chamber.
5	After three weeks, remove petri dishes from growth chamber. Observe whether or not germination has occurred. A seed is considered germinated if the cotyledon or radicle is visible without magnification.
6	Data and observations that may be taken for each treatment include (but are not limited to) number of seeds rotted, germination percentage (number of seeds germinated/total number of seeds).

Anticipated Results

For many species of plants, horticulturists do not know the germination requirements. Some native plant species have been shown to be sensitive to light (Kettenring et al., 2006). I would encourage instructors using this chapter to investigate native species being used in restorations in their area.

Questions

Compare each of the treatments with respect to germination percentage.

- What other environmental parameters could be tested?
- How effective were each of the treatments?
- What other treatments might be used to overcome dormancy requirements?
- What recommendations would you give to someone who wanted to know about overcoming dormancy of native species?

LITERATURE CITED

Kettenring, K.M., G. Gardner, and S.M. Galatowitsch. 2006. Effect of light on seed germination of eight wetland *Carex* species. *Ann. Bot.* 98: 869–874.

40 Producing Seedlings and Bedding Plants

Holly L. Scoggins

CHAPTER 40 CONCEPTS

- The four stages of plug production are based on seedling/plant growth and development and include the following: (1) radicle emergence, (2) hypocotyl and cotyledon emergence, (3) presence of true leaves, and (4) root and shoot growth until sufficient for transplanting.

- Plugs are seedlings that grow in their own tiny container—a cell in a plug tray.

- Once a plug is transplanted into the final container, the time spent in the greenhouse or nursery until shipping or sale is called growing on or finishing.

- One way to grow bedding plants is in flats—often grown and sold in cell packs—groups of four to six plants comprise an individual, separable unit. Several cell packs are joined together and supported by a plastic tray to form a sales unit called a flat.

- The number of crop cycles—one group of bedding plants that occupy the same greenhouse space for a several-week period—is a turn. The limited window of opportunity to successfully market spring bedding plants is the season.

ABOUT BEDDING PLANTS

Bedding plants are big business. The USDA National Agriculture Statistics Service (NASS) estimates there were nearly 19,000 businesses producing a wholesale value of $2.18 billion of bedding plants in 2002. This category includes both seed and vegetatively propagated herbaceous plants for the landscape, along with herb and vegetable transplants. Bedding plants make up nearly half of all floriculture industry sales in the United States. Though vegetatively propagated annuals (most are tropical perennials) are rapidly gaining share of this market, seed-grown annuals and perennials will probably always comprise the majority of sales. In fact, four of the top six bedding plants—impatiens (*Impatien walleriana* Hook. f.), wax begonias (*Begonia semperflorens-cultorum* hort.), marigold (*Tagetes patula* and *Tagetes erecta*), and pansy/viola (*Viola* x *wittrockiana*) are produced from seed. The other two are propagated by both seed and vegetative cuttings—petunias (*Petunia* x *hybrida* hort. Vilm.-Andr.) and the number-one bedding and landscape plant, geranium (*Pelargonium* x *hortorum* hort.). Vegetable transplants, though statistically lumped in with bedding plants, present their own set of challenges. Some are easy, others are difficult. And you have got to wonder why someone would buy a zucchini transplant when a seed the size of your fingernail germinates in basically the time it takes to read this.

Timing is everything. There is a limited window of sales in the bedding plant industry—most greenhouse growers that specialize in bedding plants will make 75% or more of their gross income during the spring months—March through June in most parts of the United States and Canada. The market is expanding for cool-season annuals, such as pansies (*Viola* x *wittrockiana* Gams.) for fall planting, though it is still a relatively small fraction of total bedding plant production (Figure 40.1).

CHANGES IN PRODUCTS

The primary product for sale to consumers has been flats—the traditional 11 x 22-in. trays holding plastic cell packs in various configurations. But the public has shown an increased interest in larger bedding plants, including herbs and vegetable transplants, and now 3.5-, 4-, and 6-in. pots make up over half of the bedding and landscape plant wholesale value. Also gaining in popularity are hanging baskets and "color-mixed" containers—larger planters ready for the porch or sundeck. An ongoing source of information as to new selections and cultivars

FIGURE 40.1 The spring "pack trials" feature new cultivars of bedding plants.

of bedding and vegetable plants is the nonprofit organization All-America Selections (AAS). The mission of AAS is to trial and promote superior seed-grown garden plants (both ornamental and edible) to the gardening public. The AAS Web site (www.all-americaselections.org) lists the annual winners and archives those selected since 1933. Another window to the newest and best cultivars is also an annual springtime event up and down the coast of California. Greenhouse growers flock to the "pack trials" hosted by the major seed companies. These displays feature hot new bedding plant varieties along with growing and marketing tips and techniques.

STARTING BEDDING PLANTS AND SEEDLINGS

In gardening parlance, direct seeding usually implies planting the seed directly into the landscape by the home gardener. This technique is useful for species that either resent transplanting, such as annual poppies (*Papaver rhoeas* L.), or those that have large, easy to handle seeds, such as sunflower (*Helianthus annuus* L.). But the majority of bedding plants find their way into the landscape by the two-step process of indirect seeding, where they are germinated and grown in one step and transplanted into their permanent location in the second step. This can also be accomplished by the home gardener. However, this chapter focuses on the commercial production of bedding plants, so we will examine the methods of indirect seeding as they apply to bringing the bedding plants to the retail market, finished and ready for purchase by the home gardener.

Bedding plant producers have two choices: they can purchase trays of seedlings started in tiny cells (known as "plugs") from another grower that specializes in this or they can start from scratch by germinating the seeds and growing the transplants themselves, either in plugs or seeded into open flats. In reality, most growers do some of both. The ratio is usually determined by a combination of factors—greenhouse space, labor availability, and difficulty/speed of plug production—some species are much trickier than others to grow. With the efficiency and decreasing cost of ground and air freight and advances in packaging to reduce damage and losses in shipment, many growers are opting out of the seedling business altogether, happy to buy in the propagules, ready to transplant (Figure 40.2). This, in turn, has produced a relatively new category of greenhouse businesses—those that specialize in propagative material, grown for other greenhouse operations. An example is Speedling, Inc., based in Sun City, Florida, which produces over *2 billion* transplants annually with production facilities across the United States and China.

DIRECT SEEDING TO A POT OR CELL PACK

Another option for growers who start their own seeds—growers will sometimes plant seeds directly into the pot or pack that they will be finished in. This is called direct seeding (similar to direct seeding into the landscape). This method is useful for large or irregularly shaped seeds, such as marigold (*Tagetes* species), but is very labor-intensive and not done by larger, highly automated greenhouse operations. Another disadvantage is the space taken by the flat for the entire production time.

OPEN FLATS VERSUS PLUGS

An "old-timey" way of starting bedding plant seedlings was to line them out in furrows in open flats (trays of growing media). Once the seedlings reached a certain size, it was necessary to "tease" the tangled but delicate roots apart in order to transplant them. This process was tedious and often injurious to the seedling, and it had to be done by hand. The term "transplant shock" refers to the stress any seedling can undergo when uprooted from one situation to another. The stressed seedling is slow to recover, and transplant shock can add weeks on to a production schedule. Keeping the growing media intact as a unit around the roots is an important benefit of plug production.

FIGURE 40.2 Stage 4 bedding plant plugs ready to be transplanted. (Photo courtesy of H. L. Scoggins.)

STAGES OF PLUG PRODUCTION

Floriculture researchers have divided herbaceous seedling plug production into four stages—each with its own set of optimum cultural conditions. Stage 1 is the very beginning of germination as the root (radicle) emerges from the seed. In stage 2, the germination process continues as the root grows out into the soil and the shoot emerges, consisting of the hypocotyl (stem) and cotyledons (seed leaves). True leaves begin to grow during stage 3. By the end of stage 4, the plug is well rooted and ready for shipping or transplanting.

PLUG SCHEDULING

How long does it take to grow bedding plant plugs to transplant size? Of course, it depends on the species. Some are very fast—impatiens plugs can be ready to ship or transplant within four weeks of seeding. Others take a while longer—wax begonias are slow to germinate and slow growers throughout stages 3 and 4 (though once transplanted, can be finished out nearly as fast as impatiens). The key to profitability for a bedding plant grower is the number of transplant-to-market crop cycles, called "turns." The larger the plug cell size, the bigger the transplant, and the faster the crop can be finished, or "turned." For example, it takes up to six weeks from transplanting to finish a 48-cell pack flat of impatiens from 512 plugs. By using 288s, the time to finish is reduced to four weeks. Though larger plugs cost more per cell to produce or buy in, the reduction in time spent on the greenhouse bench usually more than makes up for it. And growers maximize profits by increasing the number of turns over a growing season. Choosing a tray size may seem like six of one, half-dozen of the other, but the size of the plug greatly affects how long it takes to finish a crop. Using larger plugs as transplants can dramatically shorten bench time, allowing for more turns in the same time period. And for really quick turns, a relatively new concept in bedding plant production is the prefinished plant. The grower buys in the plants from another grower, already transplanted into their final flats. The crop can then be finished in two or three weeks and resold. Obviously, this is an expensive way to increase turns, with the cost of the flats and the shipping cutting well into profit margins. But sometimes, just having the product your customers want, and when they want it, makes up for the smaller profit with greater goodwill.

SIZES AND SHAPES OF PLUGS

Plug trays hold a number of cells—tiny chambers where a seed can germinate and produce a self-contained root system that easily pops out of the tray at transplant time. Plug trays are rectangular in shape with fairly standard overall dimensions (28 × 56 cm) (11 × 22 in.) and come in a variety of cell sizes, indicated by the number of cells per tray: 512, 384, 288, 216, 128, 102, 78, 72, 50, and 36 (Figure 40.3). Individual cells may be round, square, or octagonal, and all are perforated for bottom drainage. Cell depth also varies. The greater the cell volume, the more media it can hold, providing a greater buffer for the seedling to help protect it against drying out. Currently, "288s" seem to be the most frequently used for seed-produced bedding plants. The larger cell trays, such as 78s and 36s, are usually reserved for vegetatively propagated perennials and woody plant liners.

SEEDING AND TRANSPLANT TECHNOLOGY

Plug production lends itself to mechanization, which has revolutionized the bedding plant industry. Most commercial greenhouses that use plug trays also have an automated seeder, transplanter, or both. But with the wide variety of plug tray configurations listed above,

TABLE 40.1
Producing Seedlings of Some Commonly Grown Ornamental, Herb, and Vegetable Bedding Plants

Name	Germination[a]	Growing on as Plugs	Scheduling	Comments
Ageratum houstonianum Floss flower	Requires light to germinate. Do not cover seeds. Germinates in 8 to 10 days at 24°C to 25°C.	Reduce temperature to 18°C to 20°C for stage 4.	Ready to transplant in 3 to 4 weeks. Crop can then be finished in 6 to 8 weeks, depending on container size.	Sensitive to high salts—keep fertilizer levels low.
Begonia semperflorens-Cultorum hybrids Wax begonia	Do not cover seeds. Requires high humidity, and warm temperatures (26°C to 27°C).	Supplemental lighting in plug stages 3 and 4 can hasten development. Optimal day/night temperatures of 22/18°C.	Later stages of plug production are relatively slow—up to 8 to 9 weeks. However, once transplanted, the crop can be finished in 4 to 5 weeks.	Tiny seeds—90,000/g. Pelleted seeds are much easier to handle.
Calendula officinalis Pot marigold	Requires darkness to germinate—cover seeds. Germinates at relatively cool temperatures (20°C).	Keep temperatures moderately cool through stage 4—16°C to 21°C.	Transplantable in 4 weeks, with 8 to 14 weeks to finish, depending on time of year—faster in late spring–summer (long day plant).	Relatively large seeds (105/g) can be sown directly into packs or pots.
Capsicum annuum Bell, cayenne, and ornamental peppers	Germinates in 6 to 12 days at 21°C to 24°C. Can cover seeds or leave uncovered.	Maintain relatively warm temperatures through stage 4 (20°C to 22°C).	Transplantable in 2 to 3 weeks, grow on for 3 to 5 weeks for green plant sales (no flowers).	Transplant no later than 3 weeks after germination for fastest establishment.
Catharanthus roseus Vinca, Madagascar periwinkle	Cover seeds with media, germinate very warm (24°C to 26°C) and in darkness if possible (stage 1 only).	Drop temperature after germination to 23°C to 24°C. Maintain relatively dry media—do not overwater.	A relatively slow bedding crop, vinca is ready to transplant in 4 weeks, but requires 9 or 10 weeks more to reach saleable size.	A good crop for late spring production—grows poorly at cool temperatures.
Celosia argentea var. *plumosa* and var. *cristata* Plumed and crested celosia	Cover seed, germinates in 4 to 8 days at 24°C.	Grow through stage 4 at 18 to 20°C. Reaches transplantable size fairly rapidly—two weeks or so.	Grown on at moderately warm temperatures (19°C to 21°C), should finish 5 to 6 weeks after transplanting.	Celosia plugs often come into flower before transplanting. Pinching the bloom/bud off will hasten root establishment.
Impatiens walleriana Bedding impatiens	Rapid germination at 21°C to 26°C. High light intensity hastens germination time to 1 to 3 days.	Grow at 21°C to 24°C with moderate fertilizer levels (100 mg/L of a complete fertilizer).	Can transplant to pots/flats in 4 to 5 weeks. Impatiens are one of the fastest bedding crops—10 to 11 weeks from seed to sale.	Though often grown in shady sites in the landscape, greenhouse performance is best at moderate to high light levels.
Lavandula angustifolia Lavender	Do not cover; seeds require light to germinate, temperatures of 18°C to 24°C. Expect relatively low germination percentage (50% to 70%).	Germinates over 2 to 3 weeks, transplantable at 4 to 5 weeks. Do not overwater, even at the early plug stages.	Not a fast crop. Requires 7 to 13 weeks to finish, depending on size of pot.	Choose newer cultivars for faster germination and crop time. Perennial in USDA zones 6 to 8.
Lobelia erinus Lobelia	Sow multiple seeds per plug cell. Do not cover seed. Germinates in 5 to 15 days at 21°C to 27°C, depending on cultivar.	For stage 4 and beyond, lobelia grow best under moderately cool temperatures–16°C to 19°C day and night.	Another relatively slow bedding crop. Expect 7 to 12 weeks from transplant to finish.	Very tiny seeds (up to 45,000/g). Pelleted seed contains multiple seed per pellet and is much easier to handle.

Lycopersicum esculentum Tomato	Cover seeds and germinate at 21°C to 24°C. Usually germinates quickly—3 to 9 days.	Reduced temperatures (18 to 19 C) and high light in stage 4 reduces stretching of seedlings.	Another quick crop, transplant 2 to 3 weeks after sowing, and finish as green plants in 2 to 3 more weeks.	The top selling vegetable transplant. Watch for purplish cast to foliage—indicates phosphorus deficiency.
Ocimum basilicum Basil	Cover seeds and germinate at 21°C. Germinates in 5 to 10 days, depending on variety or cultivar	Grow on fairly warm—19°C to 20°C. Do not overwater in stage 4.	The Italian types finish 4 to 5 weeks after transplanting—"flavored" varieties like lemon basil or the purple-leaf varieties take a week or so longer.	Requires warm temperatures and abundant light for optimal growth—don't start seeds too early in spring.
Pelargonium x hortorum Garden geranium	Cover very lightly with vermiculite—light aids germination. Germinates in 5 to 10 days at 21°C to 24°C.	Spends a long time in stage 4—up to 5 weeks from sowing to transplant. Supplemental light can be beneficial in early spring.	Crop time varies as to cultivar and container size, but averages 10 to 12 weeks after transplant for a 10-cm pot. High light levels produce best quality plants.	Choose deep containers for optimal root development. Refer to resources listed for in-depth cultural information (growth regulators, photoperiod, etc.).
Petunia x hybrida Petunia	Most cultivars germinate uncovered. Many benefit from light and warm temperatures (22°C to 24°C) to hasten germination, usually in 4 to 8 days.	Reduce temperature to 17°C to 19°C. Most varieties ready for transplanting within 5 to 6 weeks. Petunias benefit from application of a weak fertilizer (75 mg/L N) as early as stage 2.	High light levels at a longer duration speeds flowering. Grows best in mid to late spring. Plants are usually marketable from 6 to 10 weeks after transplanting, depending on the season and latitude.	Small seeds (9000/g). Many companies offer pelleted seed for ease of handling.
Salvia splendens Annual salvia	Traditionally a bit tricky to germinate. Best uncovered, and not too wet. Germinates in 10 to 15 days at 24°C to 25°C.	Reduce temperature to 19°C to 21°C for stage 4. Transplant fairly rapidly (1 or 2 weeks after germination).	One of the faster crops—usually 4 to 5 weeks from transplant to market for cell packs.	Salvia plugs are very sensitive to high salts—use weak, low-ammonium fertilizer and only in stage 4.
Tagetes erecta African marigold *Tagetes patula* French marigold	Cover seeds lightly. Rapid germination (2 to 3 days) at 24°C to 27°C. Small growers may consider seeding directly to cell packs.	Lower temperatures after stage 1 to 20°C to 21°C. Plugs are ready for transplanting in 4 to 5 weeks for *T. erecta* and 5 to 6 weeks for *T. patula*.	Short day length (<12 h) hastens flower initiation in *T. erecta*. Depending on species and cultivar, crop times range from 7 to 11 weeks from sowing.	Long seeds. Use 200-cell tray or larger. Purchase de-tailed and coated seed for automated planting.
Viola x wittrockiana Pansy and viola	Darkness is necessary for germination—cover seed. Start at 17°C to 20°C with high moisture levels. Germination occurs in 5 to 10 days.	Reduce temperature to 16°C to 17°C for stage 3 (2 weeks) and 13°C to 16°C for stage 4 until transplant (2 to 3 weeks). Time from sowing to transplant-ready plugs is 6 to 7 weeks.	Supplemental lighting in both stage 4 and after transplant speeds time to flower. Crop finishes 6 to 8 weeks after transplanting. Consider finishing outside in temperate climates.	One of the challenges of growing pansies for the fall market is starting them in the late summer—warm temperatures produce lanky, weak seedlings. Consider ordering in plugs from a Northerly source.

Source: Adapted from Nau, J. 1999. *Ball Culture Guide: The Encyclopedia of Seed Germination*, 3rd ed. Ball Publishing, Batavia, IL; Dole, J.M. and H.F. Wilkins. 1999. *Floriculture: Principles and Species*. Prentice Hall, Upper Saddle River, NJ; and Styer, R.C. and D.S. Koranski. 1997. *Plug & Transplant Production. A Grower's Guide*. Ball Publishing, Batavia, IL.

[a] Temperatures given in "germination" column refers to growing media.

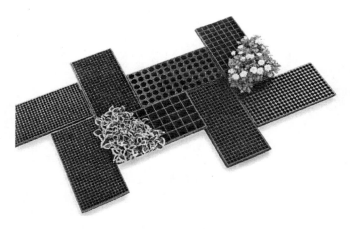

FIGURE 40.3 Bedding plant plug trays come in a wide variety of configurations. (Photo courtesy of ITML Horticultural Products, Inc.)

it is difficult to match the tray to an automated seeder or transplanter. A boon to the industry has been the formation of a "Common Element" or CE standard, cooperatively established by the major plug tray and equipment manufacturers (Figure 40.4). These standards will achieve uniformity among the most common sizes of plug trays (288 and 512 cells per sheet) and ensure compatibility between trays, automatic seeders, and transplanting equipment.

Seeding

Spend an hour or so placing barely visible petunia seeds by hand into a 512-cell plug tray, and you'll see the need for automation. A wide range of technology is available. Suitable for even the smallest greenhouse operation, handheld "wands" pick up seeds with a vacuum and deposit them a row at a time. The next step up is plate seeders—an entire tray is sown at once. A variety of plates are necessary to match seed size and tray configuration. Semiautomatic and reciprocating seeders use needles to pick up seeds of all sizes (Figure 40.5). Semiautomatic seeders are manually moved into place over the tray, and reciprocating seeders are motorized. Drum or cylinder seeders use a revolving drum to pick up and deposit seed onto a passing plug tray (Figure 40.6). The fastest of the lot, these can seed from 250 to 700 plug trays per hour.

Transplanting

Automated transplanters can transplant from 3600 to 50,000+ plugs per hour. Depending on the model, plugs are plucked, tweezed, blown, poked, or corkscrewed from their cells and planted into packs or pots. Transplanters can be "manual" (mechanical) or motorized and range in price from $3500 to $40,000. However, even moderately slow transplanters require another piece of automation for maximum efficiency—the flat or pot filler (Figure 40.7).

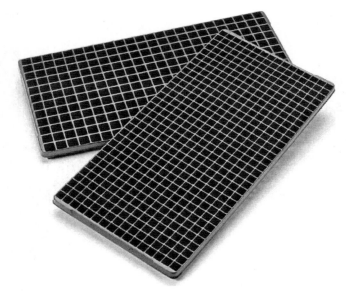

FIGURE 40.4 Common element (CE) 288 and 512 plug trays. (Photo courtesy of ITML Horticultural Products, Inc.)

FIGURE 40.5 A reciprocating needle seeder picks up marigold seeds. (Photo courtesy of J.G. Latimer.)

FIGURE 40.6 The cylinder seeder line for 128s is ready to roll at Raker's Acres, Inc., Litchfield, Michigan. (Photo courtesy of D. Steiner, Blackmore Co.)

Filling growing containers with soilless media by hand will simply not keep up with the transplanter.

Having a full plug flat with almost 100% of the cells containing a viable seedling is essential to successful automation of the transplant process. This is where "patching" comes in. Most growers employ a manual "patch line" to replace dead seedlings or empty cells with living plugs (Figure 40.8). There is new (though very expensive) technology available involving computer scanning of plug trays, robotic removal of "duds" and refilling of empty cells with viable plugs.

BEDDING PLANT SEED IMPROVEMENTS

With the ubiquity of automated seeding in commercial greenhouses, seed companies have made great strides in improving seed uniformity, germination percentage, and ease of handling. Pelleting adds mass to ridiculously tiny seeds, such as begonia. A clay coating, usually brightly colored, surrounds the seed, increasing its diameter. Several seeds can be combined into each pellet for species where multiple seeds per plug are desired, such as lobelia. Even large seeds, such as marigold, benefit from the coating to increasing uniformity (Figure 40.9). The author passes along a caveat from experience: Do not handle coated seeds with wet hands—an ounce of pelleted (and expensive) basil seed quickly turned to a sticky, unmanageable blob. Other improvements to germination and uniformity include refining and cleaning—simply removing the papery seed coat or other parts that may clog seeders and slow germination. Detailing of marigold seeds is an example.

FIGURE 40.7 A mechanical transplanter in tandem with a flat filler. (Photo courtesy of D. Steiner, Blackmore Co.)

FIGURE 40.8 Patching empty cells in a bedding flat that resulted from an incomplete plug tray. (Photo courtesy of J. G. Latimer.)

FIGURE 40.9 A freshly coated batch of seeds at Ball Seed, Inc.

Producing Seedlings and Bedding Plants

FIGURE 40.10 A fog-filled germination chamber at Ball Seed, Inc.

CULTURAL AND ENVIRONMENTAL FACTORS

Previous chapters and exercises have touched on the cultural and environmental requirements necessary for successful germination and growth of seedlings. We will elaborate further on these concepts, especially in terms of commercial greenhouse production of seed-grown bedding crops.

TEMPERATURE

The vast majority of bedding plant species germinate the fastest and with the highest percentage at relative warm media temperatures—23°C to 25°C. The temperature should then be reduced a bit, but remain fairly warm for most species throughout stages 2 and 3, from 20°C to 23°C. Heating of the root zone instead of just the air is the most efficient way to accomplish this. Some greenhouse systems grow on concrete floors with ebb and flood irrigation. The heating pipes or lines are imbedded in or under the concrete. Other growers run heating cable or pipes under the benches. Polyethylene convection tubes that normally run the length of the greenhouse near the roof can also run under benches and direct warm air under the plant. Infrared heating is another option, as it warms objects rather than air.

MOISTURE MANAGEMENT

High humidity aids germination. There are a number of ways to provide and conserve this humid environment. Germination chambers can provide a highly specialized environment—temperature and humidity are optimized for plug stages 1 and 2. Since the bright light necessary for quality stage 3 and 4 growth is not usually necessary for stages 1 and 2, plug trays can be arranged on shelves to maximize the use of space. Some growers start seeds in the greenhouse on a bench under mist irrigation (Figure 40.10). Poly tunnels or tents over the bench can be used to maintain high humidity levels. Once the plug reaches stages 3 and 4, the media should remain moist, but not saturated. Perpetually wet media prevents roots from growing to fill the plug cell and provides an ideal environment for the water molds, such as *Phytophthora*, to run amuck. However, if the media completely dries out, the tender seedling will likely become wilted beyond saving. Irrigation management for such a tiny container as a plug cell is closer to an art than a science—it takes an experienced grower to manage the needs of several species and cell sizes at once.

NUTRITION

At the time of germination, seeds really require no supplemental fertilizer, as they are "packing their own." But as early as plug stage 2, as soon as the seedling is visible, the seedlings benefit from additional nutrients. This may be accomplished if the media has a starter charge, or can be supplemented with a diluted water-soluble fertilizer. Plug-specific fertilizer formulations are available, but any complete fertilizer (macro and micronutrients) with relatively low ammonium-N (60% or more nitrate N) will suffice. Begin with 50 mg/L N every few days. Depending on the species, environment, and rate of growth, this can be increased to 100 or 150 mg/L throughout stage 4.

MEDIA TESTING

One of the most common problems associated with plug production is stunting of roots or death of the seedling from high salts. Many plant species grown as plugs are sensitive to high or low pH or high soluble salts. Soluble salts from fertilizer and/or irrigation water can build up quickly in plug cells, as there is so little media to act as a buffer. Conversely, the small volume of media results in limited nutrient reserves. Frequent monitoring via measurement of electrical conductivity (EC) and pH of the seedling's root zone environment can detect both these problems before damage to the crop is done (see chapter 5). Current media tests used by growers and analytical laboratories for greenhouse crops are the 2 water:1 substrate or 5 water:1 substrate (v/v), the saturated media extract (SME), and the pour-through. These tests are used for pH and electrical conductivity (EC) determinations and nutrient analysis. These testing methods each have their own interpretive ranges for substrate pH and EC, which, especially for soluble salts measurements, are not interchangeable. A simpler technique for plug testing is the press extraction method (PEM or Press) developed at North Carolina State University. One hour after irrigation/fertilizer application, a plug tray is selected and the grower gently presses on the top of a plug, collecting the root-zone solution (leachate) from the bottom of the plug in a container. This is repeated for four or five plugs until a sufficient volume is collected—usually 10 to 50 mL, depending on the instrument used for testing. Leachate EC and pH can then be measured. Ideally, the grower would select several trays for testing within a certain species and fertilization/irrigation regime and average the results. Testing should be done every two weeks for most crops and weekly (or even more frequently) for particularly tricky (prone to nutrient disorders) or high-value crops. The recommended pH range for plugs of most bedding crops using the press method for testing is 5.4 to 6.1. Media soluble salts levels should fall within an EC of 0.5 to 1.2 dS/m at sowing and germination and between 0.9 and 2.5 dS/m during stage 4, depending on the species.

HEIGHT CONTROL

Compact, well-rooted, and well-branched seedlings make better transplants. Tall, lanky seedlings with long internodes are more likely to be damaged in the shipping or transplanting process. Plug height can be manipulated through cultural or chemical means.

Ideal growing conditions for most species of bedding plant seedlings include high light levels, cooler temperatures, adequate, but not excessive moisture levels, and appropriate fertilizer regimens. Research has shown that an imposed phosphorus deficiency may help plugs of some species to stay compact, but this must be managed carefully. An interesting method of height control, mechanical brushing, has been used successful for years, especially in the production of tomato transplants. A boom, often the same one used for overhead irrigations, runs the length of the greenhouse several times per day. With the boom set low enough to significantly bend the plants, this action causes a release of ethylene, which inhibits apical growth. Chemical growth regulators are commonly used in the growing-out phase of the bedding plant crop cycle, but they can also be used for plug production. Extreme care must taken, as too high a concentration or amount applied can continue to limit plant growth for weeks after transplanting—greatly delaying the transplanted crop.

SUGGESTED READINGS

Ball, V. (Ed.) 1998. *Ball Redbook*. Ball Publishing, Batavia, IL. 802 pp.

Blanchette, R. (Ed.) 2001. *GrowerTalks on Structures and Equipment*. Ball Publishing, Batavia, IL. 272 pp.

Dole, J.M. and H.F. Wilkins. 1999. *Floriculture: Principles and Species*. Prentice Hall, Upper Saddle River, NJ. 613 pp.

Jerardo, A. 2002. Floriculture and nursery crops situation and outlook yearbook. Economic Research Service, U.S., Department of Agriculture, Washington, D.C.

McDonald, M. 1999. Seed quality and germination, in *Tips on Growing Bedding Plants*, 4th ed. Buck, C.A., S.A. Carver, M.L. Gaston, P.S. Konjoian, L.A. Kunkle, and M.F. Wilt (Eds.) Ohio Florists Association, Columbus, OH, pp. 13–23.

Nau, J. 1999. *Ball Culture Guide: The Encyclopedia of Seed Germination,* 3rd ed. Ball Publishing, Batavia, IL. 243 pp.

Reed, D.W. (Ed.) 1996. *Water, Media and Nutrition for Greenhouse Crops*. Ball Publishing, Batavia, IL. 202 pp.

Scoggins, H.L., D.A. Bailey, and P.V. Nelson. 2002. Efficacy of the press extraction method for bedding plant plug nutrient monitoring. *HortScience*. 37: 108–112.

Styer, R.C. and D.S. Koranski. 1997. *Plug & Transplant Production. A Grower's Guide*. Ball Publishing, Batavia, IL. 374 pp.

VanderVelde, J. (Ed.) 2000. *GrowerTalks on Plugs 3*. Ball Publishing, Batavia, IL. 197 pp.

41 Practices to Promote Seed Germination: Scarification, Stratification, and Priming

Caula A. Beyl

In nature, seeds have a varied menu of mechanisms that prevent them from germinating when there is little chance of their surviving. As a propagator, an understanding of these mechanisms and how to satisfy them is essential to obtain a high percentage of both germination and seedling establishment. To do this, various seed treatments have been developed to overcome these fail-safe mechanisms, including scarification and stratification. Scarification is any treatment that weakens or removes the seed coat, allowing the seed to take up or imbibe water. Sometimes, seeds are given a period of warm stratification (or layering seeds in a moist medium) to allow the embryo to finish its physiological and morphological development and/or until it is competent to germinate. At other times, the seed is placed into moist, cold stratification to satisfy an inherent physiological dormancy of the embryo. Stratification is also used in the case of epicotyl dormancy. In this case, the radicle or embryonic primary root will emerge without special treatments, but the seed must undergo a cold treatment before the epicotyl or embryonic shoot will emerge. No matter what the treatment, in the nursery or greenhouse, obtaining a high percentage of germination is not enough. It is also desirable that the seedlings emerge uniformly and grow vigorously as well.

Seed dormancy occurs when a seed will not germinate even if placed under ideal conditions. This may be a simple dormancy due to the seed coat physically preventing the seed from imbibing water, or it may be due to factors in the flesh of the fruit that inhibit germination. When there are two or more factors that prevent seed from germinating, such as a hard seed coat and the need for a certain number of days of cold temperatures, this is termed double dormancy. A secondary dormancy can occur when the seed is capable of germinating, but external conditions, such as high temperature, prevent it from doing so. Lettuce seeds are sensitive to high temperatures (26°C to 35°C) and undergo a heat-induced thermodormancy that prevents them from germinating even if later placed in an optimum-temperature environment.

Seeds that have hard, impervious seed coats do not germinate readily until the seed coat is weakened or removed. The seed coat acts as a barrier to water and oxygen movement into the seed. In nature, hard seed coats are broken down by different routes, such as soil microbial action, passing through the digestive tracts of birds or animals, insect activity, and even fire. Scarification is the technical term for this process of weakening or removing the seed coat. In the laboratory, the natural processes of scarification can be simulated by using concentrated sulfuric acid, sandpaper, files, hot water or even a hammer. On a large commercial scale, this can be accomplished using a rotating drum lined with an abrasive substance or sand paper (Figure 41.1). Sometimes seeds are mixed with sand and then put into the rotating drum. Immersion in hot water will often soften the seed coat sufficiently so that imbibition or water uptake can occur. Several examples of seeds that require scarification for successful germination are listed in Table 41.1. Note that many of these species are members of the legume family.

When a seed needs a prolonged moist, cold treatment to germinate and develop into a seedling, propagators mimic this by using a process called cold stratification. The term arose because seeds were arranged (or stratified) between layers of sand or soil and then subjected to chilling at temperatures ranging from 35°F to 45°F (2°C to 7°C). The amount of time and the effective temperature under which a seed must be stratified depends upon the species and its inherent requirement for chilling (often similar to the cold requirement needed by the dormant shoot buds before they will grow in spring) and the provenance or geographical source of the seeds. Seeds from individuals in northern latitudes may have a longer chilling requirement at a lower temperature than those at lower (and warmer) latitudes. The time period can be as little as three weeks for tree of heaven (*Ailanthus altissima* (Mill.) Swingle) and Norway spruce (*Picea abies* (L.) Karst.) or up to 20 weeks for Japanese maple (*Acer palmatum* Thunb.) or tuliptree (*Liriodendron tulipifera* L.). Propagators satisfy this chilling

FIGURE 41.1 Hoffman Model SC-50 commercial seed scarifier capable of administering scarification treatment to abrade seed coats of large lots of seed.

TABLE 41.1
Species Whose Seeds Benefit from Scarification or Stratification

Species	Common name	Family	Treatment
Albizia julibrissin Durazz.	Mimosa	Fabaceae	Acid scarification 30 min.
Cedrus atlantica (Endl.) Manetti ex Carr.	Atlas cedar	Pinaceae	Cold moist stratification 14 days
Cercis canadensis L.	Eastern redbud	Fabaceae	Scarification, then stratification 30 days
Chamaecyparis obtusa (Sieb. & Zucc.) Endl.	Hinoki cypress	Cupressaceae	Cold moist stratification 21 days
Cornus florida L.C. *kousa* Hance	Flowering dogwood	Cornaceae	Remove pulp and cold stratify 90 to 150 days
Cotinus coggygria Scop.	Smoketree	Anacardiaceae	Acid scarify 30 to 60 min., then stratify 3 months
Gleditsia triacanthos L.	Honeylocust	Fabaceae	Acid scarification 1 to 2 h
Gynocladus dioica (L.) K. Koch	Kentucky coffee tree	Fabaceae	Acid scarification 4 to 6 h
Hibiscus syriacus L.	Rose-of-Sharon	Malvaceae	Mechanically scarify
Koelreuteria paniculata Laxm.	Goldenraintree	Sapindaceae	Scarify 60 min, then cold stratify 90 days
Magnolia grandiflora L.	Southern magnolia	Magnoliaceae	Stratify 90 to 120 days
Olea europa L.	Olive	Oleaceae	Stratify 60 days
Pinus strobus L.	Eastern white pine	Pinaceae	Stratify 60 days
Pinus taeda L.	Loblolly pine	Pinaceae	No treatment needed
Prunus serrulata Lindl.	Oriental cherry	Rosaceae	Warm stratification 20 to 60 days, then cold for 90 to 120 days
Rhus glabra L.	Smooth sumac	Anacardiaceae	Mechanically scarify
Robinia psuedoacacia L.	Black locust	Fabaceae	Acid scarification 1 to 2 h
Zelkova sinica Schneid.	Chinese zelkova	Ulmaceae	Hot water scarify, then cold stratify 60 days
Ziziphus jujuba Mill.	Chinese date	Rhamnaceae	Hot water scarify, then cold stratify 40 days

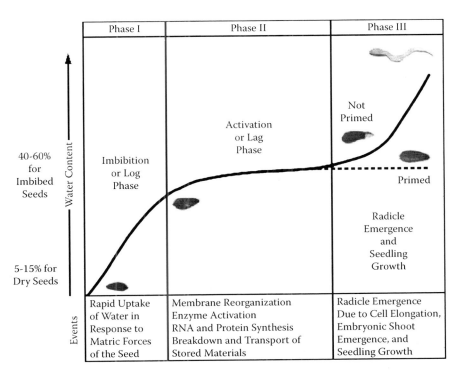

FIGURE 41.2 Water uptake over time, illustrating the three phases of germination and the effect of priming to hold germination in the activation or lag phase II.

requirement of seeds naturally by layering seed in sand or other medium in pits, bins, frames, or boxes and allowing ambient winter temperatures and rainfall to provide the moist, cold treatment. In refrigerated stratification, a variety of containers can be used as long as they do not impede aeration including simple Ziploc® bags.

Many times it is desirable to be able to hasten seed germination and seedling establishment, particularly if there is the chance of the environment not being favorable for germination. This is true even in areas of the world where agriculture is not as advanced as in the more industrialized nations. For example, if there is a narrow time window for planting seed of a crop that requires it to become well-established before a period of drought sets in, then every day spent getting the seed to germinate and the seedling to grow is crucial. Seed pregermination and seed priming are two techniques that can advance seed germination and subsequent seedling establishment. Pregermination involves allowing the seeds to imbibe, and then germinate only until incipient radicle emergence when the radicle has just appeared. The radicle or embryonic root is the first organ that emerges from the newly germinating seed. The seed is then dried down to nearly its original dry weight. When it is sown and allowed to regerminate, the germination events leading to seedling emergence from the soil are accelerated. Sometimes the seeds are not dried, but are placed in a gel matrix and fluid drilled using equipment designed to sow seeds suspended in a fluid gel.

Priming involves placing seeds in an environment where either osmotic or matric forces limit how far germination can progress. Water potential is a reflection of the free energy of water. The greater the concentration of water, the greater is the free energy and the higher the water potential. Osmotic forces are caused by solutes, which decrease the free energy and, thus, the water potential. Since the water potential of pure water is zero, decreasing the water potential causes it to be negative. In the case of priming, the osmotic forces are usually generated by either salt or polyethylene glycol added to the priming solution, with osmotic water potential values in the ranges of −0.5 to −1.5 megapascals (MPa). These are the units of water potential. Some older scientific literature expresses the water potential in bars (1 bar = 0.1 MPa). The more negative the value, the greater the ability to hold or attract water. In the case of matrix priming, seeds are imbibed in moist vermiculite, silica, or clay. The free energy of water is decreased by adsorption onto the matrix particles. There are three phases of germination (Figure 41.2). In phase I, the imbibitional or log phase, water enters the dry seed very rapidly in response to matric forces. A seed does not need to be viable to take up water in phase I.

During priming, seeds take up water in phase I, the imbibitional phase (Figure 41.2), because the driving forces for water entry into the dry seed are much stronger than those for the surrounding solution or matrix. Typical water potential values for dry seeds can range from −50 to −250 MPa. Phase II of germination is the activation phase or lag phase, so named because the rapid uptake of water slows dramatically, but even though it does not look like much is happening, this is a very busy phase for the seed. In this phase, which may last for several days, respiration starts, membranes reorganize, enzymes become activated, and protein synthesis begins from masked and newly minted mRNA. At the same time, the stored food reserves of the seed are being metabolized and mobilized so that the seed can continue germination. Because they have taken up water, imbibed seeds have a much less negative water potential than dry seeds and cannot compete with the surrounding priming solution for water. In phase III, cells in the radicle elongate, and the radicle protrudes from the germination seed. Radicle growth continues, with cell division occurring, and the embryonic shoot emerges soon afterward. During priming, seeds are held in phase II of germination and cannot proceed to phase III or radicle emergence. Holding seeds in the lag phase allows all of the important activation processes (membrane reconstitution, enzyme activation, protein synthesis, etc.) to occur at optimum temperatures. Primed seeds are then redried. When the seeds are later sown and rehydrated, germination occurs at a faster, more uniform rate than what would occur for seeds that have never been primed. Priming allows seeds to germinate faster under less than optimal temperature conditions and overcomes the problems inherent with crops that undergo thermodormancy when germinated under high temperatures, such as lettuce and celery.

The following three experiments will allow students to gain hands-on experience in the use of scarification, stratification, pregermination, and priming techniques to promote germination. Each student or small group of students can represent one replication of the experiment.

EVALUATING GERMINATION

Students should discuss and agree on criteria for measurement of germination success. What is germination? Is it when the radicle appears, when the embryonic shoot appears, or when the seedling emerges above the soil line? Once criteria for germination have been discussed and developed and students have decided what type of data should be collected, they can devise a common data sheet for recording results of the following experiments. Germination or seedling emergence should be checked frequently so that the pattern of seedling emergence over time can be documented. This is a measure of the uniformity of germination, which is also an important consideration.

EXPERIMENT 1: EFFECTIVENESS OF VARIOUS SCARIFICATION TECHNIQUES ON TWO SPECIES

There are several species of trees whose seeds have seed coats that require scarification. *Gleditsia triacanthos* L. var. *inermis* (L.) Zab. (thornless honeylocust) is a member of the legume family (Fabaceae) and like many of its relatives has a hard impervious seed coat. Honeylocust is a medium-sized tree valued for its lacy textured foliage. Although there are seedless cultivars available, many older trees are of seedling origin and have not been selected for either the thornless or seedless characteristics of many of the new cultivars. *Albizia julibrissin* Durazz. (mimosa or silk-tree), a member of the same family, is a medium-sized spreading tree with showy flowers common throughout the southern United States. Two other species that could serve as alternate choices for this exercise are *Robinia pseudoacacia* L. (black locust), a fragrant-flowered tree known for its ability to withstand harsh conditions and sites, and *Gymnocladus dioicus* (L.) K. Koch (Kentucky coffee tree), used in parks and large-scale landscape settings. If alternate species are chosen, acid scarification times must also be adjusted. It is also wise to check on the progress of scarification when using acid because the thickness of the seed coat and, thus, the time required to break it down, may vary from year to year. All of the species mentioned above (Figure 41.3) have the hard seed coat type of dormancy that acts as a barrier to germination.

The following five treatments will be compared in this exercise, which illustrates various approaches to scarification:

1. Control, with no scarification and/or water soak
2. Hot water soak
3. Cold water soak
4. Sandpaper abrasion
5. Concentrated sulfuric acid soak

Materials

The following items are needed for each student or group of three to five students:

- Seeds of *G. tricanthos* var. *inermis* (honeylocust) and *A. julibrissin* (mimosa). Since there are five treatments, each student or group will need at least 50 seeds of each species. Each student or group may act as one replication of the treatment
- Thermometer
- Large beaker and hot plate
- Cheesecloth and string
- Labels and permanent marker
- Six-pack flats for each student or group of students
- Seed germination medium, such as Jiffy Mix®
- Medium-grit sandpaper and a block of wood (a 4-in. piece of 2 × 4 works well)
- Concentrated sulfuric acid
- Face shield or eye protection and acid-resistant gloves

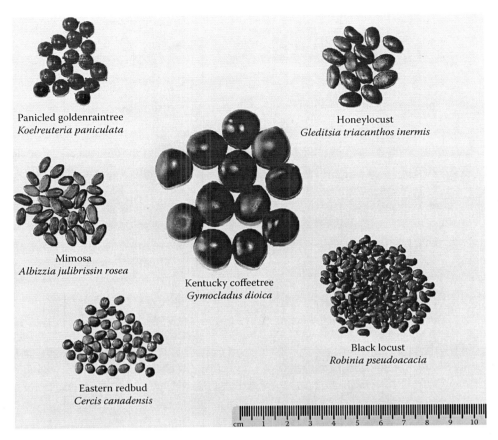

FIGURE 41.3 Seeds of *Gleditsia triacanthos* (honeylocust), *Albizia julibrissin* (mimosa), *Robinia pseudoacacia* (black locust), *Koelreuteria paniculata* (panicled goldenraintree), *Cercis canadensis* (Eastern redbud), and *Gymnocladus dioicus* (Kentucky coffeetree), prior to scarification treatment.

- Acid-impervious container, such as a large beaker, and a glass stirring rod

CAUTION: Handling concentrated sulfuric acid can be hazardous. This portion of the exercise can be done by the instructor as a demonstration. Adding water to concentrated acid will cause a sudden exothermic (heat releasing) reaction, resulting in acid splashing up onto hands, face, and clothing. Never add water to concentrated sulfuric acid. Care must also be taken to make sure that seeds are dry when put into the acid.

Follow the experimental protocols outlined in Procedure 41.1 to complete this experiment.

Anticipated Results

Scarification will not only enhance the percentage of germination, but will also result in more uniformity of seedling emergence. Results will vary from year-to-year due to differences in the thickness of the seed coats. Hot water is more effective than soaking in cold water. The control seeds should have a very low percentage of germination.

Questions

- What type of equipment is available for large-scale scarification of seeds?
- In a situation where seed has been scarified, but the resultant germination rate is still poor, what are some of the possible reasons?
- Why is uniformity of germination important as well as percentage of germination?
- What happens if seed is kept too long in the concentrated sulfuric acid?
- If you are working with a species that emerges after fire, what types of treatments could you devise to encourage germination?

EXPERIMENT 2: RESPONSE OF TWO SPECIES TO SCARIFICATION FOLLOWED BY STRATIFICATION IN DIFFERENT MEDIA

Koelreuteria paniculata Laxm. (goldenraintree) is an ornamental tree valued for its yellow flowers and brown seed panicles, which make a sound like rain in the wind. *Cercis canadensis* L. (eastern redbud) is a small tree native to the United States and valued for its flowers, which are pur-

Procedure 41.1
Comparison of Different Scarification Techniques to Enhance the Germination of Honeylocust and Mimosa

Step	Instructions and Comments
1	Each student or group should count out five lots with 10 seeds in each lot for mimosa and honeylocust.
2	They should then prepare two flats by filling them with a seed germination medium, such as Jiffy Mix®. One flat will be used for the mimosa seeds and one for the honeylocust seeds.
3	Using a label, score five trenches approximately 1/2 in. deep lengthwise along the surface of the germination medium in the flats. Each trench will hold 10 seeds from one treatment of one species. Prepare labels for each of the five treatments, and place the labels at the head of their respective trenches.
4	Since control seeds receive no treatment, they can be sown directly into their trenches in the appropriate flat for each species.
5	Place a large beaker on a hot plate, fill with water, and bring to a boil. Once the water has boiled, remove the beaker carefully from the hot plate and allow the water to begin cooling. Place the seeds for the hot water treatment in a square of cheesecloth cut large enough to hold the seeds easily, and tie with string. If the string is long enough to hang over the side of the beaker, it can be used to retrieve the cheesecloth seed bag. Check the temperature of the water, and when it has cooled to 180°F (82°C), immerse the cheesecloth bag containing the seeds (Figure 41.4). If the beaker is large enough, more than one bag of seeds can be accommodated. The seeds should be left to soak in the hot water as it cools. After 24 h, the bags are removed, opened and the seeds sown in the flats.
6	At the same time that seeds are placed in the hot water, place a second set of seeds in a corresponding beaker containing cold water. After the seeds are soaked for 24 h, they can be removed and sown in the flats in their respective trenches.
7	Seeds to be mechanically scarified are placed on one sheet of medium grit sandpaper. Use another piece of sandpaper wrapped around the piece of 2 × 4 (Figure 41.5), and use a circular motion to abrade the seeds between the two sheets of sand paper. Check the seeds periodically to determine if the seed coat has been abraded enough to just begin to see the interior tissue, and place in the flats in their appropriate trench.
8	Acid scarification requires care. The instructor may wish to conduct this portion of the exercise as a demonstration. Enough concentrated sulfuric acid should be placed in a dry beaker to constitute twice the volume of the seeds. Dry seeds are carefully placed in the concentrated sulfuric acid and allowed to remain for the recommended duration of time for the type of seed being treated (Figure 41.6). If this duration is uncertain, seeds may be checked periodically. Seeds that have received enough acid scarification are dull and may be pitted on the surface. Decant the sulfuric acid off by slowly pouring it, a small portion at a time, into a much larger volume of water. Never add water to concentrated sulfuric acid (see previous cautionary note). The seeds can then be rinsed by scraping them into a beaker containing a large volume of rinse water, decanting the water, and blotting the seeds dry. Seeds should then be sown in the flats in their appropriate place and labeled with the species name and treatment.
9	Check the flats every two to three days, and count the number of seedlings to emerge each time. Graph the number of germinated seeds over time. Note not only the final percentage of germination, but also the uniformity of emergence of seeds from each treatment. After two months, finalize the data and prepare the report.

Practices to Promote Seed Germination: Scarification, Stratification, and Priming

FIGURE 41.4 Cheesecloth bag containing seeds of goldenraintree being immersed in 180°F water for scarification treatment.

FIGURE 41.5 Seeds of goldenraintree being scarified by abrasion between two pieces of sandpaper; one serves as the base and the other is wrapped around a piece of 2 × 4-in. wood to make handling easier.

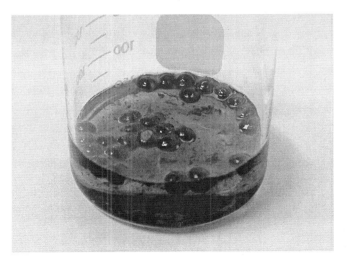

FIGURE 41.6 Appearance of goldenraintree seeds only 30 min. after being immersed in concentrated sulfuric acid for scarification treatment.

	Procedure 41.2
	Response of Two Species to Scarification Followed by Stratification in Different Media
Step	Instructions and Comments
1	Each student or group of students should take three Ziploc bags and label them with the species and the treatment (naked, peat, or vermiculite). A handful of peat that has been moistened and squeezed with hand pressure to remove excess water should be placed in the bag labeled "peat." The same process should be followed for the vermiculite treatment. Twenty scarified seeds of each species should be placed in their respective bags for the peat and vermiculite treatments and mixed thoroughly with the medium. Seeds that are destined for the naked stratification treatment should be moistened lightly and placed directly in the respective Ziploc bag. The same species used in Experiment 1 (mimosa and honeylocust) can be used in this experiment, or any other species that require scarification followed by stratification. The bags are then placed in the refrigerator for cool moist stratification.
2	The same procedure as in Step 1 should be followed for the seeds that have not been scarified to determine if scarification is necessary prior to stratification for it to be effective.
3	The stratification treatments should be observed weekly for evidence of germination or to check for fungal contamination.
4	After 4 weeks of stratification treatment, the seeds can be removed from the bags and sown in a flat containing a 1:1 mixture of peat and perlite. Water the seeds in thoroughly. At the same time, plant seeds that have received neither scarification nor stratification treatment as the controls.
5	Observe for germination weekly, and record the cumulative germination over time.

plish red in bud, later opening into a rosy pink with a purplish hue. There are also cultivars that have white flowers. Seeds of both of these species require scarification prior to stratification. Various techniques are used to stratify seeds, including stratification pits, bins, and frames, sowing in the fall and letting stratification occur naturally, or controlled stratification under refrigeration. For this experiment, seeds will be stratified either naked (with no medium), in moist sphagnum peat, or in moist vermiculite. Other media may be used as well, including hydrogels, such as TerraSorb.

Materials

The following items are needed for each group of three to five students:

- Acid-scarified seeds of goldenraintree (*K. paniculata*) and eastern redbud (*C. canadensis*). This can be done prior to the exercise by the instructor (see Experiment 1)
- Untreated seeds of the species indicated above
- Quart-sized Ziploc bags
- Vermiculite
- Sphagnum peat moss
- Standard flats (21 × 11 in.) filled with an appropriate medium (a 1:1 mixture of peat and perlite will work well)
- Labels and permanent marker

Follow the experimental protocols outlined in Procedure 41.2 to complete this experiment.

Anticipated Results

Both seeds will exhibit better germination when scarified prior to receiving stratification treatment. Eastern redbud may germinate better than goldenraintree because it only requires 30 days of stratification, but goldenraintree germinates better with a 60-day period. Fungal contamination may occur in the treatment containing sphagnum peat beyond moisture-holding capacity.

Questions

- Sometimes seeds stratified in peat moss or other organic materials mold. What can be done to minimize the likelihood of losing the seeds in this manner?
- If stratification at 5°C is effective for satisfying the cold requirement of the seed for germination, why do temperatures below freezing not work better?
- Does the source of the seeds for some species have an impact on the required length of the stratification period?
- What other materials would be appropriate choices as stratification media?

EXPERIMENT 3: EFFECT OF PRIMING OF LETTUCE AND TERRASORB™ PREGERMINATION OF SWEET CORN ON GERMINATION UNDER TEMPERATURE STRESS

All species have temperature optima for seed germination. When the temperature of the germination environment goes too much above or below these levels, germination suffers and, in some cases, is prevented altogether. The new supersweet cultivars of sweet corn (*Zea mays* L.) have genes (e.g., *su1*, *sh2*) which cause them to accumulate sugars at the expense of starches in the endosperm of the seed, making the seeds appear shrunken. Seeds of sweet corn, particularly the supersweet cultivars, are susceptible to chilling injury if germinated at temperatures below 10°C. Many crops whose origins were in tropical regions have this same susceptibility, including tomato, cotton, and okra. To gain market advantages, many farmers plant sweet corn early and thus risk exposure to chilling temperatures during germination. TerraSorb is a superabsorbent starch polymer that can absorb 3 to 400 times its weight in water, forming a hydrogel that has been used for germinating seeds, improving growing media, and as a root coating on bare root trees prior to planting. In the experiment below, it is being used as a pregermination medium for sweet corn.

Lettuce (*Lactuca sativa* L.) exposed to temperatures above at or above 30°C during germination will develop thermodormancy. Even if temperatures are then reduced to more optimum levels, the seeds will not germinate. Priming with an osmoticum, such as KNO_3 or polyethylene glycol, overcomes thermodormancy, allowing the lettuce to germinate normally.

In the experiment below, a solution of KNO_3 is being used to prime the lettuce seeds. Other salts have been used effectively, as well as polyethylene glycol solutions.

Materials

The following items are needed for each group of three to five students:

- Seeds of lettuce and a "supersweet" sweet corn, such as 'Silver Queen'
- 0.2% solution of potassium nitrate (KNO_3) for priming [alternatively, 1% potassium phosphate (K_3PO_4) or polyethylene glycol solutions of various molecular weights can be used]
- 25 g/L TerraSorb, a superabsorbent starch polymer (Industrial Services International, Bradenton, FL)
- Eight petri dishes containing Whatman #1 filter paper for each student or group
- Labels and permanent marker pen
- One Erlenmeyer flask per group
- One plastic pipette connected by tubing to a source of air to provide aeration in the priming solution for each flask
- One shallow tray or large weighing boat per group
- Fine basket strainer for rinsing seeds
- Paper towels for blotting seed dry
- Refrigerator (to impose chilling stress on sweet corn)
- Seed germinator capable of attaining a temperature of 30°C for lettuce germination

Follow the experimental protocols outlined in Procedure 43.3 to complete the experiment.

Anticipated Results

Priming will allow lettuce seed to germinate even in temperatures that would normally induce thermodormancy. Alternative treatments could be performed, comparing longer and longer times in the priming solution before the seed is dried and sown. Seeds may be held in the priming solution for as long as two weeks without negatively impacting germination. Sweet corn seeds pregerminated in TerraSorb will germinate faster and at a higher percentage than those not pregerminated.

An additional exercise that would demonstrate the rapidity of water uptake in phase I relative to phase II would be to have students measure the increase in fresh weight in the early hours of imbibition and then at a set time daily for several days thereafter. Because they are large enough to handle easily, and increases in the seed fresh weight can be measured, sweet corn seeds work well for this.

Questions

- What types of materials could be used for osmotic priming? For matric priming?
- If a seed is nonviable, will it still imbibe water in phase I of germination?
- What is a T_{50} for germination? What does calculating the T_{50} of the treatments above reveal about how effective they are?
- What other species have seeds susceptible to chilling injury? To thermodormancy?
- Sweet corn in the field takes proportionally longer to germinate as the soil temperature declines. What are other risks to a seed's survival the longer that it is in the ground? What events occur during chilling that exacerbate the risk to the seed?

\multicolumn{2}{c}{**Procedure 41.3**}	
\multicolumn{2}{c}{Evaluation of Pregermination of Sweet Corn and Priming of Lettuce for Germination outside the Optimum Temperature Range}	
Step	Instructions and Comments
1	Each student or group should count out 20 lettuce seeds each for the control and the priming treatments at two temperatures (25°C and 30°C). Each student or group should also count out 10 sweet corn seeds each for the control and the TerraSorb pregerminated treatments at two temperatures (5°C and 25°C).
2	The lettuce seeds designated for the priming treatments should be placed into an Erlenmeyer flask containing the priming solution (0.2% KNO_3). A plastic pipette inserted into the flask and connected to a source of air (Figure 41.7) will aerate the priming solution. Anoxia or lack of oxygen in the priming solution reduces the viability of the seeds. Lettuce seeds should be left in the priming solution for 48 h, after which they are removed, rinsed, and allowed to redry for 24 h. These seeds are now designated as "primed."
3	Two petri dishes lined with Whatman #1 filter paper should be prepared to receive the primed lettuce seeds, one for the 25°C and the other for the 30°C environment. Place 20 primed lettuce seeds in each, and moisten the surface of the filter paper with water until it glistens. Prepare two more petri dishes, and follow the procedure above, but using dry lettuce seeds. Place one primed and one control treatment in the 30°C incubator. The other primed and control treatment may be left out at room temperature (25°C) until germination.
3	TerraSorb should be hydrated by placing 25 g in a liter of water. Make up a sufficient quantity to create a volume two to three times that of the seeds. Mix the TerraSorb matrix with the sweet corn seeds, and spread the mixture in a shallow tray to allow the seeds to pregerminate for 24 h (Figure 41.8).
4	After 24 h, TerraSorb-pregerminated seeds of sweet corn should be rinsed using the strainer and then the seeds blotted to remove free water. Spread the seeds onto paper towels, and allow them to redry for 24 hours.
5	Place 10 sweet corn seeds, which had been pregerminated in the TerraSorb, in a petri dish lined with Whatman #1 filter paper, and moisten the surface of the paper until it glistens. As a control, prepare 10 sweet corn seeds that have received no pregermination treatment in the same way. Label both petri dishes with the treatment, and place in the refrigerator (5°C) to simulate exposure to chilling temperatures that sweet corn planted too early in the spring would experience. Now prepare two more petri dishes just like those above, and place on the bench top in the laboratory to germinate at room temperatures (25°C).
6	Place 20 primed lettuce seeds in a petri dish lined with Whatman #1 filter paper, and moisten the surface of the paper until it glistens. Do the same for 20 lettuce seeds which were pregerminated in TerraSorb. As a control, prepare 20 lettuce seeds that had received no pretreatment. Label the petri dishes with the treatment, and place in the seed germinator set to 30°C to simulate thermodormancy conditions. Now prepare three more petri dishes just like those above, and place on the bench top in the laboratory to germinate at room temperatures (25°C).
7	Examine the seeds in the petri dishes daily for the next 10 days or until additional germination has not occurred. Collect data on cumulative germination percentage. Count a seed as germinated once the radicle emerges. Graph the performance of the two seed treatments under temperature stress conditions (either chilling or thermodormancy), and compare with the performance of the two treatments conducted at room temperature.

FIGURE 41.7 Seeds of lettuce undergoing priming with 0.2% KNO_3. Note how the solution is being aerated to prevent anoxia.

FIGURE 41.8 Seeds of standard sweet corn after being pregerminated in TerraSorb superabsorbent gel in a large weighing boat for 24 h.

SUGGESTED LITERATURE

Dirr, M.A. 1998. *Manual of Woody Landscape Plants: Their Identification, Ornamental Characteristics, Culture, Propagation and Uses.* Stipes Publishing, Champaign, IL. 1187 pp.

Hartmann, H.T., D.E. Kester, F.T. Davies, Jr., and R.L. Geneve. 2002. *Plant Propagation: Principles and Practices,* 7th ed. Prentice Hall. Inc., Upper Saddle River, NJ. 880 pp.

MacDonald, B. 1986. *Practical Woody Plant Propagation for Nursery Growers.* Timber Press. Portland, OR.

Sabota, C., C. Beyl, and J. Biedermann. 1987. Acceleration of sweet corn germination at low temperatures with Terra-Sorb or water presoaks. *HortScience.* 22: 431–434.

Young, J.A. and C.G. Young. 1994. *Seeds of Woody Plants in North America.* Dioscorides Press, Portland, OR.

Part XIV

In Conclusion: Special Topics

42 Myths of Plant Propagation

Jeffrey H. Gillman

CHAPTER 42 CONCEPTS

- Over time, a variety of techniques have been used to improve propagation success, but not all of these techniques are actually beneficial.

- In any research, and especially propagation research, a control is needed to assure that observed benefits are real and not imagined.

- It is important to try new propagation techniques. Only by experimenting can we discover the best techniques to multiply the plant that we want to produce.

Up until now, most of the propagation techniques that you have been exposed to in this book will work to one extent or another. They are based on research, and most have been tested extensively by commercial propagators. This chapter is not about these established and well-tested techniques, it is about propagation techniques and ideas that are not in line with what we currently think of as mainstream.

One of the great things about propagation is the ability that we have to experiment. Throughout the course of this book you have had the chance to try your own experiments, investigating various methods and techniques. If this chapter does nothing else, hopefully it will encourage you to try new things when you propagate. New propagation methods are constantly introduced as propagators work to improve germination rate, cutting success, and speed to rooting. Some of these methods have a positive impact and so become common practice, but a number have found a place at one propagator's bench or another due, not to their usefulness, but rather to superstition and rhetoric. How could this happen? Simple. In the world of propagation, it is not unusual for a propagator to try a new technique without simultaneously testing untreated plants, cuttings, or seeds. Without testing these techniques against a control, any success that is seen may be attributed to the applied treatment instead of to a more mundane explanation, such as a healthy, well-watered stock plant. In the coming paragraphs, a number of techniques and treatments will be introduced, some of which you may want to try. If you do, be sure to test them against untreated controls, so that you can be confident in your results.

APPLYING GRAVEL TO THE BASE OF SEED TRAYS TO IMPROVE DRAINAGE

A well-drained medium is important to seed propagators for preventing newly formed roots from suffocating and for controlling soilborne diseases. Over the years a variety of methods has been used to ensure that propagation media drains appropriately, the most common of which is to add coarse materials, such as vermiculite and perlite, to the propagation medium. However, in the late nineteenth and early twentieth centuries, a technique was implemented to improve drainage that included adding coarse materials, such as oyster shells or gravel, to the base of containers rather than dispersing these materials throughout the medium. Even today it is not unheard of for propagators and gardeners to apply gravel to the base of their containers to ensure that water moves rapidly through the medium and enough air reaches the germinating seeds and roots.

DOES IT WORK?

There is no doubt at all that a container full of gravel will be better drained than a container full of soil or a commercial potting medium. The question is whether filling the bottom half of a container with gravel and then filling it the rest of the way with a more typical medium will provide better drainage to that top section. The answer to this question is no, but why? Well, the answer is that water does not move freely between media with different pore sizes, so there is not a continuous column of water between the growing medium and the gravel. Without this continuous column, water will not be drawn out of the growing medium and into the gravel, instead sitting in the media above the gravel. In other words, by applying gravel to the base of the container we end up making what amounts to a shorter, squatter container. Shorter containers have worse drainage than taller containers with a similar diameter.

SHOULD YOU TRY IT?

At this point, you realize that good drainage is important for seed propagation; however, good drainage cannot be achieved by placing a more poorly drained media on top of a well-drained media. It just doesn't work. However, if you are a skeptic, there is a quick way to test it out. Take two clear plastic cups. Punch holes in the bottom of both, and then fill one with a well-drained media and the other half way with gravel and then the rest of the way with the same well-drained media. Now water both cups. You should be able to see the water level in the media through the side of the container. What you will observe is that the water drains out of the cups unequally. In the cup with the gravel, the water level will be higher than in the cup without the gravel and will drain more slowly over time.

YOU CAN'T GRAFT AN APPLE ONTO AN ORANGE

Over the years there have been all kinds of tall tales about how one propagator or another is so good at their jobs that they can graft an apple onto an orange. Obviously, when we hear something like that, we think that the person doing the talking is making some sort of a joke or exaggerating for the sake of emphasis. But is this really such a far-fetched idea? After all, apples and oranges both have cambial layers that do essentially the same thing. Why can't you just line up the layers and—bang—there you go—an apple onto an orange or a fig onto a pear or a walnut onto a plum.

DOES IT WORK?

People have tried to graft different plants onto each other with some very interesting results, though what we usually see is incompatibility (see chapter 27). What this means is that the two plants can't work together, and the graft fails either right away, or at a later time, perhaps years in the future. When the graft lasts for a while before it fails, we say that we have delayed incompatibility. Getting around incompatibility is not easily accomplished. In fact, it is not usually accomplished at all.

Grafting an apple onto an orange will not work. But that does not mean that there are not unrelated plants that can be attached to each other successfully. It is true that grafting is typically most successful when it is accomplished using plants that are of the same species. However, it is not unusual for plants of closely related species, and even genera, to be able to be grafted together. For example, it is common practice for pears to be grafted onto quince rootstock in order to dwarf the produced tree. Lilac (*Syringa vulgaris*) was once commonly grafted onto privet (*Ligustrum ovalifolium*) because of the difficulty of rooting lilac cuttings, and trifoliate orange (*Poncris trifoliate*) may be used as a rootstock for *Citrus sp.* to confer cold tolerance. Many maples, including the full moon maple and Japanese maples, can be grafted onto each other, as can many different apples, ashes, and others.

SHOULD YOU TRY IT?

Yes, definitely. It's a lot of fun to try grafting different plants onto each other. You will certainly end up with a lot of disappointments, but you will probably discover a few surprises too. Plus, by trying to graft different plants onto each other you will have the opportunity to gain a familiarity with the plants and with grafting in general that you might not otherwise have.

GROWING PLANTS IN THE DARK TO IMPROVE ROOTING

The term etiolation refers to a plant's response to growing in darkness (see chapter 22). This response is characterized by very little chlorophyll in leaf tissue, resulting in bleached-out leaves and/or flowers, and very narrow, elongated stems. Etiolation is an old process that has been used for many years to alter the shape and even flavor of foods. Most people are familiar with cauliflower, which is commonly etiolated when it is grown, hence its white color. There are people out there who swear that growing plants in the dark will increase their likelihood of cuttings from these plants developing roots. But why would it? Intuition says that these people are wasting their time. After all, plants need nice green leaves to photosynthesize and create sugars for new and growing roots. If we etiolate, we will not have these nice green leaves, so how could this process possibly work?

DOES IT WORK?

Believe it or not, plants that are grown in the dark do tend to develop roots more readily than plants that are grown in plenty of sunlight. In fact, many plants, such as lilacs, apples, pears, and others, respond well to this technique. But why? Currently there is not a single definitive answer to this question, but there are a few characteristics of stem cuttings from plants that were etiolated that could potentially make them more likely to root than nonetiolated cuttings. First, etiolated cuttings respond to auxin much more readily than nonetiolated cuttings. Second, etiolated cuttings may have chemicals in them that can act as cofactors, aiding the activity of the rooting hormone. And third

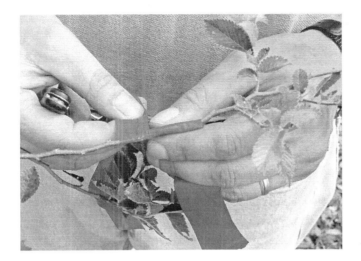

FIGURE 42.1 Wrapping a light-excluding material around a branch to prepare it for rooting (banding).

and finally, etiolated cuttings have less lignin, a structural compound, in them. It is possible that reduced lignin allows roots to penetrate out through tissues that would inhibit their extension under normal conditions.

SHOULD YOU TRY IT?

Absolutely. There are a few ways to grow plants without light. Most propagators don't keep the plants they're planning on reproducing in complete darkness the entire time they're growing them, but rather cover them with heavy shade, often in the form of a box, for a few days, or up to 10 weeks, before cuttings are taken. Another easy way to implement etiolation technique is called banding (see chapter 22) and involves wrapping a band of heavy cloth around the piece of stem where you plan on taking cuttings and leaving it there for a few weeks (Figure 42.1). This region of the stem should be cut so that the banded portion constitutes the bottom of the cutting that is taken. The cutting, and specifically the banded region, can be treated with a rooting hormone after it is removed from the plant. If you try etiolation, be sure to compare the rooting success of cuttings from plants that were etiolated with cuttings from similar plants that were not. Remember, you'll never know how well a treatment works unless you compare it to an untreated control.

USING HERBICIDES TO PROMOTE ROOT FORMATION

Once upon a time, plant propagators used herbicides to help them produce roots on cuttings. It is like a magic trick. Take a cutting and stick the base into a pan with a small amount of herbicide in it, place it in a well-drained rooting media, and put it onto a mist bench and, in a few weeks—poof. Instead of having a dead cutting you've got a plant with roots. What's going on here?

DOES IT WORK?

The herbicide of choice for rooting cuttings is 2,4-D, also known as 2,4-dichlorophenoxyacetic acid. This is the same stuff that you spray on your yard to kill dandelions while preserving the grass. This compound was first created when researchers, including Dr. Franklin Jones who received the first patent for 2,4-D, tried to create synthetic auxins during World War II to act as herbicides, perhaps against Japan's rice crop. Indoleacetic acid (IAA) was too unstable chemically to make an effective herbicide, and so something else was needed. The auxin 2,4-D was synthesized before the end of the war, but was not registered for use until 1946 because of its status as a war secret. This chemical is a strong auxin that does not break down rapidly and that tends to be more effective on what are called "broad-leaved weeds" or, in other words, most plants other than grasses. At high enough concentrations this strong auxin causes broad leaved weeds to have abnormal growth and die. Grasses are not affected to the same extent and tend to be able to tolerate a heavy dose of 2,4-D before they succumb to its effects. At lower doses, 2,4-D may not be terminal, but might instead provide the stimulation that a cutting needs for it to develop root initials. Because of its longevity, and the sensitivity of many plants to it, 2,4-D is not a compound that is commonly used to stimulate roots in cuttings anymore; however, it is still sometimes used as an auxin for micropropagation.

SHOULD YOU TRY IT?

This is a technique that is better off not tried by the casual propagator. 2, 4-D is a pesticide and is not currently labeled for use as a root-promoting compound. If you want to root a cutting by using a hormone, then you're much better off purchasing a product that contains the indolebutyric acid (IBA) or naphthalene acetic acid (NAA) that you read about in earlier chapters. If you are

FIGURE 42.2 Cutting willow branches to prepare a willow diffusate.

interested in trying to use 2,4-D as a root promoter, then do so only under the direction of a qualified instructor who is trained in the use of hazardous materials.

PLANTS THAT ROOT EASILY CAN BE USED TO PROMOTE ROOT FORMATION IN OTHER PLANTS

There is a theory that, if a tree is easy to root, then that tree must contain some sort of chemical that makes the tree easy to root. Furthermore, this chemical should be able to be extracted by soaking the stems or other portions of the plant in water. After soaking, the liquid portion of the concoction can be separated from the stems and used to treat cuttings from other difficult-to-root plants to make them more likely to develop roots (Figure 42.2).

Does It Work?

Over the years, a number of people have tried to use extracts from a variety of easy-to-root plants to promote root formation in cuttings. These extracts are created by soaking the stems of the easy-to-root plant in water for some period of time and then using the liquid that the stems were soaked in to treat cuttings from plants that they want to root (Figure 42.2). The extracts used are usually called diffusates, the theory being that important rooting compounds will diffuse from the easy-to-root plant into the water. The first person to research diffusates as a means of root promotion was C. E. Hess in the 1950s. Willow trees are the plant most commonly used for root promotion in other plants, though extracts from plants, such as ivy and black locust, have also been implicated as root promoters. In some studies, these extracts have worked, but in most they have not. There are people who will swear up and down that these extracts can coax roots from the stems of hard-to-root plants, but hard evidence is currently lacking.

Indeed, most researchers who try these compounds (such as the author of this chapter) end up unimpressed with their performance.

One argument for the use of diffusate from willow trees is the fact that willows contain the compound salicylic acid (similar to aspirin), which has been shown to increase the rooting of certain plants including faba and mung beans. Unfortunately, since these beans are extremely easy to root, showing that salicylic acid helps to promote root formation in them is not an appropriate test for whether they will promote root formation in other, more difficult-to-root species.

Should You Try It?

It is fun and easy to test whether various diffusates will help cuttings from various plants develop roots. Simply take stems from plants whose root promoting abilities you would like to test, and steep them in water for a certain amount of time. You may use the amount of time that you steep them as a variable if you like (the number of stems soaking in the water could also be a variable), perhaps allowing some stems to sit in water for an hour, some a day, and some a week. Next, take the stems out and filter the whole mess through cheesecloth, keeping the filtered water. This filtered water is the diffusate.

Place stems from the plants that you would like to root into the filtered water and let them sit for some period of time. Twenty-four hours is often noted as appropriate and should provide enough time for the cuttings to collect the compounds that they need for root formation, but certainly this can be treated as a variable, and stems can be allowed to soak in the diffusate for any length of time that you see fit. When the plants are done soaking, they should be placed onto an ordinary mist bench and allowed to sit for a few weeks as with any stem cuttings. Be sure to test

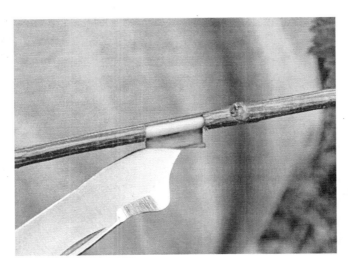

FIGURE 42.3 Girdling a stem with a budding knife.

untreated cuttings along with the diffusate treated cuttings so that you can compare the rooting success of both. You may also choose to treat some cuttings with IBA so that you can compare the success of diffusates with a synthetic auxin.

CUTTING A RING AROUND A STEM WILL BENEFIT ROOT FORMATION

The practice of cutting a ring of bark from around a stem six or so weeks before cuttings are taken is an uncommon practice, but one which some people claim is a way to encourage roots from ordinarily difficult-to-root cuttings (Figure 42.3). This cut is called a girdling cut and is one that is considered detrimental to the tree. In fact, there are certain beetles called stem girdlers who cause this type of damage and are considered pests of the trees they feed on. So why would girdling the bark around a healthy stem increase rooting? Wouldn't this just damage the plant, and particularly the stems from which the cuttings are taken?

DOES IT WORK?

Cutting a ring around a stem will disturb the physiology of the stem, no doubt about it, but this disrupted physiology is not necessarily bad for propagation. When a stem is girdled, the tissue that is removed is the phloem. When the phloem is damaged, carbohydrates manufactured in the leaves do not have a path to follow to exit the stem; hence, the base of the stem, where the girdle was made, develops an abnormally high content of carbohydrates that may be useful to a cutting when it is taken. Additionally, other compounds that are normally exported from the stem through the phloem will build up, including, potentially, auxins and auxin cofacters, which may aid in stimulating roots.

SHOULD YOU TRY IT?

If you girdle every stem on a plant you could damage that plant, perhaps even severely if the plant is small to begin with. However, if you just want to girdle a few branches and then take cuttings from them, then this is a great idea for an experiment. Make sure to wait a few weeks between the time that you girdle the stems and take the cuttings so that all of the potentially important compounds can build up in the stem, and be sure to compare the success of these cuttings with cuttings from the same plant or plants that haven't been girdled so that you can make a reasonable assessment as to whether the girdling benefited or not.

PLACING GRAIN AT THE BASE OF A CUTTING WILL BENEFIT ROOT FORMATION

When people in the Middle East and Europe propagated plants using cuttings centuries ago, they would often place grain into the base of a cutting to promote root formation. In fact, in some regions of Macedonia, it is still a common practice to embed a germinating grain of wheat in the base of a cutting to stimulate root development. Is the use of grain in this way just a superstitious waste of time or is there some practical reason why it could stimulate root development in cuttings?

DOES IT WORK?

There are few studies that investigated the usefulness of placing or embedding a germinating grain of wheat into a cutting. However, there is reason to believe that there is more to this practice than superstition. Germinating seeds tend to be higher in auxin content than other parts of the plant. Germinating grains of wheat are packed with auxin (relatively speaking) and so might act as a promoter of roots for a cutting. Another seed that has particularly high auxin content is corn, which might also serve to encourage roots from a cutting. Of course, the act of slicing the base of a cutting can promote roots in and of itself and so it might not always be clear whether the benefits of putting grain into the base of a cutting that has been sliced open is actually greater than just making the cut, unless experiments are carefully controlled.

Should You Try It?

Why not give this one a shot? All you need are some cuttings and some grain. To test whether germinating grain benefits root development, all you need to do is to slice open the base of the cutting and insert the grain, which should be soaked for 24 to 48 h before insertion to stimulate germination. Be sure to test this against other cuttings that are also slit open but which have something else inserted besides grain; perhaps a grain-sized piece of sand. It is very important when conducting any experiment to have appropriate controls, especially in a situation like this. Cuttings from certain plants respond favorably when their bases are cut open; hence, if cuttings without a slice in them are used as controls, the simple act of slicing the cutting with the grain in it may stimulate roots and lead you to the erroneous assumption that the grain promoted rooting, when in actuality it was the cut that was the culprit. Alternately, you could also test whether the slit is even necessary by just placing the grain next to the cutting as it is inserted into your propagation media just before it goes onto the mist bench. The control for this would simply be to stick cuttings into the mist bench without the grain beside it.

SUGGESTED READINGS

You Can't Graft an Apple onto an Orange

Yelenosky, G. and J.C.V. Vu. 1992. Ability of 'Valencia' sweet orange to cold-acclimate on cold-sensitive citron rootstock. *HortScience.* 27: 1201–1203.

Growing Plants in the Dark to Improve Rooting

Koukourikou-Petridou, M.A. 1998. Etiolation of stock plants affects adventitious root formation and hormone content of pea stem cuttings. *Plant Growth Regul.* 25: 17–21.

Maynard, B.K. and N.L. Bassuk. 1991. Stock plant etiolation and stem banding effect on the auxin dose-response of rooting in stem cuttings of *Carpinus betulus* L. 'fastigiata.' *Plant Growth Regul.* 10: 305–311.

Using Herbicides to Promote Roots

Delargy, J.A. and C.E. Wright. 1978. Root formation in cuttings of apple in relation to auxin application and to etiolation. *New Phytol.* 82: 341–347.

Plants That Root Easily Can Be Used to Promote Root Formation in Other Plants

Girourd, R.M. and C.E. Hess. 1964. The diffusion of root-promoting substances from stems of *Hedera helix*. *Comb. Proc. Intl. Plant Prop. Soc.* 14: 162–166.

Hess, C.E. 1959. A study of plant growth substances in easy and difficult to root cuttings. *Comb. Proc. Intl. Plant Prop. Soc.* 9: 39–43.

Sharma, A.R. and R.N. Trigiano. 1999. Rooting flowering dogwood (*Cornus florida*) microshoots. *Proc. SNA Res. Conf.* 44: 376–377.

Cutting a Ring around a Stem Will Benefit Root Formation

Cooper, W.C. 1936. Transport of root forming hormone in woody cuttings. *Plant Physiol.* 11: 779–793.

Delargy, J.A. and C.E. Wright. 1978. Root formation in cuttings of apple (cv. Bramley's seedling) in relation to ringbarking and to etiolation. *New Phytol.* 81: 117–127.

Placing Grain at the Base of a Cutting Will Benefit Root Formation

Sheldrake, A.R., 1973. Do coleoptile tips produce auxin? *New Phytolog.* 72: 433–447.

43 Intellectual Property Protection for Plants

Christopher Eisenschenk

CHAPTER 43 CONCEPTS

- Plant patents, utility patents, Plant Variety Protection Act (PVPA) certificates, and trademarks are forms of federal intellectual property protection that are available for plants and plant materials.

- Unauthorized sexual reproduction of plants can be controlled by utility patents and/or PVPA certificates.

- Unauthorized asexual reproduction of plants can be controlled by utility patents and/or plant patents.

- Of all the forms of federal intellectual property protection available for plants, a utility patent provides the strongest intellectual property protection and is the most difficult to obtain.

This chapter will address the various types of federal intellectual property protection that are available for plants and plant materials. While some forms of state intellectual property rights exist, a discussion of these rights is beyond the scope of this chapter. Additionally, this chapter is not intended to be an exhaustive discussion of the various forms of intellectual property protection available for plants and is not to be construed as legal advice regarding any form of intellectual property discussed herein.

The types of federal intellectual property protection available for plants and plant materials include patents (utility patents and/or plant patents), Plant Variety Protection Act (PVPA) certificates, and trademark protection. A discussion of copyrights is also included in this chapter to familiarize the reader with this form of intellectual property protection (primarily with respect to representations of plants produced for marketing or other purposes). Most commonly, utility and plant patents are used to protect intellectual property rights; however, the PVPA provides a form of intellectual property protection that is analogous to patent rights, albeit with a few exceptions that may impact one's decision to pursue PVPA protection for plants.

COPYRIGHT PROTECTION AND PLANTS

Federal copyright protection exists for "original works of authorship fixed in any tangible medium of expression, now known or later developed, from which they can be perceived, reproduced or otherwise communicated, either directly or with the aid of a machine or device" (see 17 U.S.C. §102(a) (1988)). Thus, where an author fixes an image of a plant or depicts the image of a plant in some form of tangible medium (an "original work"), a copyright in that original work is obtained at the moment of its fixation within the tangible medium. However, it is important to note that the scope of protection for the copyrightable material (e.g., an "original work" depicting a plant) only extends to the "original work" and not the plant itself.

17 U.S.C. §102(a) (1988) provides examples of "original works." This nonlimiting list includes the following: literary works, musical works (including lyrics), dramatic works (including accompanying music), pantomimes and choreographic works, pictorial, graphic and sculptural works, motion pictures and other audiovisual works, sound recordings, and architectural works (see 17 U.S.C. §102(a) (1988)). However, specifically excluded from copyright protection is any idea, procedure, process, system, method of operation, concept, principle, or discovery, regardless of the form in which it is described, explained, illustrated, or embodied in such work (see 17 U.S.C. §102(b) (1988)). Additionally, the term "tangible medium" refers to any physical medium, nonlimiting examples of which include paper, video/audio tapes, or electronic storage medium (e.g., computer disks).

The owner of the copyrighted "original work" has the exclusive right, subject to certain limitations, to do and/or to authorize any of the following: (a) to reproduce the copyrighted work in copies; (b) to prepare derivative works based upon the copyrighted work; (c) to distribute copies of the copyrighted work to the public by sale or other transfer of ownership, or by rental, lease, or lending; (d) in the case of literary, musical, dramatic, and choreographic works, pantomimes, and motion pictures and other audiovisual works, to perform the copyrighted work publicly; (e) in the case of literary, musical, dramatic, and choreographic works, pantomimes, and pictorial, graphic, or sculptural works, including the individual images of a motion picture or other audiovisual work, to display the copyrighted work publicly; and (f) in the case of sound recordings, to perform the copyrighted work publicly by means of a digital audio transmission (see 17 U.S.C. §106 (1988)). However, the author of an "original work" depicting a plant should also be aware of the "fair use doctrine" that allows for the limited reproduction of the copyrighted work without the consent of the author. The "fair use doctrine" allows for the reproduction of a copyrighted work for purposes including, but not necessarily limited to, criticism, comment, news reporting, teaching, scholarship, or research. Whether the reproduction of a copyrighted work is a "fair use" is determined on a case-by-case basis and includes an analysis of the following factors: (a) the purpose and character of the use, including whether such use is of a commercial nature or is for nonprofit educational purposes; (b) the nature of the copyrighted work; (c) the amount and substantiality of the portion used in relation to the copyrighted work as a whole; and (d) the effect of the use upon the potential market for or value of the copyrighted work (see 17 U.S.C. §107 (1988)).

Finally, federal copyright protection exists as of the moment that the "original work" is fixed in a tangible medium, and there is no longer any requirement for marking the first publication of a work, published as of March 1, 1998, with a copyright symbol "©," the term "Copyright" (or the abbreviation "Copr."), the author's name, and the date of publication. Additionally, one should note that there is no requirement for copyright registration with the Copyright Office in the Library of Congress in order to establish one's copyright in an "original work." However, registration of one's copyright with the Copyright Office and proper marking of the "original work" with a copyright notice provides a number of advantages in case of infringement or unauthorized reproduction of the work. For example, no lawsuit for infringement of the copyright in any United States work shall be instituted until registration of the copyright claim has been made (see 17 U.S.C. § 411). Additionally, statutory damages and attorneys' fees are not available to a copyright owner unless registration of the copyright has preceded the infringement, although a three-month grace period exists in this regard (see 17 U.S.C. §§ 412 and 504–505). Finally, the copyright statutes provide that no weight shall be given to such a defendant's assertion of a defense based on innocent infringement if a copyright notice as specified in 17 U.S.C. § 401(b)–(c) has been placed on the published copy or copies to which a defendant in a copyright infringement suit had access.

Thus, one should recognize that Web sites, photographs, marketing materials, sketches, artwork, or other forms of expression that depict a plant produced by a grower (and are fixed in a tangible medium) are a protectable form of intellectual property associated with the plant. One should also be aware of the rights that are associated with this aspect of intellectual property protection and consider enforcing these rights if infringers are identified. Additional information regarding copyrights can be accessed on the Internet at www.copyright.gov.

TRADEMARKS AND THEIR USE WITH PLANTS

The term "trademark" includes any word, name, symbol, or device, or any combination thereof (a) used by a person, or (b) which a person has a bona fide intention to use in commerce and who applies to register the trademark on the principal trademark register to identify and distinguish his or her goods, including a unique product, from those manufactured or sold by others and to indicate the source of the goods, even if that source is unknown (see 15 U.S.C. § 1127). Thus, a trademark serves to identify the source or point of origin of a particular product sold in the stream of commerce.

As would, perhaps be expected, the United States Patent and Trademark Office (USPTO) has promulgated guidelines addressing the use of varietal or cultivar names as trademarks in the *Trademark Manual of Examination Procedures* (TMEP)—4th edition (2005). As indicated in section 1202.12—Varietal and Cultivar Names (Examination of Applications for Seeds and Plants):

> Varietal or cultivar names are designations given to cultivated varieties or subspecies of live plants or agricultural seeds. They amount to the generic name of the plant or seed by which such variety is known to the public. These names can consist of a numeric or alphanumeric code or can be a "fancy" (arbitrary) name. The terms "varietal" and "cultivar" may have slight semantic differences but pose indistinguishable issues and are treated identically for trademark purposes.

> Subspecies are types of a particular species of plant or seed that are members of a particular genus. For example, all maple trees are in the genus *Acer*. The sugar maple species is known as *Acer saccharum*, while the red maple species is called *Acer rubrum*. In turn, these species have been subdivided into various cultivated varieties that are developed commercially and given varietal or cultivar names that are known to the public.

> If the examining attorney determines that wording sought to be registered as a mark for live plants, agricultural seeds, fresh fruits or fresh vegetables comprises a varietal or cultivar name, then the examining attorney must refuse registration, or require a disclaimer, on the ground that the matter is the varietal name of the goods and does not function as a trademark under §§1, 2 and 45

of the Trademark Act, 15 U.S.C. §§1051, 1052 and 1127. *See Dixie Rose Nursery v. Coe*, 131 F.2d 446, 55 USPQ 315 (D.C. Cir. 1942), *cert. denied* 318 U.S. 782, 57 USPQ 568 (1943); *In re Hilltop Orchards & Nurseries, Inc.*, 206 USPQ 1034 (TTAB 1979); *In re Farmer Seed & Nursery Co.*, 137 USPQ 231 (TTAB 1963); *In re Cohn Bodger & Sons Co.*, 122 USPQ 345 (TTAB 1959). Likewise, if the mark identifies the prominent portion of a varietal name, it must be refused. *In re Delta and Pine Land Co.*, 26 USPQ2d 1157 (TTAB 1993) (Board affirmed refusal to register DELTAPINE, which was a portion of the varietal names Deltapine 50, Deltapine 20, Deltapine 105 and Deltapine 506).

A varietal or cultivar name is used in a plant patent to identify the variety. Thus, even if the name was originally arbitrary, it "describe[s] to the public a [plant] of a particular sort, not a [plant] from a particular [source]." *Dixie Rose*, 131 F.2d at 447, 55 USPQ at 316. It is against public policy for any one supplier to retain exclusivity in a patented variety of plant, or the name of a variety, once its patent expires. *Id.*

In trademark law, a descriptive trademark conveys information regarding an ingredient, quality, characteristic, function, feature, purpose, or use of the product or service. Descriptive trademarks are, generally, not eligible for registration because the mark must become associated with the product or service among consumers (e.g., acquire secondary meaning). With respect to a view that the cultivar name is generic, trademark law holds that generic terms or common words for the products or services cannot function as a trademark because it would prevent others from rightfully using the common name for the product or service that they make. Thus, in view of the guidance provided to trademark examining attorneys by the TMEP, it is clear that cultivar names will be considered generic or, at best, descriptive and ineligible for trademark protection.

Thus, while one may not be able to trademark a particular cultivar name, it should be understood that one can still identify a particular cultivar as originating with a particular source or point of origin. In this context, a nursery or grower of a particular cultivar may trademark a different name and use it in conjunction with the cultivar name to identify the source or point of origin of the cultivar produced by the nursery or grower. This combination of a trademark and the cultivar name would then allow the consumers of the plant to identify the source or point of origin of the cultivar. With respect to marking requirements under trademark law, that portion of the combined trademark–cultivar name that constitutes the trademarked portion should be identified with a "®" or "™" symbol on the tag or label to designate a registered or common law trademark, respectively. The portion of the combined trademark–cultivar name that identifies the actual cultivar name on the tag or label should identify the cultivar using single quotation marks according to the industry norm.

THE PLANT VARIETY PROTECTION ACT

The Plant Variety Protection Act (PVPA) is a form of federal intellectual property protection that is available to growers of new sexually reproduced plants. The PVPA is administered through Plant Variety Protection Office (PVPO) of the United States Department of Agriculture and provides a breeder somewhat patent-like protection for any sexually reproduced or tuber-propagated plant variety (other than fungi or bacteria).

Thus, if a plant variety is new, distinct, uniform, and stable as defined in 7 U.S.C. § 2402 of the PVPA,* it will be eligible for a certificate under the Act. Growers should also be aware of the various statutory bars that would prevent one from obtaining a PVPA certificate. These include making the plant variety available to persons, for purposes of exploitation, within the United States more than one year prior to the filing date of the PVPA application; making the plant variety available to persons outside the United States more than four years prior to the filing date of the PVPA application (although certain exceptions may exist for tubers); or making a tree or vine available to persons outside the United States more than six years prior to the filing date of the PVPA application.

Turning now to the scope of protection for plants for which PVPA certificates have been issued, courts have held that the PVPA protects plants and seed that differ "from either of the parent plants as well as from other

---

* § 2402. Right to plant variety protection; plant varieties protectable
(a) In general
The breeder of any sexually reproduced or tuber propagated plant variety (other than fungi or bacteria) who has so reproduced the variety, or the successor in interest of the breeder, shall be entitled to plant variety protection for the variety, subject to the conditions and requirements of this chapter, if the variety is—
  (1) new, in the sense that, on the date of filing of the application for plant variety protection, propagating or harvested material of the variety has not been sold or otherwise disposed of to other persons, by or with the consent of the breeder, or the successor in interest of the breeder, for purposes of exploitation of the variety—
    (A) in the United States, more than 1 year prior to the date of filing; or
    (B) in any area outside of the United States—
      (i) more than 4 years prior to the date of filing, except that in the case of a tuber propagated plant variety the Secretary may waive the 4-year limitation for a period ending 1 year after April 4, 1996; or
      (ii) in the case of a tree or vine, more than 6 years prior to the date of filing;
  (2) distinct, in the sense that the variety is clearly distinguishable from any other variety the existence of which is publicly known or a matter of common knowledge at the time of the filing of the application;
  (3) uniform, in the sense that any variations are describable, predictable, and commercially acceptable; and
  (4) stable, in the sense that the variety, when reproduced, will remain unchanged with regard to the essential and distinctive characteristics of the variety with a reasonable degree of reliability commensurate with that of varieties of the same category in which the same breeding method is employed.

plants produced from other seeds resulting from the cross-pollination" and that "[p]lants true-to-type, although different in a strict genetic sense, are protectable under the PVPA" [see *Imazio Nursery, Inc. v. Dania Greenhouses*, 69 F3d. 1560 (Fed. Cir. 1996)]. Thus, the PVPA allows for some genetic variation within the plants and/or seeds of a variety protected by a PVPA certificate. As discussed below, such genetic variation is not permitted for plants protected under the Plant Patent Act (PPA).

With respect to the rights conferred by the owner of a protected variety, the 7 U.S.C. § 2541 indicates that:

> ... it shall be an infringement of the rights of the owner of a protected variety to perform without authority, any of the following acts in the United States, or in commerce which can be regulated by Congress or affecting such commerce, prior to expiration of the right to plant variety protection but after either the issue of the certificate or the distribution of a protected plant variety with the notice under section 2567 of this title:
>
> **(1)** sell or market the protected variety, or offer it or expose it for sale, deliver it, ship it, consign it, exchange it, or solicit an offer to buy it, or any other transfer of title or possession of it;
> **(2)** import the variety into, or export it from, the United States;
> **(3)** sexually multiply, or propagate by a tuber or a part of a tuber, the variety as a step in marketing (for growing purposes) the variety;
> **(4)** use the variety in producing (as distinguished from developing) a hybrid or different variety therefrom;
> **(5)** use seed which had been marked "Unauthorized Propagation Prohibited" or "Unauthorized Seed Multiplication Prohibited" or progeny thereof to propagate the variety;
> **(6)** dispense the variety to another, in a form which can be propagated, without notice as to being a protected variety under which it was received;
> **(7)** condition the variety for the purpose of propagation, except to the extent that the conditioning is related to the activities permitted under section 2543 of this title;
> **(8)** stock the variety for any of the purposes referred to in paragraphs (1) through (7);
> **(9)** perform any of the foregoing acts even in instances in which the variety is multiplied other than sexually, except in pursuance of a valid United States plant patent; or
> **(10)** instigate or actively induce performance of any of the foregoing acts.

Damages that one can obtain for infringement of a PVPA certificate shall be adequate to compensate for the infringement, but in no event less than a reasonable royalty for the use made of the variety by the infringer, together with interest and costs as fixed by the court (see 7 U.S.C. § 2564). It is also possible to obtain an injunction prohibiting an infringer from continuing to make and use the seeds or plants protected under the PVPA.

It is also important to recognize that the PVPA provides several exceptions/exemptions for acts of infringement. For example, it is not considered an act of infringement for a farmer to save and replant seed protected by a PVPA certificate.* Additionally, a bona fide sale of seed produced on a farm (either from seed lawfully obtained for seeding purposes or from seed produced by descent from the lawfully obtained seed and produced on the farm) for other than reproductive purposes, made in channels usual for such other purposes, shall not constitute an infringement.† A third, and potentially important, exception/exemption from infringement is a research use exception. 7 U.S.C. § 2544 states that the use and reproduction of a protected variety for plant breeding or other bona fide research shall not constitute an infringement. Thus, for both growers and academics, the use of PVPA-protected plants and seeds for plant breeding or other true research purposes does not constitute an act of infringement. As we shall see, this is not the case of plants protected by either a utility or plant patent.

PLANT PATENTS

Plant patents are granted to those who invent or discover, and asexually reproduce, any new and distinct variety of plant (see 35 U.S.C. § 161‡). Whereas the PVPA can be used to protect a sexually reproduced plant, a plant patent provides the right to exclude others from asexually reproducing the new and distinct plant variety and parts

---

* 7 U.S.C. 2543 Right to save seed; crop exemption
Except to the extent that such action may constitute an infringement under subsections (3) and (4) of section 111 [7 U.S.C. 2541(3), (4)], it shall not infringe any right hereunder for a person to save seed produced by the person from seed obtained, or descended from seed obtained, by authority of the owner of the variety for seeding purposes and use such saved seed in the production of a crop for use on the farm of the person, or for sale as provided in this section. A bona fide sale for other than reproductive purposes, made in channels usual for such other purposes, of seed produced on a farm either from seed obtained by authority of the owner for seeding purposes or from seed produced by descent on such farm from seed obtained by authority of the owner for seeding purposes shall not constitute an infringement. A purchaser who diverts seed from such channels to seeding purposes shall be deemed to have notice under section 127 [7 U.S.C. 2567] that the actions of the purchaser constitute an infringement.
† Ibid.
‡ 35 U.S.C. § 161—Patents for Plants
Whoever invents or discovers and asexually reproduces any distinct and new variety of plant, including cultivated sports, mutants, hybrids, and newly found seedlings, other than a tuber propagated plant or a plant found in an uncultivated state, may obtain a patent therefor, subject to the conditions and requirements of this title.

thereof for a 20-year period (See 35 U.S.C. § 163*). However, asexually reproduced plants can also be protected by utility patents, as 35 U.S.C. § 161, is not an exclusive form of protection which conflicts with the granting of utility patents to plants (see *Ex parte Hibberd*, 227 U.S.P.Q. 443 (Bd. Pat. App. & Int. 1985)). Additionally, plants capable of sexual reproduction (i.e., from seed) can be protected under the Plant Patent Act provided that the plant has been asexually reproduced.

As set forth in 35 U.S.C. § 161, potentially patentable plants include cultivated sports, mutants, hybrids, and newly found seedlings. Fungi and bacteria are not considered "plants" under this aspect of the patents statutes; however, utility patents can be used to protect isolated fungi and bacteria. Additionally, the single claim of a plant patent only protects the entirety of the plant, not subparts of the plant (e.g., seeds, flowers, fruit)†. It should also be noted that plants identified in the wild are not eligible for plant patent protection. Rather, only plants identified in cultivated areas are eligible for protection under the Plant Patent Act. Interestingly, neither the cultivated area in which the plant is discovered nor ownership of the plant claimed within the plant patent application need be owned by the person who identified the plant (i.e., the Applicant for the plant patent; see *Ex parte Moore*, 115 U.S.P.Q. 145 (Pat. Off. Bd. App. 1957)). Specifically excluded from protection under the Plant Patent Act are tuber-propagated plants. Examples of tuber-propagated plants include Irish potato and the Jerusalem artichoke. Because these plants are asexually propagated by the same part as is sold for food, the plants are considered ineligible for protection under a plant patent (see *Manual of Patent Examination Procedure* (M.P.E.P.), section 1601, August 2006, Revision 5).

Among the requirements that must be fulfilled to obtain a plant patent are that (1) the plant is capable of stable asexual reproduction; (2) the plant was invented or discovered and, if discovered, that the discovery was made in a cultivated area; (3) the plant is not a plant that is excluded by statute, where the part of the plant used for asexual reproduction is not a tuber food part; (4) the person or persons filing the application are those who actually invented the claimed plant, that is, discovered or developed and identified or isolated the plant, and asexually reproduced the plant; (5) the plant has not been sold or released in the United States of America more than one year prior to the effective filing date of the application; (6) the plant has not been enabled to the public, i.e., by description in a printed publication in this country more than one year before the application for patent with an offer to sale, or by release or sale of the plant more than one year prior to application for patent; (7) the plant be shown to differ from known, related plants by at least one distinguishing characteristic, which is more than a difference caused by growing conditions or fertility levels, etc.; or (8) the invention would not have been obvious to one skilled in the art at the time of invention by applicant.‡

Infringement of a plant patent results whenever someone asexually reproduces, by any means, a plant that is protected by a plant patent. It should also be noted that "experimental use" defenses are exceptionally narrow and would, likely, not be available should one be involved in litigation involving an allegedly infringed plant patent§. Should one be successful in proving infringement, the patent statutes provide for a variety of remedies. These remedies include enjoining the infringer from producing any additional plants and requiring the payment of a monetary award to the patent holder. Monetary damages can be calculated in a variety of ways including actual damages (e.g., lost sales) or a determination of what a reasonable royalty would have been for the plants sold by the infringer. In case of willful infringement, the damages can be trebled, and it may also be possible to recover attorney's fees in instances where a court finds exceptional circumstances.

UTILITY PATENTS

As noted above, utility patents are also a form of federal intellectual property protection that is available for plants and plant materials. For plants, utility patents are most often directed to transgenic plants (e.g., plants containing herbicide resistance genes or genes that confer resistance of insect pests). It is far less common for utility patents to be sought for nontransgenic plants, although a utility and plant patent can be obtained for a given plant if desired. To be eligible for patent protection, an invention must be one of the statutory classes of invention,¶ and the specification

---

* 35 U.S.C. § 163. Grant.
  In the case of a plant patent, the grant shall include the right to exclude others from asexually reproducing the plant, and from using, offering for sale, or selling the plant so reproduced, or any of its parts, throughout the United States, or from importing the plant so reproduced, or any parts thereof, into the United States.
† 37 C.F.R. § 1.164. Claim.
  The claim shall be in formal terms to the new and distinct variety of the specified plant as described and illustrated, and may also recite the principal distinguishing characteristics. More than one claim is not permitted.

‡ See U.S. Patent and Trademark Office publication "General Information About 35 U.S.C. 161 Plant Patents," accessible at www.uspto.gov/web/offices/pac/plant/.
§ *Madey v. Duke University*, 307 F.3d 1351, 1362 (Fed. Cir. 2002) (stating that regardless of whether a particular institution or entity is engaged in an endeavor for commercial gain, so long as the act is in furtherance of the alleged infringer's legitimate business and is not solely for amusement, to satisfy idle curiosity, or for strictly philosophical inquiry, the act does not qualify for the very narrow and strictly limited experimental use defense.
¶ 35 U.S.C. 101 Inventions patentable.
  Whoever invents or discovers any new and useful process, machine, manufacture, or composition of matter, or any new and useful improvement thereof, may obtain a patent therefor, subject to the conditions and requirements of this title.

that describes the inventions must be written such that it teaches those skilled in the art of how to make and use the invention that is claimed (this is commonly referred to as the "enablement requirement").* In some cases, the deposit of seed will also be required to "enable" a patent application. The specification must also provide what is referred to as the best mode of carrying out the invention that is known to the inventors at the time the application is filed, and it must also contain an adequate written description of that which the inventor considers to be the invention.† Additional requirements for patentability include a requirement for novelty‡ and nonobviousness.§ As with plant patents and the PVPA, it is possible to obtain a utility patent for a plant that has been disclosed or made available to the public, provided that a patent application is filed within one year of the date of disclosure (see 35 U.S.C. § 102(b)). Additional information regarding utility patents can be accessed at the United States Patent Office Web site www.uspto.gov/go/pac/doc/general/.

Upon the grant of a patent, the patent owner has the right to exclude others from making, using, selling, importing, or offering to sell the patented invention. Thus, it would be an act of infringement to make, use, import, offer to sell, or sell a patented plant in the United States (see 35 U.S.C. § 271). Should such an act occur, the patent owner has a right to recover damages as discussed above with respect to the infringement of plant patents. Namely, one can obtain injunctive relief that precludes the infringer from making, using, selling, offering to sell, or importing the plant. One can also obtain a monetary award of actual damages (e.g., lost sales) or a reasonable royalty for the plants sold by the infringer. As discussed above, treble damages can be obtained in some cases, as can attorney's fees in instances where a court finds exceptional circumstances.

CONCLUSION

The Supreme Court has compared and contrasted rights conferred by the PVPA, plant patents, and utility patents in a decision entitled *J.E.M. Ag Supply, Inc. v. Pioneer Hi-Bred Int'l, Inc.*, 534 U.S. 124, 140, 142 (2001). In this decision, the Supreme Court has written:

> The PVPA also contains exemptions for saving seed and for research. A farmer who legally purchases and plants a protected variety can save the seed from these plants for replanting on his own farm. See § 2543 ("[I]t shall not infringe any right hereunder for a person to save seed produced by the person from seed obtained, or descended from seed obtained, by authority of the owner of the variety for seeding purposes and use such saved seed in the production of a crop for use on the farm of the person …"); *see also Asgrow Seed Co. v. Winterboer*, 513 U.S. 179, 115 S.Ct. 788, 130 L.Ed.2d 682 (1995). In addition, a protected variety may be used for research. See 7 U.S.C. § 2544 ("The use and reproduction of a protected variety for plant breeding or other bona fide research shall not constitute an infringement of the protection provided under this chapter"). The utility patent statute does not contain similar exemptions.

The Supreme Court continues, at pages 142–143,

> To be sure, there are differences in the requirements for, and coverage of, utility patents and PVP certificates issued pursuant to the PVPA. These differences, however, do not present irreconcilable conflicts because the requirements for obtaining a utility patent under § 101 are more stringent than those for obtaining a PVP certificate, and the protections afforded by a utility patent are greater than those afforded by a PVP certificate. Thus, there is a parallel relationship between the obligations and the level of protection under each statute.

> It is much more difficult to obtain a utility patent for a plant than to obtain a PVP certificate because a utility patentable plant must be new, useful, and nonobvious, 35 U.S.C. §§ 101-103. In addition, to obtain a utility patent, a breeder must describe the plant with sufficient specificity to enable others to "make and use" the invention after the patent term expires. § 112. The disclosure required by the Patent Act is "the *quid pro quo* of the right to exclude." *Kewanee Oil Co. v. Bicron Corp.*, 416 U.S. 470, 484, 94 S.Ct. 1879, 40 L.Ed.2d 315 (1974). The description requirement for plants includes a deposit of biological material, for example, seeds, and mandates that such material be accessible to the public. See 37 CFR §§ 1.801-1.809 (2001); see also App. 39 (seed deposits for U.S. Patent No. 5,491,295).

---

* 35 U.S.C. 112 Specification

The specification shall contain a written description of the invention, and of the manner and process of making and using it, in such full, clear, concise, and exact terms as to enable any person skilled in the art to which it pertains, or with which it is most nearly connected, to make and use the same, and shall set forth the best mode contemplated by the inventor of carrying out his invention …

† Ibid.

‡ 35 U.S.C. 102 Conditions for patentability; novelty and loss of right to patent

A person shall be entitled to a patent unless—
 (a) the invention was known or used by others in this country, or patented or described in a printed publication in this or a foreign country, before the invention thereof by the applicant for patent, or
 (b) the invention was patented or described in a printed publication in this or a foreign country or in public use or on sale in this country, more than one year prior to the date of the application for patent in the United States …

§ 35 U.S.C. 103 Conditions for patentability; non-obvious subject matter
 (a) A patent may not be obtained though the invention is not identically disclosed or described as set forth in section 102 of this title, if the differences between the subject matter sought to be patented and the prior art are such that the subject matter as a whole would have been obvious at the time the invention was made to a person having ordinary skill in the art to which said subject matter pertains. Patentability shall not be negatived by the manner in which the invention was made …

By contrast, a plant variety may receive a PVP certificate without a showing of usefulness or nonobviousness. See 7 U.S.C. § 2402(a) (requiring that the variety be only new, distinct, uniform, and stable). Nor does the PVPA require a description and disclosure as extensive as those required under § 101. The PVPA requires a "description of the variety setting forth its distinctiveness, uniformity and stability and a description of the genealogy and breeding procedure, when known." 7 U.S.C. § 2422(2). It also requires a deposit of seed in a public depository, § 2422(4), but neither the statute nor the applicable regulation mandates that such material be accessible to the general public during the term of the PVP certificate. See 7 CFR § 97.6 (2001).

Because of the more stringent requirements, utility patent holders receive greater rights of exclusion than holders of a PVP certificate. Most notably, there are no exemptions for research or saving seed under a utility patent. Additionally, although Congress increased the level of protection under the PVPA in 1994, a PVP certificate still does not grant the full range of protections afforded by a utility patent. For instance, a utility patent on an inbred plant line protects that line as well as all hybrids produced by crossing that inbred with another plant line. Similarly, the PVPA now protects "any variety whose production requires the repeated use of a protected variety." 7 U.S.C. § 2541(c)(3). Thus, one cannot use a protected plant variety to produce a hybrid for commercial sale. PVPA protection still falls short of a utility patent, however, because a breeder can use a plant that is protected by a PVP certificate to "develop" a new inbred line while he cannot use a plant patented under § 101 for such a purpose. See 7 U.S.C. § 2541(a)(4) (infringement includes "use [of] the variety in producing (as distinguished from developing) a hybrid or different variety therefrom"). *See also* H. R. Rep. No. 91-1605, p. 11 (1970), U. S. Code Cong. & Admin. News 1970, pp. 5082, 5093; 1 D. Chisum, Patents § 1.05[2][d][i], p. 549 (2001).

Thus, it is apparent that the exclusionary rights conferred by various forms of intellectual property protection for a plant can have significant ramifications for both a grower and those who seek to infringe that grower's rights to the plant that has been developed. For example, a grower may wish to rely on a plant patent and a PVPA certificate to protect the plant and its progeny. However, as noted above, certain limitations exist with respect to the intellectual property rights conferred by plant patents and the PVPA. To retain complete control of the intellectual property rights associated with a plant, a utility patent may provide the most appropriate vehicle to accomplish this goal because a utility patent provides one with the right to exclude others from making, using, offering for sale, or selling the invention in the United States or importing the invention into the United States (regardless of whether the plant was asexually or sexually reproduced).

Index

A

Abelia, 263
Abies spp., 263
Abscisic acid, 146
Abutilon X hybridium, 64, 214
Acacia, 234
Acalypha, 216
Acalypha hispida, 214
Acalypha wilkesiana, 214
Acanthaceae, 221
Accelerator ® propagation container, 184
Acer spp., 152, 186
Acer griseum, 32
Acer palmatum, 277, 278, 421
Acer platanoides, 277
Acer rubrum, 277, 284, 285
Acer saccharinum, 156, 159, 277
Acer saccharum, 47, 442
Achillea, 64
Achillea filipendula, 64
Achillea millefolium, 64
Achillea ptarminca, 64
Acrylic panels, 23
Actinidia spp., 239
Actinidia deliciosa, 253
Aesculus glabra, 277
Aesculus hippocastanum, 277
Aesculus parvifolia, 7
African marigold, 415
African violet, 32, 233, 241, 245
    nematode disease, 80
Agavaceae, 17, 245
Ageratum houstonianum, 414
Aglaonema spp., 214, 243
Agrobacterium rhizogenes, 11
Agrobacterium tumefaciens, 144, 146, 265, 279
Ailanthus, 234
Ailanthus altissima, 234, 421
Albizia julibris, 234, 253
Albizia julibrissin, 234, 253, 422, 424
Algerian ivy, 156
Allamanda, 216
Allamanda cathartica, 214
Allium sp., 303, 402
Almond, 278, 284, 285
Alternation of generations, 380–381
Aluminum plant, 215
Alyssum, 92
Amaryllus, 304
American elm, 276
American Society for Horticultural Science, 11
Amoracia rusticana, 253
Anacardiaceae, 422
Anemone spp., 237, 239
Anemone hupehensis, 253
Anemone X hybrida, 235

Angel wing begonia, 64
Angiosperms, 379–389
Anthracnose diseases, 84
Anthurium andreanum, 214
Aphelandra, 191
Aphelandra squarrosa, 214
Aphids, 117, 278
Aphis gossypii, 117
Apple(s), 278, 282, 407
    blanching, 155
    juvenile, 153
    layering, 264
    mosaic virus, 76
    tolerant and resistant root stock, 279
    wooly aphid, 117
Apricot, 284
Arabian jasmine, 214
Arabidopsis, 146
Araceae, 243, 245
Aralia spinosa, 253
Arbutus spp., 69
Arbutus menziesii, 69
Arctostaphylos uva-ursi, 64, 68, 70–71
Armoracia, 254
Armullaria sp., 279
Aronia arbutifolia, 326, 327, 328
Arrowhead vine, 215
Arrowwood viburnum, 138
Ash tree, 277
Asimina triloba, 152, 234
Asparagus officinalis, 308
Aspens, 159, 253
Asplenium, 241
Asplenium rhizophyllum, 233, 242
Atlas cedar, 422
Auxins, 6, 143–145, 156, 309
    commercial powders, 197
    liquid prepared formulations, 197
Avocado tree, 155
Azaleas, 179, 264

B

Baby's breath, 253
Bacterial canker, 279
Bacterial diseases, 76–77
Balance-type electronic leaf, 38
Balfour aralia, 215
Banding, light, 226
Bark, 48
Bartlett pear, 285
Bean, 240
Bedding plants, 411
    changes in products, 411–412
    plugs, 412–413
    seed improvements, 417–419

starting, 412–413
Beech, 262
    juvenile, 152
Beefsteak plant, 214
Beet, 92, 307
Begonia spp., 47, 64
    botrytis blight, 78
    leaf propagation, 250
    nematode disease, 80
Begonia corallina, 64
Begonia grandis ssp. *evansiana*, 306
Begonia rex, 233, 239
Begonia schulziana, 32
Begonia X hiemalis, 242
Begoniaceae, 245
Bellflower, 239
Bemisia argentifolii, 117
Benching systems, 24–25, 179
    advantages, 180
    arrangement, 24
    ground beds, 25
    growing on floor, 24–25
    size, 180
Benzyladenine, 326
Beta vulgaris, 92, 240, 307
Betula spp., 159
Betula alleghaniensis, 275
Betula papyrifera, 69
Biotechnology, breeding and, 398–399
Biotic and abiotic stress, 278–279
Birch, 69, 275
Black currants, 64
Black locust, 235, 253, 422, 424, 425
Black nightshade, 163
Black root rot, 78, 80
Black rot, 265
Black walnut tree, 154
Blackberry, 253
Blanching, 155
Blanket flower, 239
Bleeding heart vine, 214
Blue rug juniper, 17
Blue spruce, 285
Boston ivy, damping-off, 92
Botrytis, 17, 80, 81, 93
    blight, 78, 82
Botrytis cinerea, 78, 91, 92
    disease assessment, 93
    inoculum, 93
Botrytis gray mold, 91
Bougainvillea glabra, 214
Bougainvillea spectabilis, 214
Boxus sp., 178
Boxwood, 59, 64, 178
Bradysia sp., 115
Brassica napus, 148
Brassica oleracea, 32
Brassinosteroids, 148
Bridal wreath spiraea, 64
Broad-leaved evergreen plants, 229
Brunfelsia pauciflora, 214
Bryophyllum spp., 241, 246
Bryophyllum daigremontianum, 233, 241, 242, 246
Bryophyllum fedtschenkoi, 246, 247
Bryophyllum pinnata, 246
Buddleia, 263
Buddleia davidii, 64, 199, 206, 207
Bulbs, 304–305
    examples, 305

Bulk density, 59–60
Burls, 157–158
Burro tail, 233, 247
Butterfly bush, 199, 207
Buxus, 265
Buxus sempervirens, 59, 64

C

Calendula officinalis, 414
California privet, 277
Callery pear, 253
Calluna, 265
Calluna vulgaris, 64, 91, 92
    cuttings, 93–95
Camellia(s), 178
    layering, 264
    scale insects, 116
Campanula spp., 239
Campsis, 263
Campsis radicans, 253
Canary grass, 143
Canary Island ivy, 64
Cape cowslip, 243
Cape grape, 215
Cape honeysuckle, 215
Cape Primrose, 233
Capsicum annuum, 414
Caricature plant, 216, 221
Carnation, 198, 199
Carpinus, 262
Carya sp., 286
Carya illinoinensis, 152
Caryophyllus, 199
Cassava, 281, 356–358
    meristem culture, 357
    plant propagation, 356–357
    in vitro propagation, 357
Castanea sativa, 156
Catharanthus roseus, 414
Cedrus, 178
Cedrus atlantica, 422
Cell pack, 412
Celosia sp., 414
Cercis canadensis, 387, 389, 425
Certification programs, 101, 104, 110
Certification tag, 103
Chaenomeles spp., 238
Chamaecyparis, 178
Chamaecyparis nootkatensis, 155
Chamaecyparis obtusa, 422
Chamaecyparis thyoides, 275–276
Chenille plant, 214
Cherry(ies), 285
    Kwanzan flowering, 156
    sweet, 285
    tart, 285
    tolerant and resistant root stock, 279
Cherry leafroll virus, 285
Chimera, 163–174
    branching and stability, 169
    classification, 167–169
    defined, 163
    examples, 169–170
    exercises, 171–174
    formation, 167
    history, 163–164
    maintaining, 170–171

Index

mericlinal, 167–168
periclinal, 168–169
propagating, 170–171
sectorial, 168
stabilizing, 171
tissue culture, 171
Chinese date, 422
Chinese zelkova, 422
Chionanthus spp., 265
Chionanthus virginicus, 32, 152
Chlorophytum, 17
Christmas cactus, 64
Chromatomyia, 116
Chrysanthemum, 47, 207, 381
    leafminers, 116
    root rot, 78
    rooting, 198, 199
Cigar plant, 64
Cinquefoil, 64
Cissus antarctica, 214
Cissus discolor, 214
Cissus rhombifolia, 214
Citron rootstick, 163
Citrus spp., 152, 279
Citrus aurantium, 163
Citrus jambhiri, 285
Citrus medica, 163
Citrus sinensis, 285
Clay pots, 184
Clematis, 240, 263
Clerodendrum thomsonae, 214
Clerodendrum X speciosum, 214
Climbing hydrangea, 263
Cockroaches, 118, 119
Coconut, 6
Cocos nucifera, 6
Cocoyam, 361–364
Codiaeum variegatum, 214
Coffee, 279
Coir, 48
Cold frames, 19, 180
Coleus sp., 47, 64, 207, 215, 216, 221, 267
    leaf propagation, 246
Coleus scutellarioides, 64
Collembola, 118, 119
Columbine, 116
Comaceae, 422
Common lilac, 265
Composts, 48–49
Compressed fiber pots, 184
Coneflower, 239
Containers, 49–55
    examples, 184
    flats and trays, 50, 184
    function, 184
    plugs and multicavity systems, 50–52
    pots, 50
    ready-to-use, 54–55
    selection criteria, 184
    ultrahigh-density systems, 52, 54
Copyright protection, 441–442
Coralberry, 57
Cordyline, 17, 267
Cordyline fruticosa, 214
Corms, 305–306
Corn, 429
Corn plant, 214
Cornus, 186
Cornus florida, 278, 379, 387, 422

Cornus sericea, 64
Corylus spp., 277
Corylus avellana, 264, 267
Costus speciosus, 214
Cotinus spp., 264
Cotinus coggygria, 422
Cotoneaster, 263
Cottage pink, 64
Cotton, 402
Cottonwood, 69
Crabapple, 277
Crape ginger, 214
Crape myrtle, 209, 238
Crassula argentea, 64, 66
Crassula ovata, 233, 242
Crassulaceae, 245
Crataegus monogyna, 163
Creeping Charlie, 215
Crepis, 147
Cresote bush, 157
Crocus, 303
Crop certification
    berry, 102
    importance of, 100
    potato, 103
    programs, 100, 101
    regulatory agencies, 100
    tag, 103
Croton, 214, 264
Crown gall, 85, 265, 279
Crown of thorns, 64
Crown rot, 279
Cryptomeria japonica, 152
Cucumber, 92, 274, 278
Cucumber mosaic virus, 76
Cucumis metuliferus, 278
Cucumis sativus, 92, 278
Cunninghamia lanceolata, 179
Cuphea ignea, 64
Cupressaceae, 422
Cupressus, 178
Currant, 102
Cutting(s)
    origin of shoots and roots from, 237–239
    preparation, 237–238
    propagation, 63–72
Cyclamen spp., 91, 93
Cyclamen persicum, 307
Cydonia oblonga, 163, 280, 284, 285
Cytisus purpureus, 163
Cytokinins, 145–146, 156

D

Dactylis glomerata, 331, 389
Daffodil, 303
Dahlia, 314
Dahlia pinnata, 6
Daisy, 116
Daktulospaira vitifoliae, 278, 279
Damping-off, 77, 91–92
    Boston ivy, 92
    Impatiens, 91
    postemergence, 92
Daphne cneorum, 265
Daucus, 147
Daylily, 79
Deciduous plants, 229

Dendradranthema, 179
Dendradranthema grandiflora, 47
Dendradranthema morifolium, 198, 199, 200
Dendranthema x grandiflorum, 64, 66, 207, 267
Deutzia, 59, 64, 263
Deutzia gracilis, 59, 64
Devil's walking stick, 253
Dianthus caryophyllus, 64, 66, 198, 199, 200
Dianthus plumarius, 64
Dianthus superbus, 64
Diffenbachia, 223, 264
Diffenbachia amoena, 214
Diffenbachia maculata, 214
Dionaea sp., 243
Dionaea muscipulata, 243
Dioscorea spp., 358
Dioscorea alata, 358
Dioscorea bulbifera, 358
Dioscorea cayenensis, 358
Dioscorea rotundata, 358
Diospyrus, 234
Disease(s), 75–85
    anthracnose, 84
    assessment, 93
    bacterial, 76–77
    chemical control of, 82
    fungal, 77–79
    leaf spot, 84
    management tactics, 80–81
    nematode, 79–80, 85, 275, 279
    rust, 79, 83
    seedling, 75, 77, 82, 92, 115
    shot-hole, 84
    virus, 76
Disinfestation, 87–89
    facilities, 87
    thermal, 87
Dogwood, 7, 64
    flowering, 79
    scale insects, 116
Doubleworking, 281–282
Downy mildew, 79, 80, 83
Dracaena, 17, 223, 267
Dracaena fragrans, 214
Dracaena marginata, 214, 222
Dracaena reflexa, 214
Droscorea bulbifera, 306
Drosera sp., 241, 243
Dumbcane, 214
Dwarf American cranberry, 138
Dwarf boxwood, 265
Dwarf brassaia, 215

E

Earwigs, 118, 119
Eastern redbud, 422, 425
Eastern white pine, 422
Echeveria, 195, 243
Echinacea spp., 239
Echinothrips americana, 117
Economic and legal liabilities, 103–104
Elderberry, 253
Elliottia racemosa, 239, 253
Elm(s)
    American, 276
    aphids, 117
    lacebark, 159, 238, 253

Elytrigia repens, 308
English ivy, 64, 91, 114, 153, 240
    juvenile, 152, 153, 158
    layering, 262
English lavender, 64
English thyme, 64
English walnut, 278
Enkianthus perulatus, 32
Environmental control systems, 27
Epiphyllum spp., 64
Epipremnum pinnatum, 214
Erica, 265
Erica arborea, 157
Ericaceae, 157
Eriosoma lanigerum, 117, 278, 279
Erisyphe pulchra, 79
Erwinia amylovora, 279
Erwinia carotovora, 76
Erwinia chrysanthemi, 76
Erysiphe, 79
Ethylene, 147–148
Etiolation, 155, 226
Eucalyptus spp., 157, 159
Eucalyptus grandis, 153, 156
Euonymus spp., 116, 209, 239
Euphorbia millii, 64
Euphorbia pulcherrima, 18, 32, 64, 76, 214, 227
European aspen, 238
European mountain ash, 238
Evergreen grape, 215
Exacum, 78

F

Fabaceae, 422
Fagus spp., 152, 262
Fall-flowering, 235
False aralia, 215
False eranthemum, 215
False spiraea, 64
Fern, 380
Fertilization, 385–386
Fiberglass-reinforced plastic, 23
Ficus sp., 164, 246, 267
Ficus carica, 177, 253
Ficus elastica, 214, 233
    layering, 264
    leaf propagation, 246
Ficus longifolia, 216
Ficus punula, 214
Figs, 177, 253. *See also Ficus* sp.
Filberts, 277
Fiorinia theae, 116
Fire spike, 214
Fireblight, 279
Firs, 289
Flats and trays, 50, 184
Florists' carnation, 64, 66
Florists' chrysanthemum, 64, 66, 267
Florists' hydrangea, 206, 398
Floss flower, 414
Flowerbulbs, 303
    propagating selected, 311–316
Flowering cherries, 277
Flowering dogwood, 379, 422
Flowering maple, 64, 214
Flowering quince, 238
Flowers, 383–384

Foliar nematode, 85
Folmeia menzeisii, 246
Forsythia, 18, 59, 191, 192
Forsythia ovata, 64
Forsythia x intermedia, 18, 59, 64, 238
Fragaria x ananassa, 100, 323, 325
Frankliniella occidentalis, 76, 117
Fraxinus sp., 277
Fraxinus angustifolia, 156
French marigold, 415
Fringetree, 152, 265
Fuchsia spp., 64, 91
Fuchsia x hybrida, 64
Fungal diseases, 77–79
Fungus gnats, 115–116
Fusarium, 115, 275

G

Gaillardia aristata, 239
Gamborg's B5 medium, 335, 336
Gametogenesis, 384–385
Garden beans, 115
Gardenia augusta, 214
Gardenia jasminoides, 214
Garland flower, 265
Genet, 234
Georgia plume, 239, 253
Geranium, 64, 66, 91, 195, 384, 411
    bacterial wilt disease, 77
    botrytis blight, 78, 91
    Rober's Lemon Rose, 173
    rust diseases, 79
    variegated zonal, 164
Gesneriaceae, 245
Gibberellins, 146–147, 156
Ginger, 214
Ginkgo biloba, 157
Gladiolus, 303
Glass, 23–24
Gleditsia triacanthos, 152, 277, 422, 424, 425
Glorybower, 216
Gloxinia, 245, 307
Goldenraintree, 422, 425, 427
Gooseberry, 102
Gossypium hirsutum, 402
Grafting, 273–292
    to achieve independent optimization of stock, 281
    approach, 277, 283
    brace, 283
    bridge, 283
    broadest application, 274
    cambial alignment, 285–286
    for clonal multiplication, 276–277
    compatibility, 284–285
    concepts, 273–275
    defined, 273
    dessication and, 287
    for diagnosis of viral infections, 282–283
    dormant or semidormant scion, 288–289
    exercises, 290–299
    history, 275–276
    indexing, 282
    to influence growth and flowering, 281
    method, 289–290
    for multiple genotypes, 281
    nurse root, 277
    plant components, 274
    pressure and, 286
    reasons for, 276
    for repair, 283
    root cutting systems vs., 277–278
    scion, 273
    seasonal considerations, 287–288
    successful, 284
    union formation, 275
    waterproofing materials and, 289
Grape, 177, 263, 278
    ivy, 214, 221
    leaves, 152
    tolerant and resistant root stock, 279
    wine, 278
Graptophyllum pictum, 214
Gray mold, 82
Greenhouse(s)
    benching systems, 24–25
    cooling systems, 27–28
        fog, 28
        pad and fan, 28
    coverings, 22–24, 179, 180
        acrylic panels, 23
        fiberglass-reinforced plastic, 23
        glass, 23–24
        polycarbonate panels, 23
        polyethylene film, 22–23
        rigid plastic panels, 23
        UV-resistant ethylene tetrafluoroethylene, 23
    defined, 179
    environmental control systems, 27
    gutter-connected, 21–22
    open-roof, 27
    retractable-roof, 22
    stand-alone, 20, 21
    structures, 19–22
        cold frames and hot beds, 19
        polytunnels, 19–20
        Quonset design, 20
    water quality, 184
Growth regulators, 143–150, 156
Gymnocladus dioica, 253, 422, 425
Gypsophila paniculata, 253

H

Hamamelis, 186
Hamamelis japonica, 156
Hawthorn, 163
Hazelnuts, 277
Heath, 265
Heather, 64, 265
Heating system(s), 25–27
    delivery, 26
    fuel, 25–26
Hebe spp., 91, 93
Hedera canariensis, 64, 156
Hedera helix, 64, 91, 114, 152, 153, 158, 240
    layering, 262
Hedging, 155
Hedychium spp., 214
Helianthemum mummularium, 91
Helianthus annuus, 116, 412
Heliothrips haemorrhoidalis, 117
Heliotrope, 64
Heliotropium arborescens, 64
Hercinothrips femoralis, 117
Hibiscus, 64, 226

Hibiscus rosa-sinensis, 64, 214
Hibiscus schizopetalus, 214
Hibiscus syriacus, 422
Hibiscus tiliaceus, 214
Hickory, 286
High temperature exposure, 156
Hippeastrum, 304
Holly, 116, 264
Honey locust, 152, 277, 422, 425
    thornless, 424
Hornbeam, 262
Horse chestnut, 277
Horseradish, 253
Horticultural plants, 391–399
    breeding of, 391–399
        biotechnology and, 398–399
        considerations in, 397
        linkage and, 397
        mutagenesis, 398
        non-Mendelian inheritance and, 397
    clonally pollinated, 396
    cross-pollinated, 396
    genetics of, 391–394
    self-pollinated, 394–395
Host plant resistance, 122
Hosta, 80
Hoya, 221, 240
    leaf propagation, 246
Hoya carnosa, 64, 214
Husker Red, 195
Hyacinthus, 303
Hyacinthus orientalis, 312
Hybrid lilies, 303
Hydrangea macrophylla, 206, 398
Hydrangea paniculata, 57, 64
Hydrangea petiolaris, 263
Hydrangea quercifolia, 159, 209, 236, 238
    root cuttings, 253
Hyoscyamus, 147
Hypericum calycinum, 238
Hypnum peat, 63
Hypoestes phyllostachya, 64

I

Ilex, 263
Ilex cornuta, 18
Ilex crenata, 228
Ilex vomitoria, 159
Ilex xattenuata, 186
Impatiens spp., 47, 64
    leaf propagation, 246
    necrotic spot virus, 75, 76, 80
Impatiens walleriana, 64, 411, 414
In-line canister filters, 16
Incompatibility, 397, 398
Indian rubber tree, 221
Indole-3-acetic acid (IAA), 185
    stability, 202
Indole-3-butyric acid (IBA), 153, 185
    preparing pure, 197–198
    stability, 202
Inorganic media, 47–48, 63
Integrated Pest Management (IPM), 113–124
    assessment of, 123
    biological control, 120–121
    chemical control, 121–122
    cultural control, 119–120
        by irrigation, 119
        by pasteurization, 119
        by sanitation, 119
        by weed management, 119
    mechanical control, 120
    quarantine and staging areas, 120
    record keeping, 122–123
    scout and monitoring techniques, 114–115
        tools for, 114
    strategies, 119
Intellectual property protection, 441–447
Intermittent mist systems, 37–40
    laboratory exercises, 38–40
IPM. *See* Integrated Pest Management (IPM)
Ipomoea batatas, 253, 307
Iris sp., 303, 308
Ivy
    Algerian, 156
    Boston, 92
    Canary Island, 64
    English, 64, 91, 114, 153, 240
        juvenile, 152, 153, 158
        layering, 262
    grape, 214, 221
    Swedish, 64
Ixora chinensis, 214
Ixora coccinea, 214

J

Jacobinia, 214
Jade plant, 64, 66, 233
Japanese anemone, 253
Japanese cedar, 152
Japanese maple, 277, 421
Japanese privet, 214
Japanese stewartia, 32
Japanese wisteria, 253
Japanese witch hazel tree, 156
Jasminum sambac, 214
Jasmonates, 148
Johnsongrass, 308
Juglans sp., 286
Juglans hindsii, 278, 285
Juglans nigra, 154
Juglans regia, 278, 285
Juniperus, 178
Juniperus chinenesis, 64
Juniperus horizontalis, 17, 64
Juniperus virginiana, 17
Justicia brandeegena, 214
Justicia carnea, 214
Juvenility, 151–161
    epigenetic changes, 152
    exercises, 158–160
    macropropagation and, 153–154
    micropropagation and, 154
    traits, 152–153

K

Kalanchoe blossfeldiana, 195
Kalanchoe tomentosa, 247
Kalmia latifolia, 157
Kentucky coffee tree, 253, 422, 424, 425
Kinnikinnick, 64, 68, 70–71
Kiwi fruit, 239, 253

Index

Koelreuteria paniculata, 238, 422, 425
Koreanspice viburnum, 128, 243
Kwanzan flowering cherry, 156

L

Laburnum vulgare, 163
Lacebark elm, 159, 238, 253
Lachenalia, 243
Lactica sativa, 429
Lagerstroemia indica, 209, 238
Larrea tridentata, 157
Lavandula spp., 91, 265
Lavandula angustifolia, 64, 414
Layering, 261–269
    air, 264
    Chinese, 264
    common, 262–263
    compound, 263
    continuous, 263
    drop, 264–265
    exercises, 266–268
    maintenance of plants, 266
    mound, 263–264
    pot, 264
    root development and, 266
    serpentine, 263
    simple, 262–263
    steps in successful, 265–266
    trench, 263
Leaf cuttings, 233, 239–241
    advantages and limitations, 239–241
    exercises, 245–251
    growth regulators and, 242
    origins of new shoots and roots, 241
    species suitable for, 242–243, 245
Leaf spot diseases, 84
Leafminers, 116
Lemon, 285
Lesion nematode, 85
Lettuce, 429
Lignotubers, 157–158
Ligustrum spp., 209
Ligustrum japonicum, 64, 214
Ligustrum ovalifolium, 64, 277, 285
Lilac, 156, 238, 253, 277
Liliaceae, 17, 243
Lilium, 17, 233, 303
Lilium longiflorum, 243, 304
Linkage, 397, 398
Liriodendron tulipifera, 421
Liriodendrum, 191
Lobelia spp., 92, 96
Lobelia erinus, 414
Loblolly pine, 422
Lobularia maritima, 92
Lonicera tatarica, 64
Lycopersicon esculentum, 92, 147, 402
Lycopersicum esculentum, 163, 415

M

Madagascar dragon tree, 214
Madagascar periwinkle, 414
Magnolia sp., 179, 186, 263
    layering, 264
    seedlings, damping-off on, 77

Magnolia grandiflora, 186, 422
Magnolia soulangiana, 156
Magnolia stellata, 156
Mahonia, 240
Malus sp., 117, 236, 239, 277
Malus domesticus, 153, 155, 278, 285, 407
Malvaceae, 422
Mandevilla, 221
Mandevilla splendens, 214
Mandevilla X amabilis, 214
Mangifera indica, 155
Manihot esculenta, 281, 356–358
Manihot glazivii, 281
Maple, 285
    flowering, 64, 214
    Japanese, 277
    juvenile, 152
    Norway, 277
    paperbark, 32
    red, 277, 284, 285
    silver, 156, 159
    sugar, 277
Maranta leuconeura, 214
Marigold, 411, 412, 415
Marsdenia floribunda, 214
Maximum tolerances, 193
Mealy bugs, 116
Media
    containers, 49–55
        flats and trays, 50, 184
        plugs and multicavity systems, 50–52
        pots, 50
        ready-to-use, 54–55
        ultrahigh-density systems, 52, 54
    for cutting propagation, 63–72
    functions, 43–44
    heat disinfestation, 5
    materials and mixtures, 47–49
        inorganic, 47–48, 63
        organic, 48–49, 60, 63
    ph, 46
    pore spaces, 44, 46, 59–60
    premixed, 49
    properties, 44–47
        biological, 46–47
        chemical, 45, 46
        physical, 44, 45, 46, 57–61
    salts and nutrients, 46
    selection, 57–61
Mediterranean shrub tree heath, 157
Medium. *See* Media
Medlar, 163
Meiosis, 381
Meloidogyne spp., 79, 279
Melons, 117, 274
Mentha spp., 267, 347–354
Mentha cardiaca, 347
Mentha spicata, 347
Mentha x piperita spp., 347
Mericloning, 321
Meristem and meristem tip culture, 320–321
Meristemming, 321
Mespilus germanica, 163
Micropropagation, 5, 6, 319–332
    cassava, 357
    cocoyam, 357, 361–364
    exercises, 347–354
    factors affecting culture establishment, 324
    mint, 347–354

purpose, 347
stages, 322–329
  donor plant selection and preparation, 322–323
  establishment of aseptic culture, 323–324
  pretransplant, 326–327
  proliferation of axillary shoots, 324
  transfer to natural environment, 327, 329
tropical root and tuber crops, 355–364
yams, 357, 360
Microsphaera, 79
Microtus, 279
Mimosa, 234, 422, 424
Mineral salts, stock solutions of, 342–343
Ming aralia, 215
Minisett technology, 358–359
Mist and fog propagation, 5, 6
Mist systems, 5, 6, 179, 180
  components, 181
  control devices, 181–182, 183
  filter, 181
  intermittent, 37–40
    laboratory exercises, 38–40
  line, 181
  nozzles, 181
  overhead, 181
  tents, 180
Mites, 118
Mitosis, 381
Mock orange, 64
Molecular markers, 398
Monterey pine tree, 155
Morus sp., 177
Moss, 380
Mother-in-law plant, 245
Mother of thousands, 233
Mountain laurel, 157
Mulberry(ies), 177
Multicavity systems, 50–52
Multiple-scion cultivars, 282
Murashige and Skoog medium, 335, 336
Muscari, 303
Mutagenesis breeding, 398
Myzus persicae, 117

N

Nandina domestica, 209
Naphthalene acetic acid (NAA), 185
  stability, 202
Narcissus, 303
Narcissus tazetta, 304
Narrow-leaved evergreen plants, 229
Narrowleaf ash, 156
Necrotic ringspot virus, 76
Nematodes, 79–80, 85, 275
  root/knot, 85, 279
Nephrolepsis, 325
Nerium oleander, 214
Ninebark, 59
Nitsch's medium, 335, 336
Nonzygotic embryogenesis, 329–332
Northern California black walnut, 278, 279
Northern spy apple, 278
Norway maple, 277
Norway spruce, 285, 421
  juvenile, 153
Nozzles, mist systems, 181

O

Oak leaf symptoms, 76
Oak rot fungus, 279
Oak tree, 277
  juvenile, 152
Oakleaf hydrangea, 159, 209, 236, 238, 253
Ocimum basilicum, 415
Octopus tree, 215
Odontonema cuspidatum, 214
Ohio buckeye, 7, 277
Olea europa, 422
Oleaceae, 277, 285, 422
Oleander, 214, 216
Oligonychus ilicis, 118
Oligonychus ungunius, 118
Olive, 422
Olive tree
  juvenile, 153
Onion, leafminers, 116
Orange, 280, 285
Orchardgrass, 331, 389
Orchid cactus, 64
Organic media, 47–48, 60, 63
Organic wastes, 48–49
Oriental cherry, 422
Oriental poppy, 239, 253
Ornamental cabbage, 32
Oryza sativa, 402
Oxalis sp., 384

P

Pachyphytum, 243
Paeonia suffruticosa, 281
Panax, 215
Panda plant, 247
Panicled goldenraintree, 238
Panonychus ulmi, 118
Pansy(ies), 78, 92
Papaver orientale, 239, 253, 254
Papaver rhoeas, 412
Paper pots, 184
Paperbark maple, 32
Paperwhite narcissus, 304
Passiflora spp., 152, 214
Passion flower, juvenile traits, 152
Passionfruit, 221
Patchouli, 240
Patents, 444–445
Pathogen testing, 101
Paulownia spp., 235, 238
Paulownia taiwaniana, 235
Paulownia tomentosa, 235, 236
Paulownia tremula, 237
Paulownia tremuloides, 236
Pawpaw, 152, 234
Pea, 92
Peach, 278, 284
  aphids, 117
  tolerant and resistant root stock, 279
Pear, 163, 264, 278, 284, 285
  tolerant and resistant root stock, 279
Peat, 48, 63
  pots, 184
Pecan, juvenile, 152
Peegee hydrangea, 57
Pelargonium sp., 64, 179, 195, 411

Index

Pelargonium x domesticum, 173
Pelargonium x hortorum, 91
Pelargonium zonale, 164
Penstemon digitalis, 195
Peperomia spp., 241, 245
Peperomia crassifolia, 214
Peperomia obtusifolia, 171, 214
Peppermint, 347
Peppers, 414
Periwinkle, 64
Perlite, 47
Persian shield, 215
Pest resistance, 122
Pest resurgence, 122
Petunia x hybrida, 147, 411, 415
PGR. *See* Plant growth regulators (PGR)
Phalaris canariensis, 143
Phaseolus vulgaris, 115, 240, 403
Philadelphus coronaries, 64
Philadelphus lewisii, 64
Philodendron, 221
Philodendron scandens var. *oxycardium,* 214
Phlox paniculata, 253
Phlox subulata, 239
Photon flux, 226
Photoperiod, 227
Photovoltaic mist controller, 38
Phylloxera, 278, 279
Physocarpus opulifolius, 59
Phytophthora sp., 75, 78, 79, 265, 279
Phytotoxicity, 122
Picea spp., 47, 289
Picea abies, 153, 285, 421
Picea pungens, 285
Pick-a-back plant, 233, 242
Piggyback plant, 245, 246
Pilea cadierei, 215
Pilea nummulariifolia, 215
Pinaceae, 422
Pines, 289
Pink spiraea, 64
Pinus spp., 178, 289
Pinus radiata, 155
Pinus strobus, 275, 422
Pinus taeda, 422
Piperaceae, 245
Pisum sativum, 92, 403
Plant(s)
    copyright protection and, 441–442
    growth regulator(s), 338–339, 342
        abscisic acid, 146
        auxins, 143–145
        brassinosteroids, 148
        cytokinins, 145–146
        ethylene, 147–148
        gibberellins, 146–147
        jasmonates, 148
        polyamines, 148
        salicylic acid, 148
    import and export, 100–101
    intellectual property protection, 441–447
    patents, 444–445
    pathogens, 6
    propagation (*See also* Propagation)
        benching systems, 24–25
        cooling systems, 27–28
        environmental requirements, 18–19
        facility design, 30–35
            exercises, 34
            initial considerations for successful, 30
            long-term issues, 33–34
            plant-centric, 31–32
            pragmatic, 32–33
        facility needs, 17–18
        future, 11
        green structures, 19–22 (*See also* Greenhouse(s))
        heating systems, 25–27
        history, 3–5
        intermittent mist systems, 37–40
        mist and fog, 5, 6
        myths concerning, 435–440
        role and importance, 6–8
        site selection, 15–16
            layout of, 15–16
            topography of, 16
        with stem cuttings, 177–188
        structures, 19–22
            cold frames and hot beds, 19
            polytunnels, 19–20
            purpose and function of, 29
            Quonset design, 20
        water guidelines, 16–17
    shoot tip meristems, 164–167
    trademarks and use, 442–443
    utility patents, 445–446
Plant propagules, 29
Plant sundew, 241
Plant Variety Protection Act, 443–444
Planting and cultural practices, 102–103
Plastic pots, 184
Platycladus, 178
Plectranthus australis, 64
Pleomele reflexa, 214
Plugs, 50–52, 412–413
    open flat *vs.*, 412
    production, 413
    scheduling, 413
    sizes and shapes, 413
Plum, 279
Podocarpus, 178
Podosphaera, 79
Pogostemon patchouli, 240
Poinsettia, 32, 64, 76, 214, 216
    botrytis blight, 78
    root rot, 78
Polka-dot plant, 64
Pollination, 385–386
Poly bags, 184
Polyamines, 148
Polycarbonate panels, 23
Polyethylene film, 22–23
Polygonium, 386
Polyphagotarsonemus latus, 118
Polyploidization, 397
Polyscias fruiticosa, 215
Polyscias guilfoylei, 215
Polyscias scutellaria, 215
Polystyrene, 47
Pomegranate, 177
Poncirus trifoliata, 280
Poplars, 253
Poppy, 412
Populus spp., 159, 234, 238, 253
Populus balsamifera, 69
Populus tremuloides, 234
Pore spaces, of media, 44, 46, 59–60
Pot marigold, 4141
Potato, 322

Potentilla fructicosa, 64
Pothos, 152, 195, 214, 221
Pots, 50
    clay, 184
    compressed fiber, 184
    paper, 184
    peat, 184
    plastic, 184
Powdery mildew, 78–79, 83
Prayer plant, 214
Priming, 423–424, 428–430
Primrose, 91
Primula spp., 91
Privet, 64, 209
    California, 277
Propagation
    bulb, 304–305
    corm, 305–306
    cuttings
        leaf, 233, 239–241, 245–251
            media for, 63–72
        stem, 177–188
    experiments, 127–140
        analysis of continuous data, 133–134
        choosing data and planning statistical analysis for, 130
        comparing treatment means of, 134
        conducting, 130–131
        counting data from, 132–133
        data analysis and interpretation in, 131–139
        evaluating data from, 127–140
        hypothesis of, 127
        multifactor, 136–139
        multiple comparison and multiple range tests, 134
        objectives of, 127–128
        orthogonal contrasts of, 135
        recording data from, 130–131
        repeating, 129–130
        selecting components of, 128–129
        standard error of mean, 134
        trend analysis in, 135–136
    flowerbulbs, 311–316
    by grafting, 273–275 (*See also* Grafting)
    leaf, 233, 239–241, 245–251 (*See also* Leaf cuttings)
    maintenance of stock plants, 187
    pests, 115–119
    plant growth regulators in, 143–150
    preventative practices for, 187–188
    rhizome, 308–309
    sanitation during, 187
    serial, 155–156
    storage organs, 303–309
    structures, 179
    substrates, 182
        characteristics of, 182–183
        chemical properties of, 183–184
        containers for, 184–185
        selection of, 183
        success of, 183–184
    tuber, 306
Prunus spp., 76, 264, 278
Prunus amygdalus, 278, 285
Prunus armeniaca, 284
Prunus avium, 280, 285
Prunus cerasifera, 59, 284
Prunus cerasus, 280, 285
Prunus dulcis, 284
Prunus persica, 278, 284
Prunus serrulata, 156, 422
Prunus subhirtella, 277

Pseudococcus longispinus, 116
Pseudomonas spp., 279
Puccinia hemerocallidis, 79
Pulvinaria, 116
Punica granatum, 177
Purple bignonia, 215
Purple heart, 64
Pyrus spp., 159, 264
Pyrus calleryana, 253, 278
Pyrus communis, 163, 278, 284, 285
Pythium sp., 75, 78, 79, 81, 115
    damping-off, 96
Pythium oligandrum, 96
Pythium ultimum, 96

Q

Quackgrass, 308
Quaking aspen, 238
Quercus spp., 152, 159, 186, 234, 277
Quince, 280, 284, 285
    rootstock, 163

R

Ralstonia solanacearum, 77
Ready-to-use containers, 54–55
Red currants, 64
Red maple, 277, 284, 285
Red raspberry, 102
    primocanes, 106, 107, 108
    root cuttings, 253
Redwood tree, 157
Reparative buds, 237
Reproduction
    cellular processes in, 381–383
    fertilization, 385–386
    gametogenesis, 384–385
    pollination, 385–386
    sexual, 379–389
Rex begonia, 233
Rhanaceae, 422
Rhaphidophora, 215
Rhaphidophora celatocaulis, 215
Rhizoctonia, 91, 92, 93
Rhizoctonia solani, 77
Rhizomes, 308–309
Rhododendron spp., 46, 91, 157, 215
    dwarf, 265
    layering, 263
    overwinter survival, 186
    soft wood cuttings, 179
Rhoicissus capensis, 215
Rhus, 234
Rhus copallina, 234
Rhus glabra, 234, 253, 422
Rhus typhina, 234, 254, 255
Ribes spp., 100
Ribes nigrum, 64
Ribes triste, 64
Rice, 402
Rigid plastic panels, 23
Rober's Lemon Rose, 173
Robinia spp., 234, 238, 254
Robinia pseudoacacia, 235, 253, 424, 425
Rock wool, 47–48
Root cutting(s), 32, 57, 114, 173

Index

advantages and limitations, 234–235
anatomical and physiological changes during, 189–194
chimera and, 235
cloning plants, 177–188
fungicide treatment, 238
growth regulators and, 237
media requirements, 49
obtaining, 235
polarity, 235
seasonality, 236–237
shoot formation, 153, 160, 233–244
size, 235
species suitable for, 238–239
sumac, 253–257
thornless evergreen, 170
tropical plant, 213–224
use of auxins for, 195
window of opportunity, 233
Root/knot nematode, 85, 279
Root rot, 78, 83, 279
Rooting competence, circumventing maturity-related loss of, 154–157
Rooting hormones, 185–187
Rooting index, 213
Rootstocks, 163
biotic stress and abiotic stress, 278–279
effects on scion vigor, 280–181
Rosa sp., 114, 177, 207, 238, 253, 281
layering, 263
Rosaceae, 422
Rose mosaic virus, 76
Rose of Sharon, 422
Rosemarinus officinalis, 64
Rosemary, 64
Roses, 115, 177
Rubber plant, 214, 233, 246, 264
Rubus spp., 239, 253, 263, 388
Rubus idaeus, 100, 254
Rubus laciniatus, 164
Rumex palustris, 147
Rust diseases, 79, 83
Rye, 115

S

Sage, 239
Sagittaria latifolia, 322
Saintpaulia spp., 241
Saintpaulia ionantha, 32, 233, 241, 245
Salicylic acid, 148
Salix spp., 114, 148, 234
Salix erythroreflexus, 207, 209
Salix fragilis, 191
Salvia spp., 92, 96, 239
Salvia officinalis, 195
Salvia splendens, 415
Sambucus spp., 253
Sanchezia speciosa, 221
Sand, 47
Sandcherry, 59
Sanitation, 80
Sanseveria spp., 233, 245
Sanseveria trifasciata, 172, 201, 233, 240, 246, 247, 248
Sapindaceae, 422
Saritaea magnifica, 215
Sassafras, 234, 253
Sassafras albidum, 253
Saucer magnolia, 156
Saxifragaceae, 245

Scale insects, 116–117
Scarification, 422, 424–428
Scatella sp., 116
Schefflera actinophylla, 215
Schefflera arboricola, 207, 215, 216, 221
Schefflera elegantissima, 215, 221
Schlumbergera truncata, 64
Sciara sp., 115
Scindapsus exotica, 220
Sclerotinia crown rot, 82
Secale sp., 115
Sedum spp., 241, 242
Sedum acre, 32
Sedum morganianum, 233, 243
leaf cuttings, 245, 247, 248
Seed(s)
analysis, 402–403
bleach test, 404–405
classification, 402
dormancy, 421
germination
evaluation of, 424–434
fertilizer regimen for, 420
light and, 408–409
moisture management for, 419
nutrition and, 419
phases of, 423
stratification and, 407–408
temperature and, 419
test for, 403
mechanical damage test, 404–405
parasites, 118–119
priming, 423–424, 428–430
processing, 402
production, 401–402
purity test, 403
scarification, 422, 424–428
tetrazolium test, 403, 404
treatments, 402
Seeding, 416
Seedling(s), 30, 32, 115, 157, 411–420. *See also* Plug(s)
chimera in, 168
clonal stands and, 234
damping off, 77, 91–92
disease, 75, 77, 82, 92, 115
ethylene and, 147
germination, 88, 407–409
carbon dioxide and, 44
nutrients and, 44
salts and, 46
grafting, 273–290
growth stimulation, 52
initial stages of growth, 18
lifting, 50
light control, 420
media testing for, 420
monitoring, 114
oak, 54
pest management, 120
plant density in trays, 18
storage organs, 303–309
streaky, 171
transplantation, 52, 54
Senicio radicans, 64
Sentinel plants, 115
Sequoia sempervirens, 157, 179
Serial propagation, 155–156
Sexual reproduction, 379–389
SH medium, 335, 336

Shearing, 155
Shingle plant, 215
Shining sumac, 234
Shoot and node culture, 321–322
Shoot apical meristems, 319–320
Shoot tip meristems, 164–167
    cell division, 166
    electron micrograph, 165
    structure, 165–166
    ultimate fate of cells, 166–167
Shore flies, 116
Shot-hole diseases, 84
Shrimp plant, 214, 216
Silene, 147
Silk tree, 253, 424
Silver maple, 156
    experiments, 159
Sinningia speciosa, 245, 307
Smoketree, 422
Smooth sumac, 234, 253, 422
Snake plant, 233, 245, 247
    leaf propagation, 246, 247
Snap dragon, 79
Solanum nigrum, 163
Solanum tuberosum, 100, 322
Solenostemon scutellarioides, 207, 215
Somatic embryogenesis, 388, 389
Song of India, 214
Sorbaria sorbifolia, 64
Sorbis aucuparia, 238
Sorghum halepense, 308
Sour orange scion, 163
Southern blight, 82
Southern magnolia, 422
Spathiphyllum, 243
Spearmint, 347
Speedwell, 64
Sphagnum peat, 63
Spider-wort, 64
Spiraea sp., 57, 239
Spiraea douglasii, 64
Spiraea X vanhouttei, 64
Springtails, 118, 119
Spruce, 47, 289
    blue, 285
    Norway, 153, 285
Staghorn sumac, 234
Star magnolia, 156
Stem/crown rot, 82
Stem cuttings, 177–188
    deciduous hardwood, 177
    herbaceous, 177, 179
    narrow-leaf evergreen, 177, 178
    semihardwood, 177, 178
    softwood, 177, 178–179
Steneotarsonemus latus, 118
Stephanotis, 214, 221
Sterile transfer hood, 344, 345
Stewartia, 186
Stewartia pseudocamellia, 32
Sticky cards, 115
Stock plant(s), 17, 129, 177, 179, 186
    area, 187
    berry, 102
    care and management, 225–230
    fertilization, 57
    field-grown, 187
    foundation, 99
    growth, 92
    infection, 91
    layering, 266
    light, 226–227
    nutrient status, 228
    ornamental, 117
    preventative practices for, 187
    timing of cutting collection, 228–229
    water status, 226
Stock solutions, 342–344
    mineral salt, 342–343
    PGR and vitamins, 343–344
Stool bedding, 155
Storage organs, 303–309
Stratification, 407–408, 422, 424–428
Strawberry, 102, 323
Streptocarpus spp., 233, 239, 241
    leaf propagation, 245, 247
Streptocarpus formosus, 233
String of pearls, 64
Striped dracaena, 214
Strobilanthes dyerianus, 215
Sugar beet, 240
Sugar maple, 277
Sumac, 234
    root cuttings, 253–257
Sunflower, 116, 412
Sunrose, 91
Swedish ivy, 64
Sweet cherry, 285
Sweet chestnut, 156
Sweet potato, 253, 307
Symphoricarpus orbiculatus, 57
Syngonium, 221
Syngonium podophyllum, 215
Synthetic foam, 48
Syringa vulgaris, 156, 238, 253, 264, 265, 277, 285

T

Tagetes sp., 41
Tagetes erecta, 411, 415
Tagetes patula, 403, 411, 415
Taraxacum officinalis, 254
Tart cherry, 285
Tatarian honeysuckle, 64
Taxus, 178
Tea scale, 116
Tecoma capensis, 215
Tecomaria capensis, 215
Tents, mist, 180
Tetranychus urticae, 118
Thielaviopsis, 81, 265
Thielaviopsis basicola, 78
Thornless evergreen blackberry, 170
Thrift, 239
Thrips, 117
Thuja, 178
Thymus, 265
Thymus vulgaris, 64
Ti, 214
Tissue culture, 122, 156, 333–346
    chimeral, 171
    laboratory equipment, 333–334
    media
        agar in, 339
        agarose in, 340
        components of, 335–338
        inorganic minerals in, 336–337

organic compounds in, 337–338
plant growth regulators and, 338–339, 342
preparation of, 340–341
sterilization of, 344
storage of, 344
surface disinfecting, 344–345
types of, 335, 336
room, 335
water, 334–335
Tobacco mosaic virus, 76
Tolmeia, 241
Tolmeia menziesii, 233, 242, 245
Tomato(es), 92, 163, 274, 415
ethylene-insensitive, 147
leafminers, 116
Tomato ringspot virus, 76, 279, 285
Trademarks, 442–443
Tradescantia, 64
Tradescantia albiflora, 64
Tradescantia fluminensis, 64
Tradescantia pallida, 64
Transplant technology, 416–417
Tree of heaven, 234, 421
Tree peonies, 281
Trialeurodes abutilonea, 117
Trialeurodes vaporariorum, 117
Trichoderma, 139
Trichoderma harzianum, 137, 138
Tristeza virus, 279
Triticum sp., 114
Triticum aestivum, 404
True-to-cultivar, 101
Trumpet creeper, 263
Trumpet vine, 253
Tuber, 306
Tuberous roots, enlarged hypocotyl and, 306–308
Tukey's honestly significant difference, 134
Tulipa, 303
Tuliptree, 421

U

Ulmaceae, 422
Ulmus spp., 117, 234
Ulmus americana, 276
Ulmus parvifolia, 159, 238
Ultrahigh-density systems, 52, 54
Umbrella tree, 215
Units of concentration, 341–342
Utility patents, 445–446
UV-resistant ethylene tetrafluoroethylene, 23

V

Venus flytrap, 243
Vermiculite, 47
Veronica incana, 64
Veronica spicata, 64
Verticillium, 115, 265
Viburnum spp., 186, 239, 263
Viburnum carlesii, 128, 131, 132, 133, 136, 243
Viburnum dentatum, 138
Viburnum lantana, 138
Viburnum trilobum, 138
Vigna radiata, 147
Vinca, 64, 414
Vinca major, 64
Vinca minor, 64
Viola spp., 92
Viola x wittrockiana, 411
Viral diseases, 76
diagnosis of, 282–283
transmission of, 282–283
Vireya rhododendrons, 215
Vitis sp., 177, 263, 278
Vitis labrusca, 278
Vitis vinifera, 152, 278
Vole, 279

W

Walking fern, 233, 242
Walking stick, 267
Walnut, 286
tolerant and resistant root stock, 279
Water, 16–17
characteristics, 16
maximum usable, 16
municipal, 17
quality guidelines, 16
well, 16, 17
Water mold, 78
Wax begonia, 32, 411
Wax plant, 64
Wayfaringtree viburnum, 138
Web blights, 77
Weeping willow, contorted, 207, 209
Weigela florida, 18, 57
Wheat, 404
White cedar, 275–276
White currants, 64
White enkianthus, 32
White fringe tree, 32
White pine, 275
Whiteflies, 117–118
White's medium, 335, 336
Wide hybridization, 397
Willow, 191
Windflower, 239
Wine grapes, 278
Winkler's graft hybrids, 164, 165
Wisteria, 263
Wisteria floribunda, 238, 253
Wood by-products, 48
Woody herbs, 265
Woody plant(s), 365–376
medium, 335, 336
Wooly aphids, 278

X

Xanthomonas campestris, 76
Xanthosoma sagittifolium, 361

Y

Yam(s)
propagation, 358–361
minisett, 358–359
in vitro, 360
Yaupon holly, 159
Yellow cedar tree, 155
Yucca, 223
Yucca elephantipes, 215

Z

Zamioculcas zamiifolia, 243
    leaf propagation, 245, 250, 251
Zea mays sp., 388, 429
Zebra plant, 191, 214
Zebrina, 64
Zebrina pendula, 64
Zelkova sinica, 422
Zingiber spp., 214
Zingiber officinale, 308
Zinnia, 76, 116
Ziziphus juju, 308
Zygotic embryogenesis, 386–388